COST ESTIMATING FOR ENGINEERING AND MANAGEMENT

PRENTICE-HALL INTERNATIONAL SERIES
IN INDUSTRIAL AND SYSTEMS ENGINEERING

W. J. Fabrycky and J. H. Mize, Editors

PHILLIP F. OSTWALD

UNIVERSITY OF COLORADO

COST ESTIMATING FOR ENGINEERING AND MANAGEMENT

Prentice-Hall, Inc., Englewood Cliffs, New Jersey

Library of Congress Cataloging in Publication Data

OSTWALD, PHILLIP F
    Cost estimating for engineering and management.

    Includes bibliographies.
    1. Engineering—Estimates and Costs. I. Title.
TA183.O84   658.1'55   73–18034
ISBN 0–13–181131–2

10 9 8 7 6 5 4 3 2

Printed in the United States of America

PRENTICE-HALL INTERNATIONAL, INC., *London*
PRENTICE-HALL OF AUSTRALIA, PTY. LTD., *Sydney*
PRENTICE-HALL OF CANADA, LTD., *Toronto*
PRENTICE-HALL OF INDIA PRIVATE LIMITED, *New Delhi*
PRENTICE-HALL OF JAPAN, INC., *Tokyo*

To Doris, Mark, Phil, Lynne

contents

Cost estimating is introduced in this text. As the first unified treatment, the book covers the philosophy, concepts, and practices of a field that is feeling a resurgence of interest and enthusiasm. This specialization is concerned with evaluation of engineering designs in economic terms. While designs certainly differ, the principles and practices used for their appraisement are remarkably similar. We state, without proof, that all designs undergo an economic appraisal. This book, then, covers those subjects that contribute positively to the successful economic attainment of the engineer's design.

Design is given a broad and liberal interpretation. Every design (1) is a new combination (2) of pre-existing knowledge (3) which satisfies an economic want. This three-part definition includes virtually every product, such as bridges, cars, chemical plants, highways, machine tools, production lines, radios, rockets, ships, systems of machines and people, and tooling. Plans, technical reports, and models that are a part of the engineering job are allowed in this definition because they undergo the reckoning. Design is a unifying term for the practice of engineering. With this as the focal point, cost estimating is the body of theory and practice which provides an economic value for the design.

The experienced cost estimator, who, after a few times "looking down the barrel" defending his estimate (management calls it probing for softness; the estimator calls it picking on his professionalism), misunderstands management's interest in this topic. The exposure of cost overruns for weapon systems and public work projects testifies to the serious embarrassment that cost estimating has faced and the importance of this specialization for management. Indeed, the well-being of firms and our country rests, in part, on cost estimating. Business firms have realized that computer management information systems, notwithstanding their unmatched ability to handle data, are helpless to overcome a lack of trust in the truth of estimated data in their cost-forecasting systems. International trade and foreign competition with our past and future trading partners present a challenge, and cost and cost estimating will be important. Productivity needs an index, and cost estimating plays a prominent role in its measurement. The engineer's design impact on changing times, employment, growth and development, pricing efficiency, income, gold and foreign trade, and the blessings of our kind of democratic society are topics of today.

Cost estimating discloses the strengths of a company, or country, or trading group for executive management—it should never hide weakness from management. Decisions, both great and small, rest in part on estimates. "Looking down the barrel" need not be an embarrassment for the cost estimator if newer techniques, professional staffing, and a greater awareness are provided for the cost-estimating function.

Cost is a nebulous term which has no standardized definition. Used in some contexts, it implies a meaning that is clearly not cost. To appreciate these distinctions, one must be prepared to understand the particular setting in which the word is used. Surprisingly, the word cost could mean profit, rate of return, or effectiveness, and dollars are one dimension for these measures. For management and many engineers, dollars are more important than amperes, foot-pounds, or mass flow. Whether the engineer is principally involved with cost-finding, profit, cost reduction, or value analysis, subtle variations of the word cost are understood.

A great number of engineers are concerned with cost engineering and estimating. Although their professional title may be something other than cost estimator, their work uses many of these concepts. These professional engineers have found that their career paths lead to economic evaluation of design. For instance, in municipal, state, and federal design sections, these matters may be handled by the designers themselves. In industry, special groups are groomed to do this exacting work. R & D design calls for a special understanding. Whatever the organization, the engineer performing estimating has a more than average amount of engineering experience. This is necessary because designs are complicated. Practicing engineers will find in this book the kinds of thinking that will help them do the cost evaluation.

Many men rise to cost estimating from the practical ranks of industry, construction, business, and government. Often they find self-study necessary to supplement their intimate grasp of practice with an appreciation of academic topics. To them this book will give a taste of the principles of a special kind of topic.

Throughout this text the terms *cost estimator* and *cost engineer* are used interchangeably. In addition, business students, practicing accountants and economists are closely identified with these activities. Their contributions are necessary in conducting a successful cost-estimating enterprise.

This text is an outgrowth of notes from a one-semester cost-estimating course and various industrial clinics and seminars. This book would be suitable for courses such as Cost Engineering, Industrial Analysis, Estimating, Manufacturing Analysis and Planning, and Economic Systems Engineering, to name a few. These courses cover practical and theoretical techniques of cost estimating/engineering for various kinds of engineering designs. It is suitable for engineering students whenever they reach their first level of specialization. For most it would be their junior or senior year, because specific programs restrict early opportunity for experimentation and broadening, and design evaluation is deferred. Schools of technology, technical institutes, and junior colleges with various programs in technology will find their students prepared to understand the material of this text. Although a great diversity of occupational and vocational problems has been included, the instructor may want to supplement the problems and design studies using his own experience. A liberal listing of chapter references helps in forming new problems and case studies.

The arrangement of the chapters and topics allows for a variety of teaching and self study approaches. Basically, the text is broken down into four areas: engineering design and the economic environment, cost-estimating methods, design estimating, and management. As design customarily precedes its calculation, this is discussed with emphasis on the art and technique of design. No reference is made to any specific specialty of designing, as that is left to other books. But on the design base, practices of cost engineering and estimating are built. The design-estimating portion, which is the largest of the four, considers the kinds of information and estimates for four categories of design. Operation, product, project, and system design contexts are constructed. Then various techniques that are pertinent to each are associated with that kind of design. After various designs are cost-estimated, the process of optimization or fine tuning is considered. The cost engineer is well suited for this task, as he has many of the economic facts at his fingertips and understands the design. Finally, a single chapter on management gives attention to the special problems of self-administration for cost engineering. It is not intended to be exhaustive.

The teacher can select those portions of the book that best meet the objectives of the class. Mathematical rigor is mostly algebra, and while some calculus is found, with suitable section selections this need not be an obstacle.

I am grateful to Lawrence E. Doyle of the University of Illinois for his encouragement and guidance. The Charles K. Kettering Foundation who funded a grant called The BUILD Program, a cooperative venture between the University of Colorado and the University of Illinois, supported tangibly many of the thoughts and underpinning ideas in this text. In a great measure the relevance of this book has been enhanced by my association with many estimators in industry, government, clinics, and seminars for over fifteen years. I hope that this text does justice to their practice.

Industrial friends, Smokey P. Call of Dana Corporation, Al Christianson of the American Paper Bottle Company, Eugene W. Groff of the New Holland Division of Sperry Rand Corporation, Ray Kincheloe of Collins Radio, Donald H. McBee of Monroe Auto Equipment, Gianni Peri of the Olivetti Corporation of America, James J. Thompson of American Lava Corporation, Donald F. Vehrs of Machine Specialties, William H. Wakerley of Ex-Cell-0, and Glen R. Wyness of the Proctor and Gamble Company provided a sense of proportion that gives balance to this text. My colleagues in the teaching of engineering design and economic evaluation have been helpful in many ways. I am indebted to the Literary Executor of the late Sir Ronald A. Fisher, F.R.S.; to Dr. Frank Yates, F.R.S.; and to Oliver & Boyd, Ltd., Edinburgh, for permission to reprint Table III from their book *Statistical Tables for Biological, Agricultural and Medical Research*. The skills of Cheryl Welsh, Lorraine Ruka, Marie Hornbostel, and Virginia Birkey persevered over some bad drafts.

Finally, I wish to thank my wife Doris for her help and encouragement, without which this book would have never been completed.

Boulder, Colorado   Phillip F. Ostwald

When shallow critics denounce the profit motive inherent in our system of free enterprise, they ignore the fact that it is an economic support of every human right we possess and without it all rights would soon disappear.

DWIGHT D. EISENHOWER, 1890–1969

IMPORTANCE OF COST ESTIMATING

1.1
PROFIT IS NECESSARY FOR BUSINESS SURVIVAL

The Winston Dictionary defines *profit* as the amount by which income exceeds expense in a given time. This notion about profit leads to unfortunate conclusions. First, profit is necessary for taxes, dividends, and capital re-investment in the firm. Taxes, whether they are national, state, or local, are the inescapable reward for successful operation—a vital contribution to continue a democratic society. If dividends, the rent on invested capital and money, were not paid, it would negate the faith of investors and jeopardize a source of money for growth. Once taxes and dividends are removed from profit, a portion referred to as plowback is necessary for equipment or other modernization needs. Successful managements do not ignore debt repayment, research funding, plant maintenance cost, salaries, or other expenses, but it is surprising that profit is sometimes overlooked. Is profit less vital than anticipated costs? Consequently, it is important that profit become a planned expense.

A new approach can be suggested: Everything is going to be spent. Thus it becomes a question of partitioning income and expense. To use a simple illustration, assume that sales revenue is going to be $1000. You expect to realize a net profit of $50. Based on your calculation the "net profit dollars after taxes" is $50 and all other costs, including income taxes, must be found within the $950. It is common at this point to hear the excuse "You can't tell until afterwards." What about prediction of sales income? Can it be safely approximated? Sales forecasts are surprisingly accurate and provide a foundation for profit estimating. The planning recognizes that what counts are current costs, not those of the previous quarter. How successfully can expenses be held to 95%? In controlling performance versus target a significant body of experience indicates that management can react to unplanned contingency costs as well as ordinary types of cost. For a long-term survival plan, the creation and assurance of profit remains a primary goal.

1.2
STEWARDSHIP NECESSARY FOR ECONOMIC SURVIVAL

Business, whether large or small, is not alone in its quest for survival. The pursuit of this objective includes government and the governed. A democratic government with its authority to impose economic laws on its citizens is not a wealth producer and has no inexhaustible source of wealth. Governmental activities such as public works, welfare, the military establishments, and a host of legislative-directed projects use the resources of its nation. Despite the nobility of cause and honest-meaning goals, governments suffer from financial bankruptcy. Curtailment of welfare programs, de-evaluation, and heavy tax loads are symptoms and manifestations of failure. Politics does not provide a shield against national ruin as the accounting ledger between nations is a reminder for long-term fiscal sobriety.

Even churches, foundations, charitable organizations, and not-for-profit trusts must have positive balances between short- and long-term debt and income. Individuals need no economic reminders. Despite credit, loans, or notes, bankruptcy or poverty is not uncommon. Unfortunately, there is no inviolate equation that will prevent financial failure. The notion that receipts and expenses must maintain a positive cash flow is an oversimplification. Benefit-cost ratios, whereby social goals are evaluated in monetary terms, provide a narrow solution. Legally imposed restrictions on credit and spending are imperfect. Knowing the profound nature of this problem, a general objective for any steward is to simply husband resources.

It has been generally assumed that competition of all kinds is increasing. This statement can be examined on pragmatic grounds. A monopolist's product must be indispensible and have no substitutes, no potential competition, and possibility of control by the government. These conditions are practically impossible to find, although they are sometimes approximated. Pure competition, on the other hand, is present when many firms provide a standard product to numerous purchasers. No single supplier or purchaser is strong enough to affect the price significantly by his actions. Pure competition does not prevail either. Rather a form of imperfect competition is the usual marketplace.

Evidence of financial failure as a consequence of increasing competition may be found by examining companies, products, governments and their programs, and individuals. The profit squeeze on companies may result in public disclosure of bankruptcy or a receivership action. Mergers or sales of assets of the company, changes in the title of a company, and interdivisional failures within a larger corporation disguise the more subtle company failures. There is a good deal of empirical evidence that products fail as well. The high rate of new products which enter the market but are withdrawn within a short time is a case in point. Curtailment or complete abandonment of various governmental programs, although politically inspired, is asserted to be a result of increasing social competition. If poverty may be accepted as a norm for the individual, evidence of that is well known.

Although evidence of all types of failure is clear, the factors causing it are not. With production exceeding public demand, particularly true in the Western world, a temperamental society cannot guarantee long-term stability in the marketplace. Shifting consumer preferences, pliable and elusive, illustrate short- and long-term effects of increasing competition. The interaction of control by governmental legislation is an obvious business factor; increasing costs of production, inflation or recession, rising nonproduction or policy costs, new inventions, and improving technology are candidates for the causes of business failure.

Inasmuch as we are concerned with cost estimating, our attention is naturally directed to the matters of invention and the pace of general technology. For our purpose invention and technology are classified into four distinct areas: operations, products, projects, and systems. The understanding and manipulation of these areas is a result of the employment of the engineering sciences, economics, and mathematics. Thus we choose to deal with the onrushing inventions and technology as factors of increasing competition within the firm and government. The reader may want to suggest other determining factors that show the importance of cost estimating within a competitive society.

One may wonder what describes the act of estimating. Stories involving teacup fortunes, foretelling, palmistry, mystics, or wandering gypsies are known by school children. The mystic and showman—fraudulent or not—have created a legend that is difficult for the legitimate act of estimating to dispel. Throw out the unknown quantities of E.S.P. and miracle workers, and consider seriously weather forecasting, control of national economics, and gambling. On the surface it appears that these three are unrelated. Examined more closely it must be admitted that weather forecasting is an inexact science, particularly at the local level; national politics has been, ineptly at times, trying to achieve economic control of the nation; and mathematics has not always won the pot in the art of gambling. All three involve estimating. For the engineering situation, the act of estimating uses concepts of engineering sciences, economics, and mathematics. Within the engineering environment there is generally an exterior problem insisting on an interior solution. Social, economic, and technical overtones exist within and without. There is always disclosed as well as undisclosed information.

Estimating is practiced by the housewife, farmer, manager, military planner, engineer, and you, the reader. Let us forever dispel the myth that estimating cannot be done. It remains, of course, to show that estimating is able to provide reliable estimates.

1.5 CERTAINTY, RISK, AND UNCERTAINTY IN ESTIMATING

With the recognition that estimating is a common art, the task beckoning the analyst becomes clearer. Few engineers enjoy the admission that their problems are subject to unknown forces. However, this is the case: The accuracy of the estimate is inversely proportional to the span of time between the estimate and the event. Thus the estimating activity is not for the short sighted. Three broad categories characterize future environmental conditions of the estimate: (1) estimates assuming certainty, (2) estimates recognizing risk, and (3) estimates admitting uncertainty.

The simplest of the states of nature is that of certainty. This simplicity seldom exists in nature or in a competitive society, but we achieve it by ignoring complications. For example, if an estimate is required, it would be easier if it were made on the basis that demand, design, product, production rate, and vendor's cost were stipulated and known with certainty. In performing the analysis, experience and wisdom lead us to believe that a set of assumptions have a high expectation of occurring. This expectation is fully warranted in many cases—labor costs, production rates, and cost indexes are sometimes stable especially in the short term. In a primitive estimate we make these assumptions to expedite a workable means of analysis. The event of certainty assumes that each action undertaken results in the same outcome and has a probability of 1.

Situations involving risk are appropriate whenever the analyst can obtain good estimates of the probability of future conditions. Research and study may be required. Risk is defined where each of several outcomes is assigned a probability and their sum equals 1. Uncontrollable factors of the future, by their very nature, can at best be anticipated. The probabilities of failure of several designs, for instance, can be estimated by experts. Sometimes these probabilities are subjective as actual measurement is either impossible or undesirable.

Engineering and physical laws, which depend to an extent on well-ordered cause and effect relationships, are unlike economic laws which depend on the action of people. The estimator does not have the good fortune of commanding the circumstances that follow his estimate. Accordingly, the state of uncertainty may be more applicable than certainty. Now we suppose that this uncertainty applies to the probabilities or relative values which describe a specific set of states of nature. For these states of nature, or competitive reactions, we have little bits and pieces of information so poorly understood that we are unable to assign any probability ranking. This is the qualification for the condition of uncertainty. It is presumed that the outcomes are identified in some context. This is a normal situation facing an estimator.

It must be admitted that in practice estimates which clearly define certainty, risk, or uncertainty are generally not found, but it remains a philosophy for dealing with future forces. Because of their importance, these theories are extended throughout this book.

When actual costs can be compared to estimated costs a deviation is usually found. It may be surprising that actual and true cost may never be known, but it is interesting to speculate on the deviation that would be revealed if they were known. This deviation would be the sum of (1) certainty, risk, and uncertainty components associated with the future; (2) mistakes, and (3) errors of belief. The certainty, risk, and uncertainty were previously introduced and form a part in the deviation. Mistakes may result from imprecision and error of technique in computation and blunders of various sorts. Typically, they may pass unnoticed, but if they occur, "nature" may be kind due to compensating effects. The solution is uncompromising arithmetic at all levels and strict attention to methods that inspire faultless computation.

Errors of belief result through ignorance or inadvertentness. Simplified illustrations include failure to recognize material price breaks, the omission of cost items, and overlooking a planned contractual increase in direct labor cost. These error are in addition to excessive or inexcessive values associated with first cost, salvage value return, costs of operation, and capitalization. Excessive profit margins, deliberately chosen, can have serious consequences. These errors of belief, as initially described here, are prevented by well-thought-out policies and practices.

Despite a variety of possible ways to make an estimate, resources for the preparation are always restricted as time, money, and the technical staff are limited. Eventually one reaches a point where the objectives must be satisfied with what time, money, and intelligence are available. The ideal policy would have the estimate coincide with the reconciled actual cost. True cost is, perhaps, never known. No procedure, mathematical technique, or policy employed in the engineering world is without its flaws and shortcomings or is able to guarantee perfect estimates. Although flaws in estimating may be obvious, these procedures and techniques are used for the simple reason that they are the best means at hand. Imperfection seldom deters usage.

1.7
ESTIMATING: WITH OR WITHOUT COMPUTER?

One solution to overcome mistakes and a limited technical staff may be the analog or digital computer. As a tireless machine it removes the burden of routine calculation. Computers, now linked by radio, telephone, and teletype, are found in the office or even in the bedroom of an engineer's home. But the computer should be more than a device for computation. Imaginations are being challenged by computer applications. Before ambitious objectives can be reached, more limited working models are needed. In view of the speed and the ability to draw on funds of information, we shall see that computer estimating models will become the accepted norm of operation. Even now the computer is broadly serving the needs of many engineering disciplines. In the area of engineering design and industrial application, the following applications are routine:

- Design of products or production tooling.
- Testing and analyzing products manufactured by automatic equipment.
- Control of chemical processes, pipelines, utility systems, and production processes.
- Accumulating data and calculating costs within process manufacturing and construction jobs.
- Evaluation of R & D projects at an early stage for profitability analysis.
- An analog simulation model for textile industries.
- A real-time analog process controller of the manufacture of soap.
- Payroll cost accounting and inventory cost control.

The list is only a fraction of the thousands of applications. A vast and seemingly inexhaustible inventory of software, hardware, applications, and unusual acronyms colors the computation picture. Computer rental or purchase is a costly expense in any budget. One unseen budgetary element is the cost of programming, sometimes the largest element in any computer operation. This book will not discuss computer languages, machines, codes, or applications, as that is left to other texts and journals. This book assumes that computer practices are widespread and that many of the cost-engineering techniques require a computer.

1.8
COST ESTIMATING

There is no trail of historical evidence that is associated with any cost-estimating title. That much is known; however, as a consequence of specialization many engineers have begun to practice collectively what might be called cost estimating within the dozen or so basic fields. The importance of the cost-estimating specialty is now an accepted fact, and increasing activity in industry and education is foreseeable. It is an activity done for engineering and management. Many professional engineering societies regularly present papers, hold meetings, and sponsor clinics devoted to the many ramifications of cost engineering or estimating. One professional group, the American Association of Cost Engineers, has gone so far as to sanctify the title *cost engineer*. These groups recognize that the engineer, unlike a scientist, has an orientation to an economic motive for the design.

A scientist is more concerned with the idea or principle and less interested in an economic justification for the principle. An engineer is more concerned with the economic results of his design. Frequently, this is the dichotomy used to separate technical and professional people into one of the two camps.

1.9
ORGANIZATIONS AND PROCEDURES

The patterns of organization for cost estimating are as varied as their environments. The estimating department, if one exists, may be structured along processing requirements or product or design lines. In developmental-directed laboratories oriented to consumer and market stimulus, the cost-estimating activity assesses potential economic worth in the early stage of development. Only a minimum amount of information is available at this particular time. Manufacturing firms, depending on their maturity, size, and products, tend to have a cost-estimating department that is organized within manufacturing rather than within engineering design. Where there are short lead-time bidding requirements, these groups tend to be centralized within design activities. For long-lead-time bidding requirements, we find separated cost-estimating functions. Plant and architectural facility groups within manufacturing perform cost-estimating activities because of the special nature of their problem. A chemical plant may staff this activity under the title of Economic Evaluation Department or Profitability Study Group. For the aerospace and defense-related industries, the research-development-build time cycle is compressed. As a consequence it is not uncommon to find parallel efforts throughout the major activities.

The choice between a decentralized or self-sustaining department rests on the size, complexity, and maturity of the organization. Generally with the more complex and mature design engineering groups, a separate cost-estimating team exists. Sometimes a committee is formed from several departments and is held accountable for cost estimating. This arrangement does depend on the firm.

Because of obsolesence in space and defense tactics, long-range cost-estimating procedures are found. Contractual awards by the government are sometimes determined on the basis of research and design competence. A need exists to recognize this type of cost analysis. Even though production cost-estimating activities are ultimately necessary, government contracts are separated into various phases for aerospace and defense firms such as conceptual research, applied research, development, preproduction, production, service, and maintenance. Within municipal, county, state, and federal governmental engineering groups, cost-estimating activities are normally dispersed throughout the organization. Frequently, every engineer is expected to understand and implement cost-estimating concepts.

In Figure 1.1 we see the schematic of the administrative cycle as a partially closed loop of information flow. Vital to the preparation of the estimate is the information which the cost estimator receives from others, verifies, and alters to meet the demands of the estimate. For any information system to be reliable, there must be audits and controls to improve its future performance.

The qualifications required for cost estimators are diversified. He or she is normally trained in one of the several engineering disciplines, having a course or two concerned with economic theory. A few economists have migrated into this activity, but the majority are engineers. Some cost estimators have received only shop training, but this is changing. A cost-estimating department staffed with estimators having practical and university backgrounds is sometimes thought superior to a fully staffed group of one kind. Within this book the title cost estimator is used synonymously with cost engineer, analyst, or engineer.

FIGURE 1.1

ADMINISTRATIVE CYCLE OF COST-ENGINEERING
WORK

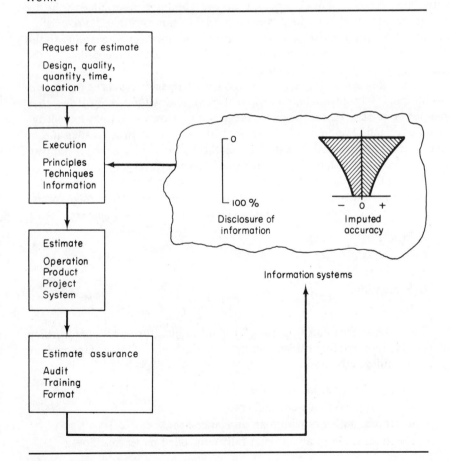

The initiation of cost-estimating work comes from a diversity of external or internal sources. The need may arise from the government, random or repeat customers, or the public-at-large for municipal, county, and state government engineering activities. Sales, marketing, engineering, and manufacturing provide many requests for cost information. Only rarely does a significant portion of the cost-estimating workload emanate from within the cost-estimating department. Consequently, a good deal of interoffice communication is required. Forms, procedures, and controls of all kinds exist to enable both accuracy and time-liness.

1.10
PRELIMINARY DEFINITIONS

At this point we shall define several terms. Cost estimating is concerned with cost determination and evaluation of engineering designs. The term *cost estimating* could well be *profit engineering* as the "making sure" of profits is a higher priority for many. However, it is recognized that cost estimating is a general title and includes engineers who may or may not see themselves causing profits directly. The estimate is the result of cost-estimating work. When used as a noun the estimate implies an evaluation of a design expressed as cost, amount, or value. When employed in its verb form, estimating means to appraise or to determine. There are four types of estimates: operation, product, project, and system. Their ordering does not suggest any ranking of difficulty.

The word *design* is given the broadest possible definition. It does not mean "board work." It implies the activities of creative engineers and is defined as follows: *Every design (1) is a new combination (2) of pre-existing knowledge (3) that satisfies an economic want.* The phrase "is a new combination" emphasizes novelty and suggests the unusual characteristics of the designer or design team and the circumstances. The designer must possess ability and resources or be the recipient of serendipity and its fortunate concentration of forces. "Pre-existing knowledge" relates to a design's (not designer now) intellectual past and to the industry on which it has been built. For the third part of the definition the engineering activity fulfills economic satisfaction.

The driving forces for design are past knowledge and wants as each alone is insufficient. Without wants, no problems exist, and without knowledge, they could not be solved. This is the chicken-egg riddle. How can a want-knowledge milieu be created? Simple answers are not possible, and to avoid reader-author disputes let us assume that demand and intellectual curiosity or sheer happenstance spurs on the design. Reasons for this become clearer later on.

A design procedure can take the shape shown in Figure 1.2. One should realize that a precise sequential process is not intended. Rather the design format is actually a hodgepodge of simultaneous continuous actions. The elements are

1. Problem.
2. Concepts.
3. Engineering models.
4. Evaluation.
5. Design.

The process of producing a change in value or quantity or a way of working establishes the content for an *operation estimate*. Examples of these man-machine or men-machines activities are

- Wrapping tape around a cable.
- Work assigned to a trenching crew.
- Secretarial activities within an insurance firm.
- The monitoring of an automatically controlled paper machine.

Consider the following: A company wants to set cost standards for labor control on a cable taping machine. This machine winds a tape on steel cables. The cable is mechanically pulled through the center of a rotating disk, the taping

FIGURE 1.2

INITIAL ENGINEERING DESIGN PROCESS

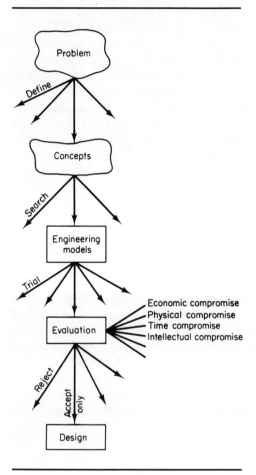

head, which carries a roll of tape. This tape unwinds via a set of rollers and adheres at an acute angle to the centerline of the cable. Work measurement methods had so far failed to provide accurate standards because of the complexity and variability in the operator's task and the uncertain effects of changes in materials. In discussion with the operators and foreman, the analyst learns that the workers select the tape width and velocity from feel based on tape breakage to machine speed. Thus we set the stage for the construction of an operation estimate.

In a *product estimate* an entire product, rather than a subassembly or a component part, is estimated. A component part refers to the smallest component of a product subassembly. A subassembly, composed of two or more parts, may be joined with other subassemblies to ultimately form a complicated product. In chemical processing industries, a subassembly would be termed an *intermediate*, while the component part would be termed raw material. The intermediate is processed further into final products. In a product estimate there is duplication and production of the same item. The following are suggested as illustrations:

- Alkyl dimethylamine intermediate production.
- Automobile production.
- Suitcase production.
- Prestressed concrete beam production.

Consider the following: In the chemical processing business conversion of alpha olefins to alkyl dimethylamines is an intermediate chemical plant process. This intermediate is raw material in another conversion to amine oxides which ends up as liquid soap formulations for home use. Natural raw materials from fats or coconut oils are available from outside the United States but have the disadvantage of price and supply fluctuation due to climatic and political forces. As a substitute to natural products, alpha olefins derived from petroleum stocks are a commercial commodity. Thus we want to to find a product estimate for the alkyl dimethylamine soap detergent.

The *project estimate* is another classification. The project, whether it is a plan, equipment, or plant expansion, is one of a thing. Normally it involves capitalization of money and is long-termed. Examples are

- High-rise apartment building.
- Purchase or lease of a lathe.
- 200,000-pound chemical plant expansion.
- Bridge construction.

As a further illustration, consider the following: The facilities design group is examining floor girder designs for a new high-rise steel building. In view of symmetry of loads, the building layout encourages a design of bays 30 × 30 feet square. A girder will be loaded with a uniform load of 12 kips per foot of length (1 kip equals 1000 pounds). On 30-foot girders this load will impose a maximum moment of 16,200 inch-pounds and a maximum shear force of 180 kips. The engineers working on this problem propose to find the design which will be the least expensive while complying with other constraints for the building.

System estimating involves elements of operations, products, or project estimates and is the most comprehensive. In its correct form a system describes configurations of hardware, a conceptual outline for the solution of a problem, a group of plans, or a large process. The motive for the system estimate may be minimum cost, maximum profit, or optimal expected gain. Examples of what we mean are

- Expansion and rearrangement of production facilities.
- Marketing, production, and distribution of a chemical plant product.
- Watershed pollution control.
- An interstate highway system.

As additional clarification, consider the following: Manufacturing firms are faced with the problem of expansion and rearrangement of production facilities. Floor space limitations impose on manufacturing facilities the question of allocation: To what extent do we store parts on the manufacturing line or in the warehouse? We are unable to store cheaply the total requirement in manufacturing space (this is the fundamental reason for warehousing). Nor can we store all materials in the warehouse, as this would normally generate prohibitive material-handling expense. Thus we have verbally described an economic balance problem between the costs of manufacturing floor space (by storing greater quantities of material on the line) and the costs of material handling (by storing a greater volume of materials in a warehouse and initiating a higher transportation cost).

This book provides the kinds of thinking that are found in cost-estimating work. The circular riddle—do problems provide the stimulus in finding solution methods, or do techniques heretofore unused discover and solve problems—is really never answered. An engineer would not seriously consider redesign of the wheel or feel any guilt in exploiting its theory and practice. An effort has been made to assimilate theories and practices which are broadly attractive to all engineers, whether the engineers are (or are to be) employed in research, development, design, production, sales, and management. While cost-estimating practices vary among the several fields of engineering, the principles do not. This becomes clearer by noting the organization of the book.

In the first five chapters we couple engineering design to its economic environment. In this text we assume that design neither leads nor lags its economic shadow. Vital to the design is the cost information on which decisions must be based. With the design and cost data at hand, the cost engineer builds a corresponding cost-design structure. In the past this consisted of columnar and recapitulation sheets. Now, however, the cost model is too involved for these simple maneuvers. Data are unfortunately out of date, demand is past history, and budgets need review; these are items that demonstrate the need for forecasting. In this book forecasting implies numerical analysis of information. How might the act of estimating be classified? In cost estimating we follow the design in a logical manner to provide a scheme of estimating. We could concoct an estimate classification according to purpose, accuracy, time, type of commitment, or information. If one were to classify estimates based on purpose, as many are, we would find estimates for the *verification of a vendor quotation*, *appropriation*, *budgeting and funding*, and *evaluation studies* or *design feasibility* in addition to *cost* or *price*. If accuracy were the determining factor, one could imagine an estimate classification as *order of magnitude*, say ±50%; as *ballpark*, say ±20%; and as *accurate*, ±5%. *Initial* and *final* are other possibilities. The classification scheme adopted in this book is associated with information. Estimates are classed as either *preliminary* or *detailed*. With information as the essential ingredient, the designs, whether they are operation, product, project, or system, provide the identifying feature. Formats, procedures, and a host of ramifications vary for these types of designs. The methods of estimating do not. Regardless of the type of design, the environments are remarkably similar. Figure 1.3 depicts the environment of cost estimation. This common touchstone illustrates the forces that play on the estimating art and that occur for all types of designs.

Chapters 6 and 7 consider preliminary and detailed methods of estimating, while Chapters 8–11 transfer these methods to the operation, product, project, and system design.

The use of judgment and experience will forever remain a vital factor in the engineering field. Certainly without it chaos results; however, judgment and experience are not really *optimization*. Designs are improved by optimization. To effect optimization, cost-estimating activities become crucial because the cost analyst is aware of cost or profit facts and business conditions. For optimization it is necessary that we formulate mathematical models representative of design and cost. The cost estimator is able to aid the improvement of the existing design process as illustrated by Figure 1.4. Optimization and the

FIGURE 1.3

ENVIRONMENT OF COST ESTIMATION

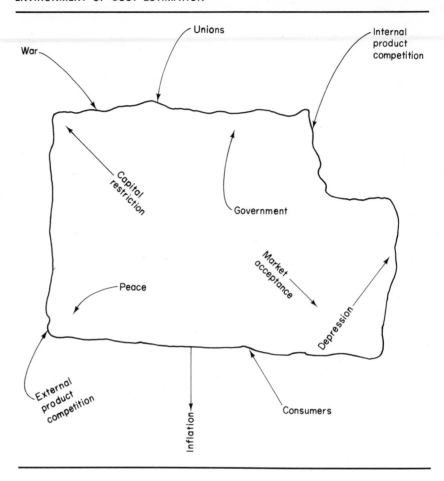

relationship of cost estimating to design is introduced in Chapter 12. Other books in operations research are necessary for more detail.

It should be clear that optimization is a mathematical approach requiring invented mathematical models. It originated in calculus, and the theories of maxima and minima remain cogent today. The development of computers is

FIGURE 1.4

IMPROVEMENT BY OPTIMIZATION

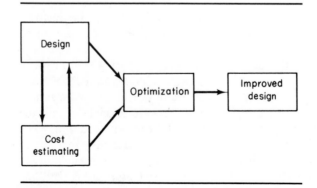

responsible for the interest in optimization, and many old theories and practices were rediscovered as a consequence. As a result of this renewed activity other discoveries and methods of optimization have been advanced in recent years.

Administrative practices are presented in the final chapter. The thrust of the last chapter is dominated more by a discussion of goals rather than by rigid rules.

1.12
AHEAD: COST ESTIMATING

With a liberal interpretation, *cost* may mean cost, profit, income, expense, or any economical measure and is an important engineering dimension. Many believe that a radio, car, or rocket design has economic value as the first and last requirement. The thought is this: Given a design (aerospace, agricultural, chemical, civil, electrical and electronic, industrial, manufacturing, marine, mechanical, metallurgical, mining, petroleum) physical and real-world restrictions are its companion. Engineering, production, marketing, sales, and finance conform to the engineering design, as the drawings and specifications are the authority for construction and operation. The salesman sells the design; the service engineer maintains this design; the accountant classifies costing points about this design; the manager plans production schedules to build the design; and the construction engineer selects processes and equipment to construct the design. Used in its broadest context, *design* causes a long chain reaction.

If all these factors have their design constraints and if information regarding income, expenses, and revenues are known or can be estimated, a mathematical model can be constructed which optimizes the design. The engineering student who brings to his job an understanding of the economic consequence of his design is valued in industry, business, and research. This man becomes a designer, a project engineer, a supervisor of engineering activities, or a business man who works closely with design, development, and the research team, for not only is cost estimating essential in order to thrive in our economy—it is also necessary for the survival of the economy.

SELECTED REFERENCES

For the discussion about decision theory, see

STARR, MARTIN KENNETH: *Product Design and Decision Theory*, Prentice-Hall, Inc., Englewood Cliffs, N.J., 1963.

For history of engineering, see

KRANZBERG, MELVIN, and W. PURSELL CARROLL, JR.: *Technology in Western Civilization*, Vols. I and II, Oxford University Press, Inc., New York, 1967.

OLIVER, JOHN W.: *History of American Technology*, The Ronald Press Company, New York, 1956.

WALKER, CHARLES R.: *Technology, Industry, and Man. The Age of Acceleration*, McGraw-Hill Book Company, New York, 1968.

For discussion of invention and design see

SCHMOOKLER, JACOB: *Invention and Economic Growth*, Harvard University Press, Cambridge, Mass., 1966.

Methods for the design process are presented by

DIXON, JOHN R.: *Design Engineering: Inventiveness, Analysis, and Decision Making*, McGraw-Hill Book Company, New York, 1966, pp. 1–52.

KRICK, EDWARD V.: *Methods of Engineering Design and Measurement of Work Methods*, John Wiley & Sons, Inc., 1962, pp. 1–70.

RUDD, DALE F., and CHARLES C. WATSON: *Strategy of Process Engineering*, John Wiley & Sons, Inc., New York, 1968, pp. 1–29.

QUESTIONS

1. How would you define profit? Discuss fully. What are the consequences of negative profit (loss)?
2. How can profit be made appealing to the individual?
3. How many not-for-profit organizations can you name? What do governments do to protect their interests?
4. Distinguish between product and technology competition. Which affects your life more?
5. What distinguishes the act of estimating? Will estimating become a science?
6. How do economic laws and physical laws differ? Are the well-ordered cause and effect relationships separable in economics and engineering fields?
7. Describe how computers have improved on back-of-envelope techniques in estimating.
8. Relate the role of cost engineering to design engineering. Contrast this role.
9. Distinguish among an operation, product, project, and system estimate. Cover mutually exclusive descriptions applicable to these four estimates. How are they similar, and how are they different?
10. What are the broad ramifications involved in certainty, risk, and uncertainty?
11. A proposal is advanced for the stated purpose of employing people to dig a very deep hole and then to fill it. Consider this proposition as it relates to (a) design engineering, (b) cost engineering, (c) political effects, and (d) physical laws that apply.
12. What ideas of feedback can you add to the figure on engineering design?
13. Do you agree that the dimension dollar is equally as important as other engineering dimensions?
14. Consider another rationale for the classification of the estimate. Would labor estimating, material estimating, and tools and machine estimating be complete?
15. In your own words, define *optimization*.
16. Do you agree with the statement "not only is cost estimating essential in order to thrive in our economy—it is also necessary for the survival of the economy"?

CASE PROBLEM

The Vacuum Chamber Environmental Test Unit. A new vacuum chamber environmental test unit was required by an aerospace firm to supplement existing units because of an increased workload. The units were already scheduled to capacity during the contract time period. The test units are necessary as all components of a spacecraft require testing in a vacuum to simulate the altitude at which they will be operating in orbit. Existing vacuum units are a small 42-inch-diameter chamber with a high pumping capacity and fast vacuum pulldown time and a large 10-foot-diameter chamber for large-component testing. After a study of probable contract and component testing schedules and environmental requirements, it was decided that another unit was required. A satisfactory commercial system cannot be purchased, and a number of other solutions were proposed:

1. Repeat the existing design and construction of the small chamber, thereby saving the engineering design costs. This design sacrifices space (inside dimensions).
2. Design and build an intermediate-sized chamber with a larger 32-inch diffusion pump. This should relieve some of the load of the larger chamber.
3. Design and build an intermediate-sized chamber with the same pump stack as the small chamber, sacrificing some pumping capacity and pump time but saving the cost of the larger pump stack and its engineering.
4. Design and build an intermediate-sized chamber that is the same as described in solution 3 but add an additional port for another pump in the future, if required.

Although a solution to this case problem cannot be provided because pertinent information remains undisclosed, consider the initial design process facing the engineer. What steps of the (a) problem, (b) concepts, (c) engineering models, (d) evaluation, and (e) design are most significant here? Why? What kinds of information are necessary for estimating to accompany these plausible designs? What general elements of information would you classify as certain? As risk? As uncertain?

chapter 2

. . . There was a certain rich man, which had a steward; and the same was accused unto him that he had wasted his goods. And he called him, and said unto him, How is it that I hear this of thee? Give an account of thy stewardship. . . .

Luke 16:1, 2
The Holy Bible
KING JAMES VERSION

ENGINEERING DESIGN AND MODELING

In this chapter we shall identify engineering design as the dominant springboard for cost-estimating practices. After design has been started and information sources recognized, cost-estimating activities begin. Emphasis in this chapter is given to design—not to teach design but to broadly illustrate designs and show how cost estimating relates.

As a new combination of pre-existing knowledge, design satisfies an economic want. Cost estimating attempts to ensure this compliance. Despite the notion that cost estimating precedes design, cost estimating acts as a shadow and requires some sort of design even before a preliminary estimate is started. In view of this relationship, design will be examined on the basis of an operation design, product design, project design, and system design. This artificial classification is tendered only to show compatibility with estimating. It only suggests a method for teaching of estimating. It does not describe a new classification for design.

2.1
THE DESIGN AND OPTIMIZATION PROCEDURE

The design process has been divided into problem, concept, engineering models, evaluation, and design. Optimization simply adds the cost notion to bring continuity leading to the improved and optimized design. Note again Figures 1.2 and 1.4.

Problem. The initial description of a problem is a vague representation satisfying some want. It is necessary to transform this into a more useful shape. By presuming the existence of a primitive problem, information (technical or nontechnical), costs, and other data are superficially gathered to give form to the problem. Suppliers, customers, competitors, standardization groups, safety and patent releases, and laboratories are sources for ideas. The reader may wonder why all the fuss over a simple problem statement. To give heed to a raw and imperfect problem is not unknown; engineering students are not the only ones guilty. But a thoughtful and reasoned statement specifying the problem leads, per chance, to a more efficient result. Questions may guide the formation of a problem statement such as "Does this fit the company's needs, interests, and abilities? Are the people connected with the problem capable of carrying it to completion, or can suitable people be hired?"

Concept. The stage is set for the concept search after a problem with subsidiary restrictions has been defined. The quest may start with idealism such as the perfect gas laws, frictionless rolling, or perpetual motion, for example. It is a searching, learning, and recognizing; it is not application, as that comes later. A timely and fortunate search may uncover unapplied principles. When one considers that a million scientific and technical articles, more or less, are published annually, it should be clear that a listing, ignoring study for the moment, of all information even within a narrow field is a hopeless cause. Here is a recognized defeat in the face of overwhelming odds. Nonetheless, the chance of finding the basic idea for a new development may be found through patent disclosures, new texts, or journals in the field. Instead of the unobtainable goal of completeness, there are other goals capable of being achieved in the search. Knowing where to start and when to stop are lessons of experience.

Engineering Models. The engineering model involves application of creditable concepts which were uncovered earlier. The formation of this model may range from a casual back-of-envelope model to a vast and complicated physical shape. The formation of a model is an engineering trait and distinguishes the engineering pattern of thought. Models, whether they are experimental or rational, permit manipulation for theories or testing. Using physical mockups, laboratory testing is able to provide numerical answers. Data are obtained, results are noted, and conclusions are stated. An engineer will choose the cheapest of the methods to state and understand his model.

Evaluation. Ultimately the engineering model reaches the evaluation point. A compromise forced on the engineer by economics, physical laws, politics, social mores, ignorance, and the human fault of stupidity discolors our engineering model. Even the moral questions may be debated. The wisest man alive cannot foresee all the future effects of the design, but it is bold to ask. Others may cooperate at this point: The stylist may abridge and direct the progress, the manager may foresee other problems, and the marketing man may be useful. But here is an appropriate place to stop, pause, and evaluate. It is at this juncture that cost estimating provides a preliminary estimate. This is discussed in Chapter 6.

The cost estimator has a responsibility that parallels the design engineer. The cost-estimating process proceeds along lines similar to those which have been discussed. After the problem is wisely stated (couched in the design engineering model), estimating ideas are considered, a cost-estimating model is selected, evaluation trials are determined (shortcomings with time, money, staff, programs, or information impede the model), and finally the cost estimate is completed.

Actually, the proper evaluation of engineering models is a continuous process. Experience says that it takes a long time to pass from the idea stage to the design stage. For instance, in the evaluation of new product ideas, one study uncovered the mortality of new products. Twenty companies with 540 possibilities in the idea stage at the research level distilled to only one which was placed into regular production.

There are questions that we ask in doing an evaluation: What will be the total cost of developing this design up to and including the sales promotion? What will be the profit of the total investment during the first few years of production and sales? How long will it take the initial investment to be returned? Does a new product coincide with the abilities and experience of our company? In considering these questions a number of factors should be noted, so many in fact that they can be best summarized by Table 2.1.

Design. Design is the execution of the plan into being and shape. In fulfilling functional requirements design involves computing, drafting, checking, specifying, and the like. It answers the question "How shall it be built?" rather than "How will it work?" By emphasizing the word design as a term in the *design process* we do not intend to overinflate its importance to the depreciation of other steps, as they are equally important; nonetheless, a greater proportion of the designers' time is tied to this work.

Detailed estimates are undertaken during and following the design act

TABLE 2.1

CONSIDERATIONS IN EVALUATING NEW
PRODUCT POSSIBILITIES

Factors Affecting Sales	Factors Affecting Production
1. Nature of product Physical properties Specifications Packaging Storage Safety factors	1. Raw materials Cost (past, present, and future) Extent and adequacy of present and future sources

2. Market factors
Uses
Volume
Price range
Character, number, and
location of customers
	a. Number of producers
	    b. Location of sources
	    c. Collection and transportation
	    d. Seasonal factors
	    e. Waste material

3. Competing products Properties and market Manufacturers	Specifications and impurities Containers Storage Safety factors Legal requirements Patents
4. Sales, service	a. Product patents b. Use patents
5. Advertising	
6. Legal requirement	Tax, tariff, etc. Possible future developments
7. Patents Product patents Use patents	2. Manufacturing process Basic character Type of equipment
8. Trademarks	Labor requirements Power requirements
9. Tax, tariff, etc.	Optimum production rate Patents (manufacture)
10. Possible future developments	Production legal matters Safety factors Environmental factors Possible future development Automation

using methods of detailed estimating (Chapter 7). A price, cost, quotation, budget, financial or operating cost policy, or measure of return is a natural result of a detailed estimate.

Practicing designers and students are sometimes convinced that the engineering process, by definition, ends with design. However, there is a growing understanding that design is one starting point for optimization. The design process provides an accurate and quantitative understanding of how the design variables interact. Optimization then finds a measure of design effectiveness, expressible in terms of the design variables, and selects the values for the design variables for an optimum level of efficiency. Optimization is not the incidental by-product of an informal design model; indeed, a specific mathematical approach in quantitative terms is required. The place of cost engineering in this optimization partnership is to fine-tune the design. More is said in Chapter 12 about fine-tuning devices.

In the design and optimization enigma, many authors discuss the creation of alternatives using terms such as brainstorming, idea generators, fantasy, empathy, and set-breaking experiences. These are terms describing various rituals purporting to enlarge the number of solutions. It is well known that methods such as these are possible to get broad solutions to sticky problems. To overlook these elegant agents would be a disservice. However, their use might be overpublicized for there is no panacea for the engineer and his confrontation with design. Generally speaking, the crux of the design is close attention to detail. Often times relationships are abstruse and are not immediately apparent to a team approach.

The word *model* has many meanings. We define it as a *representation* to explain some aspect.

It is seldom that we are able to manipulate reality, as it may be either impossible or uneconomical. The market in a free economy is an illustration of the former, while an expensive nuclear reactor for electrical generation illustrates the latter. The prediction of reality through a form of mathematical abstraction is the cause for interest in models. Engineers apply these ideas as a means to scale larger problems. For example, many times the analyst is unable to comprehend the actual system, and with limited mental powers of perspective, a model is often a satisfactory substitute. Discovery of which variables are pertinent, rejection or confirmation of prototypes, comparison to a standard, or the application of the principles of similitude give importance to modeling.

Models are frequently classified as numeric or nonnumeric, by subject, or by situation. We choose to segregate models according to physical, schematic, or mathematical notions. All three are important to the engineer and are used in the design process. The scale of abstraction would proceed from the physical to the mathematical extreme.

Physical models principally involve change of scale. The globe looks like the earth, for instance.

When one set of properties is used to represent a second set of properties, a schematic model results. It may or may not have a look-alike appearance to the real-world situation it represents. Coding processes may be used, such as the chalk-board demonstration of a football play appearing as crosses ($\times$) and circles ($\bigcirc$). Hydraulic systems are of benefit in understanding electrical systems and vice versa. Organization, flow-process, and man-machine charts are other examples of schematic models. In all cases the schematic model captures the critical feature of the real thing and ignores the unimportant in order to facilitate a solution.

Mathematical models operate with numbers, letters, or other symbols in their imitation of variables and relationships. Although they are more difficult to comprehend, it is conceded that they are the most general and precise in application. In an approximate way mathematical models explain and control the real situation. It is customary to manipulate mathematical models according to the conventional rules of mathematics.

Mathematical models are desirable in cost estimating not only because they are easy to manipulate but also because they yield more accurate results than do either physical or schematic models. The unit cost formula is an example of a model found in operation estimating. Whenever an engineer uses a *recapitulation sheet* where labor, material, and overhead are summed we are employing a mathematical model in a procedural sense. The discounted cash flow model used by engineering economists to describe a rate of return of an asset is the mathematical method for dealing with project estimates.

For cost estimating we make two distinctions within the mathematical model: *functional and variational models*. The former deals with problem-solving formulas, while the latter uses optimum solving approaches.

There are precautions to guard against in modeling. The rule of thumb is to use the fewest number. Models should be flexible to permit repeated applications. The mathematical routines of manipulation should be as simple as is necessary for a solution. Arithmetic is preferred over algebra, algebra is preferred over calculus, calculus is preferred over vectors, and so forth. The model that is simple improves its saleability to management, and increased understanding and confidence results from its use. We should point out that the analyst does not attempt to force a special technique on a particular set of circumstances; rather the problem hunting for a solution is a preferable course of action.

2.3
OPERATION DESIGN

An operation design has its economic value evaluated on the basis of immediate cost. The length of time is relatively short and the long-term consequences are assumed to be unimportant. Designs of this type abound. Their initiation is from an internal decision of management. Many engineers are concerned with designs of this nature. The economic evaluation may be done by the designer rather than by other specialists.

Activities in an enterprise are sometimes grouped under the heading *operations*. Included in this title are activities associated with engineering, finance, marketing, and industrial relations. In evaluating operations, the purpose is the attainment of an objective through the selection of operational alternatives that are most favorable from an economic point of view. Frequently, that objective is the economy of performance as it relates to operations. The economic loading of equipment above or below normal capacity, the economic load distribution between equipment and machines, the economic purchase quantity, the economic production quantity, the economical size of repair crews, when to produce and when to purchase, and the economy of maintenance require engineering analysis and are coupled with operation estimates.

The basis for economic evaluation follows many forms. In all these approaches microanalysis is necessary. A facility for breaking a design into its smallest element—both physical and economic—is a characteristic. To do this the risk involved is minimal, as the task is going to be done. If it were not minimal, many of these designs would not be started.

Before an operation design is started some authority must request and initiate this undertaking. A work order, production planner, traveler, foreman's request, work-simplification savings, employee suggestion, architect's request, and public declaration are typical reasons for an operation design. These requests precede the approval to begin a design activity.

Manufacturing engineering design is concerned with the solution to physical problems of production. The goal is the highest productive efficiency and is secured by providing the required amount of material, proper machines and tools, necessary personnel, and adequate instruction at the right time and place. Usually several processes, or at least a number of variations, can be selected for doing a job. The process design contributed by the manufacturing engineer is a master plan for production. The process design serves as the basis for manufacturing cost in many instances. This design may develop from a sketch, physical model, or complete set of engineering drawings and specifications. To illustrate, consider the isometric sketch of the invention coined the "dynamic chip breaker," shown in Figure 2.1. This device is added to an engine lathe to break chips into short lengths for convenient disposal. A bill of materials and a complete set of drawings would customarily be available. The specific problem is to evaluate several processes suitable for the fabrication of the connecting rod, one of several parts that go to the final assembly. In fabricating the connecting rod the engineer believes that the principal shape can be forged, cast, or machined. In his thinking, he may choose to outline the fabrication steps for each of these major processes, as given by Table 2.2.

FIGURE 2.1

DEVICE FOR BREAKING CONTINUOUS CHIPS
RESULTING FROM MACHINING

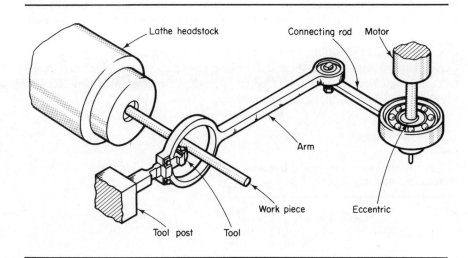

The engineer cannot rush to judgment and select one of the three methods, as each is at one time or another superior. In the case of extremely low production, the machining method would be the one adopted; for intermediate quantity

TABLE 2.2

PROCESS DESIGN FOR ALTERNATIVE
MANUFACTURING METHODS FOR
PRODUCING THE CONNECTING ROD
FOR THE DYNAMIC CHIP BREAKER

1. Forging method
   Cut barstock to length
   Rough forge billet
   Trim flashing
   Heat treat
   Tumble to remove scale
   Machine I.D. of bore and size faces of bosses
   Machine 0.375 dimension
   Deburr

2. Sand-casting method
   Prepare mold and place cores
   Cast connecting rod
   Relieve stress
   Tumble to clean
   Machine I.D. of bore and face of bosses
   Machine 0.375 dimension
   Deburr

3. Machining method
   Band saw basic shape
   Mill bore and $2\frac{3}{4}$-in. dimension
   Mill $\frac{3}{8}$ arm height
   Contour mill $\frac{1}{2}$ arm width and $3\frac{1}{8}$ dimension
   Drill 0.375 dimension
   Deburr sharp edges

requirements casting would be selected over the forging method; while for large quantities of production, forging or other precision casting methods could be adopted. The engineer knows this sort of thing intuitively. But to depend on fragmentary judgment results in inconsistencies and violates practices of good design. Assuming that the analysis is for 200 units, it is seen that the machine method proved to be superior to the other two. The labor cost was higher for this method, but the tooling cost is less than for casting and forging. Examination of Table 2.3 shows that the standard hours per piece are less for the rejected methods than for the method accepted. In other circumstances the decision could be in favor of either casting or forging.

TABLE 2.3

SUMMARY ANALYSIS FOR 200 UNITS FOR
PRODUCING THE CONNECTING ROD

Machine method	
1. Tooling costs	None
2. Perishable tools, cost/piece	$ 0.45
3. Hourly labor rate, direct labor	3.65
4. Standard hours/piece	1.050
Unit total cost	$ 4.28
Total operation cost	$856.50
Casting method	
1. Tooling costs	$250.00
2. Perishable tools, cost/piece	0.35
3. Hourly labor rate, direct labor	3.15
4. Standard hours/piece	0.905
Unit total cost	$ 4.45
Total operation cost	$890.15
Forging method	
1. Tooling costs	$ 875.00
2. Perishable tools, cost/piece	0.15
3. Hourly labor rate, direct labor	3.20
4. Standard hours/piece	0.650
Unit total cost	$ 6.61
Total operation cost	$ 13.21

Only a few steps have been shown for the process design for the fabrication of the connecting rod. Additional process design is necessary for the assembly. Station and sequence charts are required to show the various assembly, inspection, and test work. The engineer considers the assembly station requirements and specifies the makeup of these work stations. A manufacturing plan for assembly must specify the assembly tools, jigs, and fixtures and under what conditions the usage has been planned. Table 2.4 indicates the assembly design used for the dynamic chip breaker.

As has been suggested, the magnitude of the production quantity requirement to a large extent stipulates production. In the case of fabrication or assembly, plant rearrangement and layout may be anticipated; utilities such as power, gas, air, water piping, and floor space requirements are subject to design. Discussion of this area is not considered here. In the cost analysis for assembly, the designer may rely on verified plant information, sometimes called assembly standard time data or engineering performance data. This and other matters are discussed in Chapters 3 and 5–7.

TABLE 2.4

Work Stations: Three required for production

Mechanical assembly
  Press small bearing in tool arm
  Press large bearing in connecting rod
  Press connection rod pin in connecting rod and tool arm
  Fit eccentric disk with eccentric

Bench assembly
  Screw in setscrews for eccentric
  Attach rubber coupling to motor
  Screw in setscrews for latch
  Adjust equipment for reference dimensions

Inspection, test, and package
  Visual inspection for quality
  Operate equipment per specifications for 3 min
  Check all screws, fits, shafts
    Wrap in Kraft paper
    Include warranty
    Pack in container

For fabrication process design, a functional estimating model can be used. This typical functional model ignores overhead, equipment, materials, and other factors. In this elementary situation we have

$$C_T = \sum (C_l + C_t + C_{tp}) \tag{2-1}$$

where C_T = total cost per operation
    C_l = labor cost per operation
    C_t = tooling cost per operation
    C_{tp} = perishable tooling cost

The summation sign, given without indices, simply requires the addition operation for the cost element.

The term *tool design* means the design of special tools for the economical production of machine-or-assembly-made components. Tool designers are practical men with wide empirical knowledge. The tool engineer is responsible for the design, acquisition, approval, and installation of the tool. While no attempt is made to discuss tool design, this activity is subject to variation, and the final design must be evaluated in terms of economy and cost. One of the typical functional models that is applied to the evaluation of tools where a given production is known with certainty is

$$C = \frac{Na(1 + t) - S}{I + T + D + M} \tag{2-2}$$

where C = first cost of tool
    N = number of units manufactured per year
    a = savings in labor cost per unit compared to another operation
    t = percentage of overhead applied on labor saved
    S = yearly cost of setup, dollars
    I = annual allowance for interest on investment, decimal
    T = annual allowance for taxes, decimal
    D = annual allowance for depreciation, decimal
    M = annual allowance for maintenance, decimal

At every stage in manufacturing, tool engineers are confronted with dollar signs. The tool designer should know enough of cost estimating to determine whether temporary tooling would suffice even though funds are provided for more expensive permanent tooling. He should be able to comprehend his design plans so as to initiate or defend decisions when writing off the tooling on a single run as opposed to write-off distributed against probable future reruns.

Construction estimating considers the preparation of estimates of probable costs, budgets, financing, and bidding. The man most likely to do this work is a cost engineer having costwise knowledge of construction. His job is to know where the money is going, what operations are costing more than estimated, where possible savings can be made, and what the total probable cost of the operation will be. Accordingly, the cost engineer plans the construction job considering the features such as transportation and hauling equipment (trucks, carryalls, tractors) and loading and hoisting machines. He must determine what on-site plant units (including size and capacity) the job will require, such as concrete mixing, sand and aggregate crushing, classifying, and conveyor units; what administrative and shop buildings will be necessary; and where construction roads, raw and fresh water systems, air, and power lines will be necessary.

The following example illustrates a construction operation design of simple hand labor in a country without mechanical equipment. The engineer is to estimate the cost of labor required to excavate earth by hand under different conditions. For the first operation the soil is a sandy loam which requires light loosening with a pick prior to shoveling. The maximum depth of the trench will be 4 feet. Climatic conditions are good with the temperature averaging about 70°F. The laborer should easily loosen 5 cubic yards of earth per hour using a long-handle round-pointed shovel; it should require about 150 loads to move a cubic yard of earth. If a laborer can handle 2.5 shovel loads per minute, he will remove 1 cubic yard per hour. These rates of production were for 2 man-hours to loosen and 1 man-hour to remove a cubic yard of earth of the trench or a total of 3 man-hours per cubic yard. For the second operation, the soil is a tough clay which is difficult to dig and lumps badly. The maximum depth of the trench would be 5 feet. The temperature is estimated to be about 100°F without shade. Under these conditions a laborer may not loosen more than 0.25 cubic yard per hour. Because of the physical condition of the loosened earth, it is estimated to require about 180 shovel loads to move a cubic yard of earth. It is estimated that a laborer can handle 2 shovel loads per minute or 120 shovel loads per hour, which is equivalent to 0.67 cubic yard per man-hour or 1.5 man-hours per cubic yard. Given these rates of production, it is necessary to have 4 man-hours to loosen and 1.5 man-hours to remove a cubic yard, or a total of 5.5 man-hours per cubic yard. Perceiving as he did varying rates of production for hand operations, the engineer computed the total number of man-hours per cubic yard. For a given labor cost per hour, including certain fringes and other factors, the cost engineer is able to compute per yard the cost of excavation using functional cost models.

The design activity associated with the planning of more effective methods is called methods engineering. Concerned with the objective of securing maximum labor effectiveness, the methods designer must consider the repetitiveness of the activity, the hourly labor rate paid, its labor content, and the anticipated life of the activity. In addition to detailing the individual motions for a particular job, there is a standard of accomplishment, such as pieces per hour, which is useful for many things. Before one is able to measure work, the controlling circumstances must be standardized and a time value established for the performance of each of the tasks of that job. If methods engineering is to be successful, it must be understood that various human factors must be given special cognizance. If the methods engineer fails to realize the importance of the human element, little hope can be given to the success of his endeavors. When a working model of an operation has been decided on, it is essential to ensure the standardization of the working method, materials, equipment, and working conditions. Unless consideration is given these factors, it is more difficult to establish a time value of the task, for standardization has a direct bearing on the time required to perform a task.

The problem of setting time standards for labor control for an operation of a taping machine to wrap tape will now be considered. This problem was initially described in Chapter 1. Due to the complexity of the operator's task and the uncertain quality of the material, standardization was difficult. Functional models were attempted and found lacking. Throughout the investigation the methods engineer learned that tension in the tape was proportional to its speed and that the ultimate strength of the material was related to width. Using these discoveries he constructs a variational model. For this type of model there is no specific formula. The cost engineer must design his model. In this case he is attempting to control cost, and it is appropriate that cost be the major dependent variable.

The problem is to identify the average unit cost for taping the cable. We assume that the cost is the sum of three factors, namely, wrapping, tape repair, and setup. The taping cost per operation is

$$\text{taping cost} = C_o t_w \tag{2-3}$$

where C_o = cost of wrapping time, dollars per minute
t_w = minutes per cable to wrap

The tape repair cost per operation is given as

$$\text{tape repair cost} = C_o t_r (t_{w/T}) \tag{2-4}$$

where t_r = tape repairing time, minutes per operation
T = average minutes before tape breakage, minutes per operation

This varies with tape width W. The ratio $t_{w/T}$ is dimensionless, and the smaller this ratio becomes, the smaller the repair cost. Contrariwise, to leave the wrapping unchallenged for its performance raises that cost.

The setup cost includes preparation, inspection, handling and teardown, and is

$$\text{setup cost} = C_o t_s \tag{2-5}$$

where t_s = setup time for cable and tape, minutes per cable

Summation of these costs leads to

$$C_u = C_o t_w + C_o t_r (t_{w/T}) + C_o t_s \tag{2-6}$$

where C_u = dollars per unit cable

Wrapping time t_w is the time that the cable is actually being wrapped and is further described by

$$t_w = \frac{L}{lN} = \frac{L\pi d}{12Vl} \tag{2-7}$$

where d = cable diameter, inch
L = axial cable length, inch
V = wrapping speed, surface feet per minute
l = lay rate, inches per revolution
N = rotary velocity, revolutions per minute

One revolution places a "lay" of tape on the cable as shown in Figure 2.2.

There are empirical relationships that express the time at which breakage occurs or

$$VT^a = K \tag{2-8}$$

FIGURE 2.2

GEOMETRIC REPRESENTATION OF TAPE BEING
WOUND ABOUT A CABLE

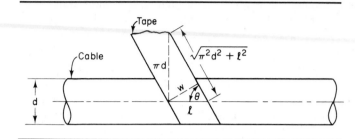

where a and $K =$ empirical parameters for a given l. Equation (2-8) is found by testing under actual conditions.

If costs in Equation (2-6) are not considered, the model yields

$$t_u = t_w + t_r(t_{w/T}) + t_s \tag{2-9}$$

where $t_u =$ time required to produce a cable, minutes

Upon substitution of Equations (2-7) and (2-8) into Equation (2-6), the cost model becomes

$$C_u = C_o\left(\frac{L\pi d}{12Vl}\right) + C_o t_r \frac{\left(\frac{L\pi d}{12Vl}\right)}{\left(\frac{K}{V}\right)^{1/a}} + C_o t_s \tag{2-10}$$

This functional estimating model is sometimes called an objective equation. The *objective* for this equation is minimum cost.

Inasmuch as the tape operation is sensitive to velocity we have constructed Equation (2-10) to have velocity as a controllable variable. It can be varied by the taping machine operator under instructions by the methods engineer.

There are production constraints that restrict the control of the operation, as

$$\theta < \frac{\pi}{2} \text{ radians} \tag{2-11}$$

because there must be an angle θ to wrap tape and

$$W \leq 5 \text{ inches} \tag{2-12}$$

because the taping head will not accept wider tape.

A particular set of information is $C_o = \$0.50$, $L = 144$, $d = 2$, $l = 1$, $t_r = 3$, $t_s = 5$, $a = 0.2$, and $K = 80$ is substituted into Equation (2-10). We can solve C_u for various values of V, or we can plot the terms as shown in Figure 2.3. A graphical sum gives C_u directly. Note that a minimum cost occurs at about 50 feet per minute.

Why did we not consider the cost of the tape material or the cable in our variational model? Inasmuch as our purpose was to standardize methods and cost, it was necessary to first fix the operating conditions. With that achieved, tape and cable material costs could be added to give a final standard cost for the operation. Tape and cable are required irrespective of finding the velocity for minimum cost, and they contribute to cost as another fixed cost factor. Their later addition does not alter the optimum velocity point for minimum cost.

Has this solution satisfied the problem for setting cost standards once and for all? Maybe not, as our empirical and theoretical variational model may have overlooked factors that were significant. The number of combinations of width,

FIGURE 2.3

OPERATIONAL COST CURVES FOR
WRAPPING TAPE AROUND A CABLE

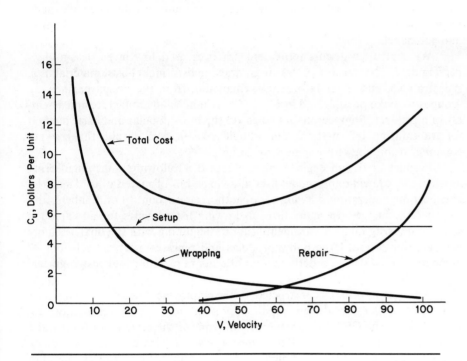

lay, diameter, material, etc., may be so vast that permanent solutions may never be found.

2.4 PRODUCT DESIGN

The factors about product design are many and varied. No doubt it is one of the most complicated and involved activities for a firm. Design acts as a governor regulating the flow of work through design, production, accounting, sales, marketing, and service. Due to its central position of importance, product design is an engineering activity concerned with phases of management, operations, manufacturing, service, and maintenance.

The objective of product design is a product. It differs from an operation or project or system design. A product design is one where there is replication. That is, there may be several or millions of similar units. Production of a single unit could qualify under special circumstances. A cohesive body of principles and practices for product estimating exists whenever several or more units are produced for sale.

Too often product-oriented companies lose sight of their central objective; the first concern is the success of the product lines. No other problem of management affects profits so squarely. Whether the products are ablating materials or zener diodes, the problems look alike in external appearance. If we consider giants such as General Motors, with a sales revenue exceeding the gross national product of all but several major world powers, with its production of over 6 million cars to the other extreme of the one-man machine shop operating from a home garage, we cannot say the problems are identical, except in a very gross way. Objectives, of course, may be alike. Executives make decisions almost every day that affect the product line in such matters as allocations of manpower, factory space, and sales efforts. They must also decide major product questions—whether to undertake a new development, to introduce a new product, or to eliminate an old one. Mistakes in any of these are usually costly and may be hazardous to long-term stability. But rarely are product decisions made entirely by executives. Often such decisions require the specialized knowledge of experts in fields of research, engineering design, manufacturing, marketing, law, finance, and personnel.

We shall now discuss terms and practices used to convey design engineering data. The purpose of drawings, drawing lists, model lists, material lists, specifications, and other information emanating from the design engineering group is to make possible and insist on the economical manufacture and assembly of a product. Engineering drawings are the most popular and basic method for transmitting information. This should remain true despite the growing computer technology for engineering design.

Product or design engineering provides the following data: engineering instructions, drawing and model lists, drawings, bill of material lists, material specifications, operation drawings, tool authorization, control of numbers, and engineering changes. In many firms a drawing list provides for an exploded series of drawing numbers for identification. Beginning with the final assembly it may cover several large bulky products and represents several echelons of numbers. It provides an orderly pattern of tying in planning and cost-engineering activities.

A sketch, informally constructed, or the employment of a cathode ray tube with a printer useful for design graphics represent the diverse range of drawings. In the case of the CRT the designer sits in front of the screen and is provided with an input keyboard that includes descriptions such as transistors, resistors, or straight lines. In this graphical man-machine system, the operator can develop complete sketches with the aid of a light pen. Drawings can be subsequently recalled from computer memory and changes readily incorporated. Whatever means are employed, the drawing promises to remain an essential ingredient of information transmittal for all phases of design and manufacturing.

The various methods for preparing and issuing bill of material lists may be partially or wholly done on the corner of the drawing or may be separately listed on other documents. Whenever there is a separate bill of material, the list may be typewritten or appear on a computer printout. It contains information such as the number of the drawing, the name of the part or assembly, the material to be used for the part, and the quantity per unit of the final product.

The control and assignment of numbers for various engineering data such as drawings, bill of material lists, and engineering instructions is a function of design engineering. Cost estimating accommodates to the design numbering system, which is hopefully comprehensive enough to serve engineering, manufacturing, and sales and yet unique enough to indicate readily specific conditions.

In the one-man shop, engineering changes present few problems. In most design activities, whether they are small or complex, product changes, no matter what their nature, must be taken care of in a systematic way. The changes instituted by design engineering and the drafting department create changes in product cost and must be evaluated by sales, marketing, and cost estimating. Certain manufacturing cost reduction practices, which in no way functionally affect the product, may alter costs. Careless disregard of revisions can lead to later chaos.

Functional models that are used to find product cost are straightforward; however, the gathering of data is not simple and represents a major task to cost engineers concerned with product costing. These functional models are algebraic in form and are a summation of separate elements. One approach would consider direct labor, direct and indirect materials, overhead, engineering costs for design and development, general administrative expenses, selling, and finally profit. Each of these categories can be broken down into subcategories. New categories peculiar to the particular situation can be isolated and often times are considered. The development of the cost associated with many of the cost elements requires use of separate forms and procedures and recap sheets.

The cost engineer armed with business and technical forecasts, prices of materials, cost and availability of labor, and the extent of competition from markets can compile these estimates into a typical model given as

$$\text{price} = M_0 e^{-k_m t} + F_0 e^{-k_f t} \tag{2-13}$$

where M_0 = price of the initial unit at inception of production, dollars
k_m = decay experience for the product
t = time, typically years
F_0 = price floor initially, dollars
k_f = decay experience of the price floor

A typical graph for a product selling price would be as shown in Figure 2.4. The floor price is the point at which the most efficient producer will make a reasonable profit. With a fairly stable floor price the margin over this floor price decays rapidly as more competitors attempt to capture the market. If we illustrate by simple discussion, the reader may agree. The price for a product will remain

FIGURE 2.4

SELLING PRICE PROJECTION FOR THE MODEL
$M_0 e^{-k_m t} + F_0 e^{-k_f t}$

static if production and demand remain in equilibrium. Excess demand tends to increase prices, while excess supply suppresses the prices. If profit margins are high, more competitors are apt to enter the market. Given a simple competitive situation, prices tend to fall with time as long as some producer can earn a return for his capital by operating a modern manufacturing operation. The floor price is based on cost estimates for an efficient operation which allows for technological improvements as well as changes in fixed and variable costs. Our presentation about product functional models is suggestive and sketchy at this point. Additional development is deferred until Chapter 9.

Variational models are not straightforward, nor can we compile a sufficient number of techniques in this brief episode. Construction of an objective function and constraints calls for understanding. But when the cost estimator confronts a product design which he feels might be optimized, the path of thinking could be similar to this. Using the intermediate chemical plant process problem, presented initially in Chapter 1, note its flow, illustrated by Figure 2.5. To optimize product production the cost estimator needs to know how the engineering variables interact. This is an important step which cannot be avoided. With this knowledge, he invents an objective function expressed in terms of product production variables along with the several constraints.

From the flow chart, we recognize that product production has a large number of attributes which must be considered. A flow chart of the total process as shown by Figure 2.5 is too big for analysis, and we select the hydrobromination subsystem for attention.

The chemistry of hydrobromination has main reactions as

$$
\underset{\text{Alpha olefin}}{\text{R}-\overset{\text{H}}{\underset{\text{H}}{\text{C}}}-\overset{\text{H}}{\text{C}}=\overset{\text{H}}{\text{C}}-\text{H}} + \underset{\substack{\text{Hydrogen}\\\text{bromide}}}{\text{HBr}} \xrightarrow[\substack{85°\text{F}\\5\text{ psig}}]{\text{Initiator}} \underset{\substack{\text{Primary}\\\text{alkyl bromide}}}{\text{R}-\overset{\text{H}}{\underset{\text{H}}{\text{C}}}-\overset{\text{H}}{\underset{\text{H}}{\text{C}}}-\overset{\text{H}}{\underset{\text{H}}{\text{C}}}-\text{Br}}
$$

$$
\text{R}-\overset{\text{H}}{\underset{\text{H}}{\text{C}}}-\overset{\text{H}}{\text{C}}=\overset{\text{H}}{\text{C}}-\text{H} + \underset{\text{Ozone}}{\text{O}_3} \xrightarrow[\text{Atm. press.}]{85°\text{F}} \underset{\text{Ozonide}}{\text{R}-\overset{\text{H}}{\underset{\text{H}}{\text{C}}}-\text{C} \underset{\text{O}}{\overset{\text{O}-\text{O}}{\diagdown}} \overset{\text{H}}{\underset{\diagdown}{\text{CH}}}}
$$

and a side reaction as

$$
\text{R}-\overset{\text{H}}{\underset{\text{H}}{\text{C}}}-\overset{\text{H}}{\text{C}}=\text{CH} + \text{HBr} \xrightarrow{>100°\text{F}} \underset{\substack{\text{Secondary alkyl}\\\text{bromide}}}{\text{R}-\overset{\text{H}}{\underset{\text{H}}{\text{C}}}-\overset{\text{H}}{\underset{\underset{\text{Br}}{|}}{\text{C}}}-\overset{\text{H}}{\underset{\text{H}}{\text{CH}}}}
$$

To recap, we have the main reaction of forming the primary bromide and, second, the formation of the ozonide, which is the free radical initiation that is used. It is important to have a high yield of the primary bromide since the secondary bromide would lead to formation of a branched chain in the alkyl dimethylamines. These branched materials are poorer in detergency quality. A schematic and engineering flow chart of the hydrobromination step is shown by Figures 2.6 and 2.7.

The main feature in the hydrobromination system is the rising film reactor. Olefin is introduced in the annular space between the gas tube and the reaction tube. The olefin is picked up and plated out in the thin turbulent film by the fast-moving HBr stream. This film results in heat mass and heat transfer with an exothermic reaction. The high temperatures favor formation of a secondary bromide. Residence time in this reactor is about 20 seconds, and a free radical

FIGURE 2.5

FLOW CHART OF THE CONVERSION OF ALPHA OLEFINS TO ALKYL DIMETHYLAMINES

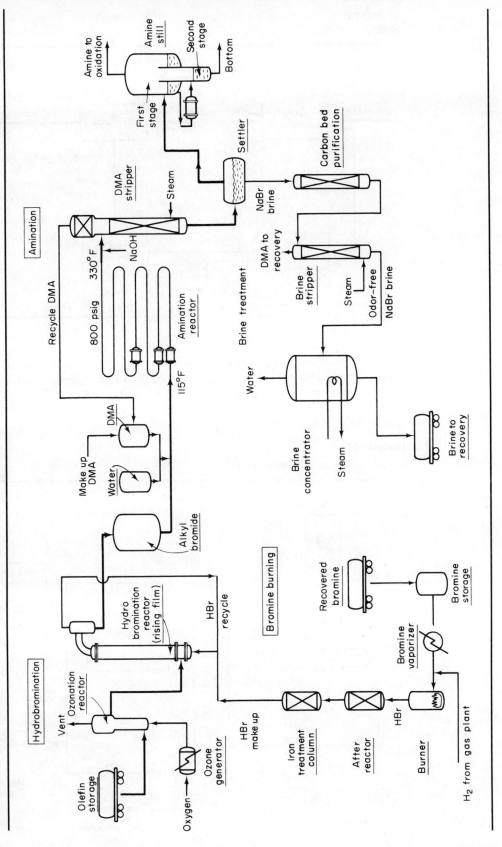

FIGURE 2.6

SCHEMATIC OF THE PRODUCT ALKYL
DIMETHYLAMINES

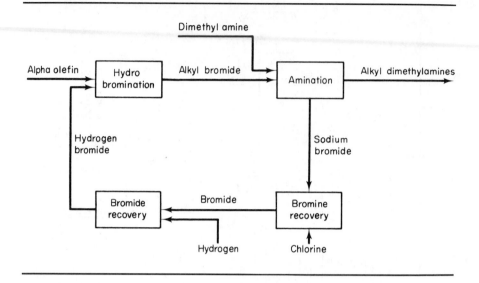

reaction mechanism gives a conversion of better than 99 % in this time. The film is maintained by blowing HBr through the tubes at about 75 feet per second with a rotary compressor. The HBr is only 96–98 % HBr, as it contains impurities

FIGURE 2.7

MODEL OF HYDROBROMINATION

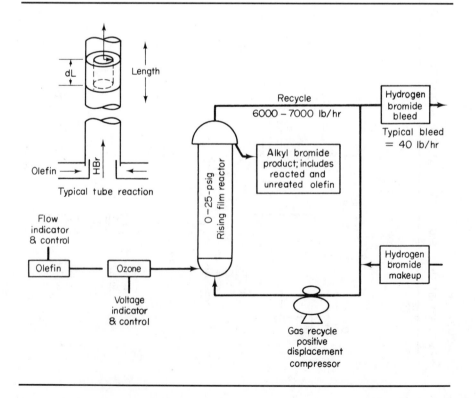

resulting from an excess of hydrogen in the burning process. By the use of a bleed stream, the HBr level in the reactor is maintained between 70 and 80%. Another main control for this system is the initiator level which is regulated by varying the voltage on the ozone generator. The 2–3% ozone in oxygen is taken up by the olefin quantitatively in the concurrent bubble column so that the ozonide level is controlled by the ozone production rate.

What has been expressed in words and figures may be related to mathematical constraints. The basic molecules of HBr into olefin may be given as

$$f_1 = \frac{dN_g}{dL} = K_g A(P_g - P_l) \qquad (2\text{-}14)$$

The reaction completeness is

$$f_2 = \frac{dC_l}{dL} = K_r C_l C_g \qquad (2\text{-}15)$$

The continuity of mass flow is

$$f_3 = \frac{dN_g}{dL} = \frac{dC_l}{dL} + \frac{dC_g}{dL} \qquad (2\text{-}16)$$

where N_g = number of moles of HBr gas
$\quad L$ = length of reaction film tube
$\quad K_g$ = constant
$\quad A$ = area
$\quad P_g$ = pressure of gas; for HBr, $P_{HBr} = P_T \times$ mole fraction HBr
$\quad P_l$ = pressure of liquid; for olefin, $P_l = K_n C_g$
$\quad C_l$ = concentration of olefin liquid
$\quad C_g$ = concentration of HBr
$\quad K_r$ = constant

The material balance around the loop is selected as

$$\text{HBr (out)} = \text{HBr (in)} - \text{HBr (reacted)} \qquad (2\text{-}17)$$

As this is a subsystem of a larger chemical plant operation, it is impossible to know exactly how profit may be affected by our optimization. Conversely, a cost model, which we minimize, would serve a better purpose at this point, and subsequently we select as the objective function

$$C = \text{cost of wasted raw materials}$$
$$= \text{unreacted olefin} + \text{HBr in bleed} + \text{ozonide} \qquad (2\text{-}18)$$

Now with the objective function and constraint equations constructed, the variational product model has been stated. The typical approach to construction of variational models requires familiarity and the ability to render a verbal description into a mathematical format. In this case differential equations have been formed which describe the production processes.

Economists choose to recognize two classes of products: consumer and producer goods. They define consumer goods as products and services that directly satisfy human wants. Color television, homes, cars, health services, and football teams would satisfy this definition. Producer goods also satisfy human wants indirectly; production machine tools, ore removal and reduction equipment, and computers are examples of producer goods. Producer goods are an intermediate step in man's effort to supply his wants. Consumer goods such as food, clothing, or shelter and our need for physical existence are reasonably stable and predictable.

If an engineer were considering the purchase of a numerically controlled machine tool in a production process being tooled up for a new product, he would be purchasing a producer good. If the factory that produced these numerically controlled machine tools were manufacturing several or more units, the factory cost engineer would associate his problem with product design and product estimating. The engineer representing his company in the exchange transaction wanting to purchase the numerically controlled machine tool would rightly approach it as a project design and use methods of project estimating.

2.5
PROJECT DESIGN

Project design is concerned with investment and is dissimilar to other designs because of the need for money appropriation. If the appropriation is a matter of immediate expense rather than a capital cost leading to depreciated expense, we are concerned with an operation design; otherwise we have a project design. A project design is for a one-of-a-kind item requiring special financing. Appropriations are more lumped-sum or first-cost type rather than operating or sequential cost. The principal concern about investment evaluation is due to its long-range impact on the financial health of the firm.

Examples of project designs are numerous: school buildings, highways, production machine tool centers, material handling systems, plants, and major equipment. It is noted that these things are physical. When one considers a task-dominated design to improve paperwork flow through an office, it is recognized as a one-time design certainly, but in the absence of capital expenditures we call this an operation design. A product design with its cost patterns is time-dependent on production of many units and is clearly of a different character. Further, a project design depends on one's perspective. To the buying engineer, the lathe is a project; to the engineer involved with design, fabrication, and construction of several or more, it is a product design. Several examples will now describe project designs.

Construction plants are frequently used in the making of roads, dams, and structures. These plants are characteristic of large construction; they are portable, and on the completion of a job their output stands fast while the plant is moved away. Construction plants are themselves the output of production plants. Construction plants vary in size and complexity ranging from wheelbarrow, pick, and shovel to plants used to build the Hoover Dam. Production plants include farming, lumbering, and mining, a great variety of factories and works, and many public work plants. The agricultural plant equipped from other production plants produces commodities from the farm. The iron mine turns out ore, from which a steel works makes steel. Production plants make equipment for themselves and for other production plants. A characteristic of the production plant is that as a whole it stands fast and its output is carried away. Service plants render service, such as transportation, storage, and communication, and provide such utilities as water, steam, gas, and electricity. In plant and building design standardization is often difficult.

The principles of cost engineering are employed to ensure that public funds obtained by various financing methods are expended in a fashion which guarantees the maximum possible benefit for many. Costs have long been a factor in decisions made by highway engineers. From the many considerations of highway transportation, its can be asserted that an economical road is achieved whenever the total cost is a minimum consistent with convenience, safety, transportation, and the ability to pay. One functional model, called the benefit-cost ratio, has been applied to public works. The benefit-cost ratio method is a comparison of the difference in annual cost to highway users when there are vehicles using an existing road in one case and an improved road in another with the annual cost of making the improvement. The applicable function equation is

$$\text{benefit-cost ratio} = \frac{r_0 - r_i}{(I_i - I_0) - (M_0 - M_i)} \qquad (2\text{-}19)$$

where $r_0 - r_i =$ decrease in road user costs after improvements per year
$I_i - I_0 =$ increase in investment costs per year
$M_0 - M_i =$ decrease in maintenance costs per year

This would simplify to a ratio of decrease in user cost per year divided by net increase in investment costs per year. A similar approach is found for irrigation, recreational development, dams, and so forth. Benefit-cost ratios are calculated for the logical alternatives and are compared with the basic condition. These matters are discussed more fully in Chapters 10 and 11.

For years the application of principles of engineering economics has aptly demonstrated the useful intermingling of design and cost engineering. The traditional course of engineering economics has been taught to a variety of engineering disciplines, and its success is evident. Many functional models exist for determining the worth of equipment. One such model is given as

$$\text{annual payment} = \frac{Pi(1 + i)^n}{(1 + i)^n - 1} \qquad (2\text{-}20)$$

where $P =$ current amount of money for paid equipment
$i =$ interest rate per period
$n =$ number of periods for the economical life of the equipment

Annual payments for several alternatives can be compared and the minimum selected. The design in this case is the equipment specification. A condensation of engineering economic theory is presented in Chapter 10.

The reader may recall the example of the design of floor girders for a new high-rise steel frame building. Due to symmetry of the building layout and floor loads, we can economize by designing a few standard sections for common use throughout the building. The building is laid out in bays, 30 by 30 feet square. We are governed by the provisions of the recent AISC Code and will design using A-36 steel. Further, we are limited to using a square or rectangular cross section for the girders where depth must equal or exceed width. The girders will be fabricated from $\frac{1}{2}$-inch plate welded into a box-beam configuration. The nature of the support system is such that the girders must be at least 6 inches wide. The girders will be loaded with a uniform load of 12 kips per foot of length. This load will impose a maximum moment of 16,200,000 in.-pounds and a maximum shear force of 180 kips. Expense of the girder installation will be the total of the costs of materials, fabrication, and erection. A-36 steel plate can be procured for a cost of \$0.12 per pound. Fabrication costs \$0.16 per inch of depth squared, and \$0.08 per inch of width squared. Erection will cost \$0.03 per pound. It should be obvious that we wish to find the design which will be the least expensive while complying with the constraints. Our cost model will be worded as

$$\text{cost} = \text{material} + \text{fabrication} + \text{erection costs}$$
$$\text{cost} = \text{weight} \times 0.12 + \text{depth}^2 \times 0.16 + \text{width}^2 \qquad (2\text{-}21)$$
$$\times\ 0.08 + \text{weight} \times 0.03$$

and consolidating material and erection costs, and expressed in terms of the variables of width W and depth D, we have

$$\text{cost} = 15.3(W + D) + 0.16D^2 + 0.08W^2 \qquad (2\text{-}22)$$

We next consider the constraint equations. The first constraint that comes to mind is the one governing the relationship of width and depth as follows:

$$f_1 = D - W \geq 0 \qquad (2\text{-}23)$$

The girders cannot be less than 6 inches wide, or

$$f_2 = W - 6 \geq 0 \qquad (2\text{-}24)$$

The AISC Code states that for A-36 steel shear stress cannot exceed 14,500 pounds per square inch on the net section. Since shear stress equals the maximum shear force divided by the net area of the cross section, we get

$$f_3 = 14,500 - \frac{V}{A} \geq 0 \qquad (2\text{-}25)$$

and when we substitute our known value of shear force and express the area in terms of W and D, we have

$$f_3 = 14,500 - \frac{180,000}{W + D} \geq 0 \qquad (2\text{-}26)$$

The code limits the tensile stress due to a bending moment to 24,000 pounds per square inch. In this beam the tensile stress will equal the moment times half the depth of the member divided by the moment of inertia of the section, or

$$f_4 = 24,000 - \frac{MD}{I \times 2} \geq 0 \qquad (2\text{-}27)$$

and again substituting known factors we get

$$f_4 = 24,000 - \frac{97.2 \times 10^6 D}{4D^3 + W(3D^2 - 3D + 1)} \geq 0 \qquad (2\text{-}28)$$

To recap, we have the following variational system of equations which we wish to minimize:

$$\text{cost} = 15.3(W + D) + 0.16D^2 + 0.08W^2 \qquad (2\text{-}29)$$

subject to

$$\begin{aligned}
f_1 &= D - W \geq 0 \\
f_2 &= W - 6 \geq 0 \\
f_3 &= 14,500 - \frac{180,000}{W + D} \geq 0 \\
f_4 &= 24,000 - \frac{97.2 \times 10^6 D}{4D^3 + W(3D^2 - 3D + 1)} \geq 0
\end{aligned} \qquad (2\text{-}30)$$

where $W, D \geq 0$.

2.6 SYSTEM DESIGN

In the context of cost estimating, a system design involves an organized design effort of operations, products, or projects in any manner and combination to satisfy an economic want. System design has several characteristics which separate it from other design practices. There seems to be an urgency to develop rational explanations about empirical systems. The empirical system, or our real world, has a distinct nonrandomness to it. A system designer attempts to anticipate this behavior and uses mathematics to explain the regularity of that system. This is an equally true statement for operation, product, and project mathematical models. What is intended now, however, is to first uncover the formulas that apply and then supply the information. For the philosophical-minded the search is for a law about laws; to the engineer, this is not an eventual reality. Accordingly, we are content to achieve a merger of design using operations, products, and projects as a sufficient description about systems design. A compatible explanation for a system estimate will be made later.

The system design, as we have proposed it, is represented by an associated system of equations. Figure 2.8 portrays a trade-off and is a classic form for many decision problems. The choice lies between the extremes of the independent variable. There are some costs which increase and some which do not. The

FIGURE 2.8

SYSTEM COST MODEL OF WAREHOUSING AND
DELIVERY OF MANUFACTURED GOODS

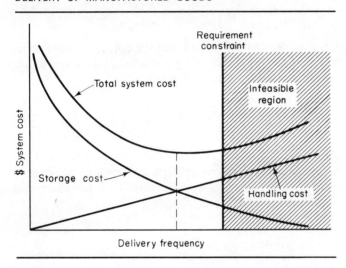

best policy is not to select an extreme alternative, but an optimum one which brings about a compromise between the increasing and decreasing costs. Now consider the problem given earlier in Chapter 1, which uses the classic trade-off illustration as a technique for establishing in-process storages and material handling policies. The method, which attempts to minimize production material handling and storage costs and reduce manufacturing space requirements, provides system cost comparisons.

We let Q_{max} be the maximum volume that will be stored during a period, and

$$Q_{max} = k\frac{U}{D} \tag{2-31}$$

where k = maximum number of deliveries stored on line during a period
U = material usage per period, cubic foot per period
D = number of deliveries per period
Then

$$C_s = Q_{max}C_d \tag{2-32}$$

where

$$C_s = k \frac{U}{D} C_d \tag{2-33}$$

and if $k = 1$ delivery, we have $C_s = UC_d/D$, where C_s = storage cost per period, C_d = cost differential between on-line manufacturing and warehousing storage, dollars per cubic foot-period. The curve decreases as the delivery frequency increases. As the frequency of delivery increases, the material handling cost for period increases. If we assume linearity, let

$$C_m = DC_l \tag{2-34}$$

where C_m = material handling cost per period
C_l = unit cost per delivery load

The objective function, or total cost, now becomes

$$C_T = \frac{UC_d}{D} + DC_l \tag{2-35}$$

In this economic balance model, as in all other types, the optimal storage policy as determined may be infeasible, for it may provide insufficient deliveries during a period. While this is not allowable, it may be stated as

$$f_1 = D - \frac{U}{L} \geq 0 \tag{2-36}$$

where L = maximum load limit of material handling device, cubic feet per delivery. This represents a constraint on our problem and must be satisfied.

Ignoring the constraint temporarily, we can solve for a minimum cost policy through calculus methods, and

$$\frac{dC_T}{dD} = -\frac{UC_d}{D^2} + C_l = 0 \tag{2-37}$$

Finally,

$$D^* = \left(\frac{UC_d}{C_l} \right)^{1/2} \tag{2-38}$$

where D^* = optimal number of deliveries per period. If feasibility is to be met,

$$D^* \geq \frac{U}{L} \quad \text{and} \quad \frac{U^2}{L^2} \leq (D^*)^2 \tag{2-39}$$

Substituting in a previous equation,

$$U \leq \frac{C_d L^2}{C_l}$$

The manufacturing storage space exactly required is

$$S_{\max} = \frac{U}{D^*} \tag{2-40}$$

What if the constraining equation appeared as in Figure 2.9? The solution is invalid if the optimal point D^* lies in the infeasible zone, and we must be satisfied by a suboptimal D which resides on the boundary. If we have a solution

FIGURE 2.9

ECONOMIC BOUNDARY RESTRICTING THE
OPTIMAL SOLUTION

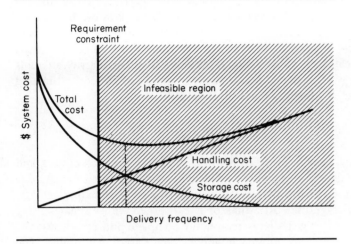

as pictured by Figure 2.9, the optimum solution is

$$C_T = \frac{UC_d}{D} + DC_l \qquad (2\text{-}41)$$

subject to

$$D = \frac{U}{L} \qquad (2\text{-}42)$$

which results in

$$C_T = C_d L + \frac{UC_l}{L} \qquad (2\text{-}43)$$

The solution is $D = U/L$. Now the systems design is not complete, for application and field testing must follow.

Before leaving this analysis, a question should be asked. Is this a law about laws? In other words, would this model apply to a chemical plant producing gasoline from a pipeline crude supply to a tank distillate storage? While the equations are a type of economic cubage quantity, the question of model *generality* is always appropriate. The reader may want to ponder this.

2.7 GUIDELINES FOR COST-ESTIMATING PERFORMANCE

Cost estimating is the rational application of quantitative methods to problems of estimating designs. The modifier *rational* suggests that the attempt will be reasonable in view of the accuracy of the numbers needed and the difficulty in obtaining them. The cost-estimating activity cannot avoid its own expense of operation, and it is fitting that we examine a notation that pertains to the effectiveness of this activity. A symbolic model is written as

$$P = f(V, c, E, n) \qquad (2\text{-}44)$$

where P = performance of the cost-estimating activity
V = value of the estimate, i.e., quotation, total cost, and so forth
c = cost of estimating per discrete component of the estimate
E = percentage error term between V and its true value
n = number of individual estimated components

The total real and potential cost inherent in making an estimate is the sum of the cost of making the estimate plus the real and potential loss resulting from the error of the estimate. If an estimate is lower than an actual realized cost, there is loss. Here the errors are of two types: those which concern the future and are prognosticated and those with uncertain past data. If the firm did not get the job, true costs remained unrealized. But we can, for the moment, postulate that potential loss exists as a result of high estimates, and an error term can be imagined for this category. Consequently, the cost-estimating activity attempts to regulate its own work such that performance is enhanced. The cost-estimating performance model will be broadened in Chapter 13.

2.8 SUMMARY

The modes of design have been structured into operation, product, project, and system. This disjunction, an admitted artificiality, was proposed for the one purpose: to see more clearly the place and time for cost-estimating activity. During the design stage the cost estimator is concerned with estimates using functional models. After a design is reasonably complete and optimization is proposed, the cost-estimating approach would adopt variational models. The purpose is to uncover a betterment of some objective, be it the operation, product, project, or system design. This chapter provides several models that intend to relate to specific design practice. Some of these models are given with little development. They suggest later ramifications. Others are given to amplify the relationship between design and cost estimator. The discussion about design techniques was intended to be skin deep for we recommend other books more suited to special engineering fields. In the construction of functional models the engineer uses existing and well-known techniques. Functional models include formulas, graphs, tables, or recapitulation sheets of information, and the approach is one of selection and choice. If betterment of design is sought, we discover the mathematical relationships. No doubt functional models will be used, but it is necessary to construct special models describing the engineering design. Remember, the models of this last kind are pretentions of reality and our confidence in them depends on a multitude of things. One of these is the determination of cost data, which is studied in the next chapter.

SELECTED REFERENCES

For discussion about the nature of models, see

ACKOFF, RUSSELL L., and MAURICE W. SASIENI: *Fundamentals of Operations Research*, John Wiley & Sons, Inc., New York, 1968, pp. 60–94.

FABRYCKY, W.J., P.M. GHARE, and P.E. TORGERSEN, *Industrial Operations Research*, Prentice-Hall, Inc., Englewood Cliffs, N.J., 1972, pp. 17–20.

MILLER, DAVID W., and MARTIN K. STARR: *Executive Decisions and Operations Research*, Prentice-Hall, Inc., Englewood Cliffs, N.J., 1960, pp. 113–70.

RUDD, DALE F., and CHARLES C. WATSON: *Strategy of Process Engineering*, John Wiley & Sons, Inc., New York, 1968, pp. 36–40.

For a general discussion about the engineering design process, see

KRICK, EDWARD V.: *An Introduction to Engineering and Engineering Design*, John Wiley & Sons, Inc., New York, 1965.

For a general discussion about systems theory, see

MESAROVIC, MIHAJLO D., ed.: *Use on General Systems Theory*, John Wiley & Sons, Inc., New York, 1964.

WILSON, WARREN E.: *Concepts of Engineering System Design*, McGraw-Hill Book Company, New York, 1965.

Some intangible features about product development may be noted in

CARLEY, H. M.: *Successful Commerical Chemical Development*, John Wiley & Sons, Inc., New York, 1955.

1. How would you categorize the design and optimization process? Identify this process and show how cost engineering might be affected.
2. Propose some primitive problems and then show the progression to ones that are more clearly formulated.
3. Examine some of the problems that you have worked on in your engineering science courses (heat transfer, dynamics, electrical theory, fluid dynamics, materials). What sort of assumptions are you required to make? What built-in assumptions make them different from real problems?
4. Can you cite some physical models, schematic models, or mathematical models?
5. Describe some trade-offs between cost and engineering design. What are some trade-offs in your personal life?
6. Discuss the classification of operation design. Name several operation designs from your practical experience.
7. What is the distinction between consumer and producer goods?
8. Assume that you are an executive of a large company requiring a product policy. Formalize this product policy.
9. What are the engineering activities generally necessary for product design?
10. List the factors relating to project design.
11. When does a short-term operation become a long-term project design?
12. Discuss the many ramifications of system design. From your other course work, what constitutes the definition for system design?
13. If a system is defined as a sequence of "executive, functional, and operating phases," how would you relate mathematics?
14. What is the role of the engineering modeler in engineering design? In cost estimating?
15. Distinguish between variational models and functional models.
16. Take a functional model and enrich it to a variational model.
17. Can you optimize variational models? Functional models? Are constraints necessary with variational models?

2-1. Equation 2-1 is a simple summation. With the operational data given, find the total cost and unit cost for each operation.

Department	Total Hours of Labor	Labor, Cost/Hr	Tooling Cost	Units of Output
Machining	4.00	$5.00	$ 200	1000
Sheet metal	0.50	$4.50	$1000	1000

2-2. Visualize another set of production assembly sequences for the Dynamic Chip Breaker (Figure 2.1) like Table 2.4. What are criteria to select an outstanding process plan?

2-3. Find the first cost of a tool where $N = 6450$ pieces, one run per year; $a = \$0.03$; $t = 50\%$; cost of each setup $= \$10$; interest for investment $= 6\%$; property taxes $= 4\%$; depreciation $= 50\%$; and maintenance allowance $= 10\%$.

2-4. What are the economic number of pieces to pay for a tool fixture if the fixture will cost $600, unit savings is $0.03, $t = 50\%$, cost of each setup $= \$10$, interest for investment $= 10\%$, taxes $= 5\%$, maintenance $= 20\%$, and annual allowance for depreciation $= 100\%$? If 7000 pieces are on the shop order, should the tool fixture be designed and built?

2-5. Construct a linear graph to portray the range of production that would make the machine method, the casting method, or the forging method of Table 2.3 most economic.

2-6. Prepare a process sheet for the aluminum 2024-T4 shaft in figure P2-6. All dimensions not shown are assumed to be $\pm\frac{1}{64}$ and the surface is 125 micro inches. Indicate operation, number, tooling, machine, and type of labor for a quantity range of 600–1000 parts. You do not need to be specific regarding information but you should indicate generally what is required. Roughly scale for missing dimensions.

FIGURE P2-6

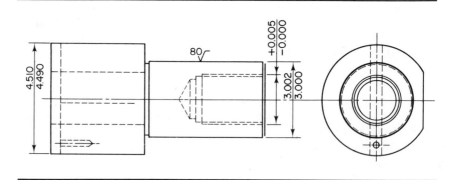

2-7. For the taping machine problem, graphically construct the curves to find C_u and t_u for $C_o = \$0.50$, $L = 144$ inch, $d = \frac{1}{2}$ inch, $l = 1$ inch, $t_r = 3$ min, $t_s = 5$ min, $a = 0.2$, and $K = 80$.

2-8. Graphically construct the two-dimensional feasible zone for the girder problem using Equations (2-23), (2-24), and (2-26) only. At what location on this space would you expect to find the optimum point? Discuss the possibility of how you would add Equation (2-28).

2-9. A revolutionary product has been developed. It is believed that novelty will sustain early sales, but long-range estimates have concluded that competing designs with a constant retail price of $3 will provide a floor. An initial price of $7, and a decay of 0.25 is expected. How many years will elapse until the new product intersects the competing price of $3 per unit?

2-10. A price dictated by modifying formula 2-13 has the following information given: initial price = $10, decay = 0.25, floor price = $6, and decay = 0.15. Using these estimates, at what year will there be an intersection of the selling price with competitive floor price? At a floor price of $3.50, what is the margin of the product's over-cost? At what year will this level be reached? Solve this question analytically. Also, solve these questions graphically by using semilog graph paper. (*Hint*: If semilog paper is not available, use your slide rule for one-log cycle to represent the vertical axis, and plot the two lines.) Discuss what your pricing policy should be after the intersection of prices.

2-11. A project has a first cost of $100,000 and an annual maintenance cost of $2500 each year over a 50-year expected life. On the average, benefits expected will be $7000 per year over the life of the project. What is the benefit-cost ratio?

2-12. A sharp bend in an asphalt road has been the scene of nonfatal accidents. The bend, some 0.7 mile in length, is analyzed with respect to safety and economy. Based on yearly destination-trip data and standard safety-accident tables, it is estimated that straightening will save $35 and $750, respectively. Buying and hauling materials, compacting and grading for the new road, and removing the old bend will cost \$28,000. Highway policy says that a road life is 20 years. Maintenance savings should amount to \$45 annually. What is the benefit-cost ratio for this project? On the basis of this calculated ratio, should this project proceed?

2-13. A project is anticipated to have a first cost of \$100,000. The interest rate for this company is 20%, and the economic life of the project is estimated at 10 years. What is the annual payment that includes capital recovery plus an interest on the principle?

2-14. A mass-production industry has on-line and off-line inventory storage. The material usage amounts to 20,000 cubic feet per week. Deliveries are observed, and 40 round trips are counted for an average week. The cost differential between on-line and off-line storage is estimated as \$0.10 per cubic foot; each delivery load costs \$10. Plot handling and storage costs, and determine the unconstrained minimum for deliveries.

2-15. A model for a hydrocracker is proposed. A simplified version has its flow illustrated by Figure P2-15(a). There are a large number of physical materials and variables such as gas-oil input, catalyst usage, hydrogen consumption,

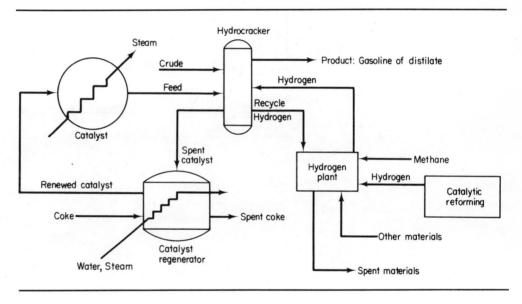

utilities required for heating and cooling, methane consumption, and temperature and pressure. Suppose that it is possible to express the pertinent factors in terms of two variables, x_1 = gas-oil input and x_2 = hydrogen input. Labor cost, incidentally, is small in refinery operations due to the automated nature of the equipment. Based on the physical characteristics of the equipment, a pressure limitation is given as $f_1 = 4x_1 - x_2^2 - 2 \geq 0$. Heat must be maintained above a certain level to assure reaction completeness in the hydrocracker and is given as $f_2 = 2x_1 + 9x_2 - 36 \geq 0$. The level of impurities such as sulfur in the gasoline output must obey $f_3 = 4x_1 + 2x_2 - 36 \leq 0$. These constraints are given in Figure 2-15(b), and the cross-hatched region identifies the zone of plausible solutions. An objective function involving the variables has been analytically found from historical records as $C = x_1^2 + x_2^2 + 1$. Study the schematic model and determine from it the materials used in the production of gasoline. Determine a word equation expressing a material balance. Plot an isocost line and determine along what boundary the minimum operating point would be.

FIGURE P2-15(b)

FEASIBLE REGION OF OPERATION OF
HYDROCRACKER AND QUALITY
CHARACTERISTICS FOR GASOLINE

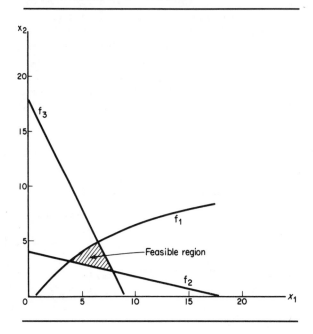

chapter 3

*O money, money, money, I'm not necessarily
one of those who think thee holy,
But I often stop to wonder how thou canst go
out so fast when thou comest in so slowly.*

OGDEN NASH
Hymn to the Thing That Makes the Wolf Go,
1934

PATTERNS OF COST INFORMATION

After the design process is underway the first concern for the cost estimator becomes the finding of costs. Its distribution to units of design or the organization and its future prediction comes later. In the personal situation, the manner of knowing the amount of money spent is a simple process—knowing the amount of cash in the billfold at two separate times provides an arithmetic and straightforward method of reckoning. The recording of cash flow in the organization requires more ingenuity, however. In a going concern the sources and kinds of cost information have been settled. Where cost estimating is emerging, a variety of methods is created to meet special needs. Generally we observe that cost estimating requires many kinds of information from many sources. After these raw data are collected their conversion to more usable forms and verification of their factual nature are undertaken.

The patterns of cost as we choose to examine them in this chapter are *historical, measured, and policy.* Although not a perfect classification, it provides an apparatus for meeting the special needs of the cost estimator.

3.1
KINDS OF INFORMATION

One of the essentials for intelligent cost finding is the continuous flow of reliable information from all activities of the organization. That this information be accurate, flexible, simple, timely, and economical is understood. Cost estimating is one of the sources that gives as well as receives information in management information systems. This basic purpose of access and transmittal of information is a dilemma that cost estimators must face. First, the amount of information is usually inadequate. Information that is common for all functions within the firm is compromised and inconsistent. There are some practices that can be advanced which benefit cost estimating, however:

1. A cost analysis which is useful for one purpose, e.g., inventory evaluation for financial statements or tax purposes, may be ill-suited to produce the information necessary for other purposes, such as a product design decision. Particular information is better.

2. Historical, measured, or policy information, whether it is about past or current costs, prices, profits, and values, is useful and indicative; however, information which is concerned with future costs, price, profits, and values is more significant. Today's decisions must relate to events which lie ahead.

3. There are times when the absolute level of cost is not as significant as the degree of variability which exists in these costs. Changes which can be made in costs or prices or profits as the result of taking one course of action rather than another are often the key elements in decision making. Estimates which are made for the short term should usually be based on short-term variations in profits and costs. Decisions for the longer term should normally be made with major thought to long-term changes in profits and costs.

A major kind of cost from internal reports is *historical*. Data as typically characterized by accounting reports are historical in nature, as they emphasize the transactions recorded through cost-controlling accounts which may be kept in some ledger system. Money is expended, and materials, labor, services, and expenses (such as power, heat, and the like) are received. Specific accounting procedures must be provided for recording the acquisition and disposition of materials, for the recording and use of labor, and for their distribution. The internal function of cost accounting as it relates to our interest here is discussed in Section 3.4. Cost accountants are primarily responsible for this kind of information.

Measured data are another kind of information. Cost engineers may find that work measurement or the economist's methods of measurement give a form of information that is amenable to certain types of estimation, either in time or dollar dimensions. Sometimes these information forms may be measured or determined by the cost engineers. Material quantities calculated from drawings and specifications are another form of measured data.

The final kind of information is *policy* data and has the property of being *fixed* for cost estimating. Accepted as factual and beyond control, the origins of policy data are varied. Union-management wage settlements or union-hall hiring of construction labor where predetermined policies dictate the wages and types of labor on equipment to be operated are commonplace. Budgets and legislative restrictions from municipal to national laws dictate certain codes of conduct and cost. The federal government requires a social security tax from the employer for the purpose of providing old-age benefits. The employee contributes an amount through the employer. This rate is subject to change by Congress. An unemployment compensation tax, sometimes called Federal Unemployment Insurance Cost, is collected by the states for the purpose of providing funds to compensate workers during periods of unemployment. The base cost of this tax is dependent on the experience and nature of the industry or contractor and is paid by the organization. This rate may be reduced by establishing a high degree of employment stability with few layoffs during a specified period of time. Employer's liability insurance is an additional form of policy cost.

3.2 SOURCES FOR INFORMATION

It is the quantifiable aspects of engineering design and cost that we seek. The statistics of cost, price, production, efficiency, performance, employment, purchases, weather conditions, time of the year, and many other variables relative to cost engineering provide the grist for the estimate. These data may emanate from internal departments within the organization, official government sources, international agencies, trade associations, trade unions, sampling organizations, or any office which gathers and divulges design and economic information. These sources may be secondary—that is, the source of data may be far removed from fundamental source—but what is more important is its reliability and timeliness. Some organizations release a vast quantity of statistics every week, month, quarter, or year on the results of economic activities falling within their category. The typical release provides what is commonly called time-series data. These time-series data provide a variety of life situations and experiences. In some situations we are willing to accept any available data and have little opportunity to be selective. Cost estimators may use their own measured data as an important source.

Internal Sources. While accounting is a major source of information, there are other departments internal to the organization that provide information. The personnel department charged with the handling of employees interprets the union contract (where unions exist), conducts labor contract negotiations, and keeps personnel records regarding wages and fringe costs. This is especially significant for future costs.

Operating departments, whether in construction, manufacturing, or crafts, to name a few, are the producing organizations. They are concerned with doing. The foreman or department manager knows the operating details at that moment. Frequently he is a direct source of information. He may often assist in the collaboration of obtaining data on special forms that report extraordinary costs of process equipment, manning, efficiency, scrap, repairs, or down time. Sometimes he is the oracle for a "guesstimate" on operations with which he is familiar.

In many organizations purchasing has the responsibility of spending money for materials. Some companies believe that purchasing is responsible for the outside manufacturer. Purchasing gets into the make-or-buy decisions and is frequently interested in the efficiency of their suppliers. Knowing about purchasing and shipping regulations, purchasing is frequently the ideal source for this class of information. Purchasing should look for less costly materials but cannot tell whether the cheaper materials will do the job (engineering decides that). Some companies require that only buyers contact vendors for trial quotes, while other companies have no provision restricting price interchange between vendor and company. A particular estimate may become so complicated that it requires several quotations. The magnitude of the quotation, the type of the organization, and particular policies dictate whether cost engineers are personally involved or whether buyers secure purchase information. The big steady-use item, big one-shot jobs, middle-sized orders, and small orders are factors in making this choice.

The contribution of sales and marketing is apparent in the pricing of products. Market demand, sales, consumer analysis, advertising, brand loyalty, and market testing are their fields of responsibility.

Government Sources. A great variety of basic economic facts and trends is available from the U.S. Government. The Bureau of Labor Statistics (BLS) provides elements of cost on the prices of materials and labor. They issue *factors*, although not a cost figure, which result from a given mixture of capital input and labor input and is called productivity. BLS surveys 80 labor markets representative of standard metropolitan statistical areas. Market data are obtained from companies on more then 60 occupations with a variety of benefits and practices. The occupations are indirect in nature, covering office clerical, maintenance, power plant, custodial, and material movement occupations, to name a few. Other indexes are published for these areas covering office clerical workers and industrial nurses, and the list goes on. With this and other measures for occupations, the BLS calculates overall movements of pay for the occupational groups. The BLS also conducts wage surveys for particular industries; the BLS is able to show wage movements over a considerable span of time. Through an integrated federal-state undertaking, they provide detailed and industry information yielding a substantial body of earning statistics. These and others are designed to assure internal comparabilities of employment, hours, and earnings data at various levels of geographical detail. These statistics average hourly earnings based on hours worked. The BLS gives attention to various outputs related to the consumer price index and the wholesale price index. The CPI measures changes in the prices of a list of goods and services which represent the items important to people and used in formal wage escalation contracts affecting many workers. The wholesale price index is a general-purpose index designed to provide a continuous monthly series showing price changes, singly and in many combinations, for all commodities sold in the primary markets of the United States. This index measures the general rate and direction of price movements in primary markets and the specific changes for individual commodities chosen to represent a wide variety of specifications. The Gross National Product (GNP) is compiled in the Office of Business Economics of the U.S. Department of Commerce. This input-to-output ratio, in its simplest form, is a relationship to productivity. The output measure is in constant dollars, and the input measure is man-hours worked. The OBE is the group responsible for determining U.S. income and product. To do this the OBE uses data from practically all the departments of the federal government as well as from many private sectors.

It should be pointed out that most cost data are based on conditions at some time in the past. Because prices or costs vary with time, some method must convert costs or prices applicable at a past time to equivalent costs or prices that are currently correct. This can be done with cost indexes. A cost index is a dimensionless number for a given year showing the cost at that time relative to a certain base year. If the cost of some time in the past is known, the equivalent cost at the current time is determined by multiplying the original cost by the ratio of the present index value to the index value applicable when the orginal cost was obtained.

International Agencies. Similar statistics available for the United States are found in other countries and are published periodically by those countries. Selected statistics measuring trade, flow of funds, international finance, and the like are published by the United Nations. Some consulting firms have international studies as a part of their business.

Business Firms. Data may be found from manufacturers' agents and jobbers, who, although they promote special areas, are willing to release information given to them by their clients. This information is usually secondary. Many firms annually report their incomes in the terms of income statements for public disclosure to their shareholders. Although obscure in some respects, diligent study may reveal elements of cost about competitors' actions. Large banks are able to aid in the fields of forecasting, business cycles, and the cost of money. Local banks are frequently compilers of local economic statistics. Economic data are made available in the monthly bulletins of the district Federal Reserve banks. Other sources include the Federal Home Loan banks, universities, Chambers of Commerce, and savings and loan institutions.

Trade Associations and Publications. Enterprises of this nature are subsidized by business, and they collect and disseminate public and proprietary data both for paying subscribers and the public at large. The National Machine Tool Builders Association, U.S. Bean Marketing Association, American Institute of Steel Construction, and Maple Flooring Manufacturing Association are typical organizations which compile published data useful for cost engineering. Some firms and associations regularly publish different types of cost indexes. The Marshall and Stevens Equipment Cost Index, the *Engineering News-Record* Construction Index, the Nelson Refinery Construction Index, and the Chemical Engineering Plant Construction Cost Index are known for such data. In the Marshall and Stevens Equipment Cost Index, an all-industry equipment index is simply the arithmetic average of the individual indexes for 47 different types of industrial, commercial, and housing equipment. Relative construction costs at various dates can be estimated by use of the *Engineering News-Record* Construction Index, which shows the variation in labor rates and material costs for industrial construction. The petroleum industry has found the

Nelson Refinery Construction Index useful, while construction costs for chemical plants form the basis for the Chemical Engineering Plant Construction Cost Index. Indexes are again discussed in Chapter 5.

Sampling Organizations. These are more noted for market research efforts, but in certain cases they do perform special services directed to cost-engineering objectives.

3.3 DEFINITIONS

Now that we are well into the book, tighter definitions of cost are appropriate. Unfortunately, the terminology of costs is in a state of confusion. Standardizing groups have parochial interests, and authors are no exception. Nonetheless, costs are used for a variety of purposes; the idea of different costs for different purposes makes it advisable to use them with an adjective or phrase to convey the shade of meaning intended. The single word *cost* is a general term for a measured amount of value deliberately released or to be released in the acquiring or creating of tangible or intangible economic resources. The dimension of cost is usually dollars. Variations of the many types of costs are distinguished in Table 3.1. This list does not cover the subject, as other terms, found in special engineering applications, do not warrant inclusion. The reader will discover that other definitions are possible even within this book, and for that shortcoming we apologize.

TABLE 3.1

MODIFYING DEFINITIONS OF COST

Capital cost	The cost of obtaining capital expressed as interest rate
Depreciated cost	A noncash tax expense deduction for recovery of fixed capital from investments whose economic value is gradually consumed in the business operation
Detailed cost	The value of the detailed estimate obtained with almost complete disclosure of engineering design data using various methods of estimating
Direct cost	That cost clearly traceable to a unit of output or segment of business operation, such as direct labor costs or direct material costs
Direct labor cost	The labor cost of actually producing goods or services
Direct material cost	The cost of raw materials or semifinished materials which can be traced directly to an operation, product, project, or system design
Engineering cost	The total of all costs incurred in a design to produce complete drawings and specifications or reports; included are the costs, salaries, and overhead for engineering administration, drafting, reproductions, cost engineering, purchasing and construction costs of prototype, and design costs
Estimated cost	Predetermined value of cost using rational methods; generally, a confusing term, and its use is not encouraged unless in a specific case its meaning would be clear
Fixed cost	That cost which is independent of the rate of output
Future cost	Cost to be incurred at a later date
Historical cost	A tabulated cost of actual cash payments consistently recorded
Indirect cost	That cost not clearly traceable to a unit of output or a segment of business operation, such as indirect labor costs and indirect material costs
Joint cost	Exists whenever from a single source or material or process there are produced units of goods having varying unit values
Manufacturing overhead cost	Includes all production costs, except direct materials and and direct labor; similar terms: overhead expense, burden, manu-

TABLE 3.1 (Cont.)

	facturing expense, indirect expense, indirect manufacturing costs, and factory expense
Marginal cost	The added cost of making one additional unit for an operation or product without additional fixed cost investment; similar terms: incremental cost, out of pocket cost, differential cost
Measured cost	A cost based on time relationships to dollars using mathematical rules; the time relationships are normally measured
Operating cost	Comprises two distinct cost elements, direct material and direct labor; materials may be computed from drawing requirements or from utilities required for a continuous process
Operation cost	Includes labor, materials used, asset value consumed, and appropriate overhead costs pursuant to the operation design
Opportunity cost	The estimated dollar advantage foregone by undertaking one alternative instead of another
Optimum cost	That operation, product, project, or system economic value for which a minimum (or a maximum, as appropriate) is uncovered for specified design variables using variational models
Period cost	Cost associated with a time period
Policy cost	A cost based on the action of others; considered fixed for the purposes of estimating
Preliminary cost	The value of a preliminary operation, product, project, or system design estimate; usually obtained quickly with a shortage of information
Prime cost	The labor and materials clearly traceable to a unit of output
Product cost	Includes operation costs, purchase materials, overhead, general and administrative expenses, and appropriate design and selling costs. It usually does not include profit.
Project cost	The investment or capital cost proposed for approval in a single evaluation of an engineered project
Replacement cost	A present cost of the design equipment or facility intended to take the place of an existing design of equipment or facility
Standard cost	Normal predetermined cost computed on the basis of past performance, estimates, or work measurement
Sunk cost	The past or continuing cost related to past decisions which are unrecoverable by current or future decisions
System cost	Usually a hypothetical cost for the evaluation of complex alternatives; it may include elements of operation, product, or project costs; in most cases it cannot be related to a total cost; its use is limited to a decision-making choice of alternatives rather than a source of total cost; less frequently it represents the absolute magnitude of all costs
Unit cost	Implies in manufacturing the sum of total material, labor, and manufacturing overhead divided by the quantity produced; for an investment it is the installed cost of the producing unit in convenient units of production
Variable cost	That cost which varies in proportion to the rate of output

3.4
HISTORICAL COST

The importance of cost accounting in the performance of the several management functions has always been recognized. Its responsibility for planning, performance, and measurement aids estimating practices by supplying economic data. In this section a superficial treatment is given with emphasis on those ideas that are complimentary to estimating.

3.4-1
Accounting Fundamentals

Accounting has been defined as the mechanism by which the money activities of an organization are recorded. Accounting systems are generally charged with the preparation of periodic balance sheets, statements of income, information to aid the control of cost, and essential data useful for the finding of product price. Although cost-accounting data may have been wisely and carefully collected and arranged to suit primary purposes for accounting, raw data are usually incompatible with forecasting, product pricing, and the like. Accounting specialties are general, public, auditing, tax, government, and cost. Cost accounting, government, institutional, and tax accounting are more important to the needs of engineering.

Cost accounting emphasizes accounting for costs, particularly the cost of manufacturing processes and of manufactured products. It deals with actual or historical costs using this base. Governmental accounting specializes in the recording of transactions of political units, such as states and municipalities. It seeks to provide accounting information about the business aspects of public administration, and it helps to control the expenditure of public funds according to law or political dictates. Tax accounting, because of the myriad laws and

TABLE 3.2

OUTLINE OF TOTAL COSTS FOR PRODUCING
MANUFACTURED PRODUCTS

Direct materials	+ Direct labor		= Prime cost
			+
Indirect materials	+ Indirect labor	+ Fixed and miscellaneous expenses	= Factory overhead

Includes:
Factory supplies
Lubricants
	Includes:
	Supervision
	Superintendence
	Inspection
	Salaries of
	    factory clerks
		Includes:
		Rent
		Insurance—fire
		    and liability
		Taxes
		Depreciation
		Maintenance
		    and repairs
		Power
		Light
		Heat
		Miscellaneous
		    factory
		    overhead
		Small tools
			=
			
			Cost of goods
			manufactured
			+

Distribution expenses	+ Administrative expenses		= Commercial expenses

Includes:
Advertising
Samples
Entertainment
Travel expenses
Rent
Telephone and telegraph
Stationery and printing
Postage
Freight and cartage out
Miscellaneous selling
    expenses
	Includes:
	Administrative and office
	    salaries
	Rent
	Auditing expenses
	Legal expenses
	Doubtful accounts
	Telephone and telegraph
	Stationery and printing
	Postage
	Engineering
	Miscellaneous administrative
	    expenses		

			+ Selling cost salaries commissions + Profit = Selling price

practices, includes the preparation of tax returns and, importantly, the consideration of the tax consequences of proposed business transactions.

A traditional cost and profit structure is shown in Figure 3.1. The terms are more accurate for manufacturing. Extension to public works would remove

FIGURE 3.1

TRADITIONAL COST AND PRICE STRUCTURE

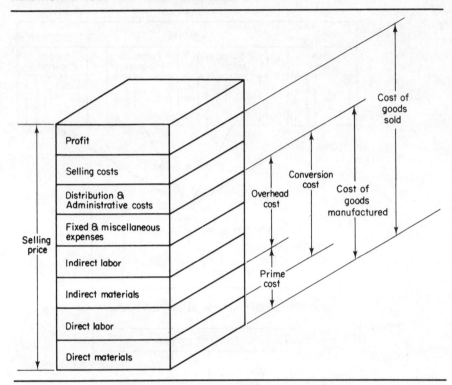

the top two layers of profit and selling costs, making the model general. That literally many variants enter into the traditional cost and profit model makes the art of accounting both interesting and confusing. This costing model is further enhanced by an examination of Table 3.2, which gives an analysis of total cost.

The accounting cycle and costing procedure is pictured by Figure 3.2, which shows the standard accounting procedure starting with the recording of the original business transactions and proceeding to final preparation and summarizing balance sheets and income statements. As daily business transactions occur they are recorded in the journal. A single journal may suffice for all entries for a small business. For the typical business, however, many types of journals abound, such as cash, sales, purchase, and general journals. The recording of these journal entries in the appropriate account headings in the ledger is the next step. Posting is the term usually applied to this process. Periodically the financial condition of the concern is stated on the balance sheet and the results of operation in the income statement. These are prepared from ledger accounts. The balance sheet shows the financial health of the business at a particular date, while the income statement is a record of the financial gain or loss during a period of time.

The cost-accounting cycle is framed about the skeleton of the manufacturing process or the physical arrangement or the service for jobs. Since cost

FIGURE 3.2

DIAGRAM OF ACCOUNTING PROCEDURE (From Max
S. Peters and Klaus D. Timmerhaus: *Plant Design and Economics
for Chemical Engineers*, 2nd Ed., McGraw-Hill Book Company,
New York, 1968, p. 342)

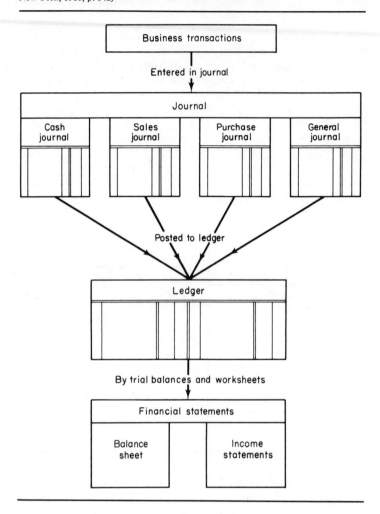

accounts are an expansion of general accounts they should, as a basic accounting procedure, be related to them. Figure 3.3 shows the relationship between general accounts and cost accounts. The bigger an organization is, the greater its span of accounting records. To illustrate: A materials account controls hundreds of different material items, the payroll account controls departmental labor costs and payroll records for each employee, and the factory overhead account controls indirect labor, supplies, rent, insurance, repairs, and many other factory expenses.

3.4-2
***The Accounting Equation
and Balance Sheet***

Assets. The assets of a manufacturing firm are those things of dollar value which it owns. These assets may be tangible as in the case of land, buildings, equipment, or inventory, or they may be intangible as in the case of trademarks, designs, and patents. For analysis, assets may be segregated into current assets, fixed assets, and intangible assets. Current assets may have three inventory accounts representing purchase goods, in-process manufactured goods, and completed products. Fixed assets include office equipment, factory equipment, buildings, and land less accumulated depreciation reserves.

FIGURE 3.3

RELATIONSHIP BETWEEN GENERAL ACCOUNTS
AND COST ACCOUNTS

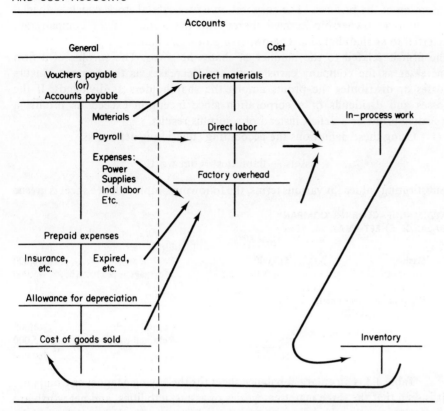

In the case of intangible assets, an engineering firm may design new products through its research and development and possibly obtain patents on them. The cost of research, engineering, and testing leading to the development of a new product may be significant and in theory could be treated as an asset in the same manner as other assets. As many research projects may be underfoot at the same time, cost can be incurred over a period of years. Some firms treat engineering costs as a part of current operating expenses.

Liabilities. The liabilities of a firm are the debts it owes. The category of liabilities is frequently broken down into current and long-term debts. Some of the more common items of business liabilities are (1) accounts payable (for example, the debts of the firms to creditors for materials and services received), (2) bank loans (the amounts the firm owes to banks for money borrowed), and (3) mortgage payable (the debt to investors for money they loaned to the business on the security of its real estate or equipment).

Net Worth. The net worth of a business is the ownership interest in the firms net assets. In certain accounting situations the use of proprietorship and capital are synonymous terms with net worth. In a simple case the net worth of a corporation consists of its capital stock and the surplus:

Net Worth

Capital stock	$40,000
Surplus (or retained earnings)..	15,000
Total	$55,000

Broadly speaking, capital stock is the portion of the net worth paid in by the owners, while surplus or retained earnings is that portion of the net worth accumulated from the excess of profits earned over the dividends paid since the inception of the business. The corporation issues capital stock which is divided into units of ownership, termed shares, and the owners of the company are referred to as shareholders. The ownership of a shareholder in the net worth of the firm is related to the number of shares he owns. The surplus of a firm increases as the company earns profit and decreases as the company incurs losses or distributes the profits among the shareholders as dividends. If the losses and dividends of a corporation since inception exceed its profits, a negative profit or a deficit instead of a surplus results.

Using these definitions the accounting equation is defined as

$$\text{assets} - \text{liabilities} = \text{net worth}$$

Substituting values for various terms, the following simple balance sheet is given:

XYZ MANUFACTURING COMPANY
BALANCE SHEET—MAY 31, 19xx

Assets		= Liabilities	
Cash	$15,000	Bank loan	$15,000
Inventory	10,000	Mortgage	15,000
Land	15,000		
Building and equipment	40,000		
		+ Net worth	
		Capital stock	45,000
		Retained earnings	5,000
	$80,000		$80,000

Table 3.3 is an enlarged balance sheet for the Flying Magnetics Company. It is seen that the three major categories of assets, liabilities, and net worth are displayed in a slightly different and enlarged format. The balance sheet is the accounting statement of financial position showing the assets, liabilities, and net worth at a given point in time.

TABLE 3.3

FLYING MAGNETICS COMPANY
BALANCE SHEET—DECEMBER 31, 19xx

Assets			
Current assets:			
Cash	$	$ 8,000	$
Customers	18,000		
Less-reserve for bad debts	1,900	16,100	
Inventories:			
Material	4,400		
Material in process	6,000		
Labor in process	4,000		
Overhead in process	5,000		
Finished products	10,000	29,400	
Total current assets			53,500
Deferred expenses:			
Insurance	1,000		
Advertising	600		1,600
Fixed assets:			
Land		10,000	
Building	28,000		
Less reserve for depreciation.	5,000	23,000	33,000
Total assets			$88,100

TABLE 3.3 (Cont.)

Liabilities and Net Worth

Current Liabilities:
Accounts payable 8,000
Bank loans 4,000
Accrued taxes 3,000
    Total current liabilities 15,000
6% Mortgage (Due in 20 years) 5,000
Net worth:
Capital stock 50,000
Retained earnings:
    Balance—January 1, 19xx 12,000
    Net profit for 19xx 6,100   18,100   68,100
        Total liabilities and net worth $88,100

TABLE 3.4

FLYING MAGNETICS COMPANY
WORKSHEET—DECEMBER 31, 19xx

Account Title	Trial Balance		Profit and Loss		Balance Sheet	
	Dr.	Cr.	Dr.	Cr.	Dr.	Cr.
Materials in process inventory	$ 6,000	$	$	$	$ 6,000	$
Labor in process inventory	4,000				4,000	
Overhead in process inventory	5,000				5,000	
Finished products inventory	10,000				10,000	
Materials inventory	4,400				4,400	
Customers	18,000				18,000	
Reserve for bad debts		1,900				1,900
Cash	8,000				8,000	
Land	10,000				10,000	
Building	28,000				28,000	
Reserve for depreciation		5,000				5,000
Prepaid insurance	1,000				1,000	
Prepaid advertising	600				600	
Accounts payable		8,000				8,000
Bank loans		4,000				4,000
Accrued taxes		3,000				3,000
6% Mortage due in 20 yr		5,000				5,000
Capital stock		50,000				50,000
Surplus		12,000				12,000
Sales		104,000		104,000		
Sales return	4,000		4,000			
Cost of goods sold	54,000		54,000			
Lease	6,000		6,000			
Salaries	22,000		22,000			
Depreciation	800		800			
Promotion, advertising	6,400		6,400			
Bad debts	600		600			
Insurance	1,600		1,600			
Sales discounts	1,000		1,000			
Purchase discount		1,500		1,500		
Underabsorbed overhead	3,000		3,000			
	$194,400	$194,400	$ 99,400	$105,500	$95,000	$88,900
Net profit to retained earnings			6,100			6,100
Total			$105,500	$105,500	$95,000	$95,000

The entries for the balance sheet are taken from the accounts in the assets and liabilities section of the ledger. The names of the balance sheet correspond with the ledger accounts. After the entries are made in the ledger and work is completed at the end of the period, usually a month, with a "worksheet" in an elementary case, the balance sheet is prepared by entering the net balance shown in each account. Table 3.4 is an example of a worksheet.

3.4-3
Profit and Loss Statement

The statement of earnings of the firm, known either as the *profit and loss* or *income* or *income and expense* statement, is a summary of its incomes and expenses for a stated period of time. The net profit or loss it discloses represents the net change in net worth during the reporting period arising from business incomes and expenses.

Definition of Profit. Profit represents the excess of revenue over cost and is an accounting approximation of the earnings of a manufacturing firm after taxes, cash and accrued expenses (representing costs of doing business), and certain tax-deductible noncash expenses such as depreciation are deducted. Loss represents the excess of cost over selling price, such as a product costing $8000 and selling for $6000 has a loss of $2000. The following example describes the effect on business net worth of profit and losses:

INVENTIONS, INC.
BALANCE SHEET—MAY 31, 19xx

Assets		Liabilities	
Customers	$ 4,000	Bank loan	$ 1,000
Gadget A inventory....	8,000	Accounts payable......	2,000
Gadget B inventory....	6,000		
		Net worth	
		Capital stock	15,000
	$18,000		$18,000

If Inventions, Inc. sold the asset gadget A for $10,000 cash, its balance sheet would change to

INVENTIONS, INC.
BALANCE SHEET—JUNE 30, 19xx

Assets		Liabilities	
Cash	$10,000	Bank loan	$ 1,000
Customers	4,000	Accounts payable......	2,000
Gadget B inventory....	6,000		
		Net worth	
		Capital stock	15,000
		Retained earnings	2,000
	$20,000		$20,000

Here it is found that net worth was increased $2000. If the business sold the asset gadget B inventory for $5000 cash, its balance sheet would look like

INVENTIONS, INC.
BALANCE SHEET—JULY 31, 19xx

Assets		Liabilities	
Cash	$15,000	Bank loan	$ 1,000
Customers	4,000	Accounts payable......	2,000
		Net worth	
		Capital stock	15,000
		Retained earnings	1,000
	$19,000		$19,000

The $6000 gadget B inventory was replaced by $5000 cash, and the net assets and the net worth was decreased $1000. From the foregoing illustrations it is seen that profits increased the net worth because they increased the net assets and losses decreased the net worth because they decreased the net assets.

Income and Expense in Business. Income represents the revenue from sales before the deduction of cost. Expenses represent costs of doing business. While income is received from the sale of merchandise or products, expenses include such common items as salaries, advertising, power and light, telephone, rent, insurance, and interest. The profit and loss statement of a firm is a summary of its incomes and expenses for a stated period of time. If the statement discloses a net profit or loss, the change represents an increment or decrement in the net worth during the period arising from business incomes and expenses and is carried to the net worth section of the balance sheet. An example of a profit and loss statement and the relationship it bears to the balance sheet follows:

SCIENCE COMMODITIES
PROFIT AND LOSS STATEMENT
MONTH ENDED JUNE 30, 19xx

Income

Fees for engineering services	$37,000	
Royalties on patents owned	3,400	
Product sales, subcontracted service........	4,000	
Interest on securities	100	
		$44,500

Expenses

Salaries	$32,600	
Rent of office	3,000	
Leasing of equipment	4,000	
Traveling	2,000	
Utilities	200	
Office supplies	500	
		$42,300
Net profit (to retained earnings)		$ 2,200

SCIENCE COMMODITIES
BALANCE SHEET—JUNE 30, 19xx

Assets		Liabilities	
Cash	$ 5,700	Accounts payable	$ 2,500
Receivables	8,500		
Bonds.............	12,000	Net worth	
Equipment	2,500	Capital stock	25,000
Patents	11,000	Retained earnings	
		Balance June 1 ⎰$10,000	
		Profit June 30 ⎱ 2,200	12,200
	$39,700		$39,700

The profit and loss statement is related to the balance sheet in that it details the profit and loss elements which caused a June 1 retained earnings of $10,000 to become a June 30 surplus of $12,200. It should be apparent that the profit amount is needed for completing the balance sheet since it is necessary to prepare the profit and loss statement first.

The Cost of Goods Made and Sold Statement. The items entering into the cost of goods made and sold statement usually appear in four inventory accounts:

1. Materials in process.
2. Labor in process.
3. Overhead in process.
4. Goods in finished inventory.

TABLE 3.5

FLYING MAGNETICS COMPANY
PROFIT AND LOSS STATEMENT
YEAR ENDED DECEMBER 31, 19xx

Net sales		$100,000
Cost of goods sold		54,000
Gross profit on sales		$ 46,000
Operating expense:		
Lease	$ 6,000	
Salaries	22,000	
Depreciation	800	
Advertising	6,400	
Bad debts	600	
Insurance	1,600	$ 37,400
Net profit from operations		$ 8,600
Miscellaneous income and expense		
Purchase discount	1,500	
Sales discount	1,000	500
Net profit before special charges		$ 9,100
Underabsorbed overhead		3,000
Net profit to retained earnings		$ 6,100

From data disclosed by inventory accounts, the cost of goods made and sold statement appears as shown in Table 3.6. Inventory accounts differences between the opening and closing inventories are cost of goods.

TABLE 3.6

FLYING MAGNETICS COMPANY
COST OF PRODUCTS MADE AND SOLD STATEMENT
YEAR ENDED DECEMBER 31, 19xx

Direct material:		$
In process—Jan. 1, 19xx	$ 7,000	
Applied in the year	9,000	
Total	16,000	
In process—Dec. 31, 19xx	6,000	10,000
Direct labor:		
In process—Jan. 1, 19xx	5,000	
Applied in the year	10,000	
Total	15,000	
In process—Dec. 31, 19xx	4,000	11,000
Factory overhead:		
In process—Jan. 1, 19xx	4,000	
Applied in the year	5,000	
Total	9,000	
In process—Dec.31, 19xx	5,000	4,000
Cost of products made		25,000
Inventory of products made—Jan. 1, 19xx		9,000
Total of products made available for sale		34,000
Inventory of products made—Dec. 31, 19xx		10,000
Cost of products sold		$24,000

So far the discussion has dealt with various aspects of accounting of costs that have actually been incurred. The principal questions that have been raised were concerned with the classification, recording, and reporting of actual cost data. This is a form of cost bookkeeping that involves the measurement of cost without the critical appraisal of the amounts so measured or the predictive parameters that are useful. These functions are concerned with what costs have been incurred in the past. With respect to the uncertainty of the future, management expects from cost-engineering analysis and accounting the notion "what costs may reasonably be expected" in addition to "what costs ought to be." The accountant's standard cost is concerned with this idea.

The engineer, when referring to *standards*, thinks in terms of a rigid specification, but analysts from other fields have dissimilar viewpoints. A *standard cost* as discussed here provides a dollar amount which is a "should be" amount and is not an immutable natural law. An attitude is necessary in viewing standard costs. Standard costs may be classified as perfection-level standards, and management encourages attainment of these standards as goals. Some businesses contend that perfection standards are preferable to attainable standards because they provide a stimulus or incentive to workers and to management alike to achieve the best possible performance.

A standard unit cost of a labor operation, part, or product is a predetermined cost that may be computed even before operations are started. In constructing the standard unit cost of an item it is necessary to study the kind and grade of materials that should be used, how each labor operation ought to be performed, how much time each labor operation should take, how the indirect services should be best administered, and the entire specifications for the complete and total manufacturing operation. The aim, of course, is to specify the most effective method of making the item and then through adherence to the specifications in the actual manufacturing operations achieve the lowest practical unit cost. By now it should be realized that a system of this type may be expected to provide a calculated and anticipated cost of all products by cost elements; comparisons of the anticipated cost with actual results and the reasons for any differences; the effectiveness of all cost elements, including material, labor, and overhead; and measurement of departmental or individual performances against accepted standards.

Despite the care used in establishing and revising standard costs the actual cost as uncovered in any particular period or any particular job are very likely to deviate from the standard. These differences are known as variances and are expressed as dollar amounts or percentages. They are favorable variances when the actual costs are less than the standard costs and unfavorable when actual costs exceed standard costs. It should not be interpreted that excess of actual cost or of standard cost is adverse to the welfare of the firm. Similarly, not all favorable variances represent actual benefits to the company. The terms *favorable* and *unfavorable* when applied to variances indicate the direction of the variance from standard cost.

Two methods of costing are about to be described. Absorption or full costing, is conventional in that all fixed and variable costs are assigned to goods or services produced. Another method, called direct costing, has gained in popularity in recent years. Direct costing and absorption costing are identical with regard to the accounting for raw materials and direct labor and the assignment of these costs to products. The important difference is that in direct costing overhead costs are separated into fixed and variable components, which leads to recognizable differences in accounting methods. Absorption costs methods are mandatory for public disclosure and for tax purposes, while for internal matters direct costing is permitted.

Direct costing (not direct cost now) can be defined as the separation of manufacturing costs between fixed and those that vary with volume. Its purpose is to provide information about cost-volume-profit relationships. Under direct costing, operating costs (factory, selling, and administrative) are separated into fixed and variable components and recorded separately. Then for the next phase, variable cost elements are handled as a product cost and are charged to the product at the appropriate moment. Fixed cost is written off as a period cost in which they are incurred. Those who advocate this system suggest it is easier and more accurate for pricing decisions, budgets, and cost control. This method is associated with variable cost estimating enlarged upon in Chapter 9. The chart of accounts is modified for recording and reporting direct costing. To keep fixed overhead out of the product cost, variable and fixed expenses should be channeled into separate accounts.

Variable or direct costs, such as direct materials, direct labor, and variable factory overhead, are examples of costs chargeable to the product. Those costs which are time-dependent, rether than volume-related, are excluded initially from the cost of the product. Fixed factory overhead, such as insurance, factory, property taxes, and salaries of the executive and managerial staff, supervisors, foremen, and the like and certain factory employees such as maintenance crews and guards, is classified as fixed cost. The matter of machine depreciation is not clear. Depreciation can be based on a facility usage or time. Whether or not an expense is classified as fixed or variable may be the result of an arbitrary decision. Depreciation on a straight-line basis is a fixed expense, while depreciation on a unit-production or labor-hour basis is a variable expense. Some factory overhead items are semifixed in nature; that is, they vary with production but not directly in proportion to the rate. Those items which contain both fixed and

TABLE 3.7

QUARTERLY ESTIMATED OVERHEAD RATE
COMPUTATIONS FOR DIRECT
AND ABSORPTION COSTING METHODS

	Normal Activity	Expected Activity	Practical Capacity
Direct labor	$ 7,500	$ 8,000	$10,000
Percent of practical capacity	75	80	100
Number of units	30,000	32,000	40,000
Variable overhead costs	$15,000	$16,000	$20,000
Fixed overhead costs	12,000	12,000	12,000
Total estimated overhead	$27,000	$28,000	$32,000

A. Absorption costing overhead rate (direct labor dollar) computations:

1. Normal activity: $\dfrac{\text{overhead for normal activity}}{\text{normal activity}} = \dfrac{\$27,000}{\$7500} = 360\%$

2. Expected activity: $\dfrac{\text{overhead for expected activity}}{\text{expected activity}} = \dfrac{\$28,000}{\$8000} = 350\%$

3. Practical capacity: $\dfrac{\text{overhead for practical capacity}}{\text{practical capacity}} = \dfrac{\$32,000}{\$10,000} = 320\%$

B. Direct costing overhead rate computations:

$$\frac{\text{variable overhead}}{\text{activity}} = \frac{\$15,000}{\$7500} = \frac{\$16,000}{\$8000} = \frac{\$20,000}{\$10,000} = 200\%$$

variable elements are supervision, inspection, factory office services, compensation insurance, health and accident insurance, heat, light, and power. Thus, direct costing classifies factory overhead according to its behavior as either fixed overhead (a period cost) or variable.

In absorption costing the factory overhead rate is usually a composite rate. Once the full method (absorption costing) has combined fixed and variable costs, a capacity or activity level (normal, expected, and practical) is decided on in order to recover all costs and expense over a certain period of time. The distinction between absorption and direct costing may be seen on the basis of how overhead rates are computed, as in Table 3.7. The table assumes a plant-wide rate; however, either departmental or cost center rates could have been used, in which case for direct costing only the variable part of the service department cost would be allocated to the producing department and included in their rates. Additionally, these rates could be based on direct labor hours, machine hours, or any other basis. With direct costing, overhead rates between practical capacity, normal activity, and expected activity for the period are no longer significant because the overhead rates are unaffected.

Assume that we have a summary cost estimate for product A for the Flagstaff Manufacturing Company which at this time produces only that product. This summary, given as Table 3.8, uses three production centers and

TABLE 3.8

SUMMARY COST ESTIMATE FOR PRODUCT A

Production Center	Estimated Hours Per Unit	Production Center Hour Rate	Production Center Cost
Light machining	6.5	$6.28	$40.82
Bench work	1.1	3.78	4.16
Finishing	0.5	5.04	2.52
	Unit conversion cost		47.50
	Unit material cost		17.50
	Unit selling & distribution costs (5% of price)		5.00
	Total direct costs		$70.00
	Selling price		$100.00
	Contribution/unit		30.00
	Contribution percent		30%

the hours are estimated for each production center. The extensions to production center cost are made by the product of the estimated hours and the machine hour rate. The unit conversion cost is $47.50. Material costs are estimated from engineering drawings, and a unit selling distribution cost is known. The income and expense statement is given as Table 3.9. From net sales we subtract the direct cost of sales at standard and the variances on direct cost. Thus, we have a profit contribution of $1492 or 27.1% of net sales. The period costs are subtracted at this point, leaving a net profit of $452.

While this illustration was for one product, consider now the case where Flagstaff will produce three products. The income and expense statement is given in Table 3.10. A similar treatment is followed where the actual profit contribution is computed for each of the three products. We know which of the

TABLE 3.9

DIRECT COSTING METHOD FOR INCOME AND
EXPENSE STATEMENT FOR ONE PRODUCT

FLAGSTAFF MANUFACTURING COMPANY
INCOME AND EXPENSE STATEMENT
MONTH OF JANUARY 19xx

Net sales (55 units at $100)		$5500
Direct cost of sales at standard		
Materials (55 units @ $17.50)	$ 962.50	
Manufacturing labor & expense (55 units @ $47.50)	2612.50	
Selling & distribution (55 units @ $5)	275.00	
		3850
Profit contribution at standard		$1650
Percent of net sales		30.0%
Variances on direct costs		
Labor	$(128.50)	
Material usage & scrap	(45.00)	
Purchase price	15.50	
Total variances on direct costs		(158)
Profit contribution at actual		$1492
Percent of net sales		27.1%
Period costs		
Budgeted amount	$1025	
Less variances on period costs	(15)	
Total period costs		$1040
Net profit		$ 452
Percent of net sales		8.2%

TABLE 3.10.

DIRECT COSTING METHOD FOR INCOME AND
EXPENSE STATEMENT FOR MULTIPLE PRODUCTS

FLAGSTAFF MANUFACTURING COMPANY
INCOME AND EXPENSE STATEMENT
MONTH OF FEBRUARY 19xx

	Total	*A*	*B*	*C*
Sales	$6000	$5000	$200	$800
Direct cost of sales at standard				
Materials	960	850	30	80
Manufacturing labor & expense	2790	2350	120	320
Selling & distribution	300	250	10	40
Total direct cost at standard	$4050	$3450	$160	$440
Profit contribution at standard	1950	1550	40	360
Percent of net sales	32.5	31.0	20.0	45.0
Variances on direct costs				
Manufacturing labor & expense	(130)	(100)	15	(45)
Materials & scrap	(20)	(25)	(10)	15
Selling & distribution	(15)	15	(20)	(10)
Total variances on direct costs	(165)	(110)	(15)	(40)
Profit contribution at actual	1785	1440	25	320
Percent of net sales	29.8	28.8	12.5	40.0
Period costs				
Budgeted amount	$1300			
Variances on period costs	(45)			
Total period costs	$1345			
Net profit	440			
Percent of net sales	7.3			

products is most and least favorable. Period costs, lumped together as $1345, are subtracted to give a profit for the period.

The method of direct costing is generally accepted as the internal method; external costing and reporting by this method is generally unacceptable to the Internal Revenue Service and the Securities and Exchange Commission. It should not be presumed, however, that this in any way reflects on the reliability of this method. Mathematical reconciliation to *absorption* methods is possible.

There is a school of thought that asserts that "time is the measure of cost." Although the slogan is debatable, there is an element of truth when applied to the needs of detailed cost estimating. For this activity there is an almost unquestioned dependence on an objective measure of time. Measured cost is defined as a time-dollar relationship where direct observational processes and mathematical rules are followed. Historical cost uncovers costs about labor, supervision, materials, and a whole host of endeavors. Measured cost, on the other hand, is limited mostly to costs of work. The two categories of work are direct and indirect. Most types of work can be segregated into one or the other category inasmuch as the direct-indirect category is determined by practice and definition. Discussed in this section are techniques of measuring these two classes of labor. Time is measured, but it is the cost that is ultimately useful for functional models.

Although the cost estimator may not be directly concerned with the measurement of labor, he does depend on work measurement. He is satisfied if such labor measurements are objective, as far as that is possible, and he is willing to use the information provided that plausible techniques were used in the determination of time. There are four methods for the measurement of time useful for estimating: time study, predetermined motion-time data, work sampling, and man-hour reports. Although one may argue in favor of a particular method, each is suitable and necessary for different occasions.

For our purpose it will not be necessary to delve into the historical background of time study. Suffice it to say that the founder of time and motion study was Frederick W. Taylor and that two of the leading pioneers were Frank and Lillian Gilbreth. Time study has many advantages, and it is useful in industry. With competition increasing in any industry, it is necessary for management to know what the various cost factors are and how much a proposed change decreases or increases our cost or improves our quality. The determination of cost comparisons and analysis of this sort is rendered by operation engineers, and time study is the backbone of many cost systems. Time study is the analysis of each operation in order to eliminate the unnecessary elements and determine the better and cheaper method of doing the necessary operations; to standardize methods, equipment, and conditions; and then, and only then, to determine by measurement the number of standard hours required for an average man to do the job.

The stop-watch procedure goes something like this: (1) Conduct analysis of methods and improve if necessary; (2) record all significant data; (3) separate the operation into elements; (4) record the time consumed by each element as it occurs each time; (5) rate the pace or tempo at which various elements of work are performed; (6) determine the allowances; and (7) convert rated elements into normal time, include allowances, and finally express the standard in common units of production. Although this practice is criticized, its role in gathering time for production elements remains significant. Its greatest testimony lies in the fact that the majority of industrial enterprises in the United States depend on time study for a substantial amount of the data used in cost estimating.

The equipment for taking a time study is simple. A clipboard and a decimal minute or hour stop-watch is all that is required. The first and most important phase of taking a time study is its preparation. Is the job ready for timing? The time-study technician resolves this question by answering the following points: Is the proper tooling being used, and is it laid out correctly? Is the material laid out in an economical manner? Are proper machines or tools being used? Is the quality of the finished part or operation up to the inspection standards? Is the motion pattern employed the most economical that can be devised at this time? The second phase of the time study is to record on the time study form all the information pertinent to the job. A good rule to follow is that you cannot get too much information on a study. After the information is recorded, the elements of work must be separated and written down in sequence. The technician keeps in mind when breaking down the operation that the elements should be as short as possible but long enough for timing accuracy. Wherever possible, manual operator time is separated from machine time. Wherever possible elements that are constant (or nearly so) are separated from variable elements. Elemental start and stop times should be easy to identify. In the example of the assembly of an electrical receptacle, we see, first, the layout of the present method, given by Figure 3.4, a detailed description of the elements along with a

FIGURE 3.4

LAYOUT OF ELECTRICAL RECEPTACLE BENCH
ASSEMBLY-PRESENT METHOD

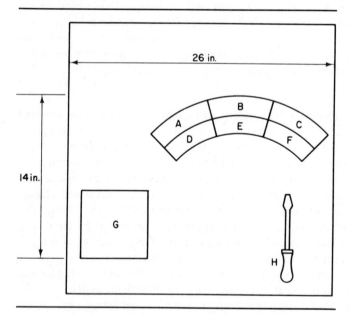

sketch of the parts involved, and finally the time study observation itself, Figure 3.5. The number of readings necessary to reflect a good average is a matter of judgment and depends primarily on the consistency of the time. During the recording of the actual times of each element, the technician has been judging the effort, or the speed, of the operator in performing the operation. He rates the speed with which motions are made and compares this mentally to a 100% average speed. The operator's effort that is reflected in the rating factor

FIGURE 3.5 (1)

Timed by R. Van Jones	Checked by Ted Davies	Workplace or mach. 0738	Mach. no. 1
Operators name Ferald Jeter	Clock No. 303-9109	Material See sketch	R.P.M. 1
Time study no.	Dept. No. Electrical assembly	Lubricant —	Strokes per min. —
Special tools used None			Feed
Part name Electrical receptacle	Part No. 1060		Operation No. 40
Remarks			

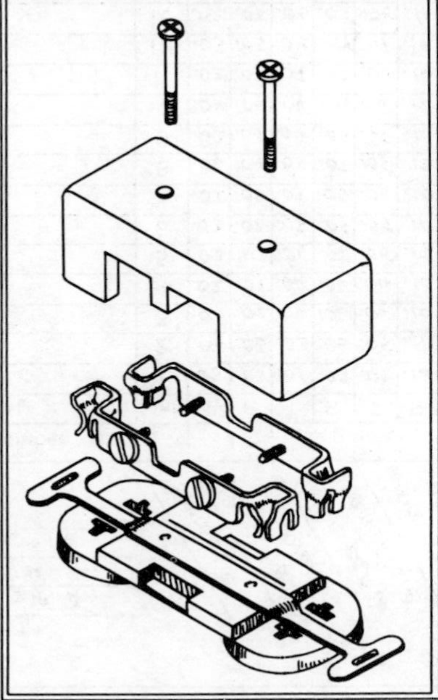

DETAILED DESCRIPTIONS OF ELEMENTS

1. Right hand reaches for back plate at bin A and grasps

2. Moves to left hand, reaches for mounting ear at B and grasps, mounting ear and inserts. Left hand holds.

3. Right hand reaches for contact and inserts (from C)

4. Right hand reaches for contact and inserts (from C)

5. Right hand reaches, grasps, transports, and places back plate over back plate. Left hand holds.

6. Reaches for A screw at F, grasps, transports, positions with right hand. Left hand holds.

7. Same as 6.

8. Right hand reaches for screwdriver at H, grasps and transports to back plate, and tightens screw. Left hand aside assembly. Right hand returns screwdriver.

FIGURE 3.5 (2)

Time study observation form

Date Sept 23
Time start 8:30 AM
Time stop 9:45 AM
Elapsed time 1:15

Sheet no. 1
No. sheets 1

Elements

1. Pick up back plate
2. Move ear to back plate
3. Contacts in back plate
4. Contacts in back plate
5. face plate over back plate
6. place plate in back plate
7. Place screw in back plate
8. Tighten screw and secure

Line	1 T	1 R	2 T	2 R	3 T	3 R	4 T	4 R	5 T	5 R	6 T	6 R	7 T	7 R	8 T	8 R
1	.03	.03	.06	.08	.08	.16	.05	.21	.06	.26	.09	.35	.04	.39	.21	.60
2	.03	.03	.03	.06	.05	.11	.03	.14	.06	.20	.04	.24	.05	.29	.17	.46
3	.02	.02	.04	.06	.04	.10	.04	.14	.08	.22	.04	.26	.05	.31	.16	.47
4	.02	.02	.04	.06	.04	.10	.04	.14	.06	.20	.04	.24	.05	.29	.16	.45
5	.02	.02	.06	.07	.05	.12	.03	.15	.05	.20	.04	.24	.05	.29	.15	.44
6	.01	.01	.02	.03	.06	.10	.04	.14	.04	.18	.06	.24	.04	.28	.16	.43
7	.02	.02	.03	.06	.05	.10	.05	.15	.06	.21	.06	.27	.06	.31	.16	.47
8	.04	.04	.03	.07	.05	.12	.06	.17	.04	.21	.04	.25	.06	.31	.16	.49
9	.02	.02	.04	.06	.06	.11	.04	.15	.05	.20	.04	.24	.08	.32	.15	.47
10	.04	.04	.04	.08	.04	.12	.04	.16	.05	.21	.06	.27	.05	.32	.16	.48
11	.01	.01	.07	.07	.07	.14	.07	.21	.05	.26	.04	.30	.09	.39	.14	.63
12	.03	.03	.07	.07	.07	.16	.07	.22	.05	.27	.04	.31	.04	.35	.21	.56
13	.02	.02	.06	.05	.05	.11	.03	.14	.08	.22	.05	.27	.05	.32	.15	.47
14	.03	.03	.07	.02	.09	.02	.06	.15	.12	.27	.06	.33	.05	.39	.18	.56
15	.03	.03	.06	.05	.11	.05	.11	.03	.14	.08	.22	.06	.27	.06	.32	.16 .49
16																

SUM = 17.36

Summary

	1	2	3	4	5	6	7	8
Total time	.39	.59	.76	.67	.92	.75	.63	2.45
No. of readings	15	15	15	15	15	15	15	15
Av. of readings	.026	.039	.051	.045	.061	.050	.055	.163
Frequency	One out of One							
Average time	.026	.039	.051	.045	.061	.060	.055	.163
El. rating factor	1.05	1.00	1.00	.95	.90	1.00	1.00	1.05
El. normal time	.027	.039	.051	.043	.056	.060	.055	.171

SUM = 0.491

Foreign elements

SYM	R	T	Description
A			
B			
C			
D			
E			
F			
G			
H			
I			

Av. cycle time $\frac{7.36}{5} = .491$
Cycle rating factor 1.05
Normal cycle time .516

Percent allowances

Pers.	Fat.	Delay	Total
5%	5%	5%	15%

Std. time per unit 0.593
Pieces per hour 101
Std. hours per 100 0.988

Allowances in minutes

Pers.	Fat.	Delay	Total

Avail. prod. min. per hr. _____
Pieces per hour _____
Std. hrs. per 100 _____

includes such intangibles as attitude, health, and interest in work as well as speed. The technician enters the stop watch reading in the R column for each element. In working up the time study back at the office, the next step is to perform subtractions for each line, 1 through 15, to determine the time for each element. These subtractions are handled to get the elemental time. Next he totals the columns labeled T to get the total time for each element. After computing the average time he multiplies by the rating factor for the normal time per element.

The observed time must be adjusted to a normal performance time. This is handled by the product of elemental time and the rating factor. The normal cycle time for a defined job is equal to the sum of the elemental normal times. After determining the normal cycle time, we apply the job allowances. The job allowances are the recognizable nonproductive minutes occurring in the performance of the defined job. This condition, which stems from the necessity to satisfy one's personal needs and to overcome fatigue and the existence of certain unavoidable delays, requires that the normal cycle time be enlarged before an acceptable job standard is determined. In many places these allowances range from 10 to 35%. The usual method of adjusting normal cycle time is through the additional percentage allowance. The standard is computed by

$$T_s = T_c(1.0 + A) \tag{3-1}$$

where T_s = standard time for a job per unit of production
T_c = normal cycle time per unit of production
A = allowance expressed as a decimal

For the electrical receptacle we have

$$T_s = 0.516(1.0 + 0.15) = 0.593 \text{ minute per unit}$$

This may be converted to pieces or units of production per time period by a reciprocal relationship. If pieces per hour are desired, then

$$60 \frac{\text{minutes}}{\text{hour}} \times \frac{\text{unit}}{0.593 \text{ minute}} = 101 \text{ units per hour}$$

Many firms prefer expressions such as pieces per hour, units per minute, or packages per week. In some businesses, it is common to express the rate per dozen or gross. Conversion to standard hours per unit or 100 units or the like is equally straightforward, or

$$H_s = T_s \times \tfrac{100}{60} = 0.593 \times \tfrac{100}{60} = 0.988 \text{ standard hour per 100 units}$$

Cost per operation can be established for the job as

$$C_j = H_s L \tag{3-2}$$

where C_j = standard job cost, dollars per quantity of units
H_s = standard hours per quantity of units
L = labor rate per hour

What is included in the labor rate L differs with the situation. In some cases only the labor rate plus direct-related fringe labor costs are selected. In other situations the rate could reflect supervision cost, machine depreciation, and so forth. This designated cost must be made clear. This receptacle assembly cost is now available as a *standard* and can be used provided the method on which it was constructed does not change.

Nonrepetitive Time Study. It should be understood what we examined previously dealt with repetitive work found in production industries. Frequently we are unable to study more than one cycle. Nonrepetitive time study, although not as accurate and free from error as other methods, does provide for a certain class of information. Sometimes called production study or all-day time study, it has been found useful for construction operations, indirect labor, production setups for highly repetitive work, allowances for repetitive work, audits, and direct-labor long-cycle type of work. This class of work is called undesignated, as it is difficult to preplan. The nonrepetitive time study differs from the repetitive motion and time study in a number of ways. The job study is not as complete, rating is frequently ignored, the preliminary investigation may be less, and preknowledge of the elements is unknown. It uses continuous timing with a stop watch and the length of the elements is based on judgment. The types of work for which this is best suited are carpentry, railroad line work, material handling, maintenance, secretarial activities, and so forth.

3.5-2
Work Sampling

Work sampling is another technique for gathering information about large segments of a work force population. It is a counting method for quantitative analysis in terms of time or percent of the activities of men, machines, or any observable state of an operation and is useful in analysis of undesignated or nonrepetitive work activities and allowances. Relatively inexpensive to obtain, and convenient to perform, it can be conducted without recourse to a stop watch. A work-sampling study consists of a number of observations taken pertaining to the specific activities of the person(s) or machine(s) at random intervals. These observations are classified into predefined categories directly related to the work situation. During the course of the work-sampling study, tally marks are made by the technician, such as "work," "idle," or "absent." The key to the accuracy is the number of observations, which may vary according to the requirements. One survey may require very broad areas to be investigated, in which case relatively few observations will be necessary to obtain meaningful results. On the other hand, to establish production standards for use with construction costs, many thousands of observations may be needed. In determining the number of observations necessary, the technician predetermines the accuracy of his results. Four thousand observations will provide more reliable results than 400. However, if accuracy is unimportant, 400 observations may be ample.

As work sampling is a statistical technique, the laws of probability must be followed to obtain accuracy of the sampling estimate. In this type of observation, an event such as equipment working or idle is instantly tallied. For this selective choice mathematicians define a binomial expression where the mean of the binomial distribution is equal to np_i with n equaling the number of observations and p_i the probability or relative frequency of event i occurring. The variance of this distribution is equal to $np_i(1 - p_i)$. As n becomes large the binomial distribution approaches the normal distribution. As work-sampling studies involve large sample sizes, the normal distribution is considered an adequate approximation. In work sampling we take a sample of size n observations in an attempt to estimate p_i or

$$p_i' = \frac{n_i}{n} \tag{3-3}$$

where p_i' = observed percentage of occurrence of an event i expressed as a decimal

n_i = number of snap observations of event i observed

n = total number of random observations taken

As shown in textbooks on probability the standard error of a sample percentage for a binomial distribution may be expressed by

$$\sigma_{p'} = \left(\frac{p'(1 - p')}{n} \right)^{1/2} \tag{3-4}$$

where σ = the standard deviation of the percentage of the binomial sampling distribution. In any sampling procedure, bias and errors may occur. This results in a deviation of p' and p. A tolerable maximum sampling error in terms of a confidence interval I commensurate with the nature and importance of the study can be pre-established. This confidence and interval may be viewed by examining Figure 3.6. The factor 1.645 is obtained from a table of probabilities for the

FIGURE 3.6

ANALOGOUS CURVE FROM "NORMAL" RELATIONSHIPS TO BINOMIAL WORK-SAMPLING PRACTICES

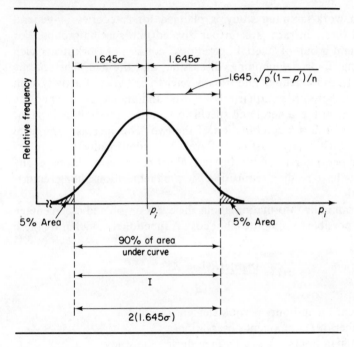

normal distribution for a confidence of 90%, which is usual for work-sampling studies. The sampling interval is given by

$$I = 2\alpha \left(\frac{p_i'(1 - p_i')}{n} \right)^{1/2} \tag{3-5}$$

where I = confidence interval obtained from study expressed as a decimal
   α = factor from normal tables for a chosen confidence

We expect that the true value of p falls within the range of $p' \pm 1.645\sigma$ approximately 90% of the time. In other words, if p is the true percentage of work estimate, the estimate will fall outside $p' \pm 1.645\sigma$ only about 10 times in 100 due to chance alone. The previous equation may be solved for the sample size when the other factors are either assumed or known, as

$$n = \frac{4\alpha^2 p_i'(1 - p_i')}{I^2} \tag{3-6}$$

Let us see how this procedure might be applied in the work-sampling study of construction labor. In view of the results of the study, values of $I = 0.04$ for a 90% confidence level seemed adequate. The event "idle or working" was to be viewed in a preliminary study of 200 observations. The idle category was found 18 times, or 9%. Based on this information the number of observations can be found for a confidence of 90% for the idle category as

$$n = \frac{10.8(0.09)(0.91)}{(0.04)^2} = 553$$

This is greater than the 200 observations which were initially made. To meet the accuracy requirements, 353 more observations are needed. In retrospect it becomes possible to determine the magnitude of the sampling error for $n = 200$, or

$$I = 2(1.645)\left(\frac{0.09(0.91)}{200}\right)^{1/2} = 0.0665$$

Normally a work-sampling study is planned for categories of general information, and these, in turn, are further divided into specific groups. For example, the general group of "clerial operations" consists of work items such as reading, writing, filing, telephoning, discussion, and the like. The specific categories would be more definite and usually refer to one area of activity such as requisitions, purchase orders, quarterly reports, and audits. Then when the observations are taken the general and specific categories are combined so that each observation is always a combination of the two. This increases the total number of items which finally must be analyzed, but, additionally, more information is received per observation bit. Instead of just having requisitions as one item, you will now have reading requisitions, writing requisitions, filing requisitions, and so forth.

With work-sampling information about the event or job and a percentage fact for each, it is possible to compute labor cost. A functional model using these data is

$$H_s = \frac{(n_i/n)HR(1 + A)}{W} \tag{3-7}$$

where H_s = standard man-hours per job element i
n_i = number of i component observations
H = total man-hours worked by men during the study
R = mean rating factor
A = allowance for unavoidable delays and personal time, decimal
W = work units accomplished during period of observing this event.

Assume a construction job where carpenters form layup walls on a dam. These concrete retaining walls are numerous and fairly well standardized. In addition the carpenters are doing other tasks which are capable of description. The study runs for 3 weeks for 14 carpenters. During the course of random work sampling, ratings similar to time study practices are made, and $R = 0.96$ is determined. A total of 17 frames are finished, and work sampling indicates that this event $n_i/n = 16\%$ of the total effort of the gang. For an allowance of 20% the time per frame may be found as

$$H_s = \frac{0.16(14 \times 3 \times 40) \times 0.96 \times 1.20}{17} = 18.215 \text{ man-hours per frame}$$

The cost per frame may be computed using the previous equation $C_j = H_s L$

with slight realignment of units. If $L = \$8.60$ and $H_s = 18.215$ man-hours per frame, we see that the labor cost for the frame is \$156.65.

Work sampling is a tool that has broad ramifications for the operation analyst. It allows him to get the facts in an easy and fast way. In summary, the following considerations should be kept in mind for work sampling:

1. Explain and sell the work-sampling method before putting it to use.
2. Isolate individual studies to similar groups of machines, operations, or activities.
3. Use as large a sample size as is practical, economical, and timely.
4. Observe the data at random times.
5. Take the observations over reasonably long periods of time, for example, 2 weeks or more, although rigid rules must bend with the situation and design of the study.

3.5-3
Man-hour Reports

In certain types of work, such as construction, crew work, or long production cycles, the foreman is frequently asked to contribute information. The time card is a common means of time reporting and cost distribution. Essentially, each foreman makes out a card for each shift covering all men under his supervision. The foreman may include badge numbers or job types and hours worked along with the description of work being done. After some period of time with the cards, foremen have been found to have an accurate knowledge of costs. A man-hour may be defined as a standard amount of work performed by one man in 1 hour. Man-hours have value if the unit hours are relatively constant and unaffected by changes in wage rates, overtime, bonuses, and so forth. Example of man-hour units are as follows:

- Number of man-hours to place 1000 bricks
- Number of man-hours to erect 100 square feet of framework
- Number of man-hours to sweep 1000 square feet of office area

If man-hour reports are to be of value as a permanent measured record, their backup data must include a variety of information to permit the engineer to consider deviations from the observation. In construction work, weather conditions, general efficiency of crews (experienced or green, native or imported), equipment, hazards, location of work (is the work located at a considerable distance and is travel time included?), material conditions, and type of construction are vital factors. As in time study, the man-hour data recognize that certain delays are incurred. The delays would fall into controllable and uncontrollable categories. Controllable delays consist of plant and equipment breakdowns, accidents to personnel, failure to coordinate material requirements and work, and so forth. Uncontrollable delays consist of extreme weather conditions or stopping work during blast.

Sometimes employees are asked to determine their own man-hours for work. This "self analysis" can be useful for estimating.

3.5-4
Predetermined Motion-Time Data

Predetermined motion-time data systems, such as Methods-Time Measurement, Work Factor, Motion-time Analysis, and others, stipulate a time for a human motion. These methods are useful for providing standards for man-time, or the manual elements, with short elements as the rule. These time systems require care in their application and in their usual form of information do not play a vital role in cost estimating. Their fault lies in the minuteness of the times which are related to small human motions. Their greatest applications are for the mass production industries.

As is pointed out in Figure 3.7, human effort is subject to a wide variety of influences and is difficult to measure. As can be seen, Figure 3.7 compares the degree of inaccuracy, the relative usefulness of the data, and the cost to obtain the data. Selection of which method to adopt is up to the user.

FIGURE 3.7

DESCRIPTIVE COMPARISON OF THE FOUR
METHODS OF TIME MEASUREMENT

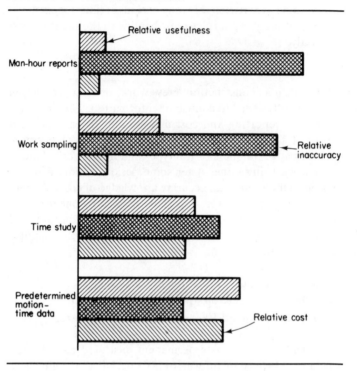

3.6
POLICY COST

Even though cost may be historical, measured, or policy, it is required for estimating purposes that the cost be for future periods. Usually, cost appears in the raw form, is generally unsuitable and requires reforming by the cost engineer. An exception would be those costs dictated by negotiation, or contractual actions which are recognized as typical policy cost and are adopted as future costs. Errors due to the uncertain future, recording mistakes, or tolerances of measurement are less of a problem for policy costs.

Policy actions within a firm may require that certain costs be established insofar as the estimator is concerned. For example, interdivisional pricing may require that division A accept all cost data resulting from division B as standard irrespective of B's performance. Wage escalator clauses in management-union contracts or the government minimum wage law are examples of where labor costs are imposed. These negotiated or legislated costs are future costs and make past costs inappropriate. Sometimes a firm commits itself to a long-term agreement for materials, labor, or other services. These costs are confirmed through a stable contractual settlement. The adoption of these policy costs for preliminary or detailed estimates is straightforward.

It is necessary to have historical, measured, or policy costs to build an estimate. Past costs are approximations of the future. Depending as it does on these types of costs, estimating is based on information analysis and speculation of these sorts of costs. Whether internal or external agencies are the origin for these data, cost engineers go to others for information. Internally it may be from management information sources, while externally many quarters would be sought out for cost data.

Accounting is a principal source for past data. Two methods of costing, absorption or direct, are the means by which cost accounts are classified. When the vital concern is planning and control and where public exposure of profit and loss data is secondary, direct costing would be a superior source for the cost engineer.

Costs may be determined by direct measurement. Information from time study, work sampling, or foremen's report can be used directly for cost estimating.

For a general approach to absorption and general accounting methods, refer to

NOBLE, HOWARD S., and C. ROLLIN NISWONGER: *Accounting Principles*, 8th ed., South-Western Publishing Company, Cincinnati, Ohio, 1961.

Cost accounting is discussed in

DICKEY, ROBERT R., ed.: *Accountant's Cost Handbook*, The Ronald Press Company, New York, 1960.

HORNGREN, CHARLES: *Cost Accounting with Managerial Emphasis*, 3rd ed., Prentice-Hall, Inc., Englewood Cliffs N.J., 1972.

MATZ, ADOLPH, J. CURRY, and GEORGE W. FRANK: *Cost Accounting*, 4th ed., South-Western Publishing Company, Cincinnati, Ohio, 1967.

The method of direct costing is discussed in

GILLESPIE, CECIL: *Standard and Direct Costing*, Prentice-Hall, Inc., Englewood Cliffs, N.J., 1962.

For an elaborate description of the measurement of work, refer to

BARNES, RALPH M.: *Motion and Time Study: Design and Measurement of Work*, 6th ed., John Wiley & Sons, Inc., New York, 1968.

A comprehensive listing of the sources of government documents significant to business, engineering, public works, and so forth is itself several major documents. The following, which can be generally found in a large public library, condenses this information:

SCHMECKEBIER, LAURENCE F., and ROY B. EASTIN: *Government Publications and Their Use*, rev. ed., The Brookings Institution, Washington, D.C., 1961.

A classified guide to current periodicals, both foreign and domestic, is

Uirich's International Periodical Directory, Vols. I and II, R. R. Bowker Company, New York, 1969.

A very practical approach to economics may be found in

DEAN, JOEL: *Managerial Economics*, Prentice-Hall, Inc., Englewood Cliffs, N.J., 1951.

KLEIN, LAWRENCE R.: *An Introduction to Econometrics*, Prentice-Hall, Inc., Englewood Cliffs, N.J., 1962.

Governmental practices and related governmental accounting procedures are discussed in

McKean, R. N.: *Efficiency in Government Through Systems Analysis*, John Wiley & Sons, Inc., New York, 1968.

Novick, David: *Efficiency and Economy in Government, the New Budgeting and Accounting Procedures*, The Rand Corporation, Santa Monica, Calif., 1954.

QUESTIONS

1. Discuss the importance of costs as they relate to the design. Consider a situation where a company has no known previous cost or design experience.

2. List several sources for information. In your judgment do you think there are ideal cost data?

3. What is the difference between direct costs (not direct costing) and indirect costs? Is it significant?

4. What are cost elements and cost units? Are they related? Explain.

5. Construct a figure similar to Figure 3.1 for public works and construction. Use appropriate language suitable to these organizations.

6. Do you believe trade associations would be overprotective of their own interests in providing cost data? What about the motives of the union in supplying certain cost information? Do you think that their objectives are consistent with the objectives of the firm?

7. Define a cost index.

8. What is direct costing? Explain fully and give its good and bad points. What is absorption costing? Explain fully and give its good and bad points.

9. Name several methods of time measurement. What is the process followed in making a time study?

PROBLEMS

3-1. Derive a functional model (see Chapter 2) that expresses the total cost of producing manufactured goods. Define the following terms by use of functional models: (a) total assets, (b) total liabilities, (c) net worth, and (d) profit.

3-2. Evaluate the effects of transactions by constructing a daily balance sheet showing an asset side and an equities side. *January 1:* John Smith starts a sheet metal business producing metal products for the home. The business is called John Smith Sheet Metal. Mr. Smith deposits $10,000 of his own money in a bank account which he has opened in the name of the business. *January 2:* The business borrows $5000 from a bank, giving a note, therefore increasing the assets and cash and the business incurs a liability to the bank. *January 3:* The business buys inventory in the amount of $10,000, paying cash. *January 4:* The firm sells material for $300 that cost $200.

3-3. Short Corporation started the year with the following balances:

Account	Balance as of January 1
Cash	$100,000
Inventory	100,000
New plant and equipment	400,000
Accounts payable	50,000
Owner equity	550,000

Transactions during the year were limited to the following: (a) Pay $100,000 for labor, purchase $150,000 worth of materials, note equipment depreciation of $50,000, adding to inventory 300,000 units costing $1 to the manufacturer. (b)

Sell 300,000 units for $2 each, cash. (c) Purchase new equipment costing $200,000. (d) Accounts payable at the end of year were the same as at the beginning of the year. Neglect income taxes. Make an end-of-year balance sheet. Make an income statement for the year just ended.

3-4. Given the following ledger from Weichman Mfg. Co., set up a balance sheet for the month of March.

Cash

March	1	Capital	$1000	March 5	Rent	$200
	10	Consulting fee	250	20	Salaries	350
	25	J. A. Wilson on acct.	500			

Customers (Accts. receivable)

March 10	J. A. Wilson	$1200	March 25	Cash on acct.	$500

Supplies on Hand

March 5	Accts. Payable	$360	March 31	Supplies used in March	$110

Equipment

March 4	Notes payable	$3200	

Accts. Payable

		March 3 Supplies	$360

Notes Payable

		March 4 Equipment	$3200

Weichman Mfg. Co. Capital

March	3	Cash	$200	March 1	Cash investment	$1000
	20	Salaries	350	10	Consulting	250
	31	Supplies used	110	10	J. A. Wilson	1200

3-5. Using the following closed ledger, construct a profit and loss statement:

Equipment

June	1	Balance	$3200

Accts. Payable

June	2	Cash Myers Co.	$360	June 1	Balance	$360
				6	Supplies	600
				28	Misc. exp.	40

Notes Payable

		June 1 Balance	$3200

Capital

		June 1 Balance	$1790
		30 From P & L	1570

Income

June	30	To P & L	$2500	June 4	Accts. pro. A. B. Jones	$2200
				15	Cash I. N. Smith	300

Lease Expenses

June	1	Cash	$200	June 30	To P & L	$200

Misc. Office Expenses

June	10	Telephone, cash	$60	June	30	To P & L	$100
	28	Elec.	40				

Salaries

June	20	Cash	$350	June	30	To P & L	$350

Supplies Expense

June	30	Supplies used	$280	June	30	To P & L	$280

3-6. Construct a balance sheet for Dynamics Corp. based on the following information:

Retained earnings	$610,000
Cash	150,000
Outstanding debt	450,000
Raw materials	100,000
Finished goods	50,000
Current liabilities	40,000
Stock ownership	400,000
Fixed assets	1,100,000
In-process materials	100,000

3-7. Develop direct-costing overhead and absorption overhead rates on the basis of direct labor dollars for the following:

Production load	80%	100%	125%
Labor	$100,000	$125,000	$156,250
Number of units	20,000	25,000	31,250
Variable cost	$ 60,000	$ 75,000	$ 93,750
Fixed costs	$120,000	$120,000	$120,000

3-8. An actual-cost audit revealed that the labor had an unfavorable 20% variance, material was favorable by $0.001, and overhead was reported unchanged. If standard part cost was $0.35 for labor, $0.075 for material, and 200% for overhead, determine the total standard cost and actual cost. What is the net variance?

3-9. A mining equipment manufacturer compiled the following data on three series of drill steel it produces.

Production Center	Hourly Rate	Hours Required Per Unit, Series			Cost Per Unit, Series		
		1700	1600	1500	1700	1600	1500
Cut off	$4.20	0.01	0.01	0.01	$0.04	$0.04	$0.04
Upsetting	5.65	0.03	0.03	0.03	0.17	0.17	0.17
Machining	6.11	0.05	0.04	0.03	0.31	0.24	0.18
Heat treat	0.40	0.20	0.18	0.16	8.00	7.20	6.40
Bench	3.90	0.5	0.5	0.5	1.95	1.95	1.95
Finishing	5.02	0.1	0.1	0.1	0.50	0.50	0.50
Unit conversion cost					10.97	10.10	9.24
Unit material cost (20¢/lb)					34.80	31.60	27.00
Unit selling & distribution cost (5% price)					3.75	3.20	2.50
Total direct costs					49.52	44.90	38.74
Selling price					$75.00	$64.00	$50.00
Contribution/unit					$25.48	$19.10	$11.36
Contribution percent					34%	30%	22.7%

Variance on direct costs are as follows:

> Mfg. labor & expense, 3% unfavorable
> Materials & scrap, 5% unfavorable
> Selling & distribution, 1% unfavorable

Make an income and expense statement similar to Table 3.10 if the manufacturer makes 200 1700-series, 100 1600-series, and 170 1500-series drill steel. If the total period costs (budgeted amount plus variance) were $5500, what are the net profit and percent of net sales?

3-10. An analyst, working with motion economy rules, had the parts bin moved closer to the belly of the operator presuming that the time per reach and move would be less. These are the data for a manual assembly operation:

- Number of parts moved: 240 per unit.
- Movements per part (reach and move): 2.
- Average savings in time to move hand 6 inches or less, both ways: 0.002 minute.
- Labor cost: $4.50 per hour.
- Working days per year: 250.
- This factory produces 175 completed sets per day.

What are the hours saved per unit, per day, and per year? If the analysis is correct, what appears to be the most serious faults in its implementation for all operators? What would you suggest as remedies to these problems?

3-11. A method has been engineered and an individual has been time-studied. The method has produced an average overall time of 2.32 minutes by actual watch-timing. The overall pace rating applied to the job is 125%. This company, which uses an incentive system, adds extra time into the production rates to cover personal time, fatigue, and unavoidable delays. The total allowance is 15%. Determine the rated time or normal minutes, the total allowed time or standard minutes, the standard hour rate for the job per unit and 100 units, and the pieces per hour. The labor rate per hour is $5.10. What is the standard labor cost per unit?

3-12. Below is a time study observation sheet. The elemental times are snapback times. Find the average, normal, and standard times for each element.

Element	Reading	Average Time	Rating (%)	Normal Time	Allowance (%)	Standard Time
1. Place bushing in jig	0.03 0.08 0.05 0.04 0.06		117		11	
2. Drill hole	0.31 0.38 0.37 0.39 0.33		100		5	
3. Remove bushing from jig and place on conveyor belt	0.07 0.08 0.09 0.12 0.11		93		13	

(a) Calculate the production per hour.
(b) By an old method, 0.834 minute standard time was required to complete this operation. Calculate the increase in output in percent and savings in time in percent.

3-13. A job is time-studied and a summary time sheet is given as

Element Description	Frequency Per Unit	Average Time	% Rating Factor
1. Get part off conveyor	1/1	0.040	100
2. Get incomplete part off	1/1	0.030	105
3. Connect length and joint	2/1	0.055	110
4. Connect 3	3/1	0.153	100
5. Assemble side	1/1	0.243	90
6. Mate two sides	1/1	0.183	90
7. Put on conveyor	1/1	0.032	100

With an allowance factor of 15% to cover P (personal), F (fatigue), and D (delay), what are the normal minutes per piece, the standard per minute, and the number of hours per 1000 units? If performance against incentive has averaged 20%, what is the incentive hourly rate if day work is paid $5.25 per hour?

3-14. The following is a raw continuous-timing two-element time study of a punch press operation:

	1	2	3	4	5
1. Handle part	0.01	0.04	0.08	0.11	0.14
2. Punch part	0.03	0.06	0.10	0.12	0.16

What is the average time for elements 1 and 2? The ratings were +0.08 and −0.09. What is the normal time for elements 1 and 2? A man and machine allowance of 20% is used in this plant. What is the standard minute per piece for this operation? How many pieces per hour? How many hours per 100 units? If only a man allowance of 20% is used, for element 1, what are the standard minutes per unit? Assume that this operation completes the part and that material costs from purchase records indicate a unit cost of $0.005. Express the labor and material cost as a manufactured standard cost if labor costs $5.40 per hour. Overhead based on direct labor dollars for the punch press is 200%.

3-15. A gang nonrepetitive time study was made of a construction crew, and the following is a tabulated summary. It is usually unrealistic to rate this kind of work. Crosses in the table indicate that the job element was necessary for the worker.

The work consisted of placing 180 yards of concrete by a pipeline. Determine a standard cost of placing a yard of concrete. Initially assume that when not assigned to this job the operator is doing something else profitable. Pay time and a half for overtime for any assigned work in this job. Assume that the union contract requires the entire crew (exclusive of concrete truck and driver) to be at the job site the entire time. What would this actual cost standard be then? What is the cost of this nonproduction?

Element	Minutes	Foreman	Pump Operator	Hopper Man	4 Laborers	2 Vibrators	Truck, Driver	Pumps & Fittings
					Crew and Equipment			
1. Make ready	60	×			×			
2. Move to job site	15	×			×			
3. Pump machine to job	45	×	×	×	×			×
4. Set up machine	90	×	×	×	×			×
5. Inspect	30	×	×	×	×		×	×
6. Adjustment	30	×	×	×	×		×	×
7. Pump concrete to forms	360	×	×	×	×	×	×	×
8. Normal delay for set of concrete	80	×	×	×	×	×		×
9. Dismantle pump	45	×	×	×	×			×
10. Put tools away	15	×	×		×			×
11. Return pump	30	×	×		×			×
12. Clean up	10	×			×			
Cost per hour		7	8	8	6	8	35	25

3-16. A work-sampling survey of an operation, which was designated into 12 categories, has the following observations:

Item	Observations	Item	Observations
1	92	7	24
2	99	8	33
3	37	9	3
4	11	10	22
5	25	11	8
6	14	12	32
			400

If this sample covered a span of 25 days for 8 hours per day, what are the percents and expected hours per item of work.

3-17. The pediatrics department in a hospital was work-sampled for 608 hours.

Work Category	Observations	Work Category	Observations
1. Routine nursing	496	8. Other	79
2. Idle or wait	263	9. Feeding	52
3. Unit servicing	183	10. Bathing	22
4. Report	129	11. Elimination	11
5. Personal time	128	12. Transporting	8
6. Intervention	102	13. Housekeeping	7
7. Unable to sample	91	14. Ambulation	7

For the 1578 observations find the % occurrence, % cumulative occurrence, element hours, and cumulative hours for each work category.

3-18. A work-sampling study is taken of a department with the following information obtained: Number of sampling days, 25; number of trips per day, 16; number of people observed per trip, 3; and number of items being sampled, 4. The sample items are broken down as A, 80; B, 320; C, 1600; and D, 2800. How many man-days and observations were sampled? What is the percentage and equivalent hours for the activities? For a confidence level of 90%, what is the accuracy of each of the items?

3-19. (a) To get $\pm 5\%$ precision on work observed by work sampling that is estimated to require 70% of the worker's time, how many random observations will be required at the 95% confidence level? For 95% $\alpha = 1.96$.

(b) If the average handling activity during a 20-day study period is 85%, and the number of daily observations is 45, what is the tolerance allowed on each day's percent activity? Use 90%.

(c) Work sampling is to be used to measure the not-working time of a utility crew. A preliminary study shows that not-working time is likely to be around 35%. For a 90% confidence level and a desired accuracy of $\pm 5\%$, what is the number of observations required for this study?

3-20. A shipping department that constructs wooden boxes for large switch gears has five direct-labor workers. A work-sampling study was undertaken, and the following observations of work elements were recorded over a 15-day, 8-hour period:

Set up and dismantle,	312
Construct crates,	264
Load switch gear in crates,	204
Move materials,	324
Idle,	96

A rating factor of 90% was found. The number of switch gears shipped during this period was 26. This firm uses an allowance value of 10% for work of this kind. Average labor cost including fringes is $6.25 per hour. What are the elemental costs? What is the standard labor cost per box? The actual cost?

CASE PROBLEM

The Endicott Iron Foundry. "We can't make any profit on that job; it has too much labor cost in it," said Dick Crawford, the foundry superintendent, to George Dobbins, cost estimator for the Endicott Iron Foundry. The Endicott Iron Foundry, like its competitors, has always estimated costs on a per pound basis for the delivered casting. Difficult castings were quoted at a higher price per pound than simple castings, but the difference in price (often based on the estimated cost) did not seem to be great enough to warrant the extra labor costs. Crawford suggested that the company was making little profit or loss on jobs that took considerable labor. What will happen if Endicott starts quoting higher prices for casting requiring extra labor? Should it recover the full cost? What devices can you suggest at this time to improve the estimates? Should the cost estimator depend on the knowledge of Dick Crawford as final? What should constitute a loss or profit for the estimate?

chapter 4

The counting of cost is the art of revealing bad answers to problems for which worse answers would otherwise be given.

Antiquity.

STRUCTURAL APPROACH TO COST

For ages men have deduced rules of conduct from fables and parables. Stripping the veneer as they do, they point up the central thought of the lesson. In our introductory thought, the admission of something less than perfect about counting of cost is made, yet there is no other way for the engineer.

Whether the source for the cost data was historical, measured, or policy, its use for conjecture is another matter. With this information consistently and objectively gathered, there must be a uniform method of distribution of costs before an engineering design can be estimated. Record keeping and distribution of costs are eventually associated with an operation cost, product price, project return, or systems effectiveness. The structure of the design estimate is fundamentally similar. The purpose of this chapter is to provide an understanding of cost data and how they are allocated for estimating.

Structuring cost information is a process of grouping like facts about a common point on the basis of similarities, attributes, or relationships. After a classification of cost accounts is undertaken, the cost information is summarized. Sometimes cost codes are given with an engineering description. In the engineering construction business the cost codes are an additional listing beyond the chart of accounts. In some cases only a chart of accounts is found, as in industry. The principles of allocation, enumerated later, concern the assigning of expenditures, as represented by original documents such as payrolls and invoices, materials, and supplies, to the cost accounts or to the engineering cost codes. It is desirable that these costs be described with terms that are compatible with the needs of a design estimate. Unfortunately, this is not always possible.

In view of his rapport with cost and its ramifications, the cost estimator is frequently charged with the preparation of budgets, cost reports, labor analyses, material studies, special economic studies, as well as cost estimates. Labor, materials, depreciation, overhead costs, and general costs constitute universal elements for these requirements. We intend to study these elements separately before uniting them into cost, price, return, or effectiveness.

4.1
EQUITABLE DISTRIBUTION OF COSTS

Uniformity and consistency in cost-accounting and cost-engineering procedures can contribute to the success of the estimate. In proration of costs (loosely meaning distribution or allocation) it is unlikely that any universal way can be suggested for all situations. Any base that is chosen for distribution purposes should meet two tests: It should result in charges to the cost center that are fair in view of the benefit that the center receives, and it should be inexpensive to administer. Distribution of costs consists essentially of determining the proper account or the cost code to which expenditures can be assigned by placing the account number or code number on all financial documents such as invoices, payrolls, materials-receiving reports, and other documents of costs. The account number is generally an accounting term, while *cost code* is more of an engineering term. Table 4.1 is a listing of a manufacturing set of cost accounts without numerical account designations. A few explanations for the table are in order: An accrued expense is an expense which has been incurred but for which payment has not yet been made. Accrual is made for taxes, salaries, interest, and other expenses incurred but unpaid at the balance sheet data. Accrued income is income which has been earned but not collected. Accrual is made for royalties, interest, and other income earned but uncollected at the balance sheet data. A deferred expense or income is one which applies to future accounting periods; an example expense would be the full payment of the premium of 3-year fire insurance policy, as of December 31 when 1 year has expired and 2 remain. Deferred expenses represent, in effect, expense inventories, and on the balance sheet they are shown as assets. Fringe benefits are employee welfare benefits including the expenses of employment not paid directly to the employee, such as group insurance and pension cost. Some fringe costs paid directly are holiday and vacation expense. A noncash expense is a manufacturing or overhead expense for accounting purposes which does not require outlay of funds, of the same amount in the same year, but which may result in a saving, for example, depreciation (tax deductible) and amortization of good will (not deductible). The distribution considered so far is simply the recording of business transactions. Sometimes this is called *primary* distribution.

In the engineering construction business and in certain manufacturing operations, a cost list is selected with the needs of design estimating as a basic consideration. Materials, supplies, equipment cost, subcontract payments, and all such expenditures are assigned from their original documents and logged against the appropriate code number which is a similar procedure to accounting. The posting may be handled by the engineering department or an experienced cost clerk. Table 4.2 is a master list of cost codes that is found in the engineering construction industry.

A primary distribution of overhead consists of assigning the various overhead costs to several departments or defined divisions within the unit, or factory, for example. Sometimes *cost centers* combine or separate departments to form homogeneous groups and are a logical point for the accumulation of costs. In making this distribution there is no distinction between a producing department (the milling machine department) and a service department (first aid). This distribution is followed by a redistribution in a secondary manner. Overhead is allocated to a designated base. A base for distribution of manufacturing overhead may be floor area, kilowatt hours, direct-labor dollars or hours, machine hours, or number of employees. The redistribution of costs originally assigned to service cost centers (first aid, for example) to the production cost centers is termed *secondary* distribution. This bookkeeping transfer technique of overhead costs to production departments is for subsequent recovery within operation or product cost. A secondary distribution, in the case of first aid, might be to prorate the total medical center costs to the production department on the basis of the number of employees within that department. If the milling machine department had two times as many employees as the lathe department, one could reason that the costs that it receives should be borne on a 2:1 ratio. Building depreciation, building insurance, building maintenance, and building taxes are often distributed on a floor area basis. Electrical power poses a confusing choice. It may be distributed on the basis of machine weight, machine-hours, horsepower hours, or even direct labor hours. Overhead costs become complicated whenever they have a joint or commonness with different levels of variability. Joint costs, or costs incurred jointly, are depreciation, insurance, property taxes, maintenance, and repairs. They are dependent on one another.

Regardless of the level of productive activity, some overhead costs are almost completely fixed per time period; others are constant within certain ranges of activity; some change as the activity fluctuates. Depreciation of a building is an example of the first class, superintendence of the second, while

TABLE 4.1

CHART OF ACCOUNTS

Assets	*Income and Expense, cont.*

Assets

1. Current assets
   - Cash in bank and petty cash funds
   - Accounts receivable
   - Notes receivable
   - Life insurance cash surrender value
   - Raw materials
   - Other direct materials
   - Work in process
   - Finished goods
   - Factory supplies
   - Fuel
   - Office supplies

2. Fixed assets, intangibles, deferred charges
   - Land
   - Building
   - Machinery and equipment
   - Autos and trucks
   - Office furniture and equipment
   - Reserves for depreciation
   - Interest prepaid
   - Taxes prepaid
   - Insurance prepaid

Liabilities

3. Current liabilities
   - Accounts payable
   - Dividends declared
   - FICA withheld
   - Federal income taxes withheld
   - Notes payable
   - Salary and wages accrued
   - Commissions accrued
   - Employees' vacations accrued
   - State and local taxes accrued
   - Interest payable accrued
   - Compensation insurance accrued

4. Valuation reserves
   - Reserves for contingencies

5. Long-term liabilities, capital stock, and surplus
   - Long-term debt
   - Capital stock
   - Capital surplus
   - Retained earnings
   - Profit or loss

Income and Expense

6. Sales, deductions, cost of sales
   - Sales
   - Returns and allowances

Income and Expense, cont.

- Commissions
- Manufacturing cost of sales
- Commercial cost of sales
- Material quantity variances
- Material price variances
- Labor variances
- Manufacturing overhead variances
- Commercial overhead variances

7. Manufacturing overhead
   - Superintendence and foremen
   - Departmental indirect labor
   - Service labor
   - Overtime and shift premiums
   - Vacations
   - Correcting defective work
   - Pensions
   - Compensation insurance & accidents
   - Group insurance
   - Social security taxes
   - Fuel used
   - Power, light, and water
   - Indirect materials
   - General factory supplies
   - Research and laboratory
   - Repairs & maintenance buildings
   - Repairs & maintenance machinery
   - General factory expense
   - Insurance
   - Taxes
   - Depreciation

8. General and administrative overhead
   - Engineering costs
   - Salesmen's compensation
   - Advertising
   - Selling expense
   - Executives' salaries
   - Office salaries
   - Executive traveling
   - Office supplies & expense
   - Company auto expense
   - Postage
   - Telephone and telegraph
   - Subscriptions
   - Donations
   - Bad debts
   - Life insurance premiums
   - Social security taxes
   - Other administrative expenses

9. Other income and charges
   - Cash discounts received
   - Interest received
   - Sundry sales and incomes
   - Interest paid
   - Cash discounts allowed
   - Federal taxes on income

TABLE 4.2

MASTER LIST OF COST CODES[a]

Project Work Accounts, 100–499		Plant and Equipment Operation Accounts, 500–699		Overhead Expense Accounts, 700–999	
100	Clearing & grubbing	500	Excavation equipment operation	700	Project administration
101	Demolition			.1	Project manager
102	Underpinning	510	Truck operation	.2	Office engineering
103	Common excavation	515	Concrete mixer plant operation	.3	Field engineering
104	Rock excavation			701	Construction supervision
105	Backfill	.1	Transportation	.1	General superintendent
115	Wood sheet piling	.2	Set up	.2	Concrete gang foreman
116	Steel sheet piling	.3	Operating labor	702	Project office
130	Excavating for caissons	.4	Maintenance & repairs	.1	Office structure
240	Concrete	.5	Fuel or power	.2	Office furniture
.01	Footings	.6	Ownership expense	.3	Office supplies
.05	Grade beams	.7	Dismantle	703	Timekeeping
.07	Slab on grade	520	Electric plant operation	.1	Timekeeper
.08	Beams	521	Hoisting equipment operation	.2	Watchmen
.11	Columns			.5	Guards
.12	Walls	.1	3-drum hoist	705	Utilities
260	Concrete forms	.11	Transportation	.1	Water
.01	Footings	.12	Set up	.2	Gas
.05	Grade beams	.13	Operating labor	.3	Power
.07	Slab on grade	.14	Maintenance & repairs	710	Storage & sanitary facilities
.08	Beams	.15	Fuel or power		
.11	Columns	.16	Ownership expense	711	Drinking water facilities
.12	Walls	.17	Dismantle	715	Temporary lighting
280	Masonry	528	Power concrete buggies	719	Payroll taxes & insurance
.01	Common brick, 8-in. wall	536	Air compressors	720	Project insurance
300	Carpentry	550	Mobile cranes	721	Performance bond
.01	Floor joist	572	Hoist tower	722	Subcontractor bonds
310	Millwork	.1	Transportation	730	First aid facilities
.01	Kitchen cabinets	.2	Set up	740	Small tools
320	Steel and misc. iron	.4	Maintenance & repairs	755	Temporary fences and enclosures
.01	Structural steel	.6	Ownership expense		
331	Blinds, drapes, & carpets	.7	Dismantle	770	Permits & fees
340	Insulation	585	Patented concrete forms	771	Photographs
342	Caulk & weatherstrip	.1	Plywood panel	775	Tests
360	Ceramic tile	.14	Repairs	776	Cutting & patching
362	Resilient flooring	.16	Ownership expense	780	Winter operation
370	Misc. metals	590	Masonry saws	785	Drayage (not charged elsewhere)
.01	Metal door frames	594	Mortar mixers		
380	Painting	595	Mortar lifts	786	Parking
430	Finish hardware			787	Storage area rental
455	Plumbing, heating, and ventilation			788	Protection of adjoining property
470	Electrical			789	General clean up
488	Clean glass			790	Plane
495	Paving, curbs, & gutter			800	Housing, feeding, & facilities for workmen
499	Allowances				

[a]Richard H. Clough, *Construction Contracting*, John Wiley & Sons, Inc., New York, 1960.

consumption of factory supplies qualifies as the third. Figure 4.1 illustrates this condition of fixed, semifixed, and variable overhead. This behavior of overhead costs leads some to suggest that it should be identified as fixed or variable and let it go at that. The fixed-variable classification is adopted with direct costing; absorption costing does not require this choice.

FIGURE 4.1

RELATIONSHIPS OF FIXED, SEMIFIXED, AND
VARIABLE COSTS

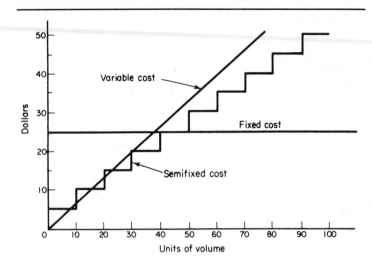

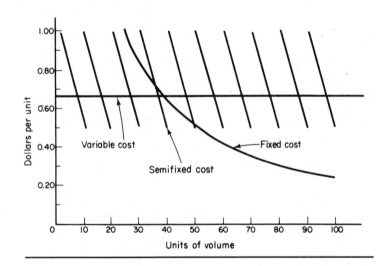

Other activities of the firm must be allocated in an arbitrary fashion. The following indicates possibilities:

Function	Possible Basis of Allocation
1. Selling	Gross sales dollar value of product sold
2. Warehousing	Size, weight, & quantity handled of product
3. Packaging & transportation	Weight, number of shipping units, size, distance separation
4. Advertising	Quantity of products sold, size of average order, kind of product advertised

Although management information systems have been initially discussed in Chapter 3, it is necessary to elaborate more from a cost-estimating point of view. Frequently, some jobs are repeats; no job should be totally estimated if information is already available. When information, tried and true, exists, a retrieval system is mandatory. Despite these needs, the structure of an information system is a compromise. Overall, it cannot accommodate cost engineers exclusively. In a practical way, a centralized cost-engineering department maintains its own library of information. The use of a particular set of data is usually low, while the volume of data generated and used in a given period of time is high. This results from a great diversity in work, and it implies that the preparation of information storage and retrieval is proportionately a great task. A management information retrieval system must serve many purposes and must cross departmental lines.

Can an information profile be developed, although not an exclusive system, to provide the general types of data that will be germane to estimating? Whether the estimator is considering an operation, product, project, or system, information is crucial to his projections. Fortunately the content of the estimates are similar. Materials, money, and assets could be listed as fundamental to the envelope of information. The classification of cost accounts or engineering estimating codes should be systematically prepared to retrieve this information. Obviously having fewer codes and subcodes reduces work, while many subcodes increase work. Inordinate detail tends to increase inaccuracies in cost distributions, while too few details limit cost control.

Cost codes and books of account vary widely in type and scope. Arabic numerals, upper and lowercase letters, and various arrangements with a sprinkling of periods, dashes, colons, and hyphens are found. Generally a simple decimal system using only Arabic numerals in progressive order is most satisfactory for general use. For example, a major building-substructure code would be divided into

110	Excavation	114	Anchor bolt inserts
111	Backfill	115	Concrete, pouring, and finishing
112	Forms	116	Dewatering
113	Reinforced steel	117	Floor surfacing

The approach is to begin with the broad cross section and subdivide it into meaningful codes. The numbering system provides a designated skeleton of the information system. Any major area, such as labor, material, and subcontracts, can be similarly structured. The estimator would use the books of accounts (generally summarized in orderly fashion and distributed to the cost-estimating department) and his own estimating cost codes along with other measured and policy information as a source for preparation of estimates.

Budgeting is a frequent cost-estimating task. A budget is a written plan covering the activities for a definite future time, and dimensions are in monetary terms for a specific period, such as a quarter or year. Budgets deal with information based on data derived from cost-estimating and accounting records and conjectures of future activities. The budgeted cost center should be the smallest unit to which a cost can be clearly traced, provided there is a balance between excessive and too little detail, consistent with the cost of preparing the budget. For example, if all cost centers within the engineering department are physically located together, heat and light should be charged to the entire department as a practical expedient.

Appropriation, fixed, and variable budgets are common classifications. An appropriation budget may be directed toward proposed expenditures for a machine tool. A fixed budget may be directed toward an operation with only one level of activity for a definite time period. This budget may not be adjusted to actual levels; this may be satisfactory if the company activities can be predetermined accurately.

Fixed Budget. As an example of the development of a fixed budget, assume an engineering department of 50 people with three cost centers, product engineering, research and development, and engineering services. Product engineering consists of engineers and draftsmen who are concerned with existing products. Improvements of present designs are responsibilities for product engineering. The research and development group is responsible for the development of new products including prototype testing. In budgeting this cost center, the engineering manager must consider the manufacturing costs involved with the prototype. The third cost center, engineering services, consists of blueprint room personnel, clerks and secretaries, and others who are assigned to the engineering department rather than a particular section. These people work part time for either or both of the other cost centers. We suppose that the support work is sufficient to warrant a separate cost center for the time and expenses. Clerks or secretaries who work exclusively for either product engineering or for research and development would be budgeted directly to those cost centers.

TABLE 4.3

EXPENSE BUDGET FOR ENGINEERING COST
CENTERS, FIRST QUARTER

Expense Account	Product Engineering	Research & Development	Engineering Services
Salaries	$50,000	$25,000	$10,000
Total salaries	$50,000	$25,000	$10,000
FICA	2,700	1,350	540
Workmen's compensation	500	250	100
Group insurance	1,000	500	200
Pension fund	1,250	625	250
Fringe benefits	1,000	500	200
Taxes	1,000	500	200
Total payroll related	$ 7,450	$ 3,725	$ 1,490
Overtime	750	600	400
Office supplies, reproductions	3,000	1,000	500
Spoiled materials	150	150	100
Freight charges	375	450	300
Traveling	750	750	500
Books	75	75	50
Memberships	75	75	50
Telephone	750	750	500
Postage	150	150	100
Professional services	1,050	600	400
General expenses	750	1,200	800
Depreciation assets	750	750	500
Rent furniture	375	600	400
Building space charges	1,125	900	600
Total expenses	$10,125	$ 8,050	$ 5,200
Engineering errors	6,000	20,000	
Manufacturing costs	1,500	30,000	
Total shop costs	$ 7,500	$50,000	0
Total expected costs	$75,075	$86,775	$16,690

Both product engineering and research and development need two budgets: an expense budget and a product budget. The expense budget, Table 4.3, is a schedule of the expenses that the manager expects his department to incur during the budget period, a quarter in this case. Complexity of the expense budget depends on the amount of information available and the degree of control desired. Four main sections are selected in the expense budget: salaries, payable overhead expenses, supplies and administrative expenses, and manufacturing expenses. There must be clear definitions to prevent ambiguous charges to improper accounts. The product budget, Table 4.4, lists current engineering

TABLE 4.4

PRODUCT ENGINEERING SECTION BUDGET

Product Identification	Variance to Date	Cost to Date	Quarters							
			1		2		3		4	
			Eng.	Mfg.	Eng.	Mfg.	Eng.	Mfg.	Eng.	Mfg.
444	+$486	$3486	$ 5,000	$10,000						
3806	+297	9297	3,000	5,000	$8000	$ 5,000	$ 1,000	$ 5,000		
443	−375	4625	3,000	5,000	5000	5,000				
2211	+22	1322	11,000	25,000						
7582					5000		5,000	5,000	$ 1,000	$ 1,000
7543					5000		5,000	5,000	5,000	10,000
7542							3,000		10,000	10,000
7547	+324	824	3,000	5,000	5000	10,000	15,000	10,000	9,000	10,000

products and estimates for several future quarters. Each product shows the future expenditures as well as the sum total of costs to date. The total variance for each product should be a part of the product budget so that comparisons may be made between expected and actual costs.

TABLE 4.5

SUMMARY ENGINEERING BUDGET

	1st Qtr.	2nd Qtr.	3rd Qtr.	4th Qtr.	Year
Product engineering					
Salaries	$50,000	$47,000	$47,000	$50,860	
Payroll-related expenses	7,450	4,700	4,700	5,080	
Expenses	10,125	1,000	22,650	8,150	
Errors in manufacturing	7,500	10,000	7,750	7,900	
	$75,075	$62,700	$82,100	$71,990	$291,865
Research & development					
Salaries	$25,000	$28,000	$29,000	$25,000	
Payroll-related expenses	3,725	2,800	2,900	2,500	
Expenses	8,050	8,000	7,850	22,350	
Shop costs	50,000	20,000	25,000	30,000	
	$86,775	$58,800	$64,750	$79,850	$290,175
Engineering services					
Salaries	$10,000	$10,000	$13,400	$13,400	
Payroll-related expenses	1,490	1,000	1,340	1,340	
Expenses	5,200	5,000	6,860	6,860	
	$16,690	$16,000	$21,600	$21,600	$ 75,890
Total for fiscal year					$657,930

After the expense and product budgets are completed, a summary, Table 4.5, for the entire department is determined. In some cases the total of the projected budgets of all activities may exceed the funds to be allocated, in which case some budgets must be cut.

Variable Budgets. In a variable budget it becomes possible to determine budgeted costs for various levels of activity. For an engineering department actually operating at $575,000, it makes little sense from a control standpoint to compare cost to a budget which previously assumed operation at $657,930. Variable budgeting is a near-cousin to the direct-costing methods previously discussed. In variable budgets and direct-costing methods, fixed costs are not translated into the product cost until later. By definition, fixed costs are constant with changes in production. They are charged against the business as a whole and not against individual products. Consider the following example, where costs are developed from knowledge of a variable, semifixed, and fixed standpoint:

	Percent capacity of Total			
Costs	*70%*	*80%*	*90%*	*100%*
1. Fixed	$2000	$2000	$2000	$2000
2. Semifixed	1200	1300	1500	1900
3. Variable	700	800	900	1000
Total	$3900	$4100	$4400	$4900

4.4
LABOR COSTS

Labor costs, which are dollars paid for wages or salaries for work performed, are a major ingredient of an estimate. In heavy construction, for example, labor costs constitute 40% or more of the bid on a building. Payrolls cover two classes of workers: first, management and general administrative employees who may or may not be on a salary basis and, second, hourly labor. General administrative employees may be management, engineers, foremen, inspectors, and clerk-typists. Hourly payroll may include carpenters, lathe operators, draftsmen, and clerk-typists. Hourly labor is sometimes classified as direct or indirect. The category direct refers to employees who can be associated with a product directly, such as milling machine operators or truck drivers. This classification presupposes that the work can be preplanned or *designated;* it is found in the manufacturing industries where the process sheet specifies the operations to follow in the manufacture of certain parts. The indirect hourly labor notion refers to workers that are generally performing undesignated work such as clerk-typists, janitors, mechanical engineers, and superintendents. In an allocation sense of cost, their work and effort is usually for a variety of tasks, making it difficult to designate precisely what portion of their work contributes to the particular operation, product, project or system. In the main the indirect operator is not clearly identifiable to a particular task.

The principle is now established that one of the costs of labor is *fringe benefits.* In the past, estimators concluded that wages or salaries that were received directly constituted the total sum of labor costs. This is not so. Fringe benefits, which are related to wages and salaries, constitute as much as 30% of the actual cost incurred for labor.

A partial listing of fringe costs may include (1) legally required payments such as payroll taxes and workmen's compensation; (2) voluntary or required payments such as group insurance and pension plans; (3) sometimes, wash-up time, paid rest periods, and travel time; (4) payment for time not worked, holidays, vacations, and sick pay; and (5) profit-sharing payments, service awards, and payment to union stewards. Fringe costs depend on local situations and must be determined individually for each case.

Federal and state laws regulate wages and salaries paid by employers. Two federal laws, the wage-hour law and the Walsh-Healey Act, are prominent. All manufacturers engaged in interstate commerce are covered by the wage-hour law. The Walsh-Healey Act covers only companies having federal contracts. Both laws require the payment of wages at time-and-a-half rates for more than 40 hours in one week; the Walsh-Healey Act adds the same requirement for more than 8 hours in one day. Some groups are exempted from the wage-hour law, such as management and engineers. A man with a 40-hour base rate of $4 must be paid $6 per hour for his overtime hours. The wage-hour law specifies a minimum per hour as the lowest paid wage. For example, the wage minimums as established by Congress constitute a floor for employed labor in most categories. There are other labor laws too numerous to mention here: Discrimination against minority races, women, or others; work hours for women and children; safety and sanitation laws; unemployment insurance, taxes, and social security taxes are typical.

The federal government requires that businesses pay a tax for old-age and medical benefits, called social security tax, or FICA (Federal Income Contribution Act). This amounts to a percentage of a fixed sum of gross earnings of an employee. The employee contributes an equal amount. This rate is subject to change by Congress and has marched steadily upward since the first law back in 1935. An unemployment compensation tax is imposed by the states and federal government for the purpose of providing funds to compensate workers during periods of unemployment. The rate is about 3%, and it may be reduced or increased due to employment experiences, stability or layoffs, and involuntary terminations during a specified period of time. All states require Workmen's Compensation and Employer's Liability Insurance as a protection for the worker. In the event of injury or death, the insurance carrier provides financial assistance to the injured person or to the survivors in the event of death.

The preceding discussion has considered a variety of possibilities with respect to the labor rate to be used in estimating cost. With these viewpoints in mind, consider the following: A labor incentive or piece rate is determined and often is used in estimating operation costs per unit of product. These rates are indicators of the actual labor cost per unit, but they must be adjusted to reflect the performance on the job and other costs to arrive at a revised piece rate per unit of production. Assume a standard measured by time study practices as 0.988 hours per 100 units and a performance factor of +1.21. This performance factor is computed using

$$\frac{\text{standard} - \text{actual}}{\text{standard}} \quad \text{or} \quad 0.21 = \frac{0.988 - \text{actual}}{0.988}$$

where actual = 0.781 hour. The increase in labor cost in this case usually begins at 100% productivity in which the percentage increase in wages is equal to the percentage of savings in hours over standard hours. In terms of the labor cost, the hours per unit of production is the reciprocal of production per unit of time, and the standard units per time are 101.2, while actual performance is $1/0.781 \times 100 = 128\%$. If the labor rate for this operation amounted to $3.75 per hour, to the operator each piece is worth a wage of $3.75/101.2 = \$0.037$. The operator is not paid $0.037 per unit, as this method of pay is rarely found. But an actual pay per hour can be computed as $0.037 \times 128 = \$4.74$. Using an adjusted figure of $4.74, labor can be determined for this operation with additions of fringe costs, workmen's compensation, or other factors.

The foregoing can be formalized algebraically. Most incentive plans do not pay additional wages until 100% standard is reached. Below that level a guaranteed wage, or *day rate*, is provided. A formula expressing these relationships is given as

$$E = RH + R(sN - H) \qquad (4\text{-}1)$$

subject to $E \geq RH$, where $E =$ total earnings per pay period, dollars

$\qquad\qquad R =$ guaranteed hourly rate of pay, dollars

$\qquad\qquad H =$ actual hours worked

$\qquad\qquad s =$ standard hours per number of units

$\qquad\qquad N =$ actual number of units produced

For purposes of illustration assume that $R = \$3.75$, $H = 1$, $s = 0.988$, and $N = 128$, and that

$$E = 3.75 + 3.75\left(\frac{0.988}{100} \times 128 - 1\right)$$
$$= \$4.74$$

Assume an hourly wage of $4.74, and to be general, let this be an established wage or the weekly average incentive wage working against incentive standards as described above:

Rate of pay per hour	$4.740
FICA (employer's $5\frac{3}{4}\%$).....................	0.273
Unemployment compensation (at $1\frac{1}{2}\%$)	0.071
Workmen's compensation (at 1%)	0.047
Labor cost	$5.131

In this case a specified selection of fringe benefits was included in the wage rate. Frequently, fringe cost are provided for in overhead calculations.

4.5
MATERIAL COSTS

Before one considers the topic of materials, it should be realized that an understanding of physical materials associated with engineering design is required. The design may call for CA-610, SAE 1020, M type HSS, or A-432 materials. While the terms are meaningless to the uninformed, materials are complicated in other respects. For instance, a casting has a variety of associated costs. There is the cost of melted metal, molding costs, core costs, cleaning costs, heat treatment costs, foundry tooling costs, and so forth. To estimate the cost for a casting, the engineer must be acquainted with the material and heat-treatment specifications, inspection requirements, and the design of the casting. Physical specifications such as tensile strength, yield strength, and elongation must be clarified as well as the chemical composition. The knowledge about engineering materials is so vast that we provide no information. Rather we consider the bill of materials and classifications of cost that relate to the needs of cost engineering.

Cost finding involved with material includes historical and policy information. In some cases, material costs are uncovered via company records, while for others the engineer must seek out the basic costs.

We define materials as the substance being altered or used in that alteration. This may involve coke and limestone to a basic blast furnace industry for converting pig iron, steel ingots to a steel rolling mill for refining and rolling into strip and coil, tin sheet to a can producer manufacturing 6-ounce tins, and cases of 24 cans for a food processer for canning frozen orange juice. The scope of what constitutes materials depends on the situation.

For engineering design the documents of importance are the engineering bill of material and specifications in addition to the drawings. Engineers' bill of material or parts list accompany the blueprints and consist of an itemized list of the materials for a design. This list may be prepared on a separate sheet or may be lettered directly on the drawing. For example, a parts list contains the part numbers or symbols, a descriptive title of each part, the quantity, material, and other information such as casting pattern number, stock size of the materials, weight of parts, or volume of materials. The parts are listed in general order of size or importance, or the special-design parts first and standard parts last. The list proceeds from the bottom upward on the drawings so that new items may be added later. In the case of a product that is to be constructed for the first time or in the case of the project where the contract must be executed, the materials must be purchased quickly to have them available when needed in the factory or on the job site. In some cases, the estimating department determines all the materials and job requirements. Inasmuch as they are conversant with the drawings and specifications and are already familiar with the details of the design and contract, they begin the job of *taking off* all material requirements. The quantity *take-off* sheets must be made up in complete detail and include all descriptions necessary to obtain the materials as specified. Special forms are available for take-offs.

Specifications are considered apart from the bill of material and engineering drawings. In the construction industry, specifications are specific statements as to construction requirements, while in manufacturing, specifications relate to the performances, or materials, or special requirements. Basically these requirements are limited to the necessary technical details pertaining to the item itself. Specifications serve as a discussion of the proposed work so that bids may be compiled or quotations may be collected, to act as a guide or a book of rules during the construction and manufacturing time period, and finally they are legal instruments to the construction industry and special requirements to manufacturing.

In addition to the type and character of the material, it is important that correct quantities such as units, weight, and volume are known. To have a correct cost such as 26 cents per pound of casting material and then improperly specify the number of pounds required for the casting is a serious flaw. In the process of determining the quantities involved, the estimator would examine the plans and specification and make the estimate, say, the number of cubic yards of 2800-pounds-per-square-inch concrete required by type of clean-and-sharp aggregate. In the construction business, contractors would then be bidding against a stated number of cubic yards and their price would be so much per cubic yard placed. In the product-oriented industries, the estimator determines the cost per unit of material after the count of materials has been established. In industry, as in construction, the estimate of materials involves extensive calculation to include allowances for waste, short ends, or losses. After these calculations are completed, often with the aid of data catalogs on weights, allowances, and the like, costs are determined with the aid of prices obtained from the purchasing or accounting departments or from other sources.

Now consider an approach for estimating the cost of equipment from material and design considerations using fundamental mathematics. In detailed estimating, the cost engineer resorts to charts and tables which show the costs of various pieces of equipment for different capacities and materials of construction. In preliminary estimating, the lack of specific designs may thwart this approach. To overcome this shortage of information, material costs are sometimes determined from simple concepts, such as the following. We can write an expression for the shell weight of a pressure vessel as

$$t = \frac{PD_m}{2SE} + C_a \tag{4-2}$$

where t = wall thickness of vessel
$\quad P$ = pressure, pounds per square inch
$\quad D_m$ = average of inside and outside vessel diameter, inches
$\quad E$ = joint efficiency, dimensionless decimal
$\quad S$ = allowable maximum working stress, pounds per square inch
$\quad C_a$ = corrosion allowance, inches of additional wall

For carbon steel, a second relation would give cost per vessel = $K\rho CtL$,

where ρ = density of material, pounds per cubic inch
$\quad K$ = cost per pound for construction material
$\quad C$ = mean circumference, inches
$\quad L$ = length, inches.

Now consider an example of the foregoing. Let a corrosion allowance for carbon steel amount to $\frac{1}{16}$ inch, $S = 10,000$ pounds per square inch, $D_m = 22$, $E = 1$ for a welded structure, $P = 25$ pounds per square inch, $\rho = 0.28$, $K = 0.245$, and $L = 18$ feet; then

$$t = \frac{25 \times 22}{2 \times 10,000} + 0.062 = 0.0275 + 0.062 = 0.090 \text{ inch}$$

Then the material cost, excluding fabrication, erection labor and a waste allowance, is

$$C_v = \text{cost per vessel} = (0.245)(0.28)(69)(0.090)(216)$$
$$= \$92.02$$

A waste allowance of 40% is not uncommon and would make the cost $1.40(92.02) = \$128.83$.

4.6 DEPRECIATION

The purpose of a discussion about depreciation is to present the generalized mechanics of depreciation calculations and some of the thinking in cost estimating. A great deal of ambiguity exists about depreciation. Forces of politics enter into the picture of depreciation. Surtax, credits, accelerated write-off, inflation or recession, obsolescence, and new technology obscure thinking about this matter. Even various accounting terms such as reserves for depreciation, allowances for depreciation, and amortization and retirement enlarge the ambiguity.

Depreciation is an accounting charge that provides for recovery of the capital that purchased the physical assets. It is the process of allocating an amount of money over the useful life of a tangible capital asset in a systematic manner. It is cost, as of a certain time, not value, that changes with time and that is allocated and recovered. There are no interest charges or any recognition of a changing dollar value. The depreciation charge is not a cash outlay. The actual cash outlay takes place at the time that the asset is acquired. Depreciation charges are the assignments of that initial cost over the life of the asset and do not involve a periodic disbursement of cash. When the rate at which the asset is depreciated increases, it does not increase the outflow of cash. In fact, it has the opposite reaction, as the depreciation charges reduce taxable income and the outflow of cash for taxes. The initial investment is a prepaid operating cost that is expensed or allocated to an operating expense account. Examples of typical fixed assets are given by Table 4.6.

Initial costs are undertaken to acquire assets that contribute to the production of revenue over long periods of time. The cost of a factory building or a machine, for example, will remain a positive factor in generating revenue provided that the building and equipment is used in the manufacture of a sellable product. As federal laws require income measurement, it is necessary that an appropriate portion of the cost of the building or equipment be charged to or matched with each dollar of revenue resulting from the sale of products produced within its walls and machines. The amount of such cost, matched with revenue, during any one period is the estimated amount of the cost that expires during the period. Thus, a fixed asset, which will not last forever, has its useful value exhausted over a period of time. In the mineral industries and forestry, depletion, somewhat analogous to depreciation, is a noncash expense representing the portion of a limited natural resource such as oil, shale, and minerals utilized in a product which is sold.

TABLE 4.6

TYPICAL FIXED ASSET ITEMS

A. Production and process equipment
    Bins and tanks for inventory and storage of raw materials
    Turret lathes
    Milling machines
    Equipment for manufacturing basic materials into finished products
    Automatic transfer equipment
    Converting-type equipment
    Piping, material-handling trucks, and other moving devices
    Industrial pumps and compressors
    Instruments and controls
    Trucks
B. Service or nonprocess facilities
    Power generation and distribution
    Furniture
    Business machines
    Motor vehicles
    First aid and safety installations
C. Buildings and land
    Factory buildings, land
    Warehousing
    Roadways, parking areas
    Shops, maintenance garages

The factors contributing to the decline in utility are considered in the categories of (1) physical wear and tear resulting from ordinary usage or exposure to weather, (2) functional factors such as inadequacy and obsolescence, and (3) governmental actions. Physical factors are commonplace such as the wear and tear on buildings and corrosion which impairs efficiency and safety. Due to technological progress, obsolescence is a frequent situation where the fixed asset is retired not because it is worn out but because it is outmoded. The superior efficiency of an asset of a later design is one reason compelling new product designs for the market. Sometimes alterations of the manufacturing design of equipment through research and development techniques make for immediate obsolescence of equipment. There is the instance of inadequacy where the asset, although neither worn out nor obsolete, is unable to meet the demands on it. An electric power company installs a hydroelectric generator and in the course of several years finds that the demands of an expanding community placed on it exceed the ability of the water and hydroelectric power to meet peak loads. Governmental or other forces may prevent operation; e.g., loss of raw material source or a new law prohibiting waste disposal are possibilities.

It should be recognized that depreciation is an approximation, for any of the methods used to determine a depreciation charge employ estimates. The significant factors that are considered regardless of the method are (1) the cost of the asset, (2) the estimated useful life of the asset, and (3) the estimated scrap value or residual value of the asset.

For consideration of new projects, engineering estimates would be required for the asset, erection, and operating capital and costs. There are in every project certain costs that can be tax expensed immediately, such as expenditures for nonphysical assets, some physical assets of extremely short life, and certain installation and start-up expenses.

The question of life is an important matter. Some firms are concerned about economic life with little regard for physical life, while others, public utilities, for example, are restricted to earning a specified amount on capital invested, and *life* takes on another ramification. The *economic life* estimate is affected by tax laws and the vagaries of human judgments.

The residual value of an asset is the amount expected to be realized on its eventual disposition. It is sometimes referred to as scrap, salvage, or trade-in value and may be an expense if there is a cost to dispose of the asset. Frequently the residual value is estimated to be nominal and is ignored in determining depreciation.

The selection of a method for depreciation is a choice by management but

TABLE 4.7

FORMULAS FOR DEPRECIATION

Depreciation Method	Including Salvage Value	With Zero Salvage Value
1. Straight line:		
Annual value	$D_{sl} = \dfrac{P - F_s}{n}$	$D_{sl} = \dfrac{P}{n}$
Cumulative depreciation	$\sum D_{sl} = \dfrac{K(P - F_s)}{n}$	$\sum D_{sl} = \dfrac{KP}{n}$
2. Declining balance:		
Annual value	$D_{db} = \dfrac{P - \sum D_{db}}{n} r$	Undepreciated investment at end of asset life is theoretically the salvage value
3. Sum-of-the-year digits:		
Annual value	$D_{sd} = \dfrac{2}{n}\left[\dfrac{n + 1 - K}{n + 1}\right](P - F_s)$	$D_{sd} = \dfrac{2}{n}\left[\dfrac{n + 1 - K}{n + 1}\right]P$
Cumulative value	$\sum D_{sd} = \dfrac{K}{n}\left[\dfrac{2n - K + 1}{n + 1}\right](P - F_s)$	$\sum D_{sd} = \dfrac{K}{n}\left[\dfrac{2n - K + 1}{n + 1}\right]P$

D = depreciation charge, dollars
P = initial investment, dollars
F_s = future salvage value of investment, dollars
n = useful life of physical asset, years
K = current year number from start of investment
r = declining-balance factor[a]

[a]*Note:* r can be 1.5 or 2 according to the U.S. Internal Revenue Service since the Revenue Act of 1954.

restricted to guidelines established by the Internal Revenue Service. The method selected should provide for a reasonable and systematic depreciation charge or allocation over the useful life of the asset. Although different methods may be used for different classes of depreciable assets, they should be used and applied consistently over the years. Depreciation methods commonly used are (1) straight line, (2) declining balance, (3) sum-of-the-year digits, and (4) units of production. Formulas for the first three are given in Table 4.7 and include the cases for an estimated salvage or zero salvage value.

Straight-Line Method. Under the straight-line method the depreciable value, which is the first cost less its estimated salvage, is divided equally among the life or accounting periods of the asset. To illustrate this method, assume the cost of the depreciable asset to be $10,000, its salvage value to be $1000, and its estimated life as 10 years; the annual depreciation is computed as follows:

$$D_{sl} = \frac{P - F_s}{n}$$

$$= \frac{10,000 - 1000}{10} = \$900 \text{ annual depreciation} \qquad (4\text{-}3)$$

If $F_s = 0$,

$$D_{sl} = \frac{10,000}{10} = \$1000 \text{ annual depreciation}$$

Sometimes the annual depreciation to the total amount to be depreciated can be expressed as a percentage of the cost of the asset. For the example the depreciation rate would be stated as 9% of the cost. The straight-line method is widely used as it is simple and provides a reasonable allocation of cost to periodic revenue provided that usage is uniform from period to period.

Declining-Balance Method. In the application of the declining-balance method of depreciation there is a steadily declining periodic depreciation charge over the estimated life of the asset. This method has a long history in England; in the United States interest in this method increased after the Internal Revenue Service code of 1954 became effective. There are several variants to this technique, and the one commonly used applies twice the straight-line depreciation rate to the declining book value of the asset. If the estimated life of an asset is 10 years, the depreciation rate is 20% (10% × 2). For the first year the rate is applied to the initial cost of the asset, while in succeeding years it is multiplied by the declining book value (cost − accumulated depreciation). The method is described by the following tabulation:

Year	Cost of Asset	Accumulated Depreciation at Year Beginning	Book Value at Year Beginning	Rate (%)	Yearly Depreciation
1	$10,000		$10,000	20	$2000
2	10,000	$2000	8,000	20	1600
3	10,000	3600	6,400	20	1280
4	10,000	4880	5,120	20	1024
5	10,000	5904	4,096	20	819
6	10,000	6723	3,277	20	655
7	10,000	7378	2,622	20	524
8	10,000	7902	2,078	20	416
9	10,000	8318	1,682	20	336
10	10,000	8654	1,346	20	269

It is seen that the estimated salvage value is not considered in determining the depreciation rate, nor is it considered in the periodic depreciation. The declining-balance method has its greatest application where the earning power of the asset is proportionately greater in the early years of its use rather than in later years.

Sum-of-the-Year Digits Method. Sum-of-the-year digits provides a declining periodic depreciation charge over an estimated life for the asset. This is achieved by applying a smaller fraction recursively each year to the cost less its salvage value. In the recursive relationship the numerator of the changing fraction is the number of remaining years of life, while the denominator is the sum of the digits representing the years of life. With an asset having an estimated life of 10 years the denominator of the fraction is $55(1 + 2 + \cdots + 10)$; for the first year the numerator is 10, for the second 9, and so forth. For a \$10,000 asset with an estimated life of 10 years and a salvage value of \$1000, the schedule of depreciation is as follows:

Year	Cost Less Salvage	Rate	Yearly Depreciation	Accumulated Depreciation at End of Year	Book Value at End of Year
1	\$9000	10/55	\$1636	\$1636	\$8364
2	9000	9/55	1473	3109	6891
3	9000	8/55	1309	4418	5582
4	9000	7/55	1145	5563	4437
5	9000	6/55	982	6545	3455
6	9000	5/55	818	7363	2637
7	9000	4/55	655	8018	1982
8	9000	3/55	491	8509	1491
9	9000	2/55	328	8836	1164
10	9000	1/55	164	9000	1000

The sum-of-the-year digits has the advantage of accelerated write-offs, which is an income tax advantage.

Unit of Production Method. Still another method of depreciation charging is based on the premise that an asset wears exclusively as demands are placed on it. A truck would assume to wear as it consumed more miles. The computation for depreciation per unit of production is as follows: $(P - F_s)/Q$, where Q is defined as the estimated number of units of output. If $P = \$10,000$, $F_s = \$1000$, and $Q = 200$ units of output, then the depreciation charge per unit of product equals \$45 per unit. This method of depreciation has the advantage that expense varies directly with operating activity. Retirement in these cases tends to be a function of use. This method is confined to those items where useful lives are determined by the factors of the extent of use. The unit of production method requires an estimate of total lifetime use or production.

Another example of the unit of production method is the case where $P = \$40,000$, $F_s = \$5000$, and a full-time life production $= 1140$ units. Then we have the following:

$$\frac{P - F_s}{Q} = \frac{35,000}{1140} = \$30.70$$

Year	Estimated Units of Production	Yearly Depreciation	Accumulated Depreciation at End of Year	Book Value at End of Year
1	90	$2763	$ 2,763	$37,237
2	125	3838	6,601	33,399
3	125	3838	10,439	29,561
4	125	3838	14,277	25,723
5	125	3838	18,115	21,885
6	125	3838	21,953	18,047
7	125	3838	25,791	14,209
8	125	3838	29,629	10,371
9	100	3070	32,699	7,301
10	75	2301	35,000	5,000

The advantages of accelerated methods of depreciation, such as those of the double-declining-balance method, and sum-of-the-year digits are compatible with the logic that the earning power of an asset is created during its early service rather than later, where upkeep costs tend to increase progressively with age. Accelerated methods offer a measure of protection against unanticipated contingency such as excessive maintenance, and they return the investment more quickly and simultaneously, decreasing the book value at the same rate. A high book value would tend to deter the disposing of unsuitable equipment even when the need for replacement is pressing. Rapid reduction of book values, provided the owner overlooks the tax benefits from capital loss, leaves the owner more free to dispose of inefficient and unsatisfactory equipment.

Depreciation of costs is collected under the control of general ledger accounts, such as accumulated depreciation or allowance for depreciation. Periodic fixed charges resulting from these accounts are analyzed and charged departmentally under the proper cost classification through the application of worksheet analysis. Distribution of depreciation of machinery and equipment is thus made to different departments based on factors such as cost of factory equipment, rates of depreciation applicable to each unit of equipment, and departmental location of each unit. In the case of plants and buildings, various property ledgers would show location of cost and accumulated depreciation. The proration of depreciation on buildings to departments may be based on the cost of the building, the total area of the building, and the area occupied by each department of the building.

Because of the clerical trouble involved, many firms use depreciation schedules that consider groups rather than specific machines or plants and other equipment. This is particularly true for many of the individual items that have a low cost, and if depreciation is a relatively minor portion of the total expense, the computation of depreciation by groups is satisfactory.

Not all equipment costs are necessarily depreciated. In the case of inexpensive tools, several methods are used to allocate the costs. At the time of the purchase, tools may be capitalized in a small tools account or charged to expenses. If the purchase of the small tools is capitalized, the tools are fixed assets with depreciation to be applied in order to establish annual or monthly amounts to charge off as expense. This is a difficult method to administer due to the variation of the length of life of many different small tools. Yet, on the other hand, many tools are as expensive as the machines which use them, and thus it would be appropriate to capitalize these items.

A *reserve for depreciation* account contains the accumulated estimated net decrease in the value of the particular asset account to which it pertains. In most industries the amount shown in the depreciation account does not appear as cash unless a special fund is set aside specifically for this purpose. To create a fund of this nature suggests that a fund is actually invested outside the company to earn an interest rate. However, interest rates found on the outside are less than the earning rate enjoyed by the company. It is wiser to employ the money for some operations. The amount equal to the depreciation will appear as other assets such as working capital, raw materials, or finished products in storage. When it becomes necessary to buy new equipment or replacements, management must convert physical assets into cash (unless sufficient cash is on hand) or use existing profit to pay for the new equipment. The appearance of a depreciation reserve on the balance sheet, as with other types of assets, represents capital retained in the business, obstensibly for the ultimate replacement of the capital asset being depreciated.

It has been suggested that a number of factors in any depreciation model are subject to estimation: salvage value and, particularly, life. If for some reason the estimates prove faulty, then it is possible to retire an investment before its capital has been recovered. In the circumstance where net income received is less than the amount invested, an unrecovered balance remains. This unrecovered balance is referred to as *sunk cost*. The term sunk cost may be defined as the difference between the amount invested in an asset and the net worth recovered by services and income resulting from the employment of the asset. As an illustration of the above statements, consider a case where a capital investment of $5000 is to be recovered in 5 years with a remaining $1000 salvage or $4000 depreciation. Based on straight-line depreciation the amount invested and to be recovered per year will be $800. As a result of excessive use, the machine was sold after 3 years, for $1400, and had actually consumed $3600 in 3 years or $1200 per year on the average. The sunk cost is equal to the difference in the actual depreciation and the depreciation charge, in this case $3600 − $2400 = $1200. Stated yet another way, sunk cost is determined to be the estimated depreciation value (or book value) minus the realized salvage of the asset. Sunk costs cannot be affected by decisions of the future and must be faced with reality.

The act of exhausting a natural resource and converting it to a saleable product is called depletion. The natural resources which are subject to depletion are oil, natural gases, metal and mineral mines, orchards, fisheries, and forests. Not all natural resources are subject to depletion, for example, soil fertility and urban land. In accounting for depletion, it is the allocation of the value of the quantity of natural resources extracted from a deposit that is considered. Like depreciation, it is a noncash expense that is an allowed tax deduction.

4.7 OVERHEAD

By definition overhead is that portion of the cost which cannot be clearly associated with particular operations, products, projects, or systems and must be prorated among all the cost units on some arbitrary basis. Broad details regarding the posting of direct labor cost, time, material, and other indirect costs have been given earlier. What will be discussed here are those overhead aspects which pertain to estimating. The key to this puzzle is the way in which indirect expenses are allocated, unitized, and charged to individual estimates; the direct costs, such as direct labor and direct materials, present little if any allocation problem. These costs do not exist unless the product is made.

The underestimating or overestimating of overhead rates is serious in view of the proportion of the total estimated cost that they play. As an illustration, consider two different operations where the labor rates are $4.25 per hour. Machine A, a numerical controlled milling machine, is initially worth $150,000, while machine B, a standard general-purpose milling machine, is worth approximately $15,000. Using an average burden rate of 200%, it would be indicated that machines A and B would each cost, on a machine-hour basis, $12.75 per hour. However, this is false machine-hour costing as the investment in machine A is 10 times that machine B. It is evident that the proper cost base and sensible allocation of overheads to handle discrepancies of this sort are necessary. Years ago machine investment per worker was lower, and it was not uncommon that overhead rates were uniformly distributed over the direct labor base. In recent decades, the ratio of fixed cost to variable cost has risen, and the simple expediency of overhead distribution via the single rate is misleading.

Overhead distribution involves the assignment of actual indirect overhead costs among the various departments so as to determine the overhead costs incurred in operating each department. On the other hand, overhead application involves the assignment of manufacturing overhead cost to products, operations, or a system costed during the period so as to determine unit cost.

The exact nature of the overhead rate differs from one company to the next, and it is not uncommon to find various types of overhead rates within one firm. A classification is given as

1. Whether the rate includes fixed costs, as in absorption costing, or not, as in direct costing.
2. The base used to apply overhead such as direct labor dollars, direct labor hours, and machine hours.
3. The activity level assumed for the absorption of fixed costs as the expected activity for the year, normal activity, or practical capacity.
4. The scope of the application of the rate to the whole plant or to a cost center.
5. Whether the rate applies to all designs (such as product lines) or to one line of the design.

Some of the methods used to distribute factory overhead costs to job costs are the following: (1) overhead costs applied on the basis of direct labor cost or time, (2) overhead cost applied on the basis of direct labor and direct material costs, (3) overhead applied on the basis of conversion costs, (4) overhead costs applied on the basis of direct material costs, (5) overhead costs applied on the unit of product basis, (6) overhead applied on a machine-hour method, and (7) overhead applied on the basis of a fixed and variable cost classification. Some of these methods can be actual rates, as they are calculated as

$$\text{rate} = \frac{\text{actual factory overhead}}{\text{actual direct labor hours or direct labor costs or prime cost}}$$

An actual burden rate has the merit of distributing among the jobs the incurred factory overhead. It is subject to certain defects, as it is unavailable until the close of the accounting period. Although it is possible to determine an overhead rate following the end of the period and to use this rate to apply actual costs for that period, historical overhead rates are at a disadvantage. Historical rates delay cost calculations until the end of the month and often later and fluctuate because of seasonal and cyclical influences acting on the actual overhead costs and on the actual volume of activity for which overhead cost is spread. In view of the dependency on predetermined overhead rates, our discussion will disregard historical rates.

In most situations overhead rates are determined from data developed from an operating budget, although overhead rates can be computed without using a budget. For the latter situation the company does not compile an operating rate; rather the estimate is made by studying behavior of recent cost experiences, and then various adjustments are made.

4.7-2
Overhead Applied on Direct Labor Cost Base

The method of applying overhead as a percentage of the direct labor dollar is one of the oldest and most popular:

$$\text{rate} = \frac{\text{overhead charge}}{\text{direct labor cost}} \qquad (4\text{-}4)$$

The numerator of the equation may consider the factory as a unit, department, or cost center of the plant activity. Both the numerator for overhead and the denominator for direct labor dollars may be expressed in terms of the expected actual, normal, or practical capacity. The steps involved in developing departmental direct labor dollar rates for Table 4.8 are as follows:

1. Make budget estimates of the direct overhead for the time period for each manufacturing overhead center. This would include those things that are used as indirect materials for that center.
2. Make budget estimates of the direct costs of service departments and of general factory expenses. This step is not shown in Table 4.8, but Table 4.11 is typical.
3. Allocate the budget expenses of the service department and general factory costs to the manufacturing overhead centers based on the logical factors relating to the services performed. This allocation for Table 4.8 is on the basis of overhead center payroll, floor space, assets, and services rendered.
4. Estimate the budgeted total direct labor on products.
5. Divide the total product overhead by the budgeted total direct labor on products to secure the departmental direct labor dollar rates.

TABLE 4.8

DEPARTMENT DIRECT LABOR DOLLAR RATES

| Overhead Manufacturing Center | Charged Directly | Overhead Center Payroll | Apportioned on the Basis of | | | Total | Total Direct Labor | Overhead Rates (%) |
			Floor Space	Assets	Services Rendered			
Press	$ 66,500	$ 11,400	$ 14,600	$ 3,600	$ 22,100	$118,200	$ 40,200	294
General machining	119,100	42,500	27,800	6,100	61,500	257,000	143,400	179
Assembly	75,400	53,600	43,100	1,900	23,700	197,700	240,200	82
Product quality	39,100	13,400	15,100	100	2,900	70,600	50,300	140
Total: mfg. overhead center	$300,100	$120,900	$100,600	$11,700	$110,200	$643,500	$474,100	136

For the example, if the aggregate summary of indirect manufacturing charges accumulates to $643,500 and the direct labor cost was found to be $474,000, the percentage overhead = 136%. Other overhead rates are shown for various cost centers in Table 4.8.

If one assumes that the time is the correct basis for overhead distribution, it follows that the percentage of direct labor cost method gives accurate results only if labor costs are proportional to labor time. The situation occurs commonly if labor is paid on an hourly basis and the rate per hour is substantially the same for all workers. The rate of overhead per direct labor hours is calculated as

$$\text{rate} = \frac{\text{overhead charge}}{\text{direct labor hours}} \qquad (4\text{-}5)$$

An aggregate summary found that \$100,000 of overhead was charged to a job having 20,000 direct labor hours. Then the rate = \$5 per direct labor hour. Considering theory alone, the rate per direct labor hour represents the ideal in methods of overhead proration, as overhead accumulates on a time basis, while the rate per direct labor hour distributes on a time basis.

Like the percentage of direct labor cost plan, another method, the prime cost method, is accurate only if it applies the same overhead to each job order as it would be applied by the use of the rate per direct labor hour. Prime cost, earlier defined as direct materials plus direct labor, is an overhead costing defined as follows:

$$\text{rate} = \frac{\text{overhead charge}}{\text{prime cost}} \qquad (4\text{-}6)$$

For the case where aggregate prime cost balances are \$500,000 and factory costs for indirect manufacturing equal \$200,000, the percentage overhead for prime cost is 40%.

The percentage of prime cost method suffers from deficiencies. It is accurate only if the method applies the same overhead to each job order as would be applied by the application of the rate per direct labor hour.

In certain limited circumstances the application of overhead cost applied to the basis of direct material cost is valid. For a few industries which have uniform products, such as cement or sugar, the method is equitable. The method is simple, as it assumes that one cost factor, i.e., direct material, is the allocation base. If products are made on facilities which may differ considerably in capital cost, it becomes a faulty cost allocation method.

Still another method of absorption costing involves the unit of product method and is determined by dividing the expected volume of product output into the total budgeted overhead to get the cost per unit. If overhead costs of \$100,000 per year are found in an output of 20,000 units per year, then a \$5 cost per unit of is allocated. This system is found more frequently in process cost systems and is used where one or a few products are made and are closely related and have common factors such as weight or volume. Process cost accounting regards production on a continuous flow basis rather than on a series of identifiable lots.

This method has the formula

$$\text{rate per machine hour} = \frac{\text{overhead charge}}{\text{machine hours}} \qquad (4\text{-}7)$$

Overhead is costed to an estimate by multiplying this rate by the number of machine hours involved in specific operation. As with other overhead rates, computation is on the basis of actual, normal, or practical activity for the coming period. Additionally, it may be computed for all machinery in the plant on a plant-wide basis or a cost center basis. However, a plant-wide machine-hour rate is seldom found. Machine-hour rates afford an accurate method of allocating overhead expenses to each job, and from the engineering point of view it is suitable for estimating the cost of the job from specifications and route sheets. It is, however, expensive and increases the cost of accounting procedures—nor is it universally applicable. It is best used whenever operations are performed by machinery which is significant in terms of the final cost of the product.

The steps in computing the machine-hour cost center rates are as follows: Table 4.9, an allocated production center budget, is constructed first and considers the physical factors associated with machines such as space occupied, horsepower, annual production hours, depreciation and direct costs (other than labor and material), and tooling expenses. These are estimated for the forthcoming year or some other period. These physical factors are used later as a basis for the final allocation of cost to each production cost center. The next budget, Table 4.10, forecasts the normal manpower requirements for manning

TABLE 4.9

ALLOCATED PRODUCTION CENTER BUDGET

Production Center	Units	Occupied Floor Space	Annual Assigned Hours[a]	Horsepower-hours	Depreciation	Tooling Expenses
Vertical boring	4	400	10,200	234,000	$ 8,000	$14,000
Light machining	20	1000	49,300	443,000	5,250	26,000
Heavy machining	2	2000	6,800	748,000	51,000	8,000
Bench work	15	375	16,000	24,000	—	1,000
Finishing	8	400	13,600	68,000	2,250	500
Total		4175	95,900	1,517,000	$66,500	$49,500

[a]Reduced by 15% for nonproductive time.

TABLE 4.10

BUDGETED MANNING FOR PRODUCTION CENTERS

Production Center	Normal Center Manning	Average Direct Hourly Rate	Average Fringe Hour Rate	Total Rate	Annual Assigned Hours	Total Direct Labor Budget
Vertical boring	6	$4.05	$1.20	$5.25	12,000	$ 63,000
Light machining	29	3.65	1.10	4.75	58,000	275,500
Heavy machining	4	4.25	1.25	5.50	8,000	44,000
Bench work	17	2.60	0.75	3.35	34,000	113,900
Finishing	8	3.20	0.95	4.15	16,000	66,400
Total						$562,800

the selected cost centers listed in the allocated production center budget. The established hourly rate per unit is found from other sources and provides the wages for the forthcoming period. The column headed Direct Labor Budget is used later as a factor in allocating costs to each cost center. In Table 4.11, an allocatable budget, is a listing of expenses which cannot be identified with the various production centers. These costs have been classified into categories to apply on the basis of a cause and effect allocation factor. For example, all those

TABLE 4.11

ALLOCATABLE BUDGET

Direct overhead		
Space:		
Rent	$ 18,000	
Repairs to factory	6,000	
Heat	1,600	
Total	$ 25,600	
Power:		
Electricity	$ 37,925	
Other utilities	4,000	
Total	$ 41,925	
Indirect labor:		
Material handlers	$ 32,000	
Inspectors	24,000	
Supervisors	45,000	
Clerical personnel	9,000	
Total	$110,000	
Indirect overhead:		
Personnel support	$ 18,000	
Time clerks	7,700	
Training	11,000	
Manufacturing engineering	34,000	
Total	$ 70,700	
Engineering and development:		
Material	$ 35,000	
Labor	248,000	
Indirect expenses	75,000	
Total	$358,000	
Equipment, tooling, services:		
Depreciation	$ 66,500	
Repairs & maintenance	15,000	
Perishable supplies	18,000	
Tooling (non-perishable)	49,500	
Repairs on tools	14,000	
Outside services	9,300	
Total	$172,300	
Packaging:		
Materials & equipment	$ 62,000	
Labor	8,725	
Supervision & consultation	3,000	
Total	$ 73,725	
Total		$852,250

costs under Space are to be allocated on the basis of occupied floor space. Power Costs are to be allocated on the basis of horsepower-hours. The production center cost per hour assignment sheet is determined in this way: Having determined the physical factors of the facilities and the manning of those cost centers and the financing to operate those centers, we are now ready to combine these data into machine-hour rates for each cost center. Table 4.12, production center cost per hour, is an exhibit of an expense assignment. The expense assignment is a sheet in which the various costs on the allocatable budget are allocated to each cost center on some selected basis. The additive costs shown on the allocated production center budget and the budgeted manning for production centers are added to get the composite machine-hour rate. The allocation factors selected for each budget category are shown in the middle of the sheet. A ratio is computed between the total cost to be allocated and the designated

TABLE 4.12

PRODUCTION CENTER COST PER HOUR

Production Center	Space	Power	Indirect Labor	Indirect Overhead	Engineering	Equipment, Tooling, Services	Packaging	Total	Annual Assigned Hours	Production Center Cost Per Hour
Vertical boring	$ 2,452	$ 6,458	$12,310	$ 7,913	$ 40,068	$32,670	$ 8,253	$110,124	12,000	$ 9.177
Light machining	6,131	12,227	53,833	34,603	175,218	46,306	36,090	364,408	58,000	6.283
Heavy machining	12,262	20,699	8,598	5,526	27,984	87,755	5,764	168,588	8,000	21.073
Bench work	2,299	664	22,284	14,318	72,500	1,485	14,920	128,470	34,000	3.779
Finishing	2,456	1,877	12,975	8,340	42,230	4,084	8,698	80,660	16,000	5.041
Total	$25,600	$41,925	$110,000	$70,000	$358,000	$172,300	$73,725	$852,250		

Allocation Factor	Floor Space	Horsepower-Hours	Direct Labor Budget	Direct Labor Budget	Direct Labor Budget	Production Center Depreciation, Tooling	Direct Labor Budget
Allocation fraction	25,600	41,925	110,000	70,000	358,000	56,300	73,725
	4,175	1,517,000	562,800	562,800	562,800	116,000	562,800
Allocation ratio	6.131	0.0276	0.1954	0.1256	0.636	0.485	0.131

factor. This ratio is used as an allocation factor by which cost is assigned to the individual cost center. For example, $25,600 of fixed space cost is to be distributed on the basis of 4175 square feet of working space. This gives a ratio of 6.121 and implies a cost of $6.131 in fixed space cost for every square foot of net working space occupied. It is multiplied by the specific working space occupied by each production center. Fixed space cost is the result. In the case of the vertical boring production center, occupied floor space amounts to 400 × 6.131 = $2452. In a similar manner all the other budgeted costs are allocated to all production centers. Adding a cost for each facility gives a total overhead cost; dividing by the annual budgeted hours produces the overhead portion of the machine-hour rate. This development has been based on a machine-hour rate where overhead has been classified as fixed.

4.7-5
Overhead Adjustment

Consider now the adjustment of overhead *points*. Assume for our example that absorption-overhead rate points are established from labor standards in effect at the moment the union-management labor contract is agreed on. Furthermore, assume that overhead is related to a practical 80% of plant capacity for each major product line. Note Table 4.13, which lists major indirect accounts considered for absorption overhead and their number of overhead points relative to the amount of direct labor to be generated, 80% of our total plant capacity for one product line. All product lines carry the entire sum of indirect charges, of course. If significant departures become apparent during the period between major overhead studies, it becomes necessary to update the overhead record for estimating. The previous discussion dealt with machine-hour determination and is representative of a periodic overhead study. Estimating absorption changes in the overhead rate are an entirely different matter.

The first two columns indicate the expected points of overhead and the dollar amount for each major account. For instance,

$$\text{overhead rate} = 450 = \frac{149,959}{33,325} \times 100$$

Now a customer wants an axle which falls within current capability range in terms of tooling and processing. This is in addition to the 80% practical business which is being planned. The expected volume is 5000 per month, and the operation estimate for the shaft is $0.2355. The total direct labor for this product will increase by (5000)(0.2355) or $1178. Some of the indirect cost accounts can be expected to increase with this new business, but not all. This remains a judgment process. Depending on the product, time, and other circumstances, the ratio of 34,503/33,325 = 103.5 is used to increase specific accounts. This adjustment process is given in the next column and a new overhead ratio of 151,808/34,503 = 440 points is computed. All direct labor costs actually established during this period would then carry this ratio.

TABLE 4.13

OVERHEAD ADJUSTMENT WORKSHEET

	Product Axle Shaft, % Points	$	Adjustment	Revised Overhead, $
Direct labor		33,325	+1178	34,503
Setup costs	11.2	3,732	×103.5%	3,863
Nonproductive labor	12.1	4,032	—	4,032
Fringe	81.5	27,160	×103.5%	28,111
Maintenance	16.8	5,599	—	5,599
Tool cost	38.9	12,963	×103.5%	13,417
Supplies	9.3	3,099	×103.5%	3,207
Scrap	7.5	2,499	×103.5%	2,586
Down time	7.2	2,399	×103.5%	2,483
Rework	2.9	966	×103.5%	1,000
Taxes & insurance	18.3	6,098	—	6,098
Depreciation	44.3	14,763	—	14,763
Gas	—	—		
Power consumption	14.3	4,765	—	4,765
Janitorial			—	
Maintenance group	16.1	5,365	—	5,365
Tool room	4.0	1,333	—	1,333
Inspection & quality	29.6	9,864	—	9,864
Cutter grind	21.5	7,165	—	7,165
Division administration	68.3	22,761	—	22,761
Balance, other items	46.2	15,396	—	15,396
Total		149,959		151,808
Overhead	450%			440%

4.7-6
Variable Overhead

Cost can be identified and recorded in the accounts as either variable or fixed with respect to changes in output. Materials and direct labor are agreed on as variable costs, and as a practical expedient it is sometimes desirable to distinguish overhead costs as variable and fixed also. To achieve this distinction each cost center has identified variable and fixed overhead. The following classification of the individual cost items is a sample of how overhead might be classified into variable or fixed:

Variable Overhead Cost	*Fixed Overhead Cost*
Indirect materials	Indirect labor
Labor-related costs	Labor-related costs
Power and light	Depreciation
	Property taxes

It will be noted that labor-related costs are included as both variable and as fixed costs. It is conceivable, if they are related to direct labor, that they are variable overhead, and conversely, if they are related to indirect labor, a fixed cost, they become fixed overhead.

4.8
GENERAL EXPENSES

In addition to the previously cited costs, there are other general expenses involved in any company's operation. These expenses may be classified as

1. Administrative expenses.
2. Distribution costs.
3. Finance costs.
4. Research, development, and engineering design expenses.

The traditional approach has been to view administration as a distinct function on the same level as manufacturing and selling. An opposite view holds that manufacturing and selling are two basic functions of a business; general administrative costs are incurred for the benefit of both. In the first view the general costs are costs of the period and not the product. For the second, part of the general and administrative cost should be assigned to the product through manufacturing overhead. Figure 3.1 describes the traditional cost and price structure. When the administration of a firm is designated as a separate function in the business enterprise, we segregate the adminstrative costs as those which have to do with phases of operation that are clearly associated with production, sales, or financing of operations.

Distribution and marketing costs refer to those costs incurred to effect the sale, the movement of goods, and associated administration wherever there are materials or products going into the hands of purchasers. The distribution of costs classification includes storing, packaging, transporting, selling, advertising, and applicable administrative expenses.

Research, development, and engineering expenses, whenever removed from the actual production effort, are frequently a general expense. These types of costs are incurred by any concern which attempts to maintain a competitive industrial posture.

Whenever a firm is required to borrow capital in order to finance its operations or extend its investment by the purchase of new equipment or the additions of plant space, interest is charged against the outstanding loans. Interest is considered to be the compensation paid for the use of borrowed capital. A fixed rate of interest is established at the time the capital is borrowed, and this becomes an obligated cost for the management of the firm.

4.9
SUMMARY

For any structural approach to cost estimating, there must be an equitable distribution of costs on which to base analysis of past costs and projections for future costs. In all organizations, budgeting is usually found. Labor and material costs are common costs. Most firms do have assets, and the topic of depreciation is applicable. Whenever there are operations or products or things physical, we find that indirect expenses are usually incurred. With this in mind, the nature of overhead and its application, determination, and allocation can be discussed. The general and administrative costs are considered and must be determined. The cost structure of many activities can be related to this background of information.

The next chapter considers forecasting. Using imperfect information we add the ingredients of judgment to provide decisions based on interpretive facts.

In these forecasting processes, the data, although always incomplete, are assumed to be the situation for the future.

For a cost-accounting understanding of this chapter, the following may be studied:

DICKEY, ROBERT I.: *Accountants' Cost Handbook*, 2nd ed., The Ronald Press Company, New York, 1960.

GILESPIE, CECIL: *Standard and Direct Costing*, Prentice-Hall, Inc., Englewood Cliffs, N.J., 1962.

NOBLE, HOWARD S., and C. ROLLIN NISWONGER: *Accounting Principles*, 8th ed., South-Western Publishing Company, Cincinnati, Ohio, 1961.

A concentrated discussion of the various approaches by industries that are concerned with the construction of cost-engineering data is given by

CLOUGH, RICHARD H.: *Construction Cost Control*, American Society of Civil Engineers, 1951.

CLOUGH, RICHARD H.: *Construction Contracting*, John Wiley & Sons, Inc., New York, 1960.

ROBIE, EDWARD H., ed.: *Economics of the Mineral Industries*, The Maple Press Co., York, Pa., 1964.

TUCKER, SPENCER A.: *Pricing for Higher Profit. Criteria, Methods, Applications*, McGraw-Hill Book Company, New York, 1968.

WASS, ALONZO: *Building Construction Estimating*, Prentice-Hall, Inc., Englewood Cliffs, N.J., 1963.

QUESTIONS

1. Why is equitable distribution of cost essential throughout the organization to cost finding, analysis, and prediction?
2. Name typical source documents used as primary information.
3. What is the purpose of the budget? How would you define a cost center for an engineering budget? What prevents the budget from being meaningless?
4. Distinguish between a cost code and a cost account. How does the nature of the organization determine whether a cost code or cost account system of classification is used? Prepare a list of cost codes for the new development of a water utility.
5. Provide classifications of budgets.
6. Distinguish between depreciation and depletion. Are there natural laws for depreciation?
7. Define overhead. What is the essential and philosophical purpose for overhead?
8. Prescribe and contrast several methods for the distribution of the indirect cost. How do absorption and direct costing differ as to the details of overhead costing?

PROBLEMS

4-1. A company has physically isolated the product manufacture of truck filters to a location away from the central plant. The desire is to determine if this new venture can compete on its own merits. The plant manager has determined monthly fixed and variable costs as $1000 and $0.25 per unit, respectively. Semi-fixed costs were estimated to be

Volume	Monthly Semi-fixed Costs
0–5000	$ 400
5000–9000	600
9000–16,000	1400
16,000–20,000	2000

Construct plots similar to Figure 4.1. If the anticipated production is about 10,000–12,000 units, what simplifications can be made for the costs?

4-2. A cost schedule of electrical rates is given as

> First 50 kwh or less per month @ $4.00
> Next 50 kwh per month   @ $0.05 per kwh
> Next 100 kwh per month   @ $0.03 per kwh
> Next 800 kwh per month   @ $0.017 per kwh
> Over 1000 kwh per month   @ $0.13 per kwh

What is the cost for 550 kwh? For 1250 kwh? Discuss how "block" costs relate to fixed, semi-fixed, and variable cost concepts. Because this is a power company committed only to rural and farm service, how is it discouraging new service to nonfarm users?

4-3. You are the manager of a newly formed drafting department. Under the old management organization the engineering cost center hired their own draftsmen, but they now work for you. There are eight draftsmen and their salaries average $14,000 per year. Using Table 4.3 as a guide, determine what expense accounts would be appropriate, and roughly prepare your first quarterly expense budget.

4-4. Consider the variable budget example in Section 4.3. Plot the percentage capacity of total cost (on the *x*-axis) versus fixed, variable, semi-fixed, and total costs. If for every 10% of capacity, $500 must be realized, at what percentage of total capacity do we find the break-even point?

4-5. Determine a company's base annual cost for a top-grade worker (description: general assembler). The base hourly rate is $5.25. Use the FICA costs of the base costs $9600 earned. State unemployment compensation runs 2%. Health insurance premiums cost $20 per month. The company carries term life insurance that costs $15,000 per year for all employees (there are 60 employees). In addition the company profit-sharing plan usually pays 5% of the base wage.

4-6. One firm decides to find a typical hourly cost for direct labor. In this instance the average work week is 48 hours, of which the final 8 hours are premium at one and a half. The wage is $5.15 per hour. Each day a total of 20 minutes is permitted as a coffee break, the final 15 minutes are cleanup, there are 2 weeks of paid vacation on basis of 40 hours worked and five paid holidays, and there is FICA at the current rate, 1.2% workmen's compensation tax, $25 per month for company-paid medical and life insurance, and a $25 Christmas bonus, all considered pertinent to the calculation.

4-7. A worker produces 56 units on an incentive plan during the 40-hour week. His base hourly rate is $5 per hour, and the standard for one unit of accomplishment is 1.10. What are his weekly earnings? If FICA is $5\frac{3}{4}$%, unemployment insurance costs the company 2.1%, the accident rate is 2.5%, 20 days of yearly vacation, and $40 per month for insurance, what is the actual direct and fringe cost per hour? Per unit?

4-8. Examine the aluminum 2024-T4 shaft in Problem 2-6 in Chapter 2. What is material cost if the density of aluminum = 0.0975 pounds per cubic inch and the price per pound is $0.60? For the (4.510/4.490) outside-diameter dimensions, the tolerance level of bar stock meets the drawing's specifications. The bar stock for this part is supplied in 12-foot lengths.

4-9. A nickel alloy with low carbon, density = 0.321 pounds per cubic inch, is the material to be used for a 125-pounds-per-square-inch storage tank. Design data are $D_m = 48$ inches, $s = 20,000$ pounds per square inch, $E = 0.85$, $C_a = 0.25$, $K = \$1.29$ per pound, and $L = 18$ feet. A waste allowance of 25% is expected. What is the unit material cost for this tank?

4-10. A foundry chooses to price its castings on the basis of delivered weight per 100 pounds. Essentially they price each new casting on past historical records of a similar design. A description of procedure is as follows:

(a) Poured metal cost per casting = (furnace labor and overhead + cost of metal charged) × casting poured weight.

(b) Cost of metal in finished casting = poured metal cost per casting less amount of remelted metal × value of remelted metal.

(c) Cost per pound of delivered metal is item (b) divided by finished weight.

For our estimate the casting poured weight is (5 pounds finished weight) 9 pounds and the cost of charged metal is $0.06 per pound. Furnace labor and overhead is $0.01 per pound. The amount of remelted metal is expected to be 3.7 pounds with a value of $0.03 per pound. Determine the resulting material cost per pound.

4-11. A design having 19.45 cubic inches of metal volume, with casting allowances added, is to be purchased from a foundry. A casting estimator who computed the volume from the various geometrical parts of the design has received quotations from a job-shop foundry:

Pounds of Total Order	Price per Pound
20,000– 50,000	$0.500
50,000–125,000	0.262
125,000–500,000	0.199
500,000–	Negotiable

If the density of iron is 0.2847 pounds per cubic inch, what are unit prices for the midpoint of each range?

4-12. Company A purchased a machine that costs $25,000. Economic life is estimated as 10 years with a salvage value of $5000. This company chooses the sum-of-the-year digits method for depreciation. Company B, in competition with company A, purchased the same machine at identical cost. The management of company B uses straight-line depreciation. Determine the yearly depreciation charges and the end-of-year book value. Compare the two methods for this competitive situation and comment on the importance of the depreciation method for cost estimating.

4-13. A manufacturer recently acquired a numerical control long-bed lathe. Due to the high initial cost and setup charges, he is in doubt as to the appropriate depreciation method to apply. The lathe will increase production on higher-quantity pieces and four methods of depreciation are available. For the following data, plot the four methods and justify your selection as to the manufacturer's best choice. Initial cost, $140,000; installation and programming, $6000; and estimated useful life, 10 years, with 20% salvage value ($r = 2$). Production outputs for the 10 years are 600, 800, 1000, 1000, 1200, 1200, 1200, 1100, 1000, and 900 consecutively.

4-14. The Federal Highway Administration says that this year's standard-sized sedan purchased for an average $4379 will cost its owner $13,553 by the time it has been driven 10 years and 100,000 miles. Besides depreciation to zero, this sum includes $2599 for repairs and maintenance; $2096 for gas and oil; $1809 for garaging, parking, and tolls; $1350 for insurance; and $1320 for state and federal taxes. Find the depreciation cost per mile. What are the yearly and per-mile costs? What are the percentages for the elements? Define a fixed cost independent of miles driven. What are the annual fixed and variable costs? If a small compact is purchased instead of the sedan and it costs $2400—and assuming that the same percentages apply—what are the yearly and per-mile costs? What are the operating advantages?

4-15. In the Federal Highway Administration report the value and upkeep costs were found to follow this schedule:

Year	Drop in Value (%)	Drop in Value	Upkeep Costs
1	28	$1226	$103
2	21	900	135
3	15	675	269
4	11	500	343
5	9	376	322
6	6	259	347
7	4	189	457
8	3	121	242
9	2	85	304
10	1	48	77
		$4379	

How many years does it take for a car to depreciate two-thirds and three-fourths of its value? If the car is driven 13,500 miles per year, what are the operating yearly costs in dollars per mile. Plot these operating costs. When is the advantageous time to trade a car assuming that the chief criterion is per-mile economy? *Discuss:* If the immediate cost of repairing an old car is less than first-year depreciation on a new one, is the best policy to buy a car and drive it until it is ready to be junked?

4-16. To supplement its rental fleet of construction equipment, a company buys three pickups for $10,000. The company considers straight-line, declining-balance, sum-of-the-year digits, and mileage as appropriate depreciation models. The recovery value is $1000 total, life is 10 years, and depreciation on the basis of mileage is $0.033 per mile per pickup. Determine the depreciation if yearly mileage is given by the model $20,000\alpha^{K-1}$, where K is the current year from time zero of the investment and α is a preference parameter of 0.8. If the criterion is to minimize the book value at the end of the fifth year, which of the four depreciation methods gives the best policy.

4-17. An assembly area has floor space allocation as follows:

(a)	Drop area	300
(b)	Conveyor 1	100
(c)	Conveyor 2	150
(d)	Bench area	800
		1350 square feet

Total overhead costs are $8000 for this area. Determine the allocation ratio and overhead costs for each assembly production center.

4-18. A company is composed of five cost centers. Each month a budget is prepared anticipating the primary distribution of certain costs. Let C_w = costs incurred within the cost center such as depreciation and supplies and indirect labor (for producing centers only) and C_m = miscellaneous costs.

Cost Center	C_w	C_m	Area (ft^2)	Direct Labor Dollars	Direct Labor Hours
Fabrication	$30,000	$1000	25,000	$6720	1600
Assembly	8,000	500	6,000	2432	640
Finishing	2,000	4000	1,900	1248	320
Engineering	4,000	9000	1,600	—	—
Administration	2,000	1000	2,100	—	—

Make a secondary distribution for engineering and administration to the producing departments, designated C_0, on the basis of space, dollars, and hours and determine cost center and plant wide overhead ratios. Discuss the merits of these allocation schemes.

4-19. Assume the following simple overhead model: $C_b = C_w + C_0 + C_m$, where C_b = overhead costs; C_w = costs incurred within department such as depreciation, indirect labor, and supplies; C_0 = costs incurred outside the department but allocated to it, such as engineering design, building depreciation, and administration; and C_m = miscellaneous overhead account.

Department	Monthly Costs C_w	C_0	C_m	Hours	Direct Labor Dollars
Fabrication	$30,000	$7000	$1000	1600	$6720
Assembly	8,000	5000	500	640	2432
Finishing	2,000	4000	500	320	1248

Determine departmental overhead rates on the basis of hours and direct labor dollars. Find the overall plant rate on the basis of hours and dollars. Now a product cost model can be defined as $C_p = C_{dm} + C_e + C_b$, where C_{dm} = direct material cost and C_e = direct labor cost. Let fabrication, assembly, and finish departmental C_e's be $2, $0.75, and $0.25, respectively, for a unit of product—also 0.48, 0.20, and 0.064 standard hours per unit. If $C_{dm} = 1 per unit, what is the product unit cost based on departmental overhead rates? If we let $C_e = 3 per unit, what is the product cost based on an overall plant rate? Discuss the reasons for the differences in product cost.

4-20. Assume that a job is routed through Machining and Finishing. Machining is heavily mechanized with costly numerical control and other automatic equipment, while Finishing has only a few simple tools. Obviously, burden costs are high in Machining and low in Finishing. Job 1 takes 1 labor hour in Machining and 10 hours in Finishing. Job 2 takes 9 labor hours in Machining and only 2 in Finishing. If a single blanket rate based on labor hours is applied to both jobs, the burden allocation would be the same in both cases (11 hours for each job). The following illustrates the previous discussion:

	Blanket Rate Machining	Finishing	Department Rate Machining	Finishing
Budgeted annual burden	$100,000	$ 8,000	$100,000	$ 8,000
Direct labor hours	10,000	10,000	10,000	10,000
Blanket rate per DLH		$5.40		
Department rates per DLH			$10.00	$0.80

Determine the overhead costs for jobs 1 and 2 using blanket and department rates. Discuss the necessity for selecting the correct base.

4-21. Management is attempting to maintain an overhead rate of 175% for an assembled product. The "problem" area is the machining process. Here the overhead rate is 225% based on $23,000 direct labor. Total direct dollars excluding machining is $89,000. Overhead charges excluding machining are $150,000. What overhead rate will the company achieve based on this information? Use direct labor dollars as a base.

4-22. Determine a machine-hour rate for the three production centers given the following monthly information:

Production Center	Area	Hours	Horse-power: hours	Deprecia-tion	Tooling	Workers	Direct Labor Budget
Fabrication	25,000	1600	2200	$27,000	$1000	10	$6720
Assembly	6,000	640	325	7,200	500	4	2432
Finishing	1,900	320	650	1,800	4000	2	1248

Item	Amount
Indirect overhead	$16,000
Power	210
Indirect labor	4,000
Engineering	13,000
Administration	3,000

Find a plant-wide rate on the basis of direct labor. Find a department rate on the basis of production hours.

4-23. The following are cost data for an automobile part:

Element	Amount	Cost Factor	Variable Overhead Rate (%)
Direct material, steel	4 lb	$ 7.25/hwt	—
Plating, chrome	0.6 lb	19.76/hwt	90
Direct labor	0.10 hr	6.15/hr	225
Tooling	$0.80/unit	—	175

Find total direct costs (assume that tooling is direct cost). If the overhead rate of the total direct costs is 75% and the variable overhead rates are given in the table, find the total cost for the automobile part or the sum of direct costs and total overhead.

4-24. The following problem illustrates variable machine hour costing. A spread sheet for five production departments is given as

Production Center	Direct Labor	Power	Packaging	Variable Indirect Material	Variable Indirect Overhead	Total
Vertical boring	$ 63,000	$ 6,458	$ 8,253	$ 2,000	$ 4,000	$ 83,711
Light machining	275,500	12,227	36,090	5,000	20,000	348,817
Heavy machining	44,000	20,699	5,764	8,000	3,000	81,463
Bench work	113,900	644	14,920		8,000	137,464
Finishing	66,400	1,877	8,698	3,000	4,000	83,975
Total	$562,800	$41,905	$73,725	$18,000	$39,000	$735,430

In addition to these variable costs there are other variable related labor costs of $20,000 which are unallocated and direct product materials of $462,500. These product materials are uniform. What is the total out-of-pocket cost or variable cost? Fixed costs are $736,000. What is the full cost? If a 10% profit can be obtained for full cost, what is the expected gross dollar revenue or sales price for this product? A profit-volume ratio (PV) or dollar contribution ratio is defined as the excess of gross dollar revenue over total variable costs divided by the gross dollar revenue. It is an indicator used for the covering of fixed costs and profits. What is this PV ratio?

Overhead Adjustment. An engineering change is incorporated into the axle estimate described in Section 4.7-5. This will require a new machine costing $35,000 and an additional $2500 for tooling. The operation of this machine will combine the new cut with some of the work done in the current process. The net effect is a $1.76-per-100 unit labor savings. A base labor adjustment of 1.76/100 × 20,000 axles/mo. = $352 per month. In the burden adjustment worksheet of Table 4.13, reduce the current base labor by this amount. With this new base we determine the related changes in specific cost areas. Use the ratio 32,973/33,325 = 0.989 to deflate chosen specific cost areas. An increase of $175 per month covers the increase in depreciation. The monthly property tax is calculated as $12 per $100 per year on one-third of the capital expenditure value, or (35,000/3) × ($12/100)/12 = $117. The insurance calculation can be made by taking 3.3% of the depreciation on capital expenditures.

$175 increase in depreciation per month × 0.033 = $6 increase in insurance. Tooling is depreciated over 4 years, and consequently the monthly depreciation is equal to about 2% of the tooling, which is added to the depreciation sum.

With these suggestions

1. Find the new overhead rate.
2. What would you expect to happen to those designs which are influenced by the new equipment?

CASE PROBLEM

chapter 5

Get your facts first and then you can distort them as you please.

Mark Twain

FORECASTING

Despite all statements to the contrary, "emotional estimating", or the other idiom "guesstimating", has not disappeared from the cost-engineering scene. Nor has its substitute, estimation by formula and mathematical models, been universally nominated as a replacement. Somewhere between these extremes is a preferred course of action. In this chapter we shall consider the ways and means by which cost estimating can be enhanced by graphical and analytical techniques—however, not according to Mark Twain's lesson.

Forecasts are of two types: business forecasting and technical forecasting. Business forecasting comprises the prediction of market demands, prices of material, cost of conversion, cost and availability of labor, and the like. Every major engineering decision is influenced by business forecasts. The usual approach to business forecasting involves extrapolation of past data into the future using linear or nonlinear curves and mathematical relationships. Future demands are expected to follow some pattern of growth and decay. Most business forecasts are made for the short-run period of up to 2 years. Medium-term forecasts cover 2 to 5 years, while long-term forecasts are for more than 5 years.

Technical forecasting includes predictions about the response to the directed actions of the firm. For example, the durability of materials within engineering construction, the need for maintenance and replacement of equipment, inventory size, and types and nature of processing to convert raw materials into finished products are factors that tend to be neutral or favorable to the control of the engineer rather than in opposition. The life of a numerical control machine, forecast at 15 years, might be extended to 25 years by maintenance activities if this fits into the operating plans. The factor of future control is partially under the direction of the firm. Technical forecasting will be discussed, with particular attention given to the Delphi technique.

5.1
GRAPHIC ANALYSIS OF DATA

The field of descriptive statistics is concerned with methods for collecting, organizing, analyzing, summarizing, and presenting data as well as drawing valid conclusions and making reasonable decisions on the basis of analysis of the data. In a more restricted sense, the term statistics is used to denote the data themselves or numbers derived from the data, as, for example, averages.

The data that are gathered for descriptive statistics and graphical presentation may be either discrete or continuous and are usually the result of a series of observations taken over time or another controllable or noncontrollable variable. Raw data, however, communicate little information. One way to communicate information develops a frequency distribution which compacts the data into manageable proportions. After the plot has been concluded, it is common to calculate a measure of central tendency, such as mean, median, or mode, and a measure of dispersion, such as the standard deviation or range.

Consider the following example of descriptive statistics. A firm is considering the construction of a manufacturing plant in a community, and it has been suggested that the labor market could be tight, thus creating a labor shortage. Employment rates for males by age groups were received from reputable sources:

RAW DATA OF UNEMPLOYMENT
FOR COMMUNITY A

Age Range	March 1st Year	March 2nd Year
14–15	20	26
16–17	87	93
18–19	636	709
20–21	206	191
22–23	202	50
24–25	81	229
26–27	15	37
28–29	13	29
30–31	19	73
32–33	25	83
34–35	38	47
36–37	53	42
38–39	36	85
40–44	89	30
45–49	101	97
50–54	86	107
55–59	111	67
60–64	117	173
65–69	144	180
70+	101	102
Total	2180	2450

As is frequently the situation, the analyst has no other recourse than to accept data on an "as is" basis. The analyst could justifiably doubt the credibility of the sample. Do the unemployment figures overlook or include transients and students who are in and out of the labor force periodically? Is the time of the year representative? What about the age 70+? Are retirements included? Despite these concerns, this information is accepted. Rearrangement of the data led to the following:

UNEMPLOYMENT RATES FOR MALES
BY AGE GROUP

Age Group	1st Year (%)	2nd Year (%)
14–19	16.7	13.8
20–24	9.4	7.8
25–34	4.3	3.7
35–44	4.0	3.4
45–54	4.2	3.4
55–64	4.9	4.0
65+	5.5	4.7

A graphic display of unemployment for males by age groups is constructed and shown in Figure 5.1. It is apparent that the data have been forced into

FIGURE 5.1

GRAPHIC DISPLAY OF UNEMPLOYMENT FOR
MALES BY AGE GROUP

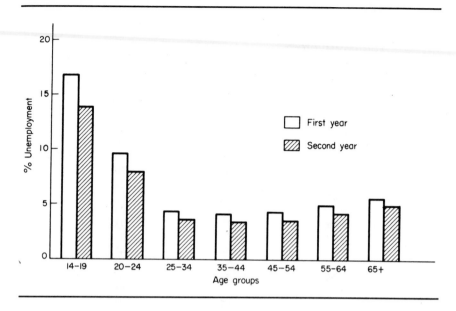

a nice-looking picture as the divisions for the age groups are not uniform. Mark Twain's advice is followed here. Data used in this illustration are composed of separate and distinct measurements and are found from a counting process and are known as discrete data.

Many engineers choose first to plot data to get a "feel" for its accuracy and association between variables, but the analysis seldom concludes with the graphical step.

5.2
EMPIRICAL DISTRIBUTIONS: CORNERSTONE FOR ANALYSIS

A market analysis has been made of the price for a standard plastic film material provided in 50-foot rolls, 4-foot widths, and 0.006-inch thickness. The data were collected through consumer reports, catalog reports, telephone queries, and advertisement checks of trade journals. The following is an illustration of 324 such observations:

Price Range $/Roll	Number of Observations	Relative Distribution	Cumulative Distribution
2.35–2.74	1	0.003	0.003
2.75–3.14	6	0.019	0.022
3.15–3.54	33	0.102	0.124
3.55–3.94	51	0.157	0.281
3.95–4.34	121	0.373	0.654
4.35–4.74	50	0.154	0.808
4.75–5.14	44	0.136	0.944
5.15–5.64	13	0.040	0.984
5.65–6.04	5	0.016	1.000
Total	324	1.000	

The relative distribution value was computed for each range of price and is equal to the count per cell divided by the total number of observations. A cumulative distribution was constructed which sums up from the lower range progressively to the maximum range. The cumulative distribution is often needed for a Monte Carlo analysis, discussed in Chapter 6. Both of these empirical distributions are given in Figure 5.2. It is evident that the mean value is near $4 per roll.

FIGURE 5.2

RELATIVE FREQUENCY CURVE AND
CUMULATIVE PROBABILITY OF OCCURRENCE
CURVE FOR DATA

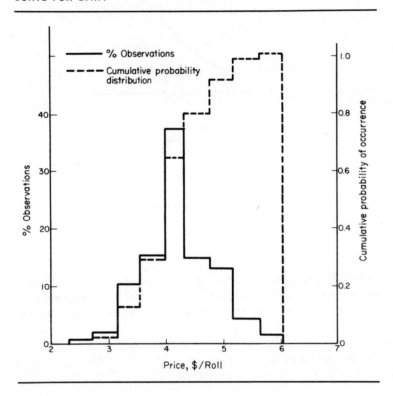

Figure 5.3 shows several types of frequency curves obtainable from analysis of data. A common type of title for the y-axis would be percentage of observations or count. All these graphical plots can be constructed in a manner similar to Figure 5.2. Curve construction calls for a trained eye. For instance, if a wealth of data exists, the bimodal or multimodal plot may be evident. But in the absence of an abundance of data, graphical conclusions of this sort are seldom found. For this and other reasons mathematical analyses are resorted to.

5.3
METHOD OF LEAST
SQUARES AND
REGRESSION

The descriptive statistical methods described so far have been concerned with a single variable and its frequency function. However, many of the problems in the estimating field involve several variables. The next several pages explain simple methods for dealing with data associated with two, three, or more variables. These methods are known as regression models and are useful tools. In regression, on the basis of sample data, the estimator wishes to know the value of a dependent variable, say y, corresponding to a given value of a variable,

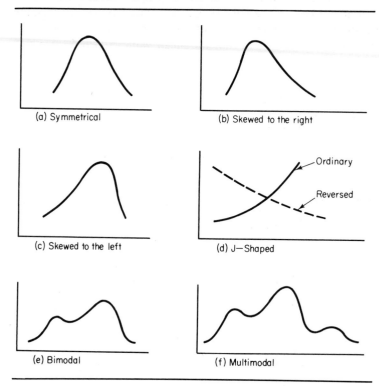

(a) Symmetrical

(b) Skewed to the right

(c) Skewed to the left

(d) J—Shaped

(e) Bimodal

(f) Multimodal

say x. This can be determined from a least-squares curve which fits the sample data. The resulting curve is called a regression curve of y on x as y is determined from a corresponding value of x. If the variable x is time, the data show the values of y at various times and are known as a time series. Regression lines or a curve y on x or the response function on time is frequently called a trend line and is used for the purposes of prediction and forecasting. Thus regression refers to the average relationship between variables.

5.3-1
Least Squares

The notion of fitting a curve to a set of sufficient points is essentially the problem of finding the parameters of the curve. The best known method is the method of least squares. Since the desired curve or equation is to be used for estimating or predicting purposes, it is useful to require that the curve or equation be so modeled as to make the errors of estimation small. An error of estimation or prediction usually means the difference between an observed value, say of the index, and the corresponding fitted curve value of the index. It will not do to require the sum of these differences or errors to be as small as possible. It is a requirement that the sum of the absolute values of the errors be as small as possible. However, sums of absolute values are not convenient mathematically. The cause of the difficulty is avoided by requiring that the sum of the square of the errors be minimized. If this procedure is followed, the values of parameters give what is known as the *best curve* in the sense of least-squares difference.

This principle of least squares states that if y is a linear function of an independent variable x, the most probable position of line $y = a + bx$ will exist whenever the sum of squares of deviations of all points (x_i, y_i) from the line is a minimum. These deviations are measured in the direction of the y-axis. The underlying assumption is that x is either free of error (a controlled assignment) or subject to negligible error. The value of y is the observed or measured quantity, subject to errors which have to be "eliminated" by this method of least squares. The value y is a random variable value from the y population values corresponding to a given x. For each value x_i we are interested in corresponding $\bar{y}_i$; suppose that our observations consist of pairs of values as

$$x_1, x_2, \ldots, x_n$$
$$y_1, y_2, \ldots, y_n$$

which are assumed to give a linear plot as in Figure 5.4. Our problem is to

FIGURE 5.4

REGRESSION LINE OF $y = a + bx$

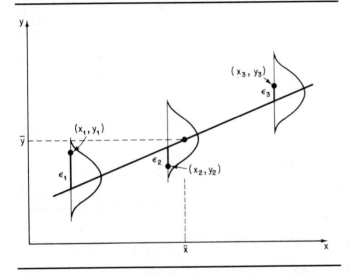

uncover a and b for a best fit. For a general point i on the line, $y_i - (a + bx_i) = 0$, but if an error ϵ_i exists, then $y_i - (a + bx_i) = \epsilon_i$. For n observations we have n equations of

$$y_1 - (a + bx_1) = \epsilon_1$$
$$\vdots$$
$$y_i - (a + bx_i) = \epsilon_i \qquad \text{(5-1)}$$
$$\vdots$$
$$y_n - (a + bx_n) = \epsilon_n$$

Summing, we can write the sum of squares of these residuals as

$$p = \sum \epsilon_i^2 \qquad \text{(5-2)}$$
$$= \sum_{i=1}^{n} [y_i - (a + bx_i)]^2 \qquad \text{(5-3)}$$

where p = sum of the squared-error term. For a minimum we insist on

$$\frac{\partial p}{\partial a} = 0 \qquad \text{and} \qquad \frac{\partial p}{\partial b} = 0 \qquad (5\text{-}4)$$

or as

$$n \sum xy - \sum x \sum y = n \sum (x - \bar{x})(y - \bar{y})$$
$$= n\left(\sum xy - \bar{x} \sum y + \bar{x} \sum y - \bar{y} \sum x\right)$$

TABLE 5.1

CALCULATION OF REGRESSION COEFFICIENTS

				Year Index		
x	y	x^2	xy	$\hat{y}$	$\epsilon = y - \hat{y}$	ϵ^2
0	87	0	0	84.875	2.125	4.516
1	89	1	89	87.264	1.736	3.014
2	90	4	180	89.653	0.347	0.120
3	92	9	276	92.042	−0.042	0.002
4	93	16	372	94.431	−1.431	2.047
5	99	25	495	96.820	2.180	4.752
6	97	36	582	99.209	−2.209	4.879
7	100	49	700	101.598	−1.598	2.554
8	101	64	808	103.987	−2.987	8.922
9	106	81	954	106.376	−0.376	0.141
10	106	100	1060	108.765	−2.765	7.645
11	109	121	1199	111.154	−2.154	4.640
12	115	144	1380	113.543	1.457	2.123
13	118	169	1534	115.932	2.068	4.277
14	122	196	1708	118.321	3.679	13.535
Totals: 105	1524	1015	11337			63.167

For $Y = a + bx$, the constants

$$a = \frac{\sum y \sum x^2 - \sum x \sum xy}{n \sum x^2 - (\sum x)^2} = \frac{(1524)(1015) - (105)(11{,}337)}{15(1015) - (105)^2} = 84.875$$

$$b = \frac{n \sum xy - \sum x \sum y}{n \sum x^2 - (\sum x)^2} = \frac{15(11{,}337) - (105)(1524)}{15(1015) - (105)^2} = 2.389$$

$Y = 84.875 + 2.389X$ or if X is year, then index $= 84.875 + 2.389(X - 1960)$. The estimated mean value for next year is $Y = 84.875 + 2.389(15) = 120.71$.

$$S_y = \left(\frac{\sum \epsilon^2}{n - 2}\right)^{1/2} = \left(\frac{63.167}{13}\right)^{1/2} = 2.204$$

$$S_{y_{\bar{i}}} = S_y\left[\frac{1}{n} + \frac{(x_i - \bar{x})^2}{\sum (x_i - \bar{x})^2}\right]^{1/2} = 2.204\left[\frac{1}{15} + \frac{(15 - 7)^2}{280}\right]^{1/2} = 1.198$$

For degrees of freedom = 13 and a 5% level of significance, $t = 2.160$. The confidence interval is $120.71 \pm 2.160(1.198) = (118.122, 123.298)$. For a single estimated value of y_i,

$$S_{y_i} = S_y\left[1 + \frac{1}{n} + \frac{(x_i - \bar{x})^2}{\sum (x_i - \bar{x})^2}\right]^{1/2} = 2.204\left(1 + \frac{1}{15} + \frac{64}{280}\right)^{1/2} = 2.508$$

The confidence interval is $120.71 \pm 2.160(2.508) = (115.293, 126.127)$. The confidence interval for the slope and intercept are

$$S_b = \frac{S_y}{[\sum (x_i - \bar{x})^2]^{1/2}} = \frac{2.204}{(280)^{1/2}} = 0.132$$

slope interval $= 2.389 \pm 2.160(0.132) = (2.103, 2.674)$

$$S_a = S_y\left[\frac{1}{n} + \frac{\bar{x}^2}{\sum (x_i - \bar{x})^2}\right]^{1/2} = 2.204\left(\frac{1}{15} + \frac{49}{280}\right)^{1/2} = 1.083$$

intercept interval $= 84.875 \pm 2.160(1.083) = (82.536, 87.214)$

and solving these two normal equations for a and b we have

$$a = \frac{\sum x^2 \sum y - \sum x \sum xy}{n \sum x^2 - (\sum x)^2}$$

and   (5-5)

$$b = \frac{n \sum xy - \sum x \sum y}{n \sum x^2 - (\sum x)^2}$$

These values would then be appropriately substituted into $y = a + bx$, which is called the line of regression of y on x. In this case we assumed that x is the assigned variable and y the observed quantity.

This least-squares equation passes through $(\bar{x}, \bar{y})$, which is the coordinate mean of all observations.

The calculation of these coefficients may be handled manually, with an electronic calculator, or by computer. Table 5.1 is an example of a tabular form in determining these values. If $Y =$ the index and $X =$ the coded year, the equation of the least-squared line is $Y = 84.875 + 2.389X$. The least-squares method can be applied whether the X-values were fixed in advance or were obtained from random samples.

The limitations of the method must be pointed out. The method of least squares is applicable when the observed values of y_i correspond to assigned (or error-free) values of x_i; the error in y_i (expressed as a variance of y) is assumed to be independent of the level of x. If inferences are to be made about regression, it is also necessary that the values of y_i corresponding to a given x_i be distributed normally, as shown on Figure 5.4, with the mean of the distribution satisfying the regression equation. It is required also that the variance of the values of y_i for any given value of x be independent of the magnitude of x. While the evidence of this statement can be statistically shown, experience shows that only a comparatively small number of the distributions met within cost engineering can be described by normal distributions. Cost data are limited at the zero end. Distributions influenced by economic, engineering, and human factors are generally skew. Transformations are required with the usual transformation via the logarithm. This stabilizes the situation, and a variable is said to have a logarithmic normal distribution if the logarithm of the variable is normally distributed. The beauty of the transformation is evident whenever a series of values at one end of the distribution is more or less unsymmetrical.

5.3-2
Confidence Limits for Regression Values and Prediction Limits for Individual Values

If variations around the universe regression are random, the method of least squares permits the computation of sampling errors and provides for the determination of the reliability of the estimate of the dependent variable from the fitted line. Furthermore, if the distribution points around the universe regression are not only random but normal in form, then the least-squares method gives the maximum likelihood estimate of the universe regression. For these reasons lines and curves of regression are commonly estimated from sample data by the method of least squares. Confidence limits for regression values can be constructed through the extension of simple statistics. The confidence limits for individual regression values and for the straight line are quadratic in form around the sample line of regression. The confidence band for the slope is fan-shaped with the apex at the mean.

The variance of an estimate permits the forming of confidence limits of the estimate. The approach, similar to the variance of a sample, in this case reckons the deviations from a line instead of a mean. The variance of y, estimated by the regression line, is the sum of squares of deviations divided by the number of degrees of freedom available for calculating the regression line, or

$$s_y^2 = \frac{\sum \epsilon_i^2}{\nu} \tag{5-6}$$

where ϵ_i is defined previously by Equation (5-1). Only two bits of information are required to determine the regression line: means $(\bar{x}, \bar{y})$ and either slope b or intercept a. With n as the number of paired observations, ν is defined as

$$\nu = n - 2 \tag{5-7}$$

Also,

$$s_y^2 = \frac{\sum \epsilon_i^2}{n - 2} \tag{5-8}$$

The variance of the mean value of y or $\bar{y}$ is

$$s_{\bar{y}}^2 = \frac{s_y^2}{n} \tag{5-9}$$

It is now possible to write the confidence limits for $\bar{y}$. A table of Student's t-distribution and of the values of t corresponding to various values of the probability α (level of significance) and a given number of degrees of freedom ν is found in Appendix II. With this t-value we state that the true value of $\bar{y}$ lies within the interval

$$\bar{y} \pm t s_{\bar{y}} \tag{5-10}$$

The probability of being wrong is equal to the level of significance of the value of t. As the regression line must pass through the mean, an error in the value of $\bar{y}$ leads to a constant error in y for all points on the line. The line is then moved up or down without change in slope.

If limits for individual values are desired, a different approach must be asserted. The statement that usually describes the *limits* for individual values goes like this: If we use the sample line of regression to estimate a particular value for y, we add to the error of the sample line of the regression some measure of the possible deviation of the individual value from the regression value. For individual values a new set of parabolic loci may be viewed as prediction limits. Figure 5.5 shows the prediction loci for individual values as well as the confidence loci for regression values. It will be noted from the figure that the prediction limits for y get wider as x deviates from its mean, both positively and negatively. This means that predictions of the dependent variable are subject to the least error when the independent variable is near its mean and are subject to the greatest error when the independent variable is distant from its mean.

If we require an estimate of the confidence limits corresponding to x_i, we calculate the limits for $\hat{y}_i$ (strictly speaking, an estimate). The variance of the estimate of this mean value is

$$s_{\bar{y}_i}^2 = s_y^2 \left[\frac{1}{n} + \frac{(x_i - \bar{x})^2}{\sum (x - \bar{x})^2} \right] \tag{5-11}$$

The confidence interval for the mean estimated value of $y_{\bar{i}}$ corresponding to specific x_i is

$$y_{\bar{i}} \pm t s_{y_{\bar{i}}} \tag{5-12}$$

For a predetermined level of significance we are able to predict the limits within which a future mean estimated value of $y_{\bar{i}}$ will lie with an appropriate chance of error.

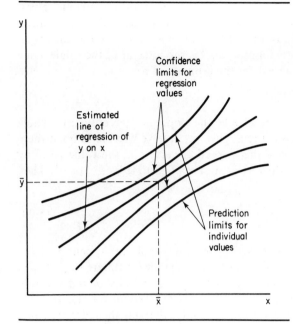

We can predict the confidence interval of a single estimated value of $\hat{y}_i$ using the variance of a single value, which has as its variance

$$s_{y_i}^2 = s_y^2 \left[1 + \frac{1}{n} + \frac{(x_i - \bar{x})^2}{\sum (x - \bar{x})^2} \right] \qquad (5\text{-}13)$$

This variance is larger than $s_{y_{\bar{i}}}^2$ because the variance of the single value is equal to the variance of the mean plus the variance of $\hat{y}$ estimated by the line or

$$s_{y_i}^2 = s_{y_{\bar{i}}}^2 + s_y^2 \qquad (5\text{-}14)$$

The confidence interval for a single value is greater too, or

$$y \pm t s_{y_i} \qquad (5\text{-}15)$$

The variance of the intercept a is a particular case of the variance of any mean estimated $\hat{y}_{\bar{i}}$. If we substitute $x_i = 0$ in Equation (5-11), the variance of intercept a is

$$s_a^2 = s_y^2 \left[\frac{1}{n} + \frac{\bar{x}^2}{\sum (x - \bar{x})^2} \right] \qquad (5\text{-}16)$$

Its confidence band is given by

$$a \pm t s_a \qquad (5\text{-}17)$$

The variance of slope b is given as

$$s_b^2 = \frac{s_y^2}{\sum (x - \bar{x})^2} \qquad (5\text{-}18)$$

Its confidence band is given by

$$b \pm t s_b \qquad (5\text{-}19)$$

The confidence band for the slope is represented by a double-fan-shaped area with the apex at the mean.

For all confidence intervals in this section, the number of degrees of freedom are $v = n - 2$.

Table 5.1 demonstrates the manual calculation of regression values, namely, the parameters a and b; the regression equation, assuming that a mean value can be determined, finds the 95% confidence interval for the mean estimated value of y_i corresponding to year 15, the interval of the single value for year 15, and finally the interval for the estimate of a and b.

5.3-3
Significance Test for Slope and Intercept

In certain situations a theoretical value of the slope or intercept exists. These values usually result from physical relationships, and we want to know if there is a significant difference between the theoretical value and the value given by the regression line. This is handled by a t-test applied to $|b - b^*|$ or b^*. Then we have

$$t = \frac{|b - b^*|}{s_b} \tag{5-20}$$

If the calculated t is greater than t given by the t-table in Appendix II for a required level of significance, we conclude a significant difference between b^* and b. The level of significance represents the probability of our drawing an erroneous conclusion. The number of degrees of freedom for selecting t is $n - 2$, where n is the number of observations used in deriving the regression line.

The intercept a^*, a theoretically correct or time-worn value, can be tested similarly, or by

$$t = \frac{|a - a^*|}{s_a} \tag{5-21}$$

The number of degrees of freedom is similar to those before.

5.3-4
Curvilinear Regression and Rectification

The world of linearity, largely an imaginary one, is a tidy and manageable world about which generalizations can be asserted with confidence. Many prefabricated tools are known that can be used with these assumptions. The function $y = a + bx$ is a characterization which is frequently employed for known nonlinear situations. It seems that electronic computers and a greater awareness are forcing a re-evaluation of linearities. A listing would show for cost-engineering purposes the following nonlinear relationships:

$y = ae^{bx}$	semilog fit[1]	(5-22)
$y = ax^b$	log-log fit	(5-23)
$y = a + \dfrac{b}{x}$	reciprocal x fit	(5-24)
$y = \dfrac{1}{(a + bx)}$	reciprocal y fit	(5-25)
$y = \dfrac{x}{(a + bx)}$	hyperbolic-type fit	(5-26)
$y = a + b_1 x + b_2 x^2 + \cdots$	polynomial fit	(5-27)

[1]The three exponential curves $y = ae^{bx}$, $y = 10^{a+bx}$, and $y = a(10)^{bx}$ are equivalent. If any two points are chosen and the constants for each curve determined so it passes through these points, the curves are equal.

Each of these functions can be statistically evaluated for the fitted data, such as

- Coefficients of the equations,
- Listing of calculated estimates,
- Standard error of y,
- Squared correlation coefficient, and
- 95% confidence limits values for y.

Computer programs do this calculation where the work is large.

Normally the *best fit* line is judged by the smallest value of the standard error, while other computed values provide an intuitive feel for how good the correlation obtained really is. Polynomial regression, that is, where for any x the mean of the distribution of y's is given by $a + b_1 x + b_2 x^2 + b_3 x^3 + \cdots + b_p x^p$, is used to obtain approximations whenever the functional form of the regression curve is a mystery. Engineers plot data a variety of ways hoping to "straighten out" a cartesian curve via semilog or log-log plots, for example. If sets of paired data straighten out on semilog paper, the analyst would conclude that the form is exponential or $y = ae^{bx}$ and he would be tempted to apply least-squares methods that utilized this form to the data.

Suppose that coded data were obtained as follows, where y, or dollars, was assumed to be related to small gas-engine horsepower as $y = a + bx^2$:

x	Observed Value, y_i	Error-Free Value, $\hat{y}$	Deviation, $y_i - \hat{y}$
0	1	a	$1 - a$
1	1.4	$a + b$	$1.4 - (a + b)$
2	1.8	$a + 4b$	$1.8 - (a + 4b)$
3	2.2	$a + 9b$	$2.2 - (a + 9b)$

The sum of the squares of the deviations is

$$P = (1 - a)^2 + (1.4 - a - b)^2 + (1.8 - a - 4b)^2 + (2.2 - a - 9b)^2$$

For P to be a minimum we are required to satisfy

$$\frac{\partial P}{\partial a} = 0 \quad \text{and} \quad \frac{\partial P}{\partial a} = 0$$

or $(1 - a) + (1.4 - a - b) + (1.8 - a - 4b) + (2.2 - a - 9b) = 0$ and $(1.4 - a - b) + 4(1.8 - a - 4b) + 9(2.2 - a - 9b) = 0$. This reduces to the normal equations

$$4a + 14b = 6.4$$
$$14a + 98b = 28.4$$

Hence $a = 1.175$ and $b = 0.122$. When substituted back into the general form, the fitted equation becomes $y = 1.175 + 0.122x^2$. A better fit may be obtained from an equation of a different form or $y = a + bx + cx^2$.

The previous little problem was straightforward, but the application of the method of least squares to nonlinear relations usually requires a good deal of computational effort, computers notwithstanding. In most cases we are able to transform or rectify a nonlinear relation to a straight-line relation. This manipulation simplifies handling of the data and permits a graphical presentation which may be revealing for certain facts. With a rectified straight line, extrapolation is simpler, and the computation of certain other supportive statistics, such as the standard deviation or confidence limits, is simpler. Several rectified cases will be identified.

The exponential function $y = ab^x$ can be rectified by log transformations as

$$\log y = \log a + x \log b$$

This straight line will have $\log y$ as the ordinate on a logarithmic division, while the abscissa will be log as well. The intercept is $\log a$ and the slope is $\log b$, and these are the fitted variables for the method of least squares.

The power function $y = ax^b$ is rectified by using logarithms:

$$\log y = \log a + b \log x$$

The fitting constants are $\log a$ and b, and the variables of concern are $\log x$ and $\log y$, which are now linearly related on a plot having log properties. Consider the second-order polynomial of the form $y = a + b_1 x + b_2 x^2$. When differentiated with respect to x, we have

$$\frac{dy}{dx} = b_1 + 2b_2 x$$

Again our straight line is obvious by plotting dy/dx versus x.

Now these previous functions indicated a regression of y on x which was linear in some fashion. Sometimes a clear relationship is not evident, and a general polynomial is selected. A predicting equation of the polynomial form [see Equation (5-27)] requires a set of data consisting of n points (x_i, y_i) and we estimate the coefficients $a, b_1, b_2, \ldots, b_p$ of the pth-degree polynomial by minimizing

$$\sum_{i=1}^{n} [y_i - (a + b_1 x + b_2 x^2 + \cdots + b_p x^p)]^2 \qquad (5\text{-}28)$$

according to

$$\frac{\partial p}{\partial a}, \frac{\partial p}{\partial b_1}, \frac{\partial p}{\partial b_2}, \ldots, \frac{\partial p}{\partial b_p} = 0 \qquad (5\text{-}29)$$

which is the least-squares criterion by minimizing the sum of squares of the vertical distances from the points to the curve. This results in $p + 1$ normal equations of the shape

$$\sum y = na + b_1 \sum x + \cdots + b_p \sum x^p$$
$$\sum xy = a \sum x + b_1 \sum x^2 + \cdots + b_p \sum x^{p+1}$$
$$\cdot$$
$$\cdot \qquad\qquad (5\text{-}30)$$
$$\cdot$$
$$\sum x^p y = a \sum x^p + b_1 \sum x^{p+1} + \cdots + b_p \sum x^{2p}$$

where summation notation has been eliminated. This is now $p + 1$ linear equations in $p + 1$ unknowns $a, b_1, \ldots, b_p$.

As an example, consider the following: An aircraft manufacturer has accumulated data on the chem-milling operation of steel panels, and paired sets of data are cost per square foot, dollars per square foot, and depth of cut,

inches—(0.01, 7.0), (0.02, 8.4), (0.03, 9.2), (0.04, 10.1), (0.05, 10.3), (0.20, 26.2)
—and resultant tabulations, very similar to previous least-squares calculations
of Table 5.1, reveal $\sum x = 0.35$, $\sum x^2 = 0.0455$, $\sum x^3 = 0.008225$, $\sum x^4 = 0.00016979$, $\sum y = 71.2$, $\sum xy = 6.673$, and $\sum x^2 y = 11.0225$. With these
values we are ready to solve the following system of three linear equations,

$$6a + 0.35b_1 + 0.0455b_2 = 71.2$$

$$0.35a + 0.0455b_1 + 0.008225b_2 = 6.673$$

$$0.0455a + 0.008225b_1 + 0.0016979b_2 = 11.0225$$

for which $a = -31.73$, $b_1 = 1660$, and $b_2 = -7025$ and the predicting equation
becomes

$$y = -31.73 + 1660x - 7025x^2$$

In practice it may be difficult to determine the degree of the polynomial
to fit data, but it is always possible to find a polynomial of degree at most $n - 1$
that will pass through each of n points, although what we want is the lowest
degree that describes our problem.

5.3-5
Correlation

It has been seen how the method of regression shows the relationship to one
independent variable which can be considered linearly related. There is a closely
related measure, called correlation, which tells how well the variables are satisfied by a linear relationship. If the values of the variables satisfy an equation
exactly, then the variables are perfectly correlated. When two variables are
involved the statistician refers to simple correlation and simple regression.
When more than two variables are involved it is referred to as multiple regression
and multiple correlation. In this section only simple correlation is considered.

Figure 5.6 indicates the location of points on a rectangular coordinate
system. If all the points in a scatter diagram appear to lie near a line as in Figure
5.6(b) or (d), the correlation is presumed to be linear and a linear equation is
appropriate for purposes of regression or estimation. If there is no relationship
indicated between the variables as in Figure 5.6(c), there is no correlation; i.e.,
the data are uncorrelated. In the case of Figure 5.6(b), the correlation coefficient
is negative linear, while for Figure 5.6(d) a positive linear correlation coefficient
is found.

With a fitted curve from data it is possible to distinguish between the
deviations of the y-observations from the regression line and the total variation
of the y-observations about their mean. A calculated difference between the two
variations gives the amount of variation accounted for by regression, and the
higher this value, the better the fit or correlation. For $y = a + bx$ no correlation
exists if $b = 0$ as in Figure 5.6(c) and the line plots as a horizontal line. Thus
x and y are independent. There is no correlation of x on y if x is independent
of y. These two statements are mathematically

$$\frac{n \sum xy - \sum x \sum y}{n \sum x^2 - (\sum x)^2} = 0 \tag{5-31}$$

and

$$\frac{n \sum xy - \sum x \sum y}{n \sum y^2 - (\sum y)^2} = 0 \tag{5-32}$$

For no correlation the product of Equations (5-31) and (5-32), or the product
of the slopes, is zero. Conversely, for perfect correlation all the points lie exactly

FIGURE 5.6

SCATTER DIAGRAMS SHOWING LEVELS OF
CORRELATION COEFFICIENTS

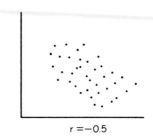

r = −0.5

(a) Negative linear correlation

r = −1.0

(b) Negative linear correlation

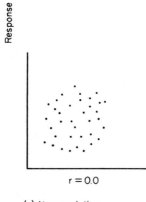

r = 0.0

(c) No correlation

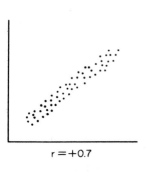

r = +0.7

(d) Positive linear correlation

on each of the two regression lines and their lines coincide, or

$$\frac{n\sum xy - \sum x \sum y}{n\sum x^2 - (\sum x)^2} \times \frac{n\sum xy - \sum x \sum y}{n\sum y^2 - (\sum y)^2} = 1$$

We call the square root of this product the correlation coefficient and denote it by r:

$$r = \frac{n\sum xy - \sum x \sum y}{\{[n\sum x^2 - (\sum x)^2][n\sum y^2 - (\sum y)^2]\}^{1/2}} \qquad (5\text{-}33)$$

which is a computation equation form. Correlation is concerned only with the association between variables, and r must lie in the range $0 \le |r| \le 1$. Using the data from Table 5.1, the correlation coefficient may be found as an example. Recollecting that $n = 15$, $\sum xy = 11,337$, $\sum x = 105$, $\sum y = 1524$, $\sum x^2 = 1015$, and $\sum y^2 = 156,500$, then

$$r = \frac{(15)(11,337) - (105)(1524)}{\{[(15)(1015) - (105)^2][15(156,500) - (1524)^2]\}^{1/2}}$$

$$= +0.981$$

which is a high score for correlation. The magnitude of the correlation coefficient r determines the strength of the relationship, while the sign of r tells one whether the dependent variable tends to increase or decrease with the independent variable. The engineer may realize that r is a useful measure of the strength of the relationship between two variables—only if, however, the variables are linearly related. The value of r will be equal to $+1$ or -1 if and only if the points of the scatter lie perfectly on the straight line, which is unlikely in cost engineering.

The interpretation of the correlation coefficient as a measure of the strength of the linear relationship between two variables is a purely mathematical interpretation and is without any cause or effect implications.

5.3-6
Multiple Linear Regression

It happens often that the method of least squares for estimating one variable by means of a related variable yields inadequate success. Although the relationship may be linear, frequently there is no single variable sufficiently related to the variable being estimated to yield good results. The extension is natural to two or more independent variables. As linear functions are simple to work with and estimating experience shows that many sets of variables are approximately linear related, or assumed so for a short period of time, it is reasonable to estimate the desired variable by means of a linear function of the remaining variables. Problems of multiple regression involve more than two variables but are still treated in a manner analogous to that for two variables. For example, there may be a cost relationship between differential profit (DP), investment (I), and market saturation (MS) which can be described by the equation $DP = a + b_1 I + b_2 MS$, which is called a linear equation in the variables DP, I, and MS. The constants of the system are noted by a, b_1, and b_2. This kind of analysis can be used to explain variations in one dependent variable by adding together the effects of two or more independent variables. It is not by any means limited to problems involving a time trend. Time is a catchall, and it takes into account gradual changes that may be due to different factors both known and suspected. For three or more variables, a regression plane is a generalization of the regression line for two variables, as previously considered. We are concerned with the linear regression function of the form

$$y = a + b_1 x_1 + b_2 x_2 + \cdots + b_k x_k \qquad (5\text{-}34)$$

where $x_0 = 1$, a is a constant, and $b_1, b_2, \ldots, b_k$ are partial regression coefficients. This is a plane in $k + 1$ dimension. We do not say that the result so obtained is the best functional relationship. We simply state that, given this assumed function and criterion, we have chosen the best estimate of the parameters.

Regression analysis requires the following assumptions:

1. The x_j-values are controlled and/or observed without error. Perfection remains a difficult requirement within cost-estimating practices, but it is nominally met.
2. The regression of y on x_j is linear.
3. The deviations $y - [j | x_j]$ are mutually independent.
4. These deviations have the same variance whatever the value of x_j.
5. These deviations are normally distributed.
6. The data are taken from a population about which inferences are to be drawn.
7. There are no extraneous variables which make the relationship of little intrinsic value.

The plane in the $k + 1$ dimension passes through the mean of all observed values, similar to the two-variable case, and

$$\bar{y} = a + b_1\bar{x}_1 + b_2\bar{x}_2 + \cdots + b_k\bar{x}_k \tag{5-35}$$

Reworking Equations (5-34) and (5-35), we have

$$a = \bar{y} - b_1\bar{x}_1 - b_2\bar{x}_2 - \cdots - b_k\bar{x}_k \tag{5-36}$$

$$y - \bar{y} = b_1(x_1 - \bar{x}_1) + b_2(x_2 - \bar{x}_2) + \cdots + b_k(x_k - \bar{x}_k) \tag{5-37}$$

As before, the coefficients are determined using the method of least squares. To illustrate we consider the case of two independent variables, and we have

$$y = a + b_1x_1 + b_2x_2 \tag{5-38}$$

with n sets (y, x_1, x_2) of triplets at this point. In each set the error is given as

$$\epsilon = y - (a + b_1x_1 + b_2x_2) \tag{5-39}$$

and the sum of squares of errors in the n sets is

$$\sum \epsilon^2 = [y - (a + b_1x_1 + b_2x_2)]^2 \tag{5-40}$$

We minimize $\sum \epsilon^2$ as before, which requires that the partial derivatives of $\sum \epsilon^2$ with respect to a, b_1, and b_2 be zero:

$$\frac{\partial(\sum \epsilon^2)}{\partial a} = \sum [y - (a + b_1x_1 + b_2x_2)] = 0$$

$$\frac{\partial(\sum \epsilon^2)}{\partial b_1} = \sum x_1[y - (a + b_1x_1 + b_2x_1)] = 0 \tag{5-41}$$

$$\frac{\partial(\sum \epsilon^2)}{\partial b_2} = \sum x_2[y - (a + b_1x_1 + b_2x_2)] = 0$$

Subscripts for summation were dropped for convenience. If we keep x_2 constant, the graph of y versus x_1 is a straight line with slope b_1. If we keep x_1 constant, the graph y versus x_2 is linear with slope b_2. Due to the fact that y varies partially because of variation in x_1 and partially because of variation in x_2, we call b_1 and b_2 the partial regression coefficients of y on x_1 keeping x_2 constant and of y on x_2 keeping x_1 constant. The normal equations corresponding to the least-squares plane for the y-, x_1-, and x_2-coordinate systems are

$$\sum y = na + b_1 \sum x_1 + b_2 \sum x_2$$
$$\sum x_1 y = a \sum x_1 + b_1 \sum x_1^2 + b_2 \sum x_1 x_2 \tag{5-42}$$
$$\sum x_2 y = a \sum x_2 + b_1 \sum x_1 x_2 + b_2 \sum x_2^2$$

The solution of this system of three simultaneous equations gives the values of a, b_1, and b_2 for Equation (5-38) and is referred to as y on x_1 and x_2. This is a regression plane, but more complicated regression surfaces can be imagined with four-, five-, . . . dimensional space. The problem given earlier as a polynomial has been solved again by multiple linear regression. Table 5.2 shows the procedure to distinguish the gross product in manufacturing, $\$10^9$, the index of output per man-hour, and the average compensation of production workers per man-hour. These were assumed to be the influencing variables.

Perhaps the true relationship does not conform to the multiple linear regression model of Equation (5-34) and is instead

$$y = (ax_1^{b_1})(x_2^{b_2})(x_3^{b_3}) \cdots (x_n^{b_n}) \tag{5-43a}$$

TABLE 5.2

MULTIPLE LINEAR REGRESSION

Gross Product in Manufacturing 10^9 Dollars, y	Index of Output Per Man-hour, x_1	Average Compensation of Production Workers Per Man-hour $1, x_2	Computations Required in Solution for y on x_1 and x_2
92.6	81.5	1.48	$\sum y = 17,563$
102.0	83.7	1.64	$\sum y^2 = 20,831,089$
105.0	86.1	1.74	$\sum x_1 = 15,201$
111.9	87.6	1.84	$\sum x_1^2 = 15,687,275$
103.8	91.2	1.89	$\sum x_2 = 32$
116.7	96.3	1.96	$\sum x_2^2 = 72$
116.4	95.0	2.07	
117.8	100.0	2.20	$\sum x_1 y = 18,064,654$
109.7	103.9	2.28	$\sum x_2 y = 3860$
121.8	107.2	2.34	$\sum x_1 x_2 = 3357$
122.0	108.8	2.44	
122.0	113.1	2.49	
134.1	118.4	2.57	
138.5	121.6	2.67	
142.0	125.7	2.76	

The normal equations are

$$17,563 = 15a + 15,201b_1 + 32b_2$$

$$18,064,654 = 15,201a + 15,687,275b_1 + 337b_2$$

$$3860 = 32a + 3357b_1 + 72b_2$$

for which

$$a = -4.9114$$

$$b_1 = 1.1563$$

$$b_2 = 1.8831$$

and the multiple linear equation becomes

$$y = -4.9114 + 1.1563x_1 + 1.8831x_2$$

and

$$\log y = a + b_1 \log x_1 + b_2 \log x_2 + \cdots + b_n \log x_n \qquad (5\text{-}43b)$$

where $\log y$ = logarithm transformation of y, which can be handled by multiple linear regression again.

5.3-7 Computer Statements

The arithmetic involved in the simple linear or nonlinear regression equations is digestible, but when an equation is to consider many variables the situation by manual methods (mechanical, electronic calculators) seems impossible. The electronic computers, capable of making short work out of massive computations, make it possible to find linear and nonlinear regression equations of 150 variables or more with all the accompanying statistical measures of reliability and to select the best equation meeting the statistical attributes. This has without a doubt created many benefits to cost engineering, but unfortunately also many pitfalls. Perhaps the greatest danger of error is encountered at the outset when the source and type of data are being selected. Despite the excellence of computation, final results are entirely dependent on the reliability of data used, the interpretation of the computations, and judgment as to reasonableness of the conclusion. Thus an idea of causation must precede statistics.

5.4
MOVING AVERAGES AND SMOOTHING

The estimator may be concerned with periodic observations of labor, material cost, overhead, product demand, and other prices. The characteristics of these observations are described as constant, variable, trend cycle, seasonal, or regular. A *trend cycle* or *seasonal* term suggests a *time series* to a set of observations taken at specific times. Examples of time series are the total annual production of aluminum in the United States over a number of years, the monthly demand for motorcycles, and the total of monthly costs of expenses in a manufacturing department. Interactions and errors of data and collection prevent the theoretical derivation of a nonempirical equation that describes a process underlying the observed data. In lieu of that, the estimator chooses an empirical graph that approximates local segments of the observed time series to forecast future events. The simplest of the time-series cases is the algebraic model using the mean. For the trivial case of constant mean or nearly so over the time interval for which the forecast is required, the dependent variable is nonsensitive. Samples of pictorial graphs where time is the abscissa and a response is the ordinate are given by Figure 5.7. The simple case, constant with no trend, has already been discussed. The linear model with a trend is found more widely than the quadratic and exponential models.

Movements are generally considered to be cyclical only if they recur after constant time intervals. An important example of cyclical movements are the so-called business cycles representing intervals of boom, recession, depression, and business recovery. Seasonal movements are well known and refer to identical or nearly identical patterns which a time series appears to follow during corresponding months of successive years. These events may be illustrated by peak summer production preceding the Christmas demand. Certain of these effects are sometimes superimposed on other effects, for example, Figure 5.7(f), where a linear and a cyclic pattern are superimposed.

Consider the following data:

PLASTIC SHEET COST DATA

Years Ago	Price ($/100 lb)
6	$60.20
5	60.50
4	68.70
3	60.24
2	60.55
1	62.32
Now	75.71

The analyst may assume that the computation of moving average at any single point in time should ideally place no more weight on current observations than those achieved some time previously. This is the major logic for the moving average. A reasonable estimate, given by the average price per plastic sheet, is $64 and the forecast for any future observation could be that same value. The actual average of N most recent observations computed at time t is given by

$$M_a = \frac{x_t + x_{t-1} + \cdots + x_{t-N+1}}{N} \qquad (5\text{-}44)$$

with the restriction that the number of terms in the numerator be equal to N and x_t be the latest term added. Another arrangement, of course, is to use the most recent three, four, or five observations and divide the sum by 3, 4, or 5. For example, the model can be arranged to another form

$$M_a = M_{t-1} + \frac{x_t - x_{t-N}}{N} \qquad (5\text{-}45)$$

FIGURE 5.7

TYPICAL TIME-SERIES MODELS

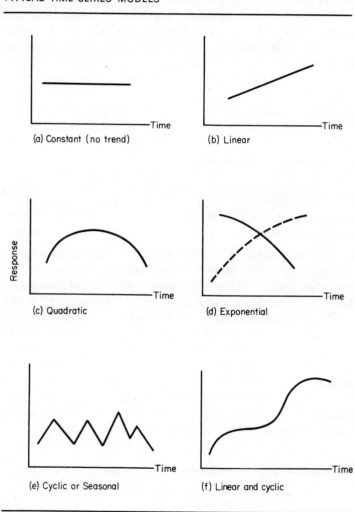

(a) Constant (no trend)

(b) Linear

(c) Quadratic

(d) Exponential

(e) Cyclic or Seasonal

(f) Linear and cyclic

for t, $N = 1, 2, 3, \ldots$, and $t > N$. If a 3-year moving average were required, the data would be arranged as follows:

Date t	Data x_t	3-Year Moving Total	3-Year Moving Average
1	60.20	.	.
2	60.50	.	.
3	68.70	189.40	63.13
4	60.24	189.41	63.14
5	60.55	189.43	63.14
6	62.32	187.30	62.43
7	75.71	192.46	64.15
8	.	.	.
.	.	.	.
.	.	.	.

It is seen that the matter of computing a moving average is simple and sufficiently straightforward for computer-processing equipment or for manual computation.

One of the difficulties with moving averages is that the rate of response is sometimes difficult to change. The rate of response is controlled by the choice of N of the observations to be averaged. If N is arbitrarily chosen large, the estimate is stable. If N is selected small, fluctuations due to random error or some other cause can be expected. The estimator is able to take advantage of these properties, for if the process is considered constant, he may wish to choose a large value of N to have accurate estimates of the mean. However, if the process is fluctuating, small values of N provide faster indications of response.

For most estimating problems some type of moving average is desired that reflects historical and current trends. A smoothing function may be defined as

$$s_t(x) = \alpha x_t + (1 - \alpha)s_{t-1}(x) \qquad (5\text{-}46)$$

where $s_t(x) =$ smoothed value of the function
$\alpha =$ smoothing constant, $0 \le \alpha \le 1$.

The α is like but not exactly equal to the fraction $1/N$ in the moving average method. Whenever this operation is performed on a sequence of observations, it is called exponential smoothing. The new smoothed value is a linear combination of all past observations. Statistically speaking, the expectation of this function is equal to the expectation of the data, which is its average. Exponential smoothing is considered accurate, computations are simple, and the computer file of historical information is shortened from $N - 1$ past observations to only one word $s_{t-1}(x)$. When the smoothing constant α is small, the function $s(x)$ behaves as if the function is providing the average of past data. When the smoothing constant is large, $s(x)$ responds rapidly to changes in trend. While no precise statements can be made regarding this smoothing, the following generally describes the effect of smoothing constant on time-series data:

Drift in Actual Data	Variation in α-Values		
	Small $\alpha \approx 0$	Little $\alpha \approx 0.5$	Large $\alpha \approx 1$
None	None	None	None
Moderate	Very small	Small	Moderate
Large	Small	Moderate	Large

As an example of the exponential smoothing for $\alpha = 0.2$, the following is presented:

EXPONENTIAL SMOOTHING

Date, t	Data, x_t	Smoothed Data, $s_t(x) = 0.2x_t + 0.8s_{t-1}(x)$
.	.	.
.	.	.
.	.	.
10	.	67.38
11	63.2	66.54
12	68.3	66.89
13	65.7	66.66
14	78.4	69.00
.	.	.
.	.	.
.	.	.

For exponential smoothing there are initial conditions that must be established, as a previous value of the smoothing function s_{t-1} is required. If data exist at the time one begins to use exponential smoothing, the best initial value is the average of the most recent end observation $s_{t-1} = M_{t-1}$. If there are no past data to average, smoothing starts with the first observation, and a prediction of the average is required. The prediction may be what the process intended to do—the reduction in labor cost of a new product or the increment of capital investment per module of long-term plant additions, for example. These predictions can also be based on similarity with other processes that have been observed for some time. If there is a great deal of confidence in the prediction of initial conditions, a small value of the smoothing constant, $\alpha \longrightarrow 0$, would be satisfactory. On the other hand, if there is very little confidence in the initial prediction, it is appropriate to have α as a larger value, $\alpha \longrightarrow 1$, so that the initial conditions are quickly discounted. This argument is the counter to the argument about flexibility of response to a change of the process. If the estimator believes that the real process is like his prediction, there is little if any reason to have a change. On the other hand, the contrary viewpoint would have a quick response between the prediction and the real process.

The estimator has a need to know time-series data, as a misjudgment about future costs, profits, sales, and the like can be expensive and make cost estimating ineffective. As an illustration of long-term time-series analysis Table 5.3 provides the early years of annual production of aluminum from 1896. The production rate from this period increased some 2900 times over initial production. The table provides a 9-year moving average and an adjusted production rate. Figure 5.8 plots the actual production and the adjusted 9-year moving average. Semilog coordinates are necessary to show the advance in the production rate. This is a long-term analysis and conjecture on it would be more certain than if only a few recent years were viewed.

Cycles are another property of time series. The cycles in the aluminum production picture are calculated by the ratio

$$\text{Normalized production} = \frac{\text{actual production}}{N\text{-year moving average}} \qquad (5\text{-}47)$$

TABLE 5.3

ALUMINUM PRODUCTION IN THE
UNITED STATES (SHORT TONS)

Year	Tons of Production	1896, 9-Year Moving Total	9-Year Moving Average	Normalized Production
1896	729			
1897	1,289			
1898	1,541			
1899	1,931			
1900	2,644	22,098	2,455	1.077
1901	2,873	27,326	3,036	0.946
1902	3,027	33,354	3,706	0.817
1903	3,562	39,976	4,442	0.802
1904	4,502	43,385	4,821	0.934
1905	5,957	55,282	6,142	0.970
1906	7,317	70,110	7,790	0.939
1907	8,163	86,281	9,587	0.851
1908	5,340	103,622	11,514	0.464
1909	14,541	122,760	13,640	1.066
1910	17,701	145,790	16,199	1.093
.	.	.	.	.
.	.	.	.	.
.	.	.	.	.

$$\text{Normalized production} = \frac{\text{actual production}}{\text{9-year moving average}}$$

FIGURE 5.8

ALUMINUM PRODUCTION IN THE UNITED
STATES FROM 1896

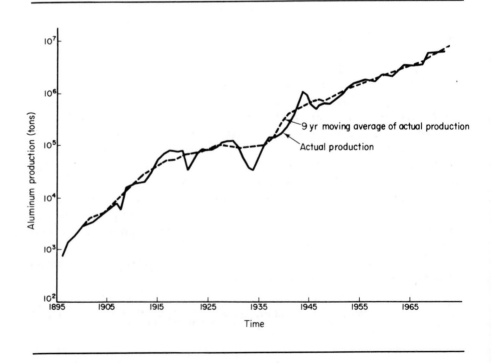

FIGURE 5.9

CYCLICAL AND IRREGULAR VARIATIONS IN
ALUMINUM PRODUCTION

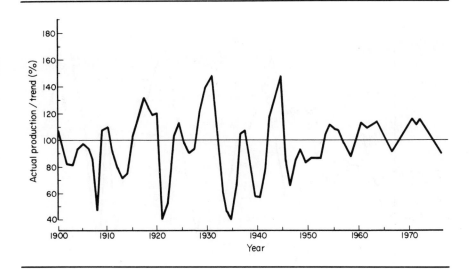

This ratio is provided in Table 5.3, and the plot of these values is given in Figure 5.9 for $N = 9$. The reader may want to associate war, peace, boom, and bust time periods to this graphical analysis and confirm this historical pattern.

Cycles may be interpreted either of two ways: as deviations from established long-term trend or as significant fluctuation due to some time-series effect. In the case where the cycles are interpreted as deviations from a trend, the peaks and troughs are normally referred to as errors from the estimate and are caused by a collection of unknown factors. In the aluminum case, cycles can be treated by an obvious historical interpretation. However, dependence on long-term data, such as industry-wide sales for a particular product, can also be faulty. For a long time the sale of motorcycles was an exclusive U.S. market, but when foreign competition introduced the lightweight scooter, sales were accelerated. Conditions of technology, competition, and advertising have altered the existing scene. Table 5.4 and Figure 5.10 picture the monthly rise

TABLE 5.4

THOUSAND MOTORCYCLE SALES IN THE
UNITED STATES

	3 Years Ago	2 Years Ago	Last Year	This Year
January	46	45	60	54
February	62	78	91	89
March	78	111	121	132
April	99	125	145	154
May	124	154	172	180
June	118	132	161	155
July	102	122	142	135
August	96	119	139	
September	75	93	111	
October	51	82	96	
November	39	51	62	
December	68	96	133	
Total	950	1208	1433	1584 (projected)

FIGURE 5.10

ACTUAL AND SHORT-TERM TREND LINE

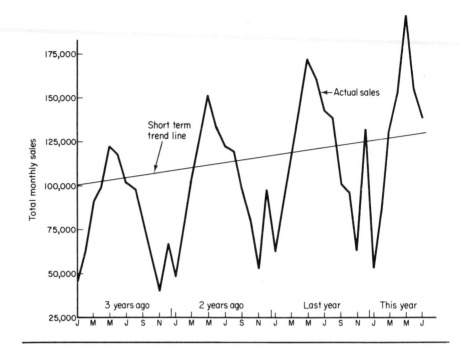

and fall of the cycles about an increasing linear demand. The high point for annual sales occurs during March (purchased for the summer months ahead) and to a lesser magnitude during the month of December. The least-squares line for these data is total monthly sales = $105,070 + 1392X$, where X is the base month 0 at October 2 years ago.

Months following this time reference would be 1, 2, 3, . . . , while months previous would be $-1, -2, -3, \ldots$. The historical data conclude with July of the present year, and the time index for the next month of August would be 22. The demand would be August sales = $105,070 + 1392(22) = 135,694$ motorcycles. Noticing the deviation of the cycles from the trend line, little faith can be given to this estimate for August. In view of the monthly variation, adjustments are called for. The following discusses one method whereby arithmetic adjustments can be made to account for the seasonal or monthly pattern. The quarterly sales are converted to percentage of the yearly sales.

PERCENTAGE OF TOTAL YEARLY SALES
BY QUARTERS

	Quarter				
	First	*Second*	*Third*	*Fourth*	*Total*
3 Years ago	19.4	35.6	28.5	16.5	100
2 Years ago	19.4	34.0	27.6	19.0	100
Last year	19.0	33.6	27.4	20.2	100
Total	57.8	103	83.3	55.7	300
Average	19.3	34.3	27.8	18.6	100
Range	$-0.3, +0.1$	$-0.9, +1.3$	$-0.4, +0.7$	$-2.1, +1.6$	

The first quarter averages 19.3% of the year's sales with a range of −0.3 and +0.1%. If the range is assumed to represent random fluctuations, the analyst may wonder if this randomness can account for the difference between the first quarter and the fourth. If the analyst compares the first quarter against the second quarter, he has greater justification to suppose a seasonal pattern of sales. The logic of a March quarter sales peak for a product that is essentially a summer vehicle could overcome timidity in view of a lack of data. However, if the range of the data were, say, to illustrate a point, 10% either way for the four quarters, the analyst would be in a weaker position to put much credence in seasonal factors. Nonetheless, the differences are assumed to be relevant, and the next step is to translate the percents into factors. For example, the first quarter averaged 19.3% of the year, while an average quarter would be 25% of the year. The percents are translated to a base of 1.0 by multiplying by the number of periods in a year. (If the comparison is on the basis of 3 months, multiply by 4; if on months of the year, by 12, and so forth.) The first quarter with an average of 19.3% of the year would have a seasonal factor of $4 \times 0.193 = 0.772$:

	Quarter			
	First	Second	Third	Fourth
Seasonal factor	0.772	1.372	1.112	0.744

The quarterly sales figures would be converted to the average quarter by dividing by 0.772, 1.372, 1.112, and 0.744, respectively. The next step is the incorporation of exponential smoothing, which uses the previous formula (5-46) of

$$\alpha(\text{adjusted sales}) + (1 - \alpha)(\text{previous smoothed value})$$

If the adjusted sales were 249,000 units for the second quarter and the previous smoothed value was actually 216,000 units, the forecast (with an α of 0.6) for the second quarter is $0.6 \times 249,000 + 0.4 \times 216,000 = 236,000$ units. Although the selection of α has been discussed, it is worthwhile for any real-world problem to forecast sales for a sample of items using different α-values and to compare the results. The value which results in the lowest absolute forecast error is superior.

After the parameters have been chosen (α-value and seasonal factors), the forecast can be made. The actual sales are posted several periods in the past to provide a basis for smoothing. They are seasonally adjusted by dividing by the seasonal factor, and the smoothing technique is applied to give the next forecast value. The forecast is then readjusted to put the seasonal factor back in by multiplying by the factor. The process is continued for each period as sales are reported. Table 5.5 uses $\alpha = 0.6$. The figure of 216,000 units for the first quarter for "3 Years Ago" was provided by the use of the least-squares equation for the previous 3 months. In starting up the system, the initial figure can be based on the average of the previous 3 or 6 months or another value acceptable to the estimator.

A typical calculation for the adjusted sales of the second quarter for "This Year" is given as

actual sales for second quarter = 489,000 units

adjusted sales = actual sales ÷ seasonal factor for
the current period

TABLE 5.5

FORECAST USING EXPONENTIAL SMOOTHING
AND SEASONAL ADJUSTMENT
(100 Units of Motorcycles)

	3 Years Ago				2 Years Ago				1 Year Ago				This Year			
	First	Second	Third	Fourth	First	Second	Third	Fourth	First	Second	Third	Fourth	First	Second	Third	Fourth
Seasonal factors	0.772	1.372	1.112	0.744	0.772	1.372	1.112	0.744	0.772	1.372	1.112	0.744	0.772	1.372	1.112	0.744
Actual sales	186	341	273	158	234	411	334	229	272	478	392	291	275	489		
Adjusted sales	241	249	246	212	303	300	300	308	352	348	353	391	356	356		
Smoothed value[a]	216	236	242	224	271	289	295	303	332	342	349	374	363	359	360	
Forecast sales	167	324	270	167	209	397	330	226	256	469	389	278	280	493	401	
Error	−19	−17	−3	+9	−25	−17	−4	−3	−16	−9	−3	−13	+5	4		

[a] $\alpha = 0.6$.

$$\text{adjusted sales} = 489,000 \div 1.372 \doteq 356,000 \text{ units}$$
$$s_t(x) = 0.6(356,000) + 0.4(363,000) \doteq 359,000$$
$$\text{forecast sales for second quarter} = \text{smoothed value} \times \text{seasonal factor}$$
$$= 359,000 \times 1.372 \doteq 493,000 \text{ units}$$

The comparison to the actual sales is fortunate at this point since the deviation approximates 400 for this period. Actually, up to this period, the calculations did not provide forecasts, as there were real-life data for a comparison. The adjusted seasonal forecast can now attend to a legitimate forecast for the forthcoming third period of "This Year":

$$\text{smoothed value} = \alpha(\text{second period smoothed value})$$
$$+ (1 - \alpha)(\text{first period smoothed value})$$
$$= 0.6(359,000) + 0.4(363,000) \doteq 360,000 \text{ units}$$
$$\text{forecast sales} = 360,000 \times 1.112 = 401,000 \text{ units}$$

In any real problem it is advisable to keep a running count of the magnitude and sign of the error to alert the estimating apparatus to deficiencies in the planning.

5.5
COST INDEXES

A *cost index* provides a comparison of cost or price changes from year to year for a fixed quantity of goods or services. It permits the estimator a means to forecast the cost of a similar design from the past to the present or future period without going through detailed costing. Provided the estimator uses discretion in choosing the proper index, a reasonable approximation of cost should result. Extrapolation through time-series analysis of cost indexes is possible for future periods. Index numbers have been used for these purposes for a long time. An Italian, G. R. Carli, devised the index numbers about 1750. He used index numbers to investigate the effects of the discovery of America on the purchasing power of money in Europe.

Index numbers are useful in other respects. With time and cost for estimating usually scarce, the cost engineer is forced to make immediate use of previous designs and costs which are based on outdated conditions. Because costs vary with time due to changes in demand, economic conditions, and prices, indexes convert costs applicable at a past date to equivalent costs now or in the future. A cost index is merely a dimensionless number for a given year showing the cost at that time relative to a certain base year. If a design cost at a previous period is known, present cost can be determined by multiplying the original cost by the ratio of the present index value to the index value applicable when the original cost was obtained. This may be stated formally as

$$C_c = C_r\left(\frac{I_c}{I_r}\right) \qquad (5\text{-}48)$$

where C_c = present or future or past cost, dollars
$\quad\;\; C_r$ = original reference cost, dollars
$\quad\;\; I_c$ = index value at present or future or past time
$\quad\;\; I_r$ = index value at time reference cost was obtained

In selecting an index to upgrade an estimate, the estimator should consider the design, region, elements of the index, and the individual elements indicated in the original estimate. If major items are ignored in the index as compared to the estimate, adjustments in the composition of the index are necessary. An index seldom considers all factors such as technology progress or local and special conditions.

Consider the following example. Construction of a 70,000-square-foot warehouse is planned for a future period. Several years ago a similar warehouse was constructed for a unit estimate of $12.50 when the index was 118. The index for the construction period is forecast as 143, and construction costs per square foot will be

$$C_c = 12.50\left(\frac{143}{118}\right) = \$15.15$$

While many indexes are published and have become accepted, their construction, alteration, and application are worthy subjects because the cost engineer needs to recognize that it may be better for the firm to develop its own index. Arithmetic development of indexes fall into several types: (1) adding costs together and dividing by their number, (2) adding the cost reciprocals together and dividing into their number, (3) multiplying the costs together and extracting the root indicated by their number, (4) ranking the costs and selecting the median value, (5) selecting the mode cost, and (6) adding together actual costs of each year and taking the ratio of these sums. The weighted arithmetic method is the most popular. Other versions incorporating refinements are also popular.

A cost *relative* of n items is

$$I_c = \frac{(C_{11}/C_{01}) + (C_{12}/C_{02}) + \cdots + (C_{1n}/C_{0n})}{n} \qquad (5\text{-}49)$$

where C_{1n} = cost of nth item for the first year. Year zero would be the base year. If three prices, say sugar, wheat, and beef, rise 4, 4, and 10%, their average rise is 6%, and the index number is 106 compared with the original price level of 100 taken as a base of comparison.

Equation (5-49) treats all items equally, and consequently the index is called simple. One item may be considered more important as though it were two or three items, thus giving it two or three times as much weight as the other item. A weighted approach is

$$I_c = \frac{W_1(C_{11}/C_{01}) + W_2(C_{12}/C_{02}) + \cdots + W_n(C_{1n}/C_{0n})}{W_1 + W_2 + \cdots + W_n} \qquad (5\text{-}50)$$

where $W_{1,2,\ldots,n}$ = weight 1, 2, and n of items 1, 2, and n. Successive years would substitute values for $C_{m1}, C_{m2}, \ldots, C_{mn}$ in the numerator, and calculation would then be relative to the base year, 0 in our case. While the formulas are simple enough, determination of engineering indexes bears little resemblance because of the variety, number, and complications involved.

Several characteristics distinguish indexes. In the construction of the index, there is a choice in the selection of the parameters for the index. Wholesale prices or retail prices, wages or volume of production, proportion of labor to materials, and the number of separate statistics used are typical alternate choices. Indexes apply to a place and time, i.e., period covered or the region considered, base year, and the interval between successive indexes, yearly or monthly. Additionally, indexes are varied as to the compiler and sources used for data. A variety of objectives create diversity for the many indexes.

While the formulas suggest a straightforward procedure, the finding of weights and their calculation is not so simple. Consider the following index:[2]

Item	Unit	Unit Cost	Quantity	Dollars	% Total
Steel	lb	$ 0.015	2500 lb	$37.50	38
Cement	bbl	1.19	6 bbl	7.14	7
Southern pine	MBF	28.50	1088 fpm	17.10	17
Labor	$/hr	0.19	200 hr	38.00	38
				$99.74 (call $100)	100

Successive years would substitute different unit costs, and a recent year exhibited the following:

Item	Unit	Unit Cost	Quantity	Dollars	% Total
Steel	lb	$ 0.062	2500 lb	$ 155.00	13
Cement	bbl	4.03	6 bbl	24.18	2
Southern pine	MBF	255.07	1088 fpm	153.04	13
Labor	$/hr	4.34	200 hr	868.60	72
				$1200.82 (call 1201)	100

During this period the material component increased by 5.38 times, while common labor rose 22.86 times. While 200 man-hours of labor might have been required in 1913 to place the quantities of lumber, cement, and steel, they were not found necessary when the index was 1200 in view of technological progress. Few, if any, jobs run 72% labor and 28% material. On the other hand, the compiler is well aware of this, but to provide a consistent service for a wide readership, the compiler chooses to report this particular index in a constant fashion. This kind of an index is keyed to materials along with a labor component, and is called a basic index.

A different approach for developing an engineering index can be suggested. An index can be developed on a materials-equipment-labor mix. Then one establishes a typical plant and develops the materials-equipment-labor index to chart the changes in costs necessary to reproduce that plant. But do various plants resemble one another? Within certain categories, the answer is yes.

[2]*Engineering News Record*, 1913. Note the labor cost of $0.19 per hour.

An ammonia plant and a polypropylene plant have in common piping, pumps, electric motors, and tanks, to name a few. The compiler[3] of this kind of index would canvas literally over 100 projects and introduce changes in productivity of about 2.5% per year. This rate is applied to all wage and salary costs.

A variety of cost indexes is available to the estimator. In building construction alone, at least 11 major indexes are compiled in the United States covering from 1 to 17 types of buildings from 1 to over 200 different locations. Some of these indexes are available at a charge from their compiler, while others are published regularly. A classification of some of the more common indexes for construction would be as follows[4]:

1. Industrial buildings in specific areas:
   a. Aberthaw (1 type—New England).
   b. Austin (1 type—central and eastern United States).
   c. Fruin-Colon (5 types—St. Louis, Mo.).
2. General buildings in specific areas:
   a. Fuller (eastern cities).
   b. Smith, Hinchman and Grylls (Detroit, Mich.).
   c. Turner (eastern cities).
3. Various types of buildings in many areas:
   a. American Appraisal (4 types in 30 cities).
   b. Boeckh (10 types in 57 areas).
   c. Campbell (5 types in 17 areas).
   d. Dow (17 types in 237 areas).
   e. Marshall and Stevens (4 types in many areas).
   f. *Engineering News Record.*

There are other special indexes covering railroad stations, airplane hangars, and utility buildings.[5]

The Marshall and Stevens Building Cost Index has been compiled since 1901. An endeavor is made to compute changing costs of four types of buildings in various parts of the United States where local modifying ratios adjust for the region. The four types of buildings covered are fireproof protected steel, fireproof reinforced concrete, masonry, and frame. The index evaluates normal costs in line with published and established prices of building materials, equipment, and labor.

A government index listing is given by *The Statistical Abstract of the United States*, a yearly publication that includes materials, labor, and construction. A yearly publication of the U.S. Department of Labor is the *Indexes of Output Per Man-Hour for Selected Industries*. This volume contains updated indexes such as output per man-hour, output per employee, and unit labor requirements for the industries included in the U.S. government's productivity measurement program. Each index represents only the change in output per man-hour for the designated industry or combination of industries. The indexes of output per man-hour are computed by dividing an output index by an index of aggregate man-hours. For an industry the index measures changes in the relationship among output, employment, and man-hours.

The Bureau of Labor Statistics publishes monthly *Wholesale Prices and Price Indexes* and covers some 2600 product groupings. The BLS may be the primary national source of information useful in the making of indexes.

[3] *Chemical Engineering*, Feb. 18, 1963, p. 143.
[4] *Cost Engineer's Notebook*, American Association of Cost Engineers, Morgantown, West Virginia, Oct. 1967.
[5] For a detailed summary of cost indexes, see *Engineering News Record, 178*, No. 11 (1967), 87–163.

5.6
TECHNOLOGY FORECASTING

So far we have examined business forecasting. What is frequently overlooked is the matter of technology forecasting, which is different from the usual methods of evaluation of past data but not entirely unassociated. A definition is in order at this point: Technology forecasting is logical analysis that leads to quantitative conclusions about future engineering qualities and properties. It is not the prediction of devices so far uninvented, but the prediction of operational performance. For instance, what will be the thrust in pounds per 1000 pounds of weight of an aircraft gas turbine 10 years from now, or what will be the highway density of cars per highway mile of urban traffic for New York City 20 years from now, or what will be the tolerance of numerical control machine tools? It assumes that progress will eventually materialize for the growth. Technology forecasting considers a very narrow field within engineering, and several analytical and nonanalytical methods are used. To illustrate technology forecasting, a specific problem of tolerance improvement of the numerical control machine tool is discussed. The reader need only generalize to other engineering applications to determine its broad applicability.

Evaluation of technology trends depends on curve plotting. It has more of a Las Vegas flavor, as little attention is devoted to standard statistical regression requirements. The trick in trend evaluation is to find the proper independent variable(s) on which the dependent variable (tolerance in our numerical control machine tool, for example) depends. The gathering of historical data, plotting, and extrapolation are the chief methods. While many dangers are evident, perhaps the major one is insufficient data.

5.6-1
Technical History

To perform technology forecasting, the estimator must first study the implication of technical history as it pertains to his subject. For instance, numerical control (N/C) may be visualized as automatic-control hardware coupled to a dynamic but rigid structure. The history that would be pertinent for our case study of N/C tolerance progress would consist of automatic controls and machine tool structures. The estimator would uncover that machine tool development followed the progress and invention of hardware essentially developed for other industries and needs. For example, cams were first used on machine tools in the late 1800's but were previously applied in the textile trade of spinning and weaving. Even paper-tape machine tool programming devices, a crude forerunner of our 5-, 8-, and 13-channel paper tape, were earlier developed in the textile industry. The player piano is an interesting sidelight of the use of paper tape to control a machine. Feedback controls, both open and closed loop, were associated with the steam engine, and radar and electronics matured to meet the requirements of World War II. The machine tool is associated with automatic controls, and to appreciate the influence of the automatic controls on N/C, it is useful that our technology estimator know these technical history details. Figure 5.11 illustrates in a gross way the growth of automatic controls. In this study the Cumulative Number of Significant Inventions has been plotted against a corresponding year. A few events have been isolated on the curve. For an invention to be significant there must be a continuing technology pattern related to past developments in some way. This historical interpretation separates that which is relevant from that which is not, and technology forecasting requires it for clearer identification of future properties that would influence the technological estimate. Figure 5.11 indicates the rapid progress of this engineering sector within the last 30 years.

FIGURE 5.11

REVIEW OF TECHNICAL HISTORY RELEVANT TO
INVENTION OF AUTOMATIC CONTROL
HARDWARE FOR NUMERICAL CONTROL
MACHINE TOOL

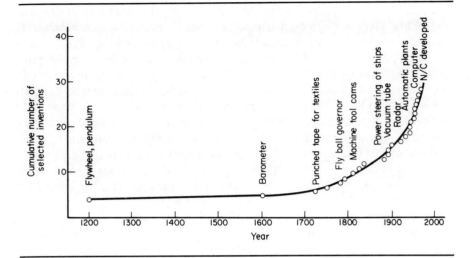

To assume that this same pattern of steep rise will continue indefinitely is beyond the speculation of the estimator at this point. Frequently, progress is subject to the S curve. Flat and slow growth in the early years, rapid middle-life development, followed by stability, decline, or a new S curve are possible life cycles.

Other types of curves could be drawn to reflect the historical trend of N/C machine tool development as influenced by automatic controls. However, it is important to note that in technology estimating the estimator is forced to consider what has happened in the past and to attempt to isolate numerically that which is significant in some quantifiable way. It may be that the accuracy of a numeric measure is imprecise, but it is necessary to try. Books that are devoted to developing specific historical patterns and charts for technological estimating are unavailable. In their stead the estimator must study general technical history books, and a few of these are listed later. Frequently the technology may be so obscure or new that a written technical history may be unavailable. In this case the estimator gathers his own history by correspondence, searching, and a general enlightenment about his field.

5.6-2
Delphi Method

After the historical patterns have been identified, our technology estimator continues to the next step. At this point he would use the *Delphi* method, which is a systematic approach of combining individual judgments to obtain a reasoned consensus. Its unique feature and potential merit is that it requires experts to consider the objections and concepts of other group members in an environment free of persuasion caused by personalities. The experts make a prolonged series of judgments subject to the objections and comments of their peers. The judgments are normally made on a questionnaire form like a multiple-question test. These judgments are analyzed by the estimator. If consistency in judgment can be found, a *significance* may be associated to the technology estimate and ultimately a curve can be drawn through the data points.

Selection and communication with a panel of experts is handled by the estimator. As a moderator he tries to provide an atmosphere free of bias and persuasion. Accordingly, he moderates the technical estimate to avoid the persuasion, prejudices, and misleading leadership that is frequently found with brainstorming. While there is a common ground between Delphi and brainstorming, the panel does not meet face to face, and direct communication is from the estimator to each panel member. This avoids the cross talk and dominant-person problems. Additionally, Delphi attempts to create a more contemplative attitude. The members of the panel need not be from the same company or from the same region or even have the same background. The qualifications of the experts are established by the estimator before he convenes the panel. Communication is consistent and uniform from the estimator to all members of the panel. Historical documents such as Figure 5.11 that the estimator has prepared are given to the panel members for their information early in the educational phase of the panel members, who must be informed of the ground rules for conducting the forecast.

The estimator prepares the statement of the technological estimate in a letter or document and distributes it to the members. Later a questionnaire form is provided to the panel members, who use it for their estimate. Space is provided for the technical estimate, and statement lines are provided on the form for the panel members to support their judgments. Statements by members of the panel supporting their judgment are optional. After the statements are returned the estimator attempts a construction of curves using all data. These curves and a compilation of remarks are returned to the members of the panel. The identity of the members of the panel is usually kept secret; there is no association of a person to a remark. At this point the members are asked if they wish to reconsider their previous technology estimate and resubmit it. This change of opinion may result from the comments of their peers. Accordingly, the members of the panel may or may not choose to alter their judgments. This process of quiet argument could continue until the estimator decides that (1) convergence has been obtained or (2) divergence of opinion will continue. In the first case, the estimator will then plot the data, according to date and the operational characteristic, and he will have completed the technical forecast. For case 2, he may start over again with a new definition of the charge or ignore the lack of correlation and plot the data anyway. Parenthetically, it should be indicated that the data cannot be treated with the usual statistical methods of correlation or variance testing. A lack of homogeneity prevents that path of analysis.

It is generally accepted that N/C machine tools are improving. To what extent this development is being immediately extended to the component is unknown. Design, manpower training, metrology, quality control, and inspection are directly concerned with this progress. A case study illustrates.

A panel of seven experts from different companies and separated by regional boundaries were convened by letter. Each member was provided selected historical data and information such as that in Table 5.6. Following several preliminaries, the question was stated as "Estimate future tolerance for the N/C machine tool for three categories: (1) total component tolerance, (2) machine positioning tolerance, and (3) machine control tolerance." Each panel member was provided a prepared and simple questionnaire for each of the three categories. The panel made a tolerance-time estimate for each category. Instructions to the panel members indicated that a total tolerance rather than a

Tolerance	Year	Statement (Optional)
.	.	.
.	.	.
.	.	.

bilateral tolerance was desired. The forecasts were returned to the estimator, who proceeded to compile the information and see if a unified viewpoint was reached. The data did not lie precisely on the drawn line, and Figure 5.12 represents the compiled opinion of the panel.

FIGURE 5.12

COMPONENT, MACHINE POSITIONING, AND
AUTOMATIC CONTROL CONTRIBUTIONS TO
NUMERICAL CONTROL MACHINE TOOL
TOLERANCE

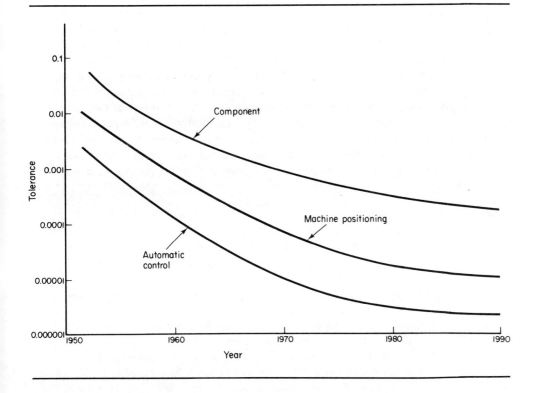

TABLE 5.6

INSTRUCTIONS TO PANEL ON FORECASTING
PERFORMANCE IMPROVEMENT FOR
NUMERICAL CONTROL MACHINE TOOLS

You are being asked to consider several questions about forecasting tolerance performance improvement for N/C machine tools. You may use any methods that you find necessary and convenient in answering the questions. Please note! There are no right or wrong answers to the questions.

The panel selected for this Delphi approach to forecasting of performance improvements in N/C consists of operating types of engineers. They do not design N/C equipment, nor are they control engineers. Operating managers and engineers would know about future innovations, but they are constrained by profit-cost relationships which mute the introduction of advanced N/C equipment. Sales engineers, design engineers, and machine tool builder types would be, perhaps, optimistic. As a consequence, if you choose to seek advice, use only working colleagues and depend mostly on your own speculative judgment.

The Delphi technique is a method of obtaining judgments. It rests on the expectation that experts can forecast, reasonably, when certain technological advances will occur. A selected panel is provided a questionnaire. To the questions the members hazard a guess and perhaps state a reason why. Later, when the several questionnaires have been received, the moderator combines all information and redistributes it again to the panel. The panel members see the written comments from the others and again respond to the questions. They may change their opinion, as a result of the comments of their peers, or they may not. The process is repeated until a consensus is reached or until it becomes the duty of the moderator to essentially average the judgments.

5.7 SUMMARY

As is now evident, forecasting is analysis using imperfect information. Estimating, when coupled with the forecasting processes, takes the analysis of imperfect information and adds to it the ingredients of judgment to provide the setting for decisions. In forecasting processes it should be remembered that data provide the information, although imperfect and incomplete, that pretend to be the situation in the future. Thus the forecasting methods of statistics and analysis are used to uncover the relationships that exist for products, production costs, prices, sales, and technology. It is chiefly these categories for which the estimator has a concern. The application of time series, single and multiple linear regression, correlation, and graphical analysis are the principal statistical techniques used in forecasting for the future. Moving averages involve a consideration of time and fluctuations that occur during these periods. Cost indexes are useful adjustments that permit estimating over lengthy periods of time. Technology forecasting is a method which is concerned with operational performance or cost and is usually dependent on the joint mature judgments of a panel of experts.

SELECTED REFERENCES

For a picture of the practical side of statistics, any of the following would be suitable:

DUNCAN, A. J.: *Quality Control and Industrial Statistics*, Richard D. Irwin, Inc., Homewood, Ill., 1965.

HOEL, P. G.: *Introduction to Mathematical Statistics*, John Wiley & Sons, Inc., New York, 1962.

JOHNSON, N. L., and F. C. LEONE: *Statistics and Experimental Design*, John Wiley & Sons, Inc., New York, 1964.

NEVILLE, A. M., and J. B. KENNEDY: *Basic Statistical Methods for Engineers and Scientists*, International Textbook Company, Scranton, Pa., 1964.

WILLIAMS, E. J.: *Regression Analysis*, John Wiley & Sons, Inc., New York, 1965.

A thorough and up-to-date treatise on moving averages can be found in

BROWN, R. G.: *Smoothing, Forecasting, and Prediction of Discrete Time Series*, Prentice-Hall, Inc., Englewood Cliffs, N.J., 1966.

An understanding of cost indexes is given in

FISHER, IRVING: *The Making of Index Numbers*, Houghten Mifflin Company, Boston, 1922.

Technology forecasting is given in depth and perhaps for the first time in

ARNFIELD, R. V., ed.: *Technological Forecasting*, Edinburgh University Press, Edinburgh, 1969.

BRIGHT, JAMES R.: *Technology Forecasting for Industry and Government*, Prentice-Hall, Inc., Englewood Cliffs, N.J., 1968.

Construction cost multipliers can be found in

Military Construction Pricing Guide, AFP 88-16, Department of the Air Force, Washington, D.C., March 1967.

Technical history books useful to canvas a wide range of topics are

KRANZBERG, MELVIN, and C. W. PURSELL: *Technology in Western Civilization*, Vols. I and II, Oxford University Press, Inc., New York, 1967.

OLIVER, JOHN W.: *History of American Technology*, The Ronald Press Company, New York, 1956.

SARTON, GEORGE: *A History of Science*, Harvard University Press, Cambridge, Mass., 1959.

USHER, ABBOT P.: *A History of Mechanical Invention*, 1st ed., McGraw-Hill Book Company, New York, 1929.

QUESTIONS

1. What are differences between estimating and forecasting? Forecasting and predicting?

2. "Statistics never lie, yet liars use statistics" is a common statement. Discuss.

3. Why are graphical plots preferred initially over mathematical analysis of data?

4. Is cost engineering more concerned with empirical evidence or theoretical data? Illustrate both.

5. Cite instances when a cumulative curve would be necessary.

6. Discuss what regression analysis is. What are its underlying assumptions?

7. What is minimized in a least-squares approach?

8. What is meant by correlation analysis? What does $r = 0$ or 1 imply?

9. Distinguish between correlation and causation. Could you have causation without correlation?

10. What is the purpose of a moving average? How does smoothing relate to a moving average?

11. If the estimator was confident of his past data, would the smoothing constant be large or small?

12. What are the differences between cycles and trend cycles?

13. Define a cost index.

14. Why would an estimator use a cost index? What qualifications are necessary before a building index is chosen?

15. Does technology forecasting deal with forecasting of devices?

16. What are the several methods used in technology forecasting?

PROBLEMS

5-1. The Federal Highway Administration analysis of a hypothetical car has determined the following driving costs per mile. Look at the data and bunch it to arrive at a neater looking curve to determine the time-to-trade decision. Based on Mark Twain's analysis rule, when is the ideal time to trade?

Year	Driving Costs Per Mile (cents)
1	16.0
2	13.8
3	14.4
4	15.0
5	13.6
6	12.4
7	13.2
8	11.0
9	12.4
10	10.4

5-2. A survey of a milling machine department found the following feed per tooth for High Speed Steel end mills cutting aluminum.

Cutter Size	Feed Per Tooth	Cutter Size	Feed Per Tooth
$\frac{3}{8}$	0.0038	$\frac{7}{64}$	0.0007
$1\frac{3}{8}$	0.0044	$\frac{1}{2}$	0.0020
$\frac{1}{2}$	0.0021	$\frac{5}{16}$	0.0040
$\frac{1}{2}$	0.0020	$\frac{1}{4}$	0.0004
$\frac{3}{8}$	0.0018	$\frac{1}{2}$	0.0012
$\frac{5}{32}$	0.0011	$\frac{3}{4}$	0.0052
$\frac{3}{8}$	0.0014	$\frac{5}{8}$	0.0012
$1\frac{1}{8}$	0.0008	$\frac{5}{8}$	0.0022
$\frac{5}{8}$	0.0012	$\frac{3}{4}$	0.0055
$\frac{1}{2}$	0.0030	1	0.0014
$\frac{1}{8}$	0.0003	$\frac{1}{2}$	0.0011
$\frac{1}{8}$	0.0006	$\frac{1}{16}$	0.0001
$\frac{1}{2}$	0.0032	$\frac{1}{4}$	0.0017
$\frac{1}{8}$	0.0003	$\frac{1}{16}$	0.0011
$1\frac{3}{16}$	0.0014	$\frac{5}{16}$	0.0017

For either the cutter size or feed per tooth, (a) Obtain a frequency table, (b) Draw a histogram and a frequency polygon, and (c) Draw a cumulative frequency diagram.

5-3. The U.S. Department of Commerce has determined the value of ship building over a 10-year period:

Year	1	2	3	4	5	6	7	8	9	10
10^6 Value	1467	1216	1360	1400	1518	1678	1818	2160	2200	2460

(a) Graph the data freehand. (b) Find the equation of a least-squares line fitting the data and plot. (c) Estimate the mean value in year 11 and the 90% confidence interval for this estimate. (d) Find the 90% confidence interval for the single estimated value for year 11.

5-4. A price index for general service-labor is given for 7 years:

Year	1	2	3	4	5	6	7
Service Index	100.0	106.0	111.1	117.2	121.3	125.2	128.0

(a) Graph the data. (b) Compute trend values and find a least-squares line fitting the data and construct its graph. (c) Predict the price index for year 8 and compare with the true value 132.6. What is the range for the individual year 8 using a 95% confidence interval?

5-5. The index for the basic union wage rate for carpenters has followed this pattern:

Year	Index
1	100.0
2	104.4
3	107.8
4	112.3
5	117.2
6	122.5
7	132.8
8	145.6
9	159.3
10	171.6

(a) Obtain the regression equation and plot this line together with the data. (b) Find a prediction interval such that the probability is 95% that the index will lie within the interval for year 11. (c) Estimate the 95% confidence interval for the trend slope and intercept values. (d) Would a better estimate result if only year 6 on were used?

5-6. From the following data, determine the values of parameters a and b using the method of least squares. Assume that the data can be plotted as $y = a + bx$.

Year	Retail Price Index, y	Change in Retail Price Index from Previous Year
0	95.1	3.2
1	97.7	2.6
2	98.4	0.7
3	100.0	1.6
4	101.1	1.1
5	102.2	1.1
6	103.5	1.3
7	104.9	1.4
8	106.6	1.7
9	109.7	3.1

Find the 90% confidence interval for the year 10 value. Estimate the 95% confidence interval for the slope.

5-7. Product cost learning has been found to follow the function $T_N = KN^s$, where T_N = unit time for Nth unit, K = man-hours estimate for unit 1, and s = the slope of the improvement rate. Transform this into a log relationship and determine the log regression line. For help, see Table 9.3.

N	Man-hours, T
5	155
8	143
13	137
17	97
25	75

5-8. The life of a tool is determined by testing a tool under standard conditions and watching flank wear. The following tool life data were obtained through laboratory experiments. Find the parameters n and k of Taylor's tool life equation $VT^n = k$ for a test of material, $A151-4140$ steel; tool, K–3H carbide; depth of cut, 0.050 inch; feed, 0.010 inches per revolution; and tool wear limit, 0.005 inch of flank wear.

Cutting Speed Surface feet per minute	Tool Life (min)
400	7, 9, 8
450	6, 8.5
500	5.5, 7.5, 6
550	4, 7, 6
600	5
650	3
700	2.5, 3, 3.4
750	3

5-9. Find the correlation coefficient of the following and assess the significance:

Production	85	99	98	98	109	127	132	134
Employees	410	450	405	400	405	435	460	450

5-10. A company which has been converting manual production methods to automatic means wants to evaluate its effort. Is there correlation for this number of employees and shipment value? What assessment do you make?

10^6 of Shipments	Number of Employees
$573	1573
606	1550
648	1530
720	1550
765	1540
798	1550
848	1560

5-11. Analyze the following data using multiple linear regression:

	Cost/1000 Units		
Weight	Length: 1.50	1.75	2.00
60.5	5070	4770	4540
52.6	4540	4215	3930
42.3	3660	3660	3120
33.3	2390	2390	2100

5-12. The productive output in man-hours is assumed to be a function of the products of the *attitude quotient* raised to some unknown power. Run a regression analysis on the following data and determine these coefficients.

Y = output, man-hours

X_1 = attitude quotient

X_2 = attitude quotient where the model is assumed to be

$$Y = (X_1)^{B_1}(X_2)^{B_2}$$

X_1	X_2	Y
8	8	20
8	7	18
8	6	16
7	8	17
7	7	15
7	6	13
6	8	17
6	7	14
6	6	12

5-13. A company uses extensive amounts of plastic film for one of its automatic processes. This purchased film constitutes the major cost for machine-hour cost.

Budget Period	Unit Price	Budget Period	Unit Price
1	$60.20	11	$63.20
2	60.50	12	68.30
3	68.70	13	65.70
4	60.20	14	78.40
5	60.50	15	81.20
6	62.30	16	82.10
7	75.70	17	82.10
8	73.10	18	84.60
9	80.10	19	83.10
10	77.40	20	82.10

Plot a four-period moving average and actual unit price. Graph normalized price. Using an exponential smoothing function, determine the smoothed data for $\alpha = 0.25$.

5-14. Aluminum production for a 10-year period commencing in 1911 is shown:

Year	Production, 10^3 Tons
1911	19
1912	21
1913	24
1914	30
1915	45
1916	57
1917	65
1918	62
1919	64
1920	69

Determine a 9-year moving total and average and indicate its normalized production. Analyze Figures 5.8 and 5.9 and indicate boom-bust and war periods.

5-15. Plot the following data:

Period	Value	Period	Value
1	7.5	11	4
2	8	12	2.5
3	9.5	13	1.8
4	9.7	14	0.8
5	10	15	0.2
6	9.9	16	0.1
7	9.8	17	0.3
8	9	18	0.9
9	7.6	19	2
10	6	20	4

Using a smoothing function with $\alpha = 0.1$, determine smoothed data and overplot the raw information. What familiar function does the information resemble?

5-16. The cost of a purchased component used in the assembly of a product was analyzed on a time series. Make a table for a 3-year moving total and average price. Plot the average price versus time and describe the movement.

Years Ago	10	9	8	7	6	5	4	3	2	1	Now
Cost (cents/each)	23.2	24.1	26.3	25.7	26.8	27.2	28.0	27.8	28.0	28.5	28.3

5-17. A warehouse construction job is anticipated in 2 years. A similar layup steel-walled warehouse was constructed for a unit price of $18.50 per square foot of wall when the index was 107. The index now is 143, but in 2 years it is expected to be 147. What will be the unit estimate for the construction? What is the bid cost of a warehouse with 700,000 square feet of wall?

5-18. A production man-hour index based on 1957–1959 = 100 is given as 1939, 41.9; 1949, 48.0; 1959, 107.6; and 1969, 205.8. Convert the index to a 1969 basis = 100.

5-19. A company receives an order for a large quantity of drill steel to be used in mining. The delivery point is equally spaced between its plants in the United States and France. Material and transportation costs are assumed to be equivalent for both countries. The efficiency index is 1.3 French man-hours = 1.0 U.S. man-hours for the same job. Indirect cost percentages are 120% France = 75% United States. The direct unit cost (labor only) is $99.65 France = $153.53 United States. Which of the two plants should this company build and ship from?

5-20. Determine the fabrication base cost for a pressure vessel. A pressure vessel is a cylindrical shell capped by two elliptical heads. The base cost estimates the vessel fabricated in carbon steel to resist internal pressure of 50 pounds per square inch with average nozzles, manways, supports, and design size. Design: diameter, 8 feet; height, 15 feet; shell material, stainless 316 solid; and operating pressure, 100 pounds per square inch. Estimating data: Construction 3 years hence with a 5% material increase per year, carbon steel material costs = $8000, factor for non carbon steel material = 3.67, factor for non-standard pressure = 1.05. What is the expected cost?

5-21. A cast-steel foundry uses indexes to price its raw materials. They are purchased over a period of time and stored in open inventory, but they are priced out on a current index basis to remain competitive. Find the indexes for periods 2 and 3 with the reference period as 1. Speculate on the next period index.

Item	1-Ton Finished Casting, Proportion, %	Price, Period		
		1	2	3
1 Pittsburgh scrap steel, No. 1 heavy	80	$37.00	$37.50	$38.00
2 Metal alloy No. 1	15	48.00	48.50	50.00
3 Metal alloy No. 2	5	57.00	56.25	55.00

5-22. A specialty nonferrous company buys metals for inventory using a unit-priced index to determine when to buy significant amounts.

Item	Reference Base-Period Price	Index Quantity	Price, Recent Periods			
			1	2	3	4
Copper, pound	$ 0.5050	4.5	$ 0.5075	$ 0.5045	$ 0.5055	$ 0.510
Lead, pound	0.155	6.0	0.160	0.150	0.155	0.159
Zinc, pound	0.180	5.0	0.180	0.180	0.179	0.182
Tin, pound	1.775	$\frac{1}{2}$	1.779	1.775	1.779	1.850
Gold, troy ounce	70.30	$\frac{1}{8}$	70.25	70.20	70.10	70.05
Silver, troy ounce	1.884	1.0	1.850	1.801	1.846	1.850

Construct an index based on the reference price. Which period would have been the ideal one to buy significant quantities for inventory? Should you buy inventory during the upcoming period?

5-23. On-the-farm costs to produce 100 pounds of pork were determined in six states:

Cost Item	N. Car.	Ill.	Mo.	Kan.	Ind.	Iowa
Feed	$15.69	$11.90	$13.29	$13.21	$11.23	$12.00
Labor	2.89	2.04	1.91	2.00	2.05	2.12
Depreciation	0.67	2.50	1.31	2.28	1.47	0.96
Taxes, insurance, & maintenance	0.40	1.40	1.48	0.48	1.69	2.06
Vet. supplies	0.34	0.78	0.28	0.57	0.23	0.63
Utilities	0.16	0.24	0.20	0.92	0.95	0.50
Miscellaneous	0.58	0.17	0.12	—	0.38	1.46
Total/cwt	$20.73	$19.03	$18.59	$19.46	$18.00	$19.73

Determine state indexes based on an all-state sample reference. Three states, North Carolina, Illinois, and Missouri, reported $22.99, $21.88, and $22.07 average hog prices for the year. Using this sample, find the margin over costs for the other states.

5-24. Basic union wage rates and indexes for major construction trades are given. What is the next year's index for carpenters? For the total trade? A new job is to be estimated, but it will have no painting component. Construct a revised index free of the painting trade and determine next year's index. Discuss: While the

Trade	Weight	Year		
		1	2	3
Carpenter	31.4	100.0	104.4	107.8
		2.04	2.13	2.20
Electrician	13.6	100.0	102.2	104.3
		2.30	2.35	2.40
Laborer	10.1	100.0	106.1	110.9
		1.47	1.56	1.63
Plumber	15.3	100.0	107.6	110.3
		2.23	2.40	2.46
Painter	11.5	100.0	105.7	108.5
		1.76	1.86	1.91
Others	18.1	100.0	105.0	107.9
		1.98	2.08	2.13
Weighted average index		100.0	105.0	107.9

rates relate to the amount paid the tradesman for 1 hour, the indexes do not encompass the efficiency with which that hour is utilized in successive years.

5-25. Construct a "bread-basket" index of these items. The base year is 1. The prices of these items were collected under similar circumstances over a 5-year period.

Item	Price, Yearly					5-Year Total
	1	2	3	4	5	
1. Milk, homogenized, ½ gal	$0.55	$0.45	$0.69	$0.58	$0.65	$ 2.92
2. Ice cream, regular vanilla, ½ gal	0.65	0.69	0.48	0.49	0.89	3.20
3. Eggs, grade A large, 1 dozen	0.41	0.45	0.43	0.34	0.83	2.46
4. Margarine, 1 lb, regular Blue Bonnet Parkay	0.29	0.33	0.34	0.30	0.35	1.61
5. White bread, 1-lb loaf, sliced	0.25	0.25	0.29	0.25	0.41	1.45
6. Instant coffee, 10-oz jar	1.21	1.29	1.49	1.58	1.18	6.75
7. Flour, 5 lb, all-purpose Pillsbury	0.62	0.56	0.58	0.68	0.63	3.07
8. Ground beef, less than 25% fat, 3-lb package	2.04	2.31	2.04	2.64	3.07	12.10
9. Potatoes, U.S., one 5-lb bag	0.40	0.39	0.49	0.59	0.69	2.56
Total	$6.42	$6.72	$6.83	$7.45	$8.70	$36.12

What is the average percentage rise over the 5-year period? What are the major and minor items contributing to this increase? Construct a time-series plot of the index. Determine a least-squares-equation fit for this time series. If these items constitute 3% of the grocery bill for a hypothetical time period and individual, what are the gross dollars lost to inflation?

CASE PROBLEM *Technology Forecasting for a Mobile Production Plant To Process Junk Cars Located in Rural Areas.* The annual number of cars discarded has been increasing with some 9 million junked yearly. It is estimated that there are over 20 million car hulks resting in piles across the nation. Of these, 16 million are in automobile graveyards or in the hands of scrap metal processors, leaving 4 million abandoned throughout rural areas.

These cars comprise over 30 million tons of scrap metal. At a time when metal resources are dwindling, this source of scrap metal becomes economically interesting.

Efforts in the past to utilize this pool of scrap metal have centered largely within the urban areas. The reasons for this are varied. The density of junk automobiles is high. Loading and transporting junked cars are expensive due mostly to the bulk of the automobile. Only about six cars can be trucked on flat-bed trailers at a time, and it is uneconomical to haul them long distances. These economic realities have caused the oversight of junk cars in rural and mountain areas.

Advances in the technology of the scrap industry have made it increasingly attractive to process junk cars in rural areas. Old-style balers are being replaced by smaller more efficient machines, such as the shredders, and new methods of separation have improved the quality of the scrap. Consider the problem of a mobile plant that processes junk automobiles in rural and mountain areas.

One alternative is to mount a plant on railroad cars and take it to rural areas, leaving the plant on a spur where it processes the scrap. This special train can be pulled from location to location as dictated by the supply of scrap. It provides a means for the processed material to be transited to the steel mill quickly and economically. The heart of the plant is a hammermill consisting of a hopper for material and a chamber holding a large rotary shaft. Attached to the shaft are hammers which are free to rotate about their own pivot point. Hammers shred the material until it is fist-sized and passes through a screen at the bottom of the chamber. The shredded material is then removed by the conveyor for further processing. A hammermill can shred some 150 tons of scrap per day. The hammermill shaft is driven by a 3000-horsepower motor, or its equivalent diesel motor-generator set.

Cars must be properly prepared before being processed. Tires, seats, motors, transmissions, and gas tanks must be removed to keep nonmetals and cast iron out of the scrap, maintaining quality. Engines and transmissions are difficult to shred, and the cast iron of the blocks and casings is undesirable for steel purity. Gas tanks are removed for safety precautions.

The production train consists of flatcars for equipment, the crane, the shredder, and conveyors; cabooses for employee accommodations; and the locomotive (supplied and rented from the railroad as needed) and gondolas for scrap. Direct-labor employees are estimated to be one man to operate the crane, one man to inspect automobiles, one man to run the power plant and conveyor motors, and two men to inspect the scrap, sort if necessary, and assist in attaching the conveyor links.

A car ready to be processed has a value of about $6 at the spur and about 30 man-minutes are necessary to prepare a car for the hopper. Each stripped car weighs about 1 ton. Transferring the shredded scrap to a steel plant costs $4 per ton for 100 miles.

Inasmuch as legislation now prevents the railroads from leasing equipment like locomotives or renting or providing a pull service to private entrepreneurs, no estimate is available for occasional movements. A rough fixed cost of the plant is

Hammermill installed	$1,000,000
Crane	50,000
3000-hp electric motor	80,000
Diesel-generator setup	50,000
Conveyor and drive equipment with separator	30,000
Total	$1,210,000

You are appointed as a moderator to forecast the development of this scheme. Use the Delphi technique, and determine the future engineering qualities and properties that can be forecast by the techniques of technology forecasting. List these significant variables. Where would you expect to find resource information for the design of this problem? What technical design features of this proposal bear watching? Consider the commercial justification. Do you think that an operation of this kind requires government subsidy, or would it be privately successful? Using estimates of available scrapped cars, the scrap prices available in your region, and the nearest steel producer, roughly determine the economic prospects for this proposal.

chapter 6

I went home, and to bed, three or four hours after midnight.... An accidental sudden noise waked me about six in the morning, when I was surprised to find my room filled with light.... Rubbing my eyes, I perceived the light came in at the windows. I got up and looked out to see what might be the occasion of it, when I saw the sun just rising above the horizon, from whence he poured his rays plentifully into my chamber.

This event has given rise in my mind to several serious and important reflections. I considered that, if I had not been awakened so early in the morning, I should have slept six hours longer by the light of the sun, and in exchange have lived six hours the following night by candlelight; and, the latter being as much more expensive light than the former, my love of economy induced me to muster up what little arithmetic I was master of, and to make some calculations, which I shall give you, after observing that utility is, in my opinion, the test of value....

In the six months between the 20th of March and the 20th of September, there are

Nights	*183*
Hours of each night in which we burn candles	*7*
Multiplication gives for the total number of hours	*1,281*
These 1,281 hours multiplied by 100,000 the number of inhabitants (of Paris) give	*128,100,000*
One hundred twenty-eight millions and one hundred thousand hours, spent at Paris by candlelight, which, at half a pound of wax and tallow per hour, gives the weight of	*64,050,000*
Sixty-four millions and fifty thousand of pounds, which, estimating the whole at the medium price of thirty sols the pound, makes the sum of ninety-six millions and seventy-five thousand livres tournois	*96,075,000*

An immense sum! that the city of Paris might save every year, by the economy of using sunshine instead of candles.

Benjamin Franklin, "An Economical Project" (apparently written March 20, 1784), in *The Complete Works of Benjamin Franklin*, Vol. VI, edited by John Bigelow, G. P. Putnam's Sons, New York, 1888, pp. 277–283.

PRELIMINARY METHODS

A design is without specific form and shape in the early stages of its evolution. For instance, the design engineers may have progressed through problem definition, concepts, engineering models, and evaluation with only the final design step remaining. The preliminary estimate is requested at some point in the initial evaluation. With a notable lack of hard facts and specific information the cost estimator is asked to provide this first estimate. Using various methods, rules of thumb, and simple calculations, a quick and relatively inexpensive estimate is provided. Obviously the accuracy of the estimate depends on the amount and quality of information and the time available to prepare the estimate. The preliminary estimate may cause the firm to take some sort of action. An estimate, such as an operation cost, product price, project return, or system effectiveness, can have serious financial overtones. More frequent, however, is the case where the preliminary estimate is used to screen designs and aid in the formulation of a budget. They are used by engineering and management to commit or stall additional design effort, or for appropriation requests for capital equipment, or for culling out uneconomic designs at an early point. Although decisions based on the preliminary estimate may not lead to legal obligations or authorization for capital spending, mistakes can be costly by eliminating potentially profitable designs.

We define a preliminary estimate as one which is made in the formative stages of design. At this time a noticeable lack of verifiable information exists. Overlooked in this definition is the accuracy involved, type of design evaluated, the nature of the organization, dollar amount, and the purpose for the estimate. A precondition of accuracy for preliminary estimates cannot be imposed, as special designs or objectives create a unique set of requirements. An estimate involving, say, pennies is no less of a challenge than an estimate involving millions of dollars, for many of the same methods are used at both ends of the dollar scale. Other terms used in practice include conceptual, battery limit, schematic, order of magnitude, and mean preliminary estimate.

In this chapter we shall discuss general estimating methods and point out their advantages and shortcomings. These methods range from experience and judgment as the dominant requirement to ones with a minimum of mathematics.

6.1
CONFERENCE METHOD

The conference method is a nonquantitative method of estimation. It provides a single value or estimate made through experience. In addition to cost or price, other motives such as savings potential, marginal revenue, and the like can be estimated. The method relies on the collective judgment of contrasting differences between previously determined estimates and their associated designs and an unknown but to-be-determined estimate with its design. The procedure, although it has many forms, involves representatives from various departments conferring with estimating in a round-table fashion and jointly estimating cost or price as a lump sum. Sometimes labor and material are isolated and estimated with overhead, distribution, selling expenses, and profit added later through various formula methods by the estimator.

The conference method may also be used within the cost-estimating department. Estimators having specialized knowledge confer on a design and determine a cost figure without counsel from other departments.

The way in which the method is managed depends on the available information. Various gimmicks can be used to sharpen judgment. A *hidden-card* technique has each of the committee experts reveal a personal value. This could provide a consensus. If agreement is not initially reached, discussion and persuasion are permitted as influencing factors. The major drawbacks are the lack of analysis and a trail of verifiable facts leading from the estimate to the governing situation. Although little faith and accuracy can be assigned to the estimate, the lack of method and procedural rigor seldom deters usage.

6.2
COMPARISON METHOD

The comparison method is similar to the previous method except that it attaches a formal logic. If we are confronted with an unsolvable or excessively difficult design and estimating problem, we designate it problem *a* and construct a simpler design problem for which an estimate can be found. The simpler problem is called problem *b*. This simpler problem might arise from a clever manipulation of the original design or a relaxation of the technical constraints on the original problem. Thus we attempt to gain information by branching to *b* as various facts may already exist about *b*. Indeed, the estimate may be in final form, or portions may exist and there need only be a minor restructuring of data to allow comparison. The alternative design problem *b* must be selected to bound the original problem *a* in the following way:

$$C_a(D_a) \le C_b(D_b) \tag{6-1}$$

where $C_{a,b}$ = value of the estimate for designs *a* and *b*
$\quad D_{a,b}$ = design *a* or design *b*

Also, D_b must approach D_a as nearly as possible. We adopt the value of our estimate as something under C_b. The sense of the inequality in Equation (6-1) is for a conservative stance. It may be management policy to estimate cost slightly higher at first and once the detailed estimate is completed with D_a thoroughly explored we comfortably find that $C_a(D_a)$ is less than the original comparison estimate.

    An additional lower bound is possible. Assume a similar circumstance for a known or nearly known design *c*, and a logic can be expanded to have

$$C_c(D_c) \le C_a(D_a) \le C_b(D_b) \tag{6-2}$$

We assume that designs *b* and *c* satisfy the technical requirements (but not the economic estimate) as nearly as possible.

    Consider now an example of comparison estimating. It is desired to decrease the rolling speed of a ball bearing relative to the rotational speed of the shaft. Among other factors the maximum operating speed of a ball bearing depends on its size and the rolling speed of the balls within the raceways. A preliminary design solution is to add an intermediate ring with raceways on its inner and outer peripheries for light radial loads. This would cut the relative rolling speed to approximately half that of the balls in an unmodified bearing. Figure 6.1 indicates a possible arrangement of design A. The adaptation of an additional outer roll of balls achieves many of the similar features required of design A and is known as design B, or Figure 6.2. Velocities are reduced and other technical advantages are achieved. The design requirements for B satisfy most of A. The cost of B can be determined as the outer roll of balls is in many

FIGURE 6.1

PRELIMINARY DESIGN OF BALL BEARING

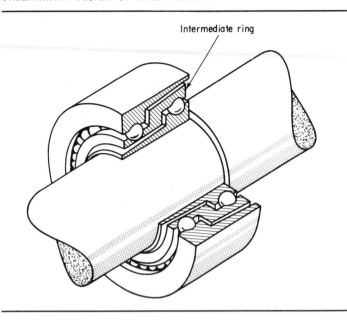

FIGURE 6.2

UPPER COST FOR ESTIMATING PRELIMINARY
DESIGN

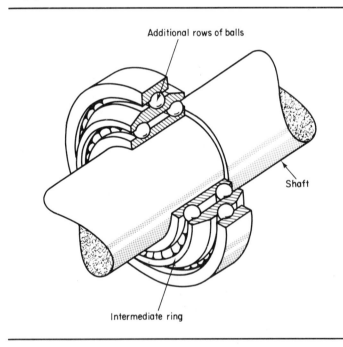

ways similar to single-roll ball bearing technology. The technology for conventional double-row ball bearings, Figure 6.3, has known costs and is our design C. With costs determined for designs B and C, comparison becomes possible.

Precautions which consider up-to-date costs, processing variations, similar production quantities, and spoilage rates are factors which call for insight in picking a specific value for design A between the B and C range.

FIGURE 6.3

CONVENTIONAL DOUBLE-ROW BALL BEARING
WITH KNOWN COST; LOWER COST

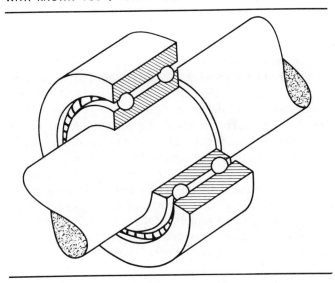

	Cost, Design B	Cost, Design C
0.3937 Bore 1.811 inside diameter Ball bearing	$0.976	$0.921

While there is no clear boundary that distinguishes the various characteristics of estimating methods, one thing is certain. The conference and comparison methods are weak, and invariably those practices can be improved by crude analysis which tends to become progressively more quantitative. Unordered ranking, exclusion chart, and band chart are the simplest methods which use limited amounts or poor-quality data for input.

In unordered ranking, grossly unrelated data are gathered for a *unit evaluation* of some sort. At this point graphing or the use of mathematical regression methods are not a concern. For example, consider the use of packaging materials for protection of electronic cards:

Design Packaging Material Choices	Cost of Square Foot of Surface Protection Per Inch of Packaging Material Thickness
Cardboard scrap	$0.72
Shredded paper	0.25
Aluminum foil and waste	0.81
Egg carton construction	0.46
Foam plastic	0.45
Sawdust	0.19
Plastic encapsulation	2.40
Cardboard layering	0.26
Jute fiber	0.20
Molded wood blocking	0.83

Some of these methods of protection would be unsuitable, such as sawdust or jute fibers, as they would cause a subsequent cleaning operation. Technical judgments naturally reduce the selection. The determination of which one to adopt rests on the various feasible actions.

An exclusion chart is a scatter plot of prices, costs, indexes, or other measured values with an area of the graph free of data points. This excluded area is a rectangular or convex polygon after lines are roughly drawn to circumscribe the data. While the unordered ranking method provided no systematic scale for economic evaluation, the exclusion chart adds specific and numeric scales. The abscissa is designated with a definable measure. The selection of one or several measures for decision making is a difficult choice. Such terms as productivity rate, worth, figure of merit, benefit, or gain are used. Others include horsepower, performance, accuracy, reliability, size, and weight. After a measure is established in a specific way (generally, uniquely quantitative) the measure does not necessarily provide the chart with universally acceptable qualities, as the ordinate may be one of several things: cost, savings, profit, rate of return, or price. The choices for the axes must lead to nontrivial solutions. A simple test of the adequacy is to question whether one design could excel in most of the criteria generated and still not be designated "best." Important criteria are missing if this is so.

Although the excluded area is hopefully free of data, an outlier or occasional points may fall within the exclusion zone. Outlier data may be uncovered on the basis of extraordinary features. The number of outliers depends on the quantity and the screening of data. Selection of data, as was seen with unordered ranking, may ignore quality, location of sold material, demand, performance, or appearance. The application of statistical regression methods to rough data of this kind is pointless. Figure 6.4 is an exclusion chart for small mini bikes where horsepower is used as the horizontal criterion and price represents the vertical axis.

Band charts are extensions of exclusion charts. Figure 6.5 shows the number of man-hours to handle and install standard pipe on a per linear foot

FIGURE 6.4

EXCLUSION CHART SHOWING REGION WHERE
PRICES DO NOT FALL

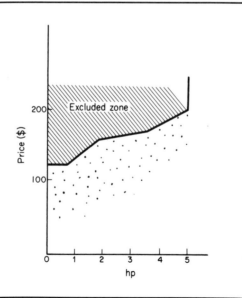

FIGURE 6.5

MAN-HOURS TO HANDLE AND INSTALL
STANDARD PIPE

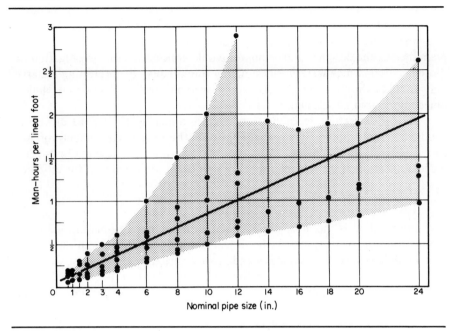

basis.[1] In constructing band charts the estimator draws a representative mean line through the scattered points. No attempt is made to disguise or ignore variation of the points, as all data are retained. There is usually no concern over too much data.

The ball-bearing example can be plotted as a band chart. For various

FIGURE 6.6

UPPER AND LOWER LIMITS ON SPECIAL
BALL-BEARING DESIGN

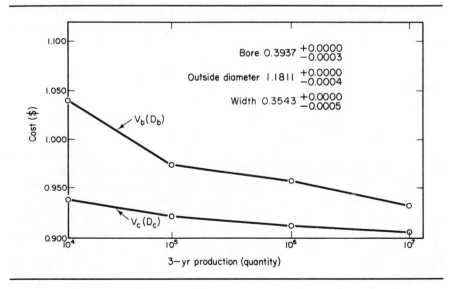

[1]*Data for Estimating Piping Costs*, American Association of Cost Engineers, Morgantown, West Virginia June 1964.

production quantities a band chart like Figure 6.6 can be constructed using the approach previously outlined.

What can be done with this kind of quantitative analysis? It calls for ingenuity to determine an estimate on which to plan, buy, sell, or conduct business using these simplest methods. A tabular display like unordered ranking is a first step. Technological factors, often subtle, hinder an easy solution. Nonetheless, the designer and estimator would concentrate the investigation on those materials which were cheapest. Other things equal, the protecting material that would be the cheapest and that would overcome technical and nontechnical objections would be selected.

As for extracting decisions from exclusion charts, it could be unwise to choose a design and its estimate that fell within the zone where competition had feared to tread. Band charts can be similarly interpreted. These methods force the identification of the variables that are significant to the value of the estimate.

6.4
UNIT METHOD

The unit method is the most popular of the preliminary estimating methods. Many other titles exist that describe the same thing—order of magnitude, lump sum, module estimating, flat rates—and involve various refinements. Extensions of this method lead to the factor-estimating method, which is discussed in the next chapter. Examples of unit estimates are found in all activities. For instance,

- Cost of house construction per square foot of livable space.
- Cost of fabricated components per pound of casting.
- Cost of electrical central power station per kilowatt generated.
- National norm cost of university education per student year.
- Chemical plant cost per barrel of oil capacity.
- Product price per hardware component.
- Factory cost per machine-shop man-hour.
- Cost per mile of highway.
- Mission cost per weapon system.

While typically vague in these contexts, the strongest assumption necessary for their application is that the design to be estimated is like the composition of the parameter used to determine the estimate. Notice that the estimate is *per* something. Data for these estimates are collected from technical literature where private cost data did not exist (an unlikely but possible event), and from government, banks, or the files of cost engineering or accounting. If accounting data are used, they must be recast to be useful for estimating costs. Because of the similarity of this method to the factor method, we defer an example until the next chapter.

Unit estimates are figured easily. Consider a residential house. Using the general contractor's actual total cost of new construction, we divide by the livable square footage or the cubic volume or the like. This average estimate lumps subcontracts, materials, fees, overhead, and profit into one rate.

Some estimators contend that the percentage error [meaning (estimate/actual $- 1$) $\times 100$] of the unit estimate is as good as an estimate using detailed methods. Their reasoning rests on the beneficial happenstance of statistical averaging where errors are compensated for by the average of averages of averages of . . . . When unit estimates are placed side by side with detailed estimates and compared with actual facts, months and years may have passed, the design may have changed between the preliminary and detailed estimate, and basic cost data may have altered. Clear-cut comparisons are difficult to make. The majority of commercial practice uses detailed methods, and this author believes in the superiority of detailed methods for improved accuracy. Indeed, the reader may be aware of examples where one can be dangerously misled by using average values, and the faulty premise of improved accuracy exhibits the following fallacy. Given an arbitrary nonlinear function $f(x_1, \ldots, x_n)$ of random variables $x_1, x_2, \ldots, x_n$, it is usually erroneous to assume that the expected value of the function $f(x_1, \ldots, x_n)$ is equal to the function of the expected values of the random variables or

$$Ef(x_1, \ldots, x_n) \doteqdot f(E(x_1), \ldots, E(x_n)) \qquad (6\text{-}3)$$

6.5
EXPECTED VALUE METHOD

As usually prepared, the estimate represents an "average" concept. It does not reveal anything about the probability of the expected values, however, and uses information which is called certain, or *deterministic*.

Much has been written on the topic statistical decision theory, also called Bayesian analysis. Despite the subject's importance, delving into details is too encompassing for a text of this sort. For this simpler discussion we assume that the estimator can give a probability point estimate to describe each element of uncertainty as represented by the economics of the design. This assignment has nonnegative numerical weights associated with possible events such that if an event is certain its associated weight equals 1. If two events A and B are mutually exclusive, the weight of the event "either A or B" equals the sum of the weights for each of the events. These weights are really a numerical judgment of future events, and the techniques for deriving probability weights are the following: (1) analysis of historical data to give a relative frequency interpretation, (2) convenient approximations like the normal or the equal likelihood distribution and (3) introspection, or what is called *subjective probability*. Subjective probabilities call for judgmental expertise and a pinch of luck. Better success is assured when past data are analyzed; on the other hand, data may be unavailable, and it should be remembered that data are past, while these probabilities should be indicators of the future. Sometimes both past data and a reshuffling of probabilities are jointly undertaken. This type of discrimination is not new to professional cost-engineering practice, as that is what estimating is all about.

Certainty, risk, and uncertainty were concepts introduced in Chapter 1. For the most part estimators have preferred to deal with the simplest case of certainty. Despite this practice, it seldom exists. The category involving risk is appropriate whenever it is possible to estimate the likelihood of occurrence for each condition of the design. These probabilities describe the true likelihood that the predicted event will occur. Formally, the method incorporates the effect of risk on potential outcomes by means of a weighted average. Each outcome of an alternative is multiplied by the probability that the outcome will occur. This sum of products for each alternative is entered in an expected value column, or mathematically for the discrete case,

$$C(i) = \sum_{j}^{n} p_j x_{ij} \qquad (6\text{-}4)$$

where C = expected value of the estimate for alternative i
p_j = probability that x takes on value x_j
x_{ij} = design event

The p_j's represent the independent probabilities that their associative x_{ij}'s will occur with $\sum_{j=1}^{n} p_j = 1$. The expected value method exposes the degree of risk when reporting information in the estimating process.

Consider the following example: An electronics manufacturing firm is evaluating a portable TV. Market research has indicated a substantial market available for a small lightweight set if priced at $85 retail. This implies that the set will have to be sold for approximately $65 to wholesalers. Before the decision can be made to enter the market, several questions need to be answered. Three important ones are: What will be the first year's sales volume in units? How much will the sets cost to produce? What will be the profit? To answer these questions, marketing was asked to furnish an estimate of the first year's sales in units. Subsequently, the cost estimators are requested to provide a total cost per unit. After consumer research, marketing presented the following forecast:

Annual Sales Volume	Probability of Event Occurring
100,000	0.1
150,000	0.1
200,000	0.2
250,000	0.6[a]

[a]Most frequent case.

After inspecting the sales forecast, the cost-estimating group provided the following estimates based on 250,000-unit production:

Cost Per Unit ($)	Probability of Event Occurring
40	0.1
45	0.7[a]
50	0.1
55	0.1

[a]Most frequent case.

The illustration can now be divided into (1) risk not visible and (2) risk visible. Assume that marketing and estimating use the most probable figure from their studies and do not report any uncertainty. The profit is calculated as

$$\text{profit} = (65 - \text{cost})\text{volume}$$

where cost = $45
volume = 250,000 units
profit = (65 − 45)250,000 = $5,000,000

On the other hand, assume that the organization encourages a policy of reporting risk in estimates. The probability numbers are used without the effects of editing. Since there are 4 possibilities for cost and 4 for volume, we can calculate 16 profit possibilities:

$(65 - Cost)Volume$	Joint Probability of Occurrence	Expected Value Profit
$(65 - 40)100,000 = 2,500,000$	0.01	25,000
$(65 - 40)150,000 = 3,750,000$	0.01	37,500
$(65 - 40)200,000 = 5,000,000$	0.02	100,000
$(65 - 40)250,000 = 6,250,000$	0.06	375,000
$(65 - 45)100,000 = 2,000,000$	0.07	140,000
$(65 - 45)150,000 = 3,000,000$	0.07	210,000
$(65 - 45)200,000 = 4,000,000$	0.14	560,000
$(65 - 45)250,000 = 5,000,000$	0.42	2,100,000
$(65 - 50)100,000 = 1,500,000$	0.01	15,000
$(65 - 50)150,000 = 2,250,000$	0.01	22,500
$(65 - 50)200,000 = 3,000,000$	0.02	60,000
$(65 - 50)250,000 = 3,750,000$	0.06	225,000
$(65 - 55)100,000 = 1,000,000$	0.01	10,000
$(65 - 55)150,000 = 1,500,000$	0.01	15,000
$(65 - 55)200,000 = 2,000,000$	0.02	40,000
$(65 - 55)250,000 = 2,500,000$	0.06	150,000
Total	1.00	Profit = $4,085,000

The reader may argue that simple mathematics need not be so complicated. Both cost estimating and marketing could have computed an expected value as $215,000(100,000 \times 0.1 + 150,000 \times 0.1 + 200,000 \times 0.2 + 250,000 \times 0.6)$ and $\$46(40 \times 0.1 + 45 \times 0.7 + 50 \times 0.1 + 55 \times 0.1)$ and $215,000 \times (65 - 46) = \$4,085,000$. In the first case of risk not visible, the total profit was overstated. The calculations for this example require that the cost per unit be independent of volume and that cost be inversely related to volume, which are generally not the case.

6.6 COMPUTER SIMULATION TECHNIQUES

Widely employed in engineering, science, and business, simulation is used to estimate the cost of systems. Simulation is defined as the manipulation and observation of a synthetic model representative of a real design which for technical or economic reasons is not susceptible to direct experimentation. This synthetic model ideally represents the essential characteristics of the real system with the frills excluded. The computer is mandatory in analysis of this sort.

Simulation models are classified into several groupings:

1. Real versus abstract. A real system would be a prototype aircraft or a pilot plant, for example, versus a system of mathematical or logical statements.
2. Continuous versus discrete.
3. Deterministic versus probabilistic. All systems are probabilistic to some degree; however, these systems may be conveniently described by constant value models. In probabilistic situations it is possible to introduce random events such that the various parameters of operation are affected, while in deterministic, or constant value systems, we assume that the value used for the parameter is an ideal approximation for that particular system. Sometimes probabilistic simulation models are termed *Monte Carlo*, which is a time-related simulation tool and mathematically simple for convenient use with digital computers.

4. Machine versus man-machine. The "machine" referred to here is a pure computer simulation for which it is necessary that all eventualities be programmed into the computer. The man-machine simulation allows program interruption by the human and permits him to intervene at strategic points. This division of labor allows the computer to do what it can do well and encourages the ability of man to interpret qualitatively rather than quantitatively.

5. Steady state versus transient.

Although simulation uses mathematics, it is not mathematics per se. In simulation, you *run* problems, not *solve* them as you do in mathematics. The intent is the collection of pertinent data from the experiment as one runs and watches the outcome of many simulation trials. Figure 6.7 is a description of the

FIGURE 6.7

SIMULATION-DIRECT SOLUTION APPROACH TO
COST-ENGINEERING PROBLEMS

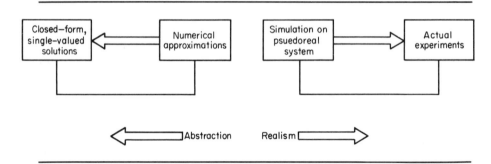

methods that are useful to engineering-economic systems analysis. On the one hand, the actual experiment of a cost-engineering design provides realism. On the other, the orthodox mathematical solution to economic and engineering systems problems remains an abstraction. This "running" and "watching" by simulation is further described by Figure 6.8.

Suppose that two systems are to be judged and that cost is the criterion. Figure 6.9 illustrates four cases where the cost estimates are shown as probability distributions. For case (a) all probable costs of system A are lower than B and there is no difficulty in choosing A. The situation for (b) has a possibility that the actual cost of A will be higher than B. Given that this is a small chance, the estimator would select system A. As the amount of overlap increases, the expected cost estimate C_a may not be given a clear choice. For (c) the average cost estimates are equal, although their distributions are not. Certainly the cost distribution of B is greater and there is an associated risk of having a lower cost than A. On the other hand, A is less variable; the estimator's assessment of cost return and risk will serve as the guide here. In (d) the expected cost estimate of system B is lower but less certain than A. If only the expected value estimate is used in this case, the estimate would likely choose the more desirable alternative B.

Presume that a system cost can be expressed as cost $A = x + y$, where x and y are probability distributions. The distribution of $f(x)$ is found by field data methods and is given as

FIGURE 6.8

TYPICAL FLOW CHART FOR SIMULATION
STUDIES

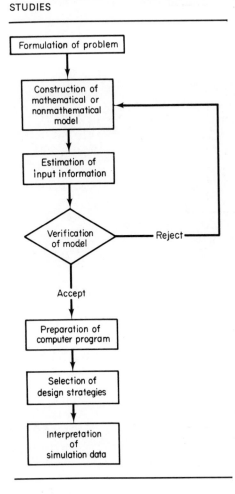

10^6, Cost of x	Frequency of Occurrence
$1.1	0.05
1.2	0.10
1.3	0.15
1.4	0.20
1.5	0.20
1.6	0.15
1.7	0.10
1.8	0.05

These data could be plotted as a discrete frequency curve as shown in Chapter 5; however, for Monte Carlo methods it is expressed as a cumulative probability distribution as in Figure 6.10.

FIGURE 6.9

VARIATIONS OF COST DISTRIBUTIONS

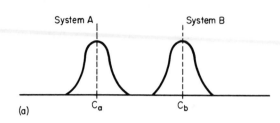

(a)

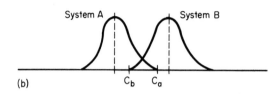

(b)

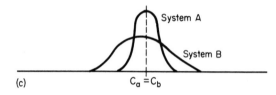

(c)

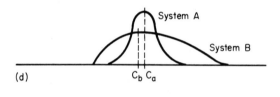

(d)

Assume that $f(y)$ is given by a theoretical distribution with a functional form of

$$f(y) = \tfrac{1}{8} e^{-y/8}$$

To find the cumulative probability distribution we use

$$F(y) = \int_0^{C_y} f(y)\, dy = \int_0^{C_y} \frac{1}{8} e^{-y/8}\, dy$$

$$= 1 - e^{-C_y/8}$$

and
$$C_y = -8 \ln(1 - F(y)) \tag{6-5}$$

Our example requires that the random variable x be added to the random variable y. This requirement is established by the model of the system under study. If a random number is found from a random number table or a computer file and has the properties of being between 0 and 1, it is set equal to the cumulative probability distribution. A random number 0.73 is drawn for design x, and projecting horizontally from the point on the vertical axis corresponding to the random decimal to the cumulative curve, the corresponding value of the random variable is $1,600,000. In a similar way we supply a random number 0.18 to the

FIGURE 6.10

CUMULATIVE FREQUENCY DISTRIBUTION OF A
COST ELEMENT

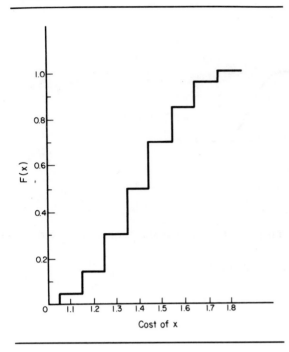

Cost of x

functional model for y:

$$C_y = -8 \ln(1 - 0.18)$$
$$C_y = 1.587 \quad \text{or} \quad \$1,587,000$$

Our solution is cost $A = 1,600,000 + 1,587,000 = \$3,187,000$ and is accepted as a sample value of our cost A. This procedure is repeated many times, any finally a distribution of cost A can be found. This distribution can be tested for its cost estimate average, range, standard deviation, and other statistical properties. In using Monte Carlo techniques to estimate design cost, it is usually required that the input relationships be independent.

6.7
PROBABILITY ESTIMATING

It was seen in the previous two sections that cost was treated as a single-point value with a probability or as a random variable. The single-valued estimate assumed conditions of certainty. Estimators knowing the weaknesses of information and techniques recognize that there are probable errors. The indication that cost is a random variable opens up the topic of probability cost estimating. A *random variable* in statistical parlance is a numerical-valued function of the outcomes of a sample of data. Another approach to single-valued estimating involves making a single estimate and bracketing this estimate for each cost element. This forms the basis for probability cost estimating.

The following procedure is based on a method developed for PERT (Program Evaluation and Review Technique). It involves making a most likely cost estimate, an optimistic estimate (lowest cost), and a pessimistic estimate (highest cost). These estimates are assumed to correspond to the beta distribu-

FIGURE 6.11

LOCATION OF ESTIMATES FOR PERT-BASED
BETA DISTRIBUTION

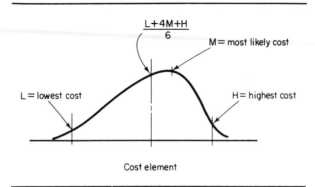

Cost element

tion shown by Figure 6.11. This particular figure is skewed left. Symmetric and skewed-right distributions are also possible.

With the three estimates made, a mean and variance for the cost element can be calculated as

$$E(C_i) = \frac{L + 4M + H}{6} \qquad (6\text{-}6)$$

and

$$\mathrm{var}(C_i) = \left(\frac{H - L}{6}\right)^2 \qquad (6\text{-}7)$$

where $E(C_i)$ = expected cost for element i
L = lowest cost
M = modal value of cost distribution
H = highest cost
$\mathrm{var}(C_i)$ = variance of element i

If several elements are estimated this way and are assumed to be independent of each other and are added together, the distribution of the total cost is approximately normal. This follows from the *central limit theorem*: The mean of the sum is the sum of the means, and the variance of the sum is the sum of the variances. The distribution of the sum of costs will be normal despite the individual shape of cost elements. There must be four or more elements to satisfy the conditions of the central limit law:

$$E(C_T) = E(C_1) + E(C_2) + \cdots + E(C_n) \qquad (6\text{-}8)$$

and

$$\mathrm{var}(C_T) = \mathrm{var}(C_1) + \mathrm{var}(C_2) + \cdots + \mathrm{var}(C_n) \qquad (6\text{-}9)$$

where $E(C_T)$ = expected total cost
$\mathrm{var}(C_T)$ = variance of total cost

Table 6.1 shows an application of this approach. The expected cost is $174.16 and the variance representing probable error is 7.89. Now various statements regarding the interval for the future value can be made, such as some cost will not be exceeded and probability boundaries can be established. While it may be meaningful to build up these refinements, little in the way of a practical algorithm can be shown for the following reasons. The central limit theorem requires the condition of having mutually independent random variables. Costs from within an organization or firm are seldom independent. The elements themselves are bounded at the zero end. The number of elements that can be structured to overcome the underlying assumptions may be large. According to PERT practices, the optimistic and pessimistic costs would be bettered or exceeded only 1 time in 20 if the activity were to be performed repeatedly under the same conditions. The making of preliminary estimates, even allowing for a greater margin of error than is hoped for with these requirements, is difficult at best for most practices of estimating. Only in system estimating have special techniques of probability estimating found much value.

Exercises dealing with probability estimating are given in the chapter problems. Values of the standard normal distribution function, given in Appendix I, are necessary for a solution.

TABLE 6.1
CALCULATION OF EXPECTED COST AND
VARIANCE USING PERT-BASED APPROACH

Cost Element	Lowest Cost, L	Most Likely Cost, M	Highest Cost, H	$E(C_i)$	$var(C_i)$
1. Direct labor	$37	$39	$ 43	$39.33	1.00
2. Direct material	91	94	103	95.00	4.00
3. Indirect expenses	20	21	27	21.83	1.36
4. Fixed expenses	17	18	19	18.00	0.11
				$174.16	6.47

6.8 AN ORDINAL SCALE METHOD

Preliminary designs can be evaluated using a simple scale measure such as first, second, third, . . . or good, better, best. While hardly quantitative, these ranking ideas may someday be developed into a systematic approach to identify the relative position of preliminary designs or intangible things.

An ordinal number identifies the position of an element in the set. The letter b is the second position in the alphabet; thus its ordinal number is 2 or 2nd. The number of elements in a set is called a cardinal number, which is 26 for the alphabet. Each item in simple-order scale ranks above or below every other item. If this ordering can be arranged, we have a solution. In the method of *unordered ranking* a measure existed and when ranked "sawdust" would be a preferred solution on a cost basis.

Ordinal scale methods are interesting. Eventually, research may codify an ordinal scale method useful for preliminary estimating. Now consider an elementary approach to this method.

We first rank the alternatives by comparing their expected value outcome, and then assign an interval value between each alternative. This ranking of alternatives uses ordinal numbers, and the assigning of interval values between the ranks denotes magnitude, which uses cardinal numbers. In one method we assign a value from 1 to 0 to each of the designs in relation to their preference. Each value is to be greater than the sum of the lower values. For instance, suppose that there are three designs, A, B, and C, and that the estimator assigns values of $A = 0.95$, $B = 0.80$, and $C = 0.50$ according to their attractiveness. Now as $A > B + C$ we require B and C to have lower values. The estimator now changes the values of B and C to 0.60 and 0.30 to have $A > B + C$ and $B > C$. In this manipulation we assume A to be more attractive than B and C combined; also, B is better than C.

As an example assume that the estimating department wants to evaluate an idea of a computer service for performing N/C machine tool calculations. The computations would be performed for the parent company and additionally be sold as a computer service to customers. The returns from this venture are savings within the company and a revenue from the sale of the computer service to other firms. An estimate of savings is available within the company, but the company lacks information on how well a program like this will sell. The programming service will be provided for three applications, plastic, sheet metal, and machining. Each type of service will have relative importance to potential customers. The task is to evaluate the ranking of the three basic services and determine which one is least risky. The cost of developing the computerized service and savings was found to be

Application for Numerical Control Machine Tool Computation	Preparation Cost	Savings Per Year Over Conventional Method	Subjective Probability of Program's Internal Success
Plastics	$ 3,000	$2500	0.9
Machining	12,000	7000	0.6
Sheet metal	5,000	2000	0.8

This company requires that ventures of this sort pay back within a discounted 3 years. The basis would have savings minus development cost, and ignoring the time value of money for simplicity, we have

	Application					
	Machining		Plastics		Sheet Metal	
Development cost		−$12,000		−$3000		−$5000
Expected savings	0.6(7000)3 =	12,600	0.9(2500)3 =	6750	0.8(2000)3 =	4800
Net savings		+$600		+$3750		−$200

Marketing information for selling costs and revenues was lacking, although an idea of the ranked success of the three alternatives could be stated. In terms of market potential,

Application	Success Rank	Interval Value
Machining	1st	1.0
Plastics	2nd	0.6
Sheet metal	3rd	0.5

The ranking is adjusted to

Application	Success Rank	Adjusted Interval Value
Machining	1st	1.0
Plastics	2nd	0.5
Sheet metal	3rd	0.4

where $1.0 > 0.5 + 0.4$ and $0.5 > 0.4$.

A revenue estimate for machining is fixed as $20,000 and the product of this base figure with the adjusted interval value gives

Application	Potential Return
Machining	$20,000 \times 1 = \$20,000$
Plastics	$20,000 \times 0.5 = \$10,000$
Sheet metal	$20,000 \times 0.4 = \$8000$

A subjective probability evaluation of the potential return is established as 0.5, 0.9, and 0.8, and the expected marketing revenue is determined relative to the machining application:

Application	Subjective Probability	Expected Marketing Revenue
Machining	0.5	$20,000 \times 0.5 = \$10,000$
Plastics	0.9	$10,000 \times 0.9 = \$9000$
Sheet metal	0.8	$8000 \times 0.8 = \$6400$

The recap of cost and savings and marketing revenues would give

	Machining	Plastics	Sheet Metal
Net savings	$600	$3,750	$−200
Marketing revenues	10,000	9,000	6400
Net revenues after 3 years	$10,600	$12,750	$6200

It is seen that cardinal numbers have been used to assign cost and saving values to the three alternatives where marketing revenues were denoted by their ordinal numbers, machining 1st, plastics 2nd, and sheet metal 3rd, in order of their potential. The final tally of net revenues indicates a gross 3-year revenue of approximately $29,000. Plastics is considered to be the best bet, and if management chooses, it could try this one first.

The method is developmental inasmuch as speculative estimates are made throughout the analysis. However, these estimates are stated openly, and if a management decision were made to improve the estimate, the apparent weaknesses could be remedied. While methods such as these are arbitrary, the benefit of ordinal ranking will create future interest inasmuch as subjective ranking is an ordinary human trait. Wide agreement on how various designs rank in ordinal scale for function and cost is possible.

6.9 SUMMARY

Estimates are constructed on available information. If there is no information, there can be no estimate. Conversely, if the information is complete, an estimate is not needed because actual values are available. The estimator operates within these limits. In this in-between region we have chosen to separate estimating methods into classes of preliminary and detailed.

Preliminary methods are less numeric than detailed methods. Accuracy of the estimate is improved by attention to detail, but balancing this is the speed and cost of preparation which favor preliminary methods. The methods are preliminary because they correspond to a design which is not well formulated.

The conference method, with and without the counsel of others, is useful to determine a cost figure. While seldom accurate, it leads to progressive improvement by simple methods of crude ranking and charting. For systematic evaluation, ranking and charting methods require identification of gross parameters that are sensitive but narrowly successful. The expected value method is concerned with notions about probabilities which must be known in the process. Ordinal scale methods are introduced to show how specialized techniques can fit into the preliminary estimating scheme. Simulation is a computer technique useful in preliminary estimating.

Here we return to the question, How do you select the right method for a particular application? Regrettably, no textbook can supply an infallible set of rules—you will have to rely on experience and continuing analysis.

SELECTED REFERENCES

For a discussion about statistics and expected value concepts, see

> MILLER, IRWIN, and JOHN FREUND: *Probability and Statistics for Engineers,* Prentice-Hall, Inc., Englewood Cliffs, N.J., 1965.

The mathematics of ordinal and cardinal numbers are explained in

> BRANDT, R. N. et al.: *Elementary Mathematics of Sets with Applications,* Malloy Inc., Ann Arbor, Mich., 1958.

For manipulations of ordered numbers and ranking, see

> CHURCHMAN, C. W., and R. L. ACKOFF: "An Approximate Measure of Value," *Operations Research 2* (1954).
> RIGGS, JAMES L.: *Economic Decision Models for Engineers and Managers,* McGraw-Hill Book Company, New York, 1968.

Simulation is discussed in

> DIENEMAN, PAUL F.: *Estimating Cost Uncertainty Using Monte Carlo Techniques,* RM 4854-PR, The Rand Corporation, Santa Monica, Calif., Jan. 1966.
> NAYLOR, THOMAS H., J. L. BALINTFY et al.: *Computer Simulation Techniques,* John Wiley & Sons, Inc., New York, 1966.

A discussion of the methods of probabilistic cost estimating may be found in

ABRAHAM, C. T., R. PRASAD, and M. GHOSH: *A Probabilistic Approach to Cost Estimation*, Report 68-10-001, IBM Corporation, Armonk, N. Y., 1968.

HUSIC, FRANK, J.: *Cost Uncertainty Analysis*, Paper RAC-P-29, Research Analysis Corporation, McLean, Va., 1968.

ZUSEMAN, MORRIS: *The Use of Tchebycheff-Type Inequalities To Bound the Upper Limit of a Cost Estimate*, Research Paper P-478, Institute for Defense Analysis, 400 Army-Navy Drive, Arlington, Va., 1969.

QUESTIONS

1. Discuss the timing of preliminary estimates. Is the timing a precise point in a well-ordered organization?
2. Use the conference method to estimate
   (a) The price for a clean-air car using a turbine drive.
   (b) The cost of Figure 1.2 for a sale of 1000 units.
   (c) The ticket price for a 5000-mile SST one-way trip.
   (d) The cost of a year's college education in 1985.
   (e) The time to dig a trench $2 \times 4 \times 10$ feet in soft clay.
   (f) The improvement in the efficiency of handling mail of a central post office near you after converting to automated methods.
   (g) The price of the standard FHA house in 1985.
3. What advantages can you cite for the conference method? Disadvantages?
4. Assume that design A can be redesigned into designs B and C with known costs. What safeguards can you suggest to assure that A is properly estimated?
5. Referring to Chapter 5, what statistical methods do you think initially appropriate to analyze data used in unordered ranking methods? What is the mean cost and standard error for the packaging material? Is there much or little variability?
6. Construct an exclusion chart for
   (a) The price of used Volkswagen bugs (see your newspaper).
   (b) The cost of various types of home workshop table saws.
   (c) The price of Volkswagen tires.
7. Discuss some substitute unit measures for "cost per square foot of residential construction." What are your local values?
8. When is the "average of averages" a safe measure?
9. What are the ways in which probabilities are determined? What is the distribution of the rolls of one six-sided die considering only the face-up event? Consider two dice, and determine the distribution of numbers from 2 to 12.
10. Point up the human frailities in determining subjective probabilities. Would you think that one would under or overestimate these point probabilities?
11. Construct an ordinal scale for
    (a) Your several girl friends (be careful now!).
    (b) Spending a windfall from an inheritance (which you never expected to receive) for various items where you are uncertain of the amount.

PROBLEMS

6-1. Use the conference method to estimate some product or sample statistic which is familiar to the entire team but whose exact value is unknown, for instance, cost of interstate highway construction per mile or number of automobiles in your town. Then check its value.

6-2. A manufacturer of factory-built houses has collected raw data for flooring and has converted them to a common square-foot base:

Material	Size	Material Cost	Labor Cost
Laminated oak blocks	$\frac{1}{2} \times 9 \times 9$ in.	$0.46	$0.17
Parquet	$\frac{5}{16} \times 12 \times 12$ in.	0.42	0.33
Strip flooring, oak	$\frac{25}{32} \times 2\frac{1}{4}$ in.	0.52	0.25
Resilient tile A Grade	$9 \times 9 \times \frac{1}{8}$ in.	0.12	0.14
B Grade		0.17	0.14
C Grade		0.18	0.14
Vinyl asbestos	$9 \times 9 \times \frac{1}{16}$ in.	0.21	0.14
Softwood, C grade	S4S, 1×6 in.	0.33	0.16
Linoleum	$\frac{1}{8}$ in., plain	0.30	0.15
Slate, irregular flags	Irregular flags	1.20	0.80
Terrazzo	$\frac{1}{4}$ in. thick	0.95	0.79

Order this table. Using only your snap judgment, where would you cost cork tile? Where would you cost carpet? These houses are delivered to the site by company trucks. What are the nonfeasible alternatives?

6-3. A garage mechanic has fashioned a new mailbox design with an interesting feature. When the mailman closes the hinged cover, a spring-loaded flag pops up at the back of the box telling the owner that he has received mail. A survey of popular catalogs has revealed the following data:

Size	Weight	Material	Cost	Features
$14 \times 7 \times 4$	4 lb, 4 oz	Steel	$ 5.90	Holds magazine, wall mount
$13 \times 7 \times 3$	3 lb, 6 oz	Steel	3.90	Holds magazine, wall mount
$14 \times 6 \times 4$	4 lb, 15 oz	Aluminum	8.99	Holds magazine, wall mount
$14 \times 7 \times 4$	4 lb, 12 oz	Steel, aluminum	6.90	Holds magazine, wall mount
$6 \times 2 \times 10$	4 lb, 8 oz	Forged iron	8.10	Letter-sized
$6 \times 2 \times 10$	2 lb, 11 oz	Steel	4.40	Letter-sized
$5 \times 2 \times 11$	2 lb, 11 oz	Steel	2.55	Liberty bell emblem
10×3	1 lb, 8 oz	Brass	5.90	Mail slot
10×3	1 lb	Steel	1.77	Mail slot
10×3	1 lb, 5 oz	Aluminum	5.90	Mail slot
$18 \times 7 \times 6$	8 lb, 8 oz	Steel	6.69	Red signal flag, mount post
$19 \times 7 \times 6$	10 lb	Steel	9.79	Black wrinkle finish
$18 \times 6 \times 9$	10 lb	Steel	9.29	Name holder
$18 \times 7 \times 6$	8 lb, 1 oz	Steel	2.37	Size no. 1
$18 \times 7 \times 6$	8 lb, 1 oz	Steel	4.90	Rural mailbox
$18 \times 7 \times 6$	4 lb	Aluminum	3.85	Rural mailbox
$17 \times 9 \times 10$ unit		Steel	50.00	Apartment type

Examine the data and plot an exclusion chart. What price range or limit would you advise? What do you estimate as the manufactured cost for a 25, 50, or 100% full markup based on your rough estimate?

6-4. A study of past records of installed insulation cost (including labor, material, and equipment to install) for central steam-electric plants revealed the following data:

10^5 Equipment Cost	10^4 Insulation Cost	10^5 Equipment Cost	10^4 Insulation Cost
$ 5.5	$ 3.5	$14.1	$ 9.2
10.7	5.5	14.8	9.3
34	28	15.1	14.1
2	1.4	15.3	13.8
6	6.4	21.3	15.0
1.5	2.1	34.0	15.8
8.1	7.2	24.1	9.8
10.1	6.4	26.0	15.8

Plot a band chart of this information on cartesian coordinates. On log-log coordinates. Which do you think is more suitable for cost-estimating purposes? Major equipment costs have been estimated as $300,000 for a new project. What is the estimate for the installed insulation cost?

6-5. The cost of a residential home with 2000-square-foot livable space and a basement and garage is $36,500. The house dimensions are such that 1200 square feet are on the first floor. The volume including house, garage, and basement is 30,200 cubic feet. Determine unit estimates.

6-6. The delivered cost of major equipment for a fluid process plant is $2 million. Plant cost ratios, sometimes called *Lang factors*, have been developed, and a sample is

3.10 for solid process plants
3.63 for solid-fluid process plants
4.74 for fluid process plants

Estimate the total cost of the plant.

6-7. (a) A company sells two different designs of one item. A study discloses that 65% of its customers buy the cheapest design for $75. The remaining 35% pay $110 for the expensive model. What is the expected purchase price?

(b) A salesman makes 15 calls without a sale and 5 calls with an average sale of $200. What is his expected sales per call?

(c) A machine tool builder takes old lathes as trade-ins for new models and sells the returned lathes through a second-party outlet. Analysis shows that the markup is $2500 on 70% of the lathes and $4000 on the rest. What is the expected markup?

(d) An insurance company charges $20 for an additional $50 increment of insurance (from $100 to $50 deductible). What is their assessment of the risk for the increment of insurance?

6-8. A student is interested in selling his car instead of trading it in. His estimating model, he reasons, is the sale price of a new car − (depreciation + major maintenance cost). Other costs for driving are the same regardless of whether he drives a new car or not. The original sticker price is $4379 and a major maintenance cost is $500. His subjective probability for a major maintenance cost is given as

Life	Cumulative Probability of Major Maintenance	Cumulative Decline in Depreciation
2	0.2	0.49
3	0.4	0.64
4	0.7	0.75
5	1.0	0.83
6	0.1	0.89
7	0.2	0.93
8	0.4	0.96
9	0.7	0.98
10	1.0	0.99

When should he sell his car? Initially assume that the next car, whenever he buys it, will be equal to his first car's price. Next assume that the new car's price increases by 1% compounded per year.

6-9. A small company with assets of $250,000 is considering the possibility of a redesign of one of its basic products. The engineering cost and the manufacture of new tooling will cost $25,000. Three alternatives are to be evaluated, and the estimators are to determine the profit and subjective probability of success. Production costs for designs A and B are about the same:

	No Change		Design A		Design B	
Year	Profit	Probability	Profit	Probability	Profit	Probability
1	$20,000	0.4	$40,000	0.3	$30,000	0.2
2	25,000	0.3	45,000	0.5	45,000	0.3
3	30,000	0.3	50,000	0.2	60,000	0.5

Determine the expected value of the total assets for each of the 3 years for the three alternatives.

6-10. In a hospital an investigation concerns the use of linen in the operating room. The hospital is considering four alternatives: (1) Buy bulk linen and have the desired sizes made within the hospital, (2) buy factory-made linen, (3) buy a disposable paper linen, and (4) buy a disposable synthetic linen. The following information is given:

	Price to Make (or Buy) Per Piece	Cost of Handling or Upkeep Per Piece Per Usage	$1000 Per Month			
Option			1	2	3	4
1	0.50	0.20	5	5	6	6.5
2	0.60	0.20	6	7	8	8.5
3	0.80	0.05	7	7	6.5	6.5
4	0.80	0.05	8	8.2	8.1	8.9

(a) List several factors you would want to know before making an appraisal of the above unordered ranking.

(b) The options underwent a trial period within the hospital, and the total cost of usage was found for a period. Construct a band chart showing the region of total cost over the 4-month period.

(c) Assume that the hospital adopted the disposables in place of the linen. The disposable suppliers, companies A and B, were asked to make studies of the total cost to use disposables during the next year:

Company	Annual Usage and Probability	Cost Per Unit and Probability
A	2000–0.2	$0.70–0.2
	2500–0.7	0.75–0.2
	3000–0.1	0.80–0.6
B	2500	$0.80

Uncover the difference in the two estimates and make a supplier recommendation.

6-11. Work the problem in Section 6.6 using these random numbers:

Random Number x	Random Number y
0.07	0.59
0.33	0.30
0.16	0.10
0.75	0.96
0.58	0.88
0.23	0.98
0.89	0.14
0.08	0.43
0.10	0.76
0.96	0.70

Determine the mean cost A. Plot the distribution of cost A and determine its standard deviation. How does this deviation generally compare with the variations of distributions A in Figure 6.9?

6-12. Work the problem in Section 6.6 using these random numbers:

Random Number x	Random Number y
0.24	0.64
0.82	0.98
0.83	0.25
0.18	0.94
0.66	0.03
0.76	0.23
0.07	0.96
0.62	0.80
0.61	0.64
0.96	0.99

Determine mean cost A and its standard deviation.

6-13. (a) A is distributed exponentially with a mean of 3 and B is distributed from a continuous distribution as $f(x) = \frac{2}{9}x$, where $0 \leq x \leq 3$. Find the cost of the distribution where $C = A + B$. (*Hint*: Integrate the two functions and assume random rectangular variates for substitution.)

(b) A simulation model is defined as $c = x + y$, where x is given by

$$f(x) = \begin{cases} \dfrac{1}{b-a}, & 2 \leq x \leq 8 \\ 0, & \text{elsewhere} \end{cases}$$

and the continuous variable y is given by a frequency:

Cost of y	Occurrence
$1	0.15
2	0.25
3	0.40
4	0.15
5	0.05

Determine cost c after five simulation trials.

(c) Distribution A has the form $f(x) = \frac{1}{4}e^{-x/4}$, where $0 \le x$, and distribution B has the following estimated data:

B_i	f	B_i	f
1	4	5	18
2	15	6	8
3	22	7	2
4	30	8	0

If $C = A/B$, find the mean of C after five trials.

6-14. Use the technique of probability cost estimating to find the expected total mean cost and variance:

Cost Element	Optimistic Cost	Most Likely Cost	Pessimistic Cost
1. Direct labor	$79	$95	$95
2. Direct material	60	66	67
3. Indirect expenses	93	93	96
4. Fixed expenses	69	76	82

6-15. A five-element cost program has been summarized:

Cost Item	Optimistic Cost	Most Likely Cost	Pessimistic Cost
1	$ 4	$ 4.5	$ 6
2	10	12	16
3	1	1	1.5
4	4	8	12
5	2	2.5	4

Determine the elemental mean costs, the total cost, and the elemental and total variances.

6-16. A probability cost estimate found that the mean cost of 10 independent elements was $169 and that the variance was $127. Using the central limit law and the table of the normal distribution function in Appendix I, find the probability that the cost will exceed $185. Use

$$Z = \frac{\text{upper limit cost} - \text{mean cost}}{\text{standard deviation}}$$

6-17. A probability cost estimate of 16 elements determined that the grand mean cost was $12 and that the variance was $4. Find the probability that the cost will exceed $15.

6-18. Order the estimates for the protection of electronic cards in Section 6.3. What is their cardinal number? What is the second ordinal choice? The third?

6-19. Use the ordinal scale method for this information:

Application	Preparation Cost	Savings Per Year over Conventional Method	Probability of Success	Success Rank	Interval Value
Plastics	$2000	$7800	0.1	2nd	0.5
Machining	9000	3700	0.4	1st	1.0
Sheet metal	6000	3000	0.5	3rd	0.4

Revenue estimates for machining amount to $20,000. A 3 year's savings on cost reduction schemes is policy. A subjective probability evaluation of the potential return is 0.5, 0.9, and 0.8 for machining, plastics, and sheet metal. Find the net revenue. Which application is best?

CASE PROBLEM

Uranium Ore Processing. Chamberlin Mining Ltd. has discovered a rich ore deposit in Saskatchewan, Canada. Prospector Chamberlin hires consultant Ralph Light to provide him with preliminary estimate facts and flow-sheet data.

Light reports that three process methods are used: acid-leach countercurrent decantation solvent type, acid-leach resin, and an alkaline leach. Operating costs are affected greatly by the grade of ore, mill capacity, and reagent consumption for a particular operation. Although sulfuric acid costs are decreasing, labor costs are not, and the general trend is upward. Construction costs are increasing with new plants requiring pollution controls.

Because Chamberlin is unwilling to disclose his proprietary facts such as profits above costs, transportation, depreciation rates, overhead rates, and assets, Light is forced to submit only preliminary data for his client:

Flow-sheet Type of Plant	Plant Capacity (tons of ore feed per day)	Capital Cost installed ($ per ton)	Direct Operation Cost ($ per ton)	Uranium Purity
Acid-leach,	500	$11,000	$5.70	0.90
CCD	1000	7,500	4.55	0.93
	2000	5,500	3.80	0.95
Acid-leach,	500	9,000	5.35	0.88
resin	1000	6,000	4.60	0.90
	2000	4,250	3.80	0.94
Alkaline-	500	12,000	6.10	0.93
leach	1000	8,000	4.95	0.95
	2000	6,000	4.15	0.96

Now the prospector determines that a plant runs 300 days per year and he figures a model that incorporates the full cost of capital and operation will be good for a starter. Based on these preliminary estimating data, what will be the direct operating dollars per recovered pound of U_3O_8 if the ore grade of feed is 0.20% U_3O_8? What conclusion as to capacity and flow sheet do you recommend on this basis?

From a separate exploration report Chamberlin knows that the ore body is limited to 10–12×10^6 tons before primary grades are exhausted. New processing equipment will then be required. What is the capital investment cost for the nine plants? Ignoring depreciation, which type of plant minimizes total capital and operating costs over the life of the ore? Which flow sheet should be analyzed by detailed estimating methods?

Nothing succeeds like excess.
OSCAR WILDE

DETAILED METHODS

The previous chapter presented approximation methods for finding estimates. Their purpose was to screen and eliminate unsound proposals without extensive engineering cost. If these estimates lead to a continuation rather than a dismissal decision, additional methods are required. Now we confine our attention to methods that are more thorough in preparation and accurate in results as well as costly in design and estimating. At this point in time the designer would have extended his own preparations, and the estimator can construct an estimate on enlarged quantities of verified information. In some cases the detailed estimate may be a re-estimate, as only limited updating needs to be done. Accordingly, we go along with Oscar Wilde's tongue-in-cheek advice on excess for success.

The man-hours devoted to estimating the operation, product, project, or system design naturally varies with circumstances, and we have no rules that can guide management. The case for increased accuracy from a detailed estimate is often made. Some practitioners claim that detailed estimates are within $\pm 5\%$ about a future actual value. Whether the particular value is $\pm 5\%$ or $\pm 50\%$ is not significant now; however, we want to assert that methods that generally are more accurate have a commensurable increase in cost. Thus, data are purified, design has increased detail, and in actual estimating situations management stipulates that the estimate be within an interval about the future actual or standard value.

Detailed estimating methods are more quantitative. Arbitrary and excessive judgmental factors are suppressed, although they are never eliminated, and emphasis shifts to the mathematical model. The model may range from a casual back-of-envelope model to a complicated computer program. These ideas were first broached in Chapter 2. Whether the model is a recapitulation columnar sheet, a set of computational rules, or a functional model, the intent is the same: We want to find the value of the estimate through formal and rigid rules.

Techniques that are presented in this chapter include the factor method, standard time data, power law and sizing factor, marginal analysis, and composite relationships. Actually, marginal analysis is not a method of detailed estimating; rather detailed estimating permits this kind of analysis. It is included because of its importance. We like to think that these unrelated methods are appropriate for any design, although in specific situations some would be preferred to others. For example, a cost-volume relationship would be more useful for an operation or product design, while the power law and sizing factor is relevant to a project design.

7.1
FACTOR METHOD

The factor method is a basic and important method for detailed estimates. Other terms such as ratio, parameter, and percentage methods are about the same thing. Essentially the factor method determines the estimate by summing the product of several quantities, or

$$C = [C_e + \sum_i f_i C_e](f_I + 1) \qquad (7\text{-}1)$$

where C = value (cost, price, and so forth) of design being evaluated

C_e = cost of selected major equipment

f_i = factor for the estimating of buildings, instrumentation, and so forth

f_I = factor for the estimating of indirect expenses such as engineering, contractor's profit, and contingency

$i = 1, \ldots, n$ factor index

The factors f are uncovered by historical, measured, or policy methods of information previously reported on in Chapters 3 and 4. Data from internal reports, from the industry at large, or from the government may be the principal source for the factors.

A natural simplification leads to the preliminary unit estimating model $C = fC_e$, where one factor is used to find the composite cost. This factor-estimating formula has variants such as $C = \sum_i f_i D$, where D is the particular design parameter, e.g., material machining time, combat force, and capital investment, and f_i is the factor in units compatible to the design.

The unit cost-estimating method was limited to a single factor for calculating overall costs. The factor method achieves improved accuracy by adopting separate factors for different cost items. For example, the approximate cost of an office building can be estimated by multiplying the area by an appropriate unit estimate such as the dollars-per-square-foot factor. As an improvement, individual cost-per-unit-area figures can be used for heating, lighting, painting, and the like, and their value C can be summed for the separate factors and designs.

Plant cost can be estimated by the factor method. Reconsider the project hydrobromination of alkyl bromide introduced in Chapter 1 and enlarged on in Chapter 2. Recall that the hydrobromination process is but one step in the sequence of making an intermediate product for a liquid soap formulation. The hydrobromination flow chart, given by Figure 7.1, is a microprocess plant.

The flow chart and the specification sheet are input data to the project estimator. With the flow chart, the basic item (or items) of the process is identified. This basic item should be a major item for a building such as the structural shell, or tons of concrete for a highway, or process equipment in a chemical plant. After the cost of the basic item has been determined, the next step is to find the cost relationship of other components as a percentage, a ratio, total cost, or factor of the basic item. For the building structural shell, the correlating factors would be brick, concrete, masonry, architectural and reinforced concrete, carpentry, finish trim, electrical work, plumbing, and so forth. For the chemical process project, equipment erection, piping and direct materials, insulation, instrumentation, and engineering can be correlated to process cost. These correlations can be statistically found using the methods of Chapter 5. In some instances the variations in the cost of components being analyzed are independent of the variation in the cost of the basic item. These independent components, e.g., roads, railroad siding, and site development for a new plant, must be estimated separately by other methods.

There are practical considerations in applying the factor method. For the chemical plant problem, variations between estimates are due to

- Size of the basic equipment selected.
- Materials of construction.
- Operating pressures, temperatures.
- Technology such as fluids processing, fluids-solids processing, or solids processing.

FIGURE 7.1

FLOW CHART FOR HYDROBROMINATION
PROCESS

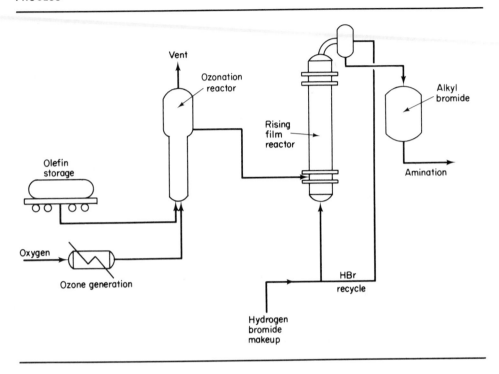

- Location of plant site.
- Timing of construction.

Practically speaking, if the physical size of the selected basic item becomes larger and therefore more costly, the factor relating, say, the engineering design cost is marginally smaller. If the basic item is constructed with more expensive materials such as stainless steel or glass-lined materials, the factors become nonproportional to the item being estimated. This factor behavior is described by Figure 7.2. This is nothing more than the law of diminishing returns. Usually factor data are collected and normalized to some past year. Frequently 1957 is chosen since this year serves as a benchmark for many indexes. First, the basic item is submitted for bid or estimated internally at today's price. With one or more bids for the basic item an average value is deflated to a reference point such as 1957 for finding the factors. Next it is indexed forward or inflated to the time of anticipated construction. Figure 7.3 shows the time scale adjustments. Regional indexes may account for differences in labor and material cost. An example now illustrates what has been said.

In the hydrobromination process, the rising film and ozonation reactor are the basic items chosen. These two are selected to avoid a time-consuming bulk take-off of materials. Several fabricators are asked for bids, and the average or a reasonable value gives

Major Process Item	*Cost*
1. Rising film reactor	$220,000
2. Ozonation reactor	70,000
	$C_e = \$290,000$

FIGURE 7.2

GENERAL EFFECTS OF INCREASING BASIC ITEM
COST UPON FACTOR

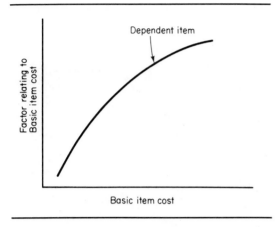

FIGURE 7.3

TIME SCALE ADJUSTMENTS FOR ESTIMATING
COSTS

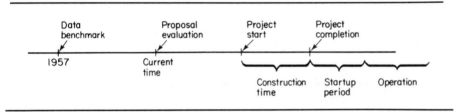

These current costs are adjusted to the factor base of 1957–1959. In this instance assume an engineering plant construction cost index is chosen. The components of this index are weighted as to equipment, machinery, and supports, 61%; erection and installation labor, 22%; buildings, materials, and labor, 7%; and engineering and supervision manpower, 10%. Each of these categories is further divided into smaller groups. In this way the estimator can make comparisons and confirm whether his projected plant is similar to the index.

Using Equation (5-48), we revise it to show

$$C_r = C_c \left(\frac{I_r}{I_c} \right) \qquad (7\text{-}2)$$

where C_r = benchmark cost, dollars
  C_c = current cost, \$290,000
  I_r = 100 for 1957–1959
  I_c = current index as 114.1 for equipment
  C_r = \$290,000(100/114.1) = \$254,163

With this adjusted major component cost, from a curve plotted for the base year, Figure 7.4 yields engineering costs and fees, 1.1; erection, 1.7; and direct materials, 4.1. Each of these items has its own index to the time of project start. Let us further state that the data of Figure 7.4 are Gulf Coast information, while our hydrobromination plant is to be constructed in the Midwest where direct materials are 8% more expensive, erection costs are 13.2% more costly,

FIGURE 7.4

TYPICAL FACTOR CHART BASED ON 1957–1959
PROCESS EQUIPMENT COSTS

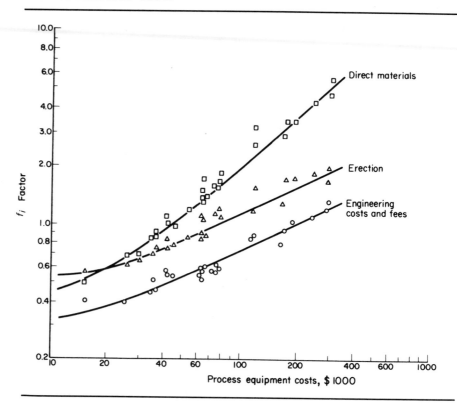

and engineering costs and fees are identical. Furthermore, these regional adjustment factors take into account the mix of the components of the index. These regional factors are applied to current values. This approach now provides Table 7.1. Process buildings, site development, utilities, and a railroad siding are independent of the basic reactors and must be estimated by other means.

TABLE 7.1

FACTOR ESTIMATE FOR A CHEMICAL PLANT PROCESS

Estimated Item	Current Time Cost	Bench Mark Index	Deflated 1957–1959 Cost	1957–1959 f_i	1957–1959 Cost	Project Start Index*	Gulf Coast Project Start Cost	Midwest Regional Index	Project Start Plant Site Cost
1. Equipment:									
Major Process Items	$290,000	114.1	$254,163	1.	$ 254,163	118.0	$ 299,912	—	$ 299,912
Erection				1.7	432,077	189.0	816,625	1.132	924,420
Direct Materials				4.1	1,042,068	127.0	1,323,426	1.08	1,429,300
2. Engineering Costs and Fees				1.1	279,000	145.0	405,389	1.0	405,389
3. Building Site Development:									
Site Development									125,000
Process Building									210,000
Railroad Spur									4,000
Utilities									65,000
								Total	$3,463,021

*1957–9 Index for all categories = 100

Finally, the capital investment, $3,463,021 is provided as the estimate. A factor for indirect costs, contingency, and profit can be applied to this raw capital investment. See page 351 for another example.

The example dealt with plant equipment estimating, but the factor method is used more broadly than indicated by this example. It is used to estimate labor, materials, utilities, and indirect costs as a multiple of some other estimated or known quantity. Operating costs as a percentage of plant investment or as a percentage of the product selling price are popular.

The power law and sizing model is frequently used for estimating equipment. This model is concerned with designs varying in size but similar in type. The unknown costs of a 200-gallon kettle can be estimated from data for a 100-gallon kettle provided both are of similar design. No one would expect that the 200-gallon kettle would be twice as costly as the smaller one. The law of economies of scale assures that. The power law and sizing model, similar to the index model [Equation (5-48)], is given as

$$C = C_r \left(\frac{Q_c}{Q_r}\right)^m \tag{7-3}$$

where C = total value sought for design size Q_c
C_r = known cost for a reference size Q_r
Q_c = design size
Q_r = reference design size
m = correlating exponent, $0 < m < 1$

An equation expressing unit cost C/Q_c can be found, or

$$C\left(\frac{Q_r}{Q_c}\right) = C_r\left(\frac{Q_r}{Q_c}\right)\left(\frac{Q_c}{Q_r}\right)^m = C_r\left(\frac{Q_c}{Q_r}\right)^{-1}\left(\frac{Q_c}{Q_r}\right)^m$$
$$\frac{C}{Q_c} = \left(\frac{C_r}{Q_r}\right)\left(\frac{Q_c}{Q_r}\right)^{m-1} \tag{7-4}$$

As total cost varies as the mth power of capacity in Equation (7-3), C/Q_c will vary as the $(m-1)$st power of the capacity ratio. If we let $m = 1$, we have a strictly linear relationship and deny the law of economies of scale. For chemical processing equipment m is frequently near 0.6, and for this reason the model is sometimes called the *sixth-tenth* model. The units on Q are required to be consistent as it enters only as a ratio.

Let us assume that in 1964 an 80-kilowatt diesel electric set, naturally aspirated, cost $16,000. The plant engineering staff is considering a 120-kilowatt unit of the same general design to power a small isolated plant. If the value of $m = 0.6$, we have

$$C = 16,000(\tfrac{120}{80})^{0.6} = \$20,400$$

The model can be altered to consider changes in price due to inflation or deflation and effects independent of size, or

$$C = C_r\left(\frac{Q_c}{Q_r}\right)^m \frac{I_c}{I_r} + C_1 \tag{7-5}$$

where C_1 = constant unassociated cost. The Marshall and Steven's price index for this class of equipment was 187 in 1964 and now is 194. Assume that we want to add a precompressor, which when isolated and estimated separately costs $1800, or

$$C = 16,000(\tfrac{120}{80})^{0.6}(\tfrac{194}{187}) + 1800 \doteq \$23,300$$

The determination of m is important to the success of this model, and methods of curvilinear regression and rectification given in Chapter 5 are cogent. If the statistical analysis assumes constant dollars, then the index ratio I_c/I_r is used for increases or decreases for inflation or deflation effects.

Model (7-3) is usually applicable to project estimates for purchased assets. Erection costs, on-site labor estimates, pre-operation charges, training of operators, use taxes, and FOB charges are items not included. When encountering these exceptions, the term C_1 is employed.

The model usually does not cover those situations where the estimated design Q_c is greater or less than Q_r by a factor of 10. Some typical exponents for equipment cost versus capacity are given in Table 7.2.

TABLE 7.2

TYPICAL EXPONENTS FOR POWER LAW
AND SIZING MODEL[a]

Equipment	Size Range	Exponent
Blower, centrifugal (with motor)	1–3 hp	0.16
Blower, centrifugal (with motor)	$7\tfrac{1}{2}$–350 hp	0.96
Compressor, centrifugal (motor drive, air service)	20–70 hp drive	1.22
Compressor, reciprocating (motor drive, air service)	5–300 hp drive	0.90
Dryer, drum (including auxiliaries, atmospheric)	20–60 ft^2	0.36
Dryer, drum (including auxiliaries, vacuum)	16.5–40	0.20
Heat exchanger, shell and tube, floating head, c.s.	100–400 ft^2	0.59
Heat exchanger, shell and tube, fixed sheet, c.s.	50–400 ft^2	0.44
Kettle, cast iron, jacketed 100 psi	250–800 gal	0.24
Kettle, glass-lined, jacketed	200–800 gal	0.31
Motor, squirrel cage, induction, 440 V, explosionproof	1–20 hp	0.53
Motor, squirrel cage, induction, 440 V, explosionproof	20–200 hp	1.00
Pump, centrifugal, horizontal, s.s.	3–7.5 hp	0.61
Pump, centrifugal, horizontal, cast iron	2–7.5 hp	0.21
Reactor, glass-lined, jacketed (without drive)	50–300 gal	0.41
Reactor, glass-lined, jacketed (with drive less motor)	50–300 gal	0.81
Tank, flat head, c.s.	300–1400 gal	0.66
Tank, flat head, c.s., glass-lined	100–1000 gal	0.57

[a]From Max S. Peters and Klaus Timmerhaus, *Plant Design and Economics for Chemical Engineers*, 2nd ed., McGraw-Hill Book Company, New York, 1968.

7.3
STANDARD TIME
DATA METHOD

The previous two methods were primarily concerned with plant and equipment estimating. The standard time data method estimates labor.

We have discussed several methods whereby work measurement data are collected. Time data in their raw form are not usable by the cost engineer.

Frequently these data include bad methods, unwarranted conditions, or non-average operators. Sometimes it is lacking due to the limitations of a narrow range of observed work. The engineer uses regression analysis to extend these raw data into a more digestible form. For it is not the original time measurements that the estimator ultimately desires; rather it is a set of engineering performance data, or standard time data, or more briefly standard data, that he uses for estimating. Standard time data may be defined as a catalog of standard tasks that are used for performing a given class of work. Parenthetically, it should be indicated that standard time data are arranged in a systematic order and are used over and over again. The advantages over direct observation methods such as time study, predetermined motion time data, work sampling, and man-hour reports are lower cost and greater consistency. Description is provided in advance of the need for the data, and, more estimators can use standard time data. Standard time data may be divided into preliminary or detailed. The estimator is more likely to be concerned with a preliminary standard data early in estimating; later, detailed data become more important. Standard time data are ordinarily determined from any of the various methods of observing work. In manufacturing, time-study and predetermined motion data are the major sources. In construction and hospitals, work-sampling and man-hour reports are the principal means of information. In certain government work, such as that of the post office or armed forces, work-sampling and man-hour methods are used. Whatever the source of work data, the development of standard data is similar.

Standard time data have been defined as the expression of time values in a concise final form from which a standard time can be obtained reflecting the cost of a given operation performed under usual conditions. The key word in this definition is *usual*. It is unwise to divise standard time data without first knowing conditions, specifications, methods, and procedures. If these precautions are heeded, the time values are based on the local conditions in which the standard is to apply.

Although the number of basic studies used for standard time data is a statistical question, it is realized that accuracy is a function of the number of observations. Assuming that the basic data were correctly and uniformly determined, one of the decisions that faces the analyst is to determine whether a particular element is *constant* or a *variable*. A constant is accepted by convenience and assumed to have less than an arbitrary percentage S slope between the limits of work scope. It is admitted that very little work is ever constant, but in order to get things done we overlook some variability. Consider the element "handle" with the independent variable given as weight. If the time slope, say a positive 2%, from the minimum to the maximum weight is found, the analyst arbitrarily chooses to classify this element as a constant if the S limit value amounted to 5%. In this case a constant value "handle" would be found. If the increase of slope between the limits of the defined element is greater than 5%, the element is classified as a variable. The 5% value is for illustration purposes. The dependent variable time can be related to weight either by a graph, a table, or regression formulas. A particular element may be one of many elements that define a set of standard time data for some operation, i.e., "painting." Moreover, if the element "handle" is $Y\%$ or less of the standard time of the total expected mean operation time, it could be classified as a constant irrespective of its elemental variability. In this operation of spray painting, it was determined that the element "handle" constituted better than 37% of the expected mean time for the operation. To have classified the element as a constant would have destroyed operation sensitivity.

The calculation of a constant element is a straightforward procedure after it is decided that the element is a constant. A constant is simply the mean of the elemental time observations. In the analysis of variable elements, however, the technique of handling the times must be considered on the basis of each variable element. What is adequate and accurate for one element could be too expensive for the next. The simplest method of expressing the variable is to divide the work into classes and to select a time value for each class, for example, small, medium, and large. This accuracy may be insufficient and the variable could be plotted as a curve on a graph. In plotting observed points a scatter is likely, but if the proper independent variable has been determined and the data collected consistently and sufficiently, the points show the trend of the curve.

Generally curves are not used as the form or final expression for standard time data, as there is a tendency to incorrect interpolation, they are slow to read, and they are subject to faulty extension. Consequently, charts replace curves as the final expression of information because they are faster and are consistently applied when compared to curves. The data for "handle part" has been graphed in the form of the straight line in Figure 7.5. This line was then

FIGURE 7.5

PLOT OF ELEMENT INFORMATION FOR DETAILED
STANDARD TIME DATA

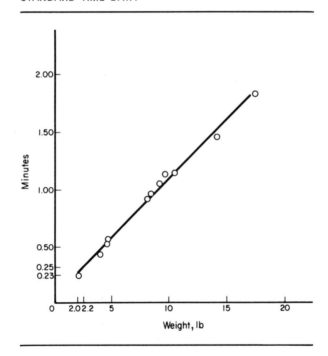

formed into a table. Starting at 2 pounds and projecting 10 % steps of the dependent variable time the variable is transformed into a table. For example, at 2 pounds, the dependent time would be read as 0.23 minutes. Next at 2.2 pounds the time would be 0.25 minutes, and so forth. This tabular representation is a device to provide essential steps uniformly throughout the range of weight. The transformation is included along with other elements that make up one set of detailed standard time data for an operation known as "paint." Table 7.3 provides the standard time data for the operation and includes the setup time (normally given in hours), the handling time, and the process or spray painting

TABLE 7.3

SAMPLE OF DETAILED STANDARD TIME DATA
FOR A PAINTING OPERATION

1. Setup Hours	0.10 hr

2. Handling Elements (Load and Unload), normal minutes per piece

A. Part Handling			B. Tray Handling				Special Elements		
Weight	Chassis	Other Parts	$L + W + H$	Pieces Per Tray	Get Aside-Loaded Tray	Get, Load, Aside-Loaded Tray	C. Fixture Load and Unload		
2.0	0.23	0.23	1.5	86	0.002	0.038	Type Part	Load	Unload
2.2	0.25	0.25	3.0	72	0.003	0.041	Dials, Knobs, Handles	.054	.060
2.5	0.33	0.27	4.5	60	0.003	0.044	Screwheads, Bolts	.022	.014
2.8	0.36	0.29	5.5	50	0.004	0.047			
3.2	0.40	0.31	6.5	42	0.005	0.049			
3.8	0.44	0.34	7.5	35	0.006	0.051			
4.1	0.48	0.37	8.5	29	0.007	0.053	D. Load or Unload Tray		
4.5	0.53	0.40	9.5	24	0.009	0.058			
5.0	0.58	0.43	10.5	20	0.011	0.061		Pieces Per Tray	Load Tray	Turn Over Piece	Unload Pieces
5.8	0.64	0.47	11.5	17	0.013	0.066	$L + W + H$				
6.2	0.70	0.51	12.5	14	0.014	0.071	1.5	86	0.045	0.003	0.035
7.0	0.77	0.55	14.0	12	0.020	0.077	3.0	70	0.048	0.004	0.037
7.5	0.85	0.60	14.5	10	0.024	0.083	4.8	50	0.053	0.005	0.040
9.0	0.94	—	16.5	8	0.031	0.094	6.5	36	0.059	0.007	0.044
9.4	1.03	—	20.0	6	0.044	0.12	8.0	20	0.078	0.012	0.055
10.6	1.13	—	22.5	5	0.053	0.13					
11.5	1.24	—	25.0	4	0.069	0.15					
12.8	1.36	—	27.5	3	0.096	0.19					
14.1	1.50	—	30.0	2	0.15	0.25					
15.5	1.65	—	32.0	1	0.30	0.42					

3. Visual Inspection

$L + W + H$	36	56	74	90	109	122	138
Minutes per piece	0.08	0.10	0.12	0.15	0.18	0.21	0.26

4. Turn Piece Around or Hook

$L + W + H$	18	27	41	61	81	110	138
Minutes per piece	0.09	0.11	0.14	0.20	0.29	0.45	0.61

5. Spray Paint	Minutes Per Square Inch
Enamel	0.004
Epoxy	0.026
Wrinkle	0.022

time. For detailed data, operation or run time is usually given in normal minutes, while for preliminary data, the units are standard hours. Whenever the word *standard* is used with estimates, one understands that allowances (including personal and delays) have transformed normal time into a form ready for the manipulation of time into dollars.

Table 7.3 can be used to estimate manual and paint processing time for direct labor. With a print, the estimator establishes the size for handling and area for spraying. This is illustrated by one problem.

This is one set of data. The job estimator would have at his disposal sets of standard data that cover all job descriptions or work stations. As a particular job was estimated, he would use the appropriate data for its estimation.

Indirect labor in fabricating- and assembly-type plants is supplemental to machine and assembly operators who work directly on the product. In many businesses the number of indirect employees of all kinds exceeds the number of direct employees. Indirect employees include truckers, schedulers, shipping clerks, and stock attendants.

In job-estimating indirect employees, who do not perform carefully designated and repetitive-type work, modification of the standard data methods is required.

Indirect labor departments have a workload that can be measured. It may be the number of engineering specifications written, number of purchase orders typed, number of cards filed, or the like. In measuring an existing production control department, work elements can be identified and their average frequency noted and prorated to a production variable such as units of production, number of machine setups, number of shifts, or number of direct employees. Thus we have data structured for the job estimating of production control clerks as given by Table 7.4.

TABLE 7.4

DATA TO DETERMINE INDIRECT WORK MANNING

Description of Work for Fabrication Operation Production Control Record Clerk	Observed Minutes Per Occurrence	Average Frequency Per Job Setup	Normal Minutes Per Setup
1. Post supply requisitions	2.2	$\frac{1}{1}$	2.20
2. Type purchase orders	2.5	$\frac{1}{10}$	0.25
3. Type purchase vouchers	2.8	$\frac{1}{15}$	0.19
4. Post travelers entries— in and out	3.0	$\frac{1}{45}$	0.07
5. Trips for travels	5.0	$\frac{1}{25}$	0.20
6. Help accounting & production answer questions	3.0	$\frac{1}{25}$	0.12
7. Phone calls	1.5	$\frac{1}{4.5}$	0.33

Total normal minutes per setup   3.36

Personal allowance: 1.122

Standard job factor per posting of setups: 3.76 min

These data fail to include normal absenteeism. Performance against this 3.76-minute job factor for posting of setup was found to be 90%. With this as a correction, it is a simple matter to determine the number of production control record clerks, and for an assumed 720 fabrication setups per week in a new plant the number of production control clerks equal

$$3.76 \times \frac{1}{0.9} \times 720 \times \frac{1}{5 \times 480} = 1.26$$

or, conservatively, one employee.

When job estimating new or enlarged processing plants, the cost engineer determines from his designs and flow sheets the number of operators required. Equipment-operator tables can be previously tabulated by work-sampling studies of existing plants and transferring that information, or suppliers can be asked for manning requirements. Published equipment-operator tables also exist.

The factor and standard time data methods demonstrate graphical and tabular estimating relationships. This information, if it were mathematically fitted, could be called *cost estimating relationships* or CERs. CERs are popular jargon in system design for the Department of Defense and other national political organizations. Simply, a CER is a functional model that mathematically describes the cost of an item or activity as a function of one or more independent variables. The power law and sizing model is a typical CER. CERs as practiced in system estimating within the Department of Defense are of two basic types: (1) those used to estimate physical quantities such as the numbers of aircraft and number of personnel, and (2) rates of activity expressing support personnel as a function of the number of direct operating personnel. Second, CERs are used to estimate the dollar cost impact such as the cost of a turbojet airframe as a function of airplane gross weight and speed. An example of turbojet engine development cost would be

$$C = 0.13937 x_1^{0.74356} x_2^{0.07751}$$

where C = cost in 10^6
    x_1 = maximum thrust, pounds
    x_2 = production quantity milestone

Their suitability and credibility rest on the input information and the goodness of the estimate. In the realm of long-range planning (5–15 years), an assessment of a CER accuracy must wait until many years have passed. Perhaps no analysis is ever made of the CER. Because of this shortcoming, emphasis shifts to the quality of the study and the analysis undergirding it. As in all functional estimating models there must be a logical or theoretical relationship of the variable to cost, a statistical significance of the variables' contribution, and independence of the variables to the explanation of cost.

Cost estimating provides estimates about designs to make future decisions. There is the premise that the action recommended by the estimate will add to the benefits of the enterprise to make it worth the trouble. Typical situations in which estimating weighs the proposition "will it add to the benefits" include the following: A firm is considering a new plant producing an intermediate soap product, a plant enlargement requires indirect labor manning, and a special wrench has been designed and will be marketed shortly. Not all situations assure that the originator will be better off after than he was before. Consider this case. An electric utility serving several separated geographical regions employs transmission repair crews for high-voltage line failures. The manager has permission to add another maintenance operator to a crew. In district A the average of the cost of repair per maintenance operator is estimated as $780, while in district B the unit estimate is $470. Our manager chooses to assign the operator to district B. But it is possible that the difference in the estimates occurred because the size of the repair crew in B was better adapted. If so, the new line mechanic may add little to cost reduction, while in A he may lead to a greater decrease in cost. The reasoning that would lead to the right choice is based on marginal cost and is a method of analysis that is important to cost engineering. Too often decisions are made on the basis of average cost or average return rather than marginal cost or marginal return.

    Marginal costs are the added costs incurred as a result of adopting a change in operations or making an engineering-change order for a product. The change in operations usually implies increasing or decreasing production by one unit. Marginal cost is used interchangeably with differential cost or incremental cost. In a broad sense all estimates are marginal estimates inasmuch as they are concerned with creating changes from a current course of action. Only after a detailed estimate is made can the engineer realize and exploit the advantages of marginal analysis.

We can explain the arithmetic of marginal analysis with a simple illustration such as

Estimated Quantity	Estimated Total Cost	Analyzed Marginal Cost	Analyzed Average Cost
0	0	0	—
1	$1000	$1000	$1000
2	1900	900	950
3	2700	800	900

Before any production the hypothetical cost is zero. Marginal cost, by convention, is also zero and average cost is undefined at this point. At two units, marginal cost is the amount which is added by the production from one to two units, or $1900 - 1000 = \$900$. Average cost (arithmetic mean) is as usually determined. The principles of marginal cost are better explained by calculus, which is the more precise way to cope with these matters. Differential calculus, which is our concern, involves the maxima, minima, first and second derivative, and the arithmetic sign and magnitude of these derivatives.

The brute-force way to explain marginal analysis is through an extensive table, where total cost is enumerated for successive units, i.e., 61, 62, 63, . . . in the region of interest. However, the approach recommended is to have a scattering of cost-estimated points and then proceed to fit linear and nonlinear regression lines. To illustrate we fit a first-, second-, and third-order polynomial regression line through the estimating data for a special wrench. Total cost data are given in Table 7.5. We analyze the marginal properties of these data from fitted regression models.[1]

TABLE 7.5

TOTAL COST ESTIMATE FOR A SPECIAL
WRENCH PRODUCT

Six-Month Production	Total Cost, C_T
5,000	$ 6,875
10,000	10,460
15,000	13,260
20,000	16,400
25,000	21,125
30,000	29,250

[1]Higher-ordered polynomials sometimes create problems. Usually it is acceptable practice to fit a polynomial regression equation of low order. The number of data points for the equation may be 3 to 8 times the order of the equation.

A simple model relating variable and fixed costs to production rate n is given as

$$C_T = nC_v + C_f \qquad (7\text{-}6)$$

where C_T = total cost dollars per period
$\quad n$ = number of production units per period
$\quad C_v$ = lumped variable cost per unit
$\quad C_f$ = lumped fixed cost per period

If at various values for n, C_v is the same, we have the familiar linear cost model; if C_v is permitted to vary, we are concerned with nonlinear models. For the constant C_v, nC_v is a straight line increasing at a constant rate per unit, and C_v is the slope of the variable cost line. The linear model statistically fitted for the wrench data is $C_T = 0.84n + 1527$, where $C_v = 0.84$. The fixed cost is \$1527 and is the vertical axis intercept.

The average cost model is given as

$$C_a = \frac{nC_v + C_f}{n} = \frac{C_T}{n} \qquad (7\text{-}7)$$

where C_a = average cost, dollars per unit. If C_v is constant, we have the linear case again. Equation (7-7) is still valid if C_v is nonconstant and we have the nonlinear case, which is the usual fare.

For linear cost situations marginal cost equals C_v and is a constant. For the general case we define marginal cost as

$$C_m = \frac{dC_T}{dn} \qquad (7\text{-}8)$$

where C_m = marginal cost, dollars per unit
$\quad dC_T/dn$ = derivative of total cost function with respect to quantity n

By similar reasoning, if the output n is increased by an amount dn from an established level n and if the matching increase in cost is dC_T, then the increase in cost per unit increase in output is $dC_T/\Delta n$. Marginal cost is the limiting value of this ratio as Δn gets smaller, i.e., marginal cost as the derivative of the total cost function. It measures the rate of increase of total cost and is an approximation of the cost of a small additional unit of output from the given level. Sometimes marginal cost is called the slope or tangent of the total cost curve at the point of interest.

For the data of Table 7.5 fitted to a second-order polynomial or $C_T = 6594. + 0.08n - 0.0000217n^2$, marginal cost dC_T/dn becomes

$$C_m = 0.08 - 0.0000434n$$

The marginal cost is a linear line inclining downward with a gradient of 0.0000434. A third-order polynomial fit of the same data is

$$C_T = 840 + 1.527n - 7.418 \times 10^{-5}n^2 + 1.827 \times 10^{-9}n^3$$

and the marginal cost function becomes

$$C_m = 1.527 - 14.836 \times 10^{-5}n + 5.481 \times 10^{-9}n^2$$

from which a marginal curve is directly plotted. The marginal cost curve is a parabola.

Plotting the average cost and marginal cost curve, we notice that the marginal cost line intersects the lowest point of the average cost curve. At this

FIGURE 7.6

AVERAGE, MARGINAL COST AND RETURN
RELATIONSHIPS

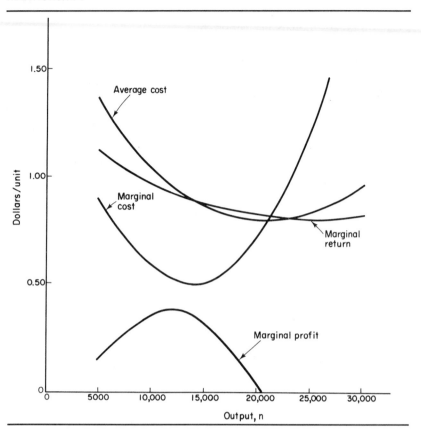

particular value of *n*, average cost is minimum and equal to marginal cost. This is shown by Figure 7.6.

The decision for the maintenance operator was made on the average cost estimate. It should have been made on the basis of the marginal yield which it promises. If district A has a greater marginal cost reduction by the addition of the maintenance operator, it is the preferred location. The choice between the two districts would be based on the greatest negative C_m.

The parallel to marginal cost is *marginal return*. It associates with many of the same concepts. In an operation design, marginal return would be marginal savings; in a product design we use the popular term marginal revenue; in a project design the word marginal rate of return connotes some of the same concepts; finally, in system design the term marginal effectiveness is frequently employed. It is usually assumed that a specific economic measure can always be chosen that will adequately reflect important differences among different values of the design variables. In a market situation total return may be thought of as the total money revenue of the producers supplying the demand and the total money outlay of the consumers providing the demand. While one can treat revenue curves on the basis of demand elasticity, estimators usually assume demand elasticity as given information and proceed.

In the case of an on-going operation or where production and sale of products are reasonably stable, demand projections are an easier thing to estimate. Whenever perturbations upset this stability, marginal analysis is prone to weakness in data. It is between these extremes that our estimator operates.

For our purpose we assume that a total return function is estimated with knowledge about input price and demand and that it increases or decreases with increasing output according to whether the demand is elastic or inelastic. Elastic demand implies a demand which increases in greater proportion than the corresponding decrease in price and vice versa.

For continuation of our special wrench problem the total return or revenue is given in Table 7.6. A zero-sale estimate is a trivial requirement, but it aids the construction in a least-squares sense. The total return column contains no reference to the price of the wrench. It can be computed by dividing total return by the estimated sales volume. In this sense price is regarded as average return per unit of output. We tacitly require for any further analysis that production equal sales volume. In a similar fashion we can fit polynomial regression curves to these data and proceed to analyze these determined models on the basis of marginal principles.

TABLE 7.6

ESTIMATED TOTAL REVENUE FOR
WRENCH PRODUCT

Estimated Sales Quantity	Total Revenue
0	0
5,000	$ 6,000
10,000	11,125
15,000	15,200
20,000	19,300
25,000	23,500
30,000	27,000

For a product design we can construct a model of the form

$$R_T = nR_v + R_f \tag{7-9}$$

where R_T = total return dollars per period
n = number of units per period
R_v = lumped variable return per unit
R_f = lumped fixed return per period

This is a linear or nonlinear model as either R_v is a constant multiplier or a nonlinear term dependent on other circumstances. If we are considering the sale of a product, no receipts are obtained until the sale of at least one product or $nR_v > 0$. For a no-sale condition $R_f = 0$, too. If the total return curve is linear, the marginal return is a constant and equals R_v. For a product sold commercially this is the sales price.

For an average return the model becomes

$$R_a = \frac{nR_v + R_f}{n} = \frac{R_T}{n} \tag{7-10}$$

where R_a = average return dollars per unit.
Marginal return is defined in a way similar to marginal cost, or

$$R_m = \frac{dR_T}{dn} \tag{7-11}$$

where R_m = marginal return, dollars per unit

dR_T/dn = derivative of total return function with respect to quantity n

For the data of Table 7.6 fitted to a second-order polynomial,[2] or $R_T = 90 + 1.138n - 8.13 \times 10^{-6}n^2$, the marginal return becomes

$$R_m = 1.138 - 16.26 \times 10^{-6}n$$

and thus marginal return is a linear line inclining downward with a gradient -16.26×10^{-6}. For a third-order polynomial fit,

$$R_T = -48 + 1.23n - 1.64 \times 10^{-5}n^2 + 1.833 \times 10^{-10}n^3$$

the marginal return function would be $R_m = 1.23 - 3.28 \times 10^{-5}n + 5.499 \times 10^{-10}n^2$ from which a marginal return parabolic curve is immediately found. This curve has been plotted in Figure 7.6. In the usual circumstance the second-order marginal model is considered more accurate than a first-order model.

Marginal estimating depends heavily on calculus. Rules for functions which are dependent on one variable are given in Tables 7.7 and 7.8. The marginal models are for the most part derived functions from fitted data. Occasionally natural physical functions are employed. For either case the functions C_T or R_T are assumed continuous and differentiable. A good deal of information can be uncovered from these basic models. Returning to the maintenance operator problem, the decision should be made on the basis of the marginal yield which it promises. Indeed, the addition of maintenance operators can proceed until the net yield is zero. This is the standard calculus procedure of finding the minimum cost and is dictated by rule 7 in Table 7.7 for C_m. If the size of the crew is to the

TABLE 7.7

TESTS OF TOTAL COST AND
TOTAL RETURN FUNCTION

Function	Definition	Cost Function	Return Function
1. Increasing function	Function increases as n increases	$\dfrac{dC_T(n)}{dn} > 0$	$\dfrac{dR_T(n)}{dn} > 0$
2. Decreasing function	Function decreases as n increases	$\dfrac{dC_T(n)}{dn} < 0$	$\dfrac{dR_T(n)}{dn} < 0$
3. Concave upward curve	Curve's derivative is increasing function	$\dfrac{d^2 C_T(n)}{dn^2} > 0$	$\dfrac{d^2 R_T(n)}{dn^2} > 0$
4. Concave downward curve	Curve's derivative is decreasing function	$\dfrac{d^2 C_T(n)}{dn^2} < 0$	$\dfrac{d^2 R_T(n)}{dn^2} < 0$
5. Stationary point	Point where tangent is horizontal	$\dfrac{dC_T(n)}{dn} = 0$	$\dfrac{dR_T(n)}{dn} = 0$
6. Return curve has a relative maximum at a point	Stationary point		$\dfrac{dR_T(n)}{dn} = 0$ $\dfrac{d^2 R_T(n)}{dn^2} < 0$
7. Cost curve has a relative minimum at a point	Stationary point	$\dfrac{dC_T(n)}{dn} = 0$ $\dfrac{d^2 C_T(n)}{dn^2} > 0$	

[2]This second-order polynomial does not pass through the origin. It might be assumed that at $n = 0$ there should be no revenue, which is correct; however, the data of Table 7.6 are least-squares approximations, and a displaced origin is not uncommon with estimated data. Furthermore

$$R_T = \begin{cases} 0 & \text{if } n \leq 0 \\ R_v n + R_f & \text{if } n > 0 \end{cases}$$

TABLE 7.8

STANDARD CALCULUS TESTS FOR
OPTIMIZATION OF FUNCTION

Name	*Definition*	*Test*	*Figure 7.7*
1. Increasing function $f(x)$	$f(x)$ increases as x increases	$f'(x) > 0$	Arcs *AB*, *DH*
2. Decreasing function $f(x)$	$f(x)$ decreases as x increases	$f'(x) < 0$	Arc *BD*
3. Concave upward curve, $c = f(x)$	Slope $f'(x)$ is an increasing function	$f''(x) > 0$	Arcs *CDE*, *FG*
4. Concave downward curve, $c = f(x)$	Slope $f'(x)$ is a decreasing function	$f''(x) < 0$	Arcs *ABC*, *EF*, *GH*
5. Stationary point of curve, $c = f(x)$	Point where tangent is horizontal	$f'(x) = 0$	Points *B*, *D*, *F*
6. Critical value of function $f(x)$	Value of $f(x)$ at stationary point		
7. $f(x)$ local maximum at $x = x_B$	$f(x_B) > f(x_B \pm \delta)$ for small values of δ	$f'(x_B) = 0$ $f''(x_B) < 0$	Point *B*
8. $f(x)$ has a local minimum at $x = x_D$	$f(x_D) < f(x_D \pm \delta)$ for small values of δ	$f'(x_D) = 0$ $f''(x_D) > 0$	Point *D*
9. Point of inflection of curve $c = f(x)$	Changes concavity for $f'' = +$ to $-$ or $f'' = -$ to $+$	$f''(x) = 0$	Points *C*, *E*, *F*, *G*

In exceptional cases an inflection point may occur when $f''(x)$ does not exist. The local maxima and minima of differentiable functions are always stationary points; that is, they satisfy the equation $f'(x) = 0$. These points separate regions of rise and fall of the function; however, other points (for example, point *F*) may also be stationary. At an inflection point, direction of concavity usually changes.

10. Finding local maxima and minima of one variable starts with solving for $f'(x) = 0$. If X_0 is any root so that $f'(x_0) = 0$, then
   a. $f(x_0)$ is max if $f''(x_0)$ is negative.
   b. $f(x_0)$ is min if $f''(x_0)$ is positive.
   If $f''(x_0) = 0$, then as x increases from $x_0 - \delta$ to $x_0 + \delta$ where δ is an infinitesimal step
   c. $f(x_0)$ is max if $f(x) - f(x_0)$ is negative or if $f'(x_0)$ changes $+$ to $-$,
   d. $f(x_0)$ is min if $f(x) - f(x_0)$ is positive or if $f'(x_0)$ changes $-$ to $+$,
   e. And has neither a max nor min at $x = x_0$ if $f'(x)$ does not change sign.

Note: The single prime mark, $f'(x)$, indicates the first derivative while two prime marks, $f''(x)$, indicate the second derivative or the derivative of the derivative.

left of the minimum point, then increasing the crew quantity will have a negative sign (rule 2) and improvement is possible until $dC_T(n)/dn = 0$. With comparative and competitive operations between districts A and B, the procedure is to examine the magnitude and sign of C_m. The greatest negative marginal cost indicates the design with greatest cost reduction for least input resources (a negative cost change is a gain). Given adequate resources, both crews should be expanded until the yields are zero. If resources are not present, the next strategy to adopt is to achieve the same marginal cost for both districts, and this may be done by reassigning operators from one district to another district.

Whenever we have a functional estimating model, say either a cost or profit one, and our interest is to find its optimum, the classic methods of calculus are powerful. This method belongs to the indirect classification of optimization because it solves equations and indirectly finds the optimum through solution of one or more equations. While we cannot hope to duplicate the knowledge of a course in calculus (particularly the differentiation part),

FIGURE 7.7

CORRESPONDING CURVE ARC FOR TABLE 7.8

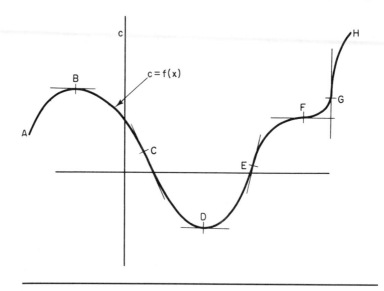

Table 7.8 and Figure 7.7 summarize some of the tests used to identify the optimum for a single-variable model.

With curves and mathematical models, like those described for the wrench, it is possible to optimize sales revenue and cost. The question should be asked, What shall we optimize? Maximizing sales revenue may not guarantee maximum profit, nor does minimizing cost balance other factors for maximum profit. In some cases these actions may in fact reduce profits. One of the traditional ways to study this interaction is by *break-even* analysis. A break-even point is defined as the point of production or operation at which there is neither a loss nor profit nor savings. Several points of neutrality are possible, and the location of these points is a straightforward exercise after the curves and models have been formulated from the data.

Using models (7-6) and (7-9) we assume that the linear approximation of a point of indifference occurs whenever $C_T = R_T$. Ignoring profits tax and solving for n, we have

$$n = \frac{C_f - R_f}{R_v - C_v} \tag{7-12}$$

Model (7-12) is for the class of problems where R_f is not zero; for a product that is commercially sold no income is received until products are sold and R_f as a fixed residual income does not exist, so $R_f = 0$. In the case of fitted models to estimated data R_f may exist to satisfy the best fit in a least-squares sense; moreover, when using linear models $C_T = 0.84n + 1527$ and $R_T = 0.894n + 1106$ the break-even point is

$$n = \frac{1527 - 1106}{0.894 - 0.84} = 7796 \text{ units}$$

This is pictured in Figure 7.8. The denominator of the foregoing is in reality unit profit without the consideration of fixed cost or return. But as fixed cost must be absorbed in the production and sales volume, the net unit profit when divided into fixed dollars yields the break-even point in terms of dollars and units sold.

FIGURE 7.8

LINEAR BREAK-EVEN ANALYSIS

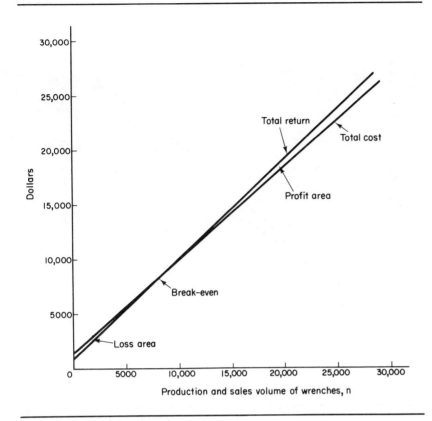

Several comments about linear break-even analysis are worth noting at this point. These relationships are valid only for the short-term period. *Short term* for one activity may be only weeks, while another firm with a more stable product may view *short term* over an extended period of months and even years. To establish arbitrary rules about what is short or long term is dangerous unless specific cases are analyzed. As was demonstrated by the example, the mathematics and graphics involving linear (or nonlinear) costs and incomes are capable of providing information for break-even production volume to meet existing cost and profit parameters for the immediate period. These price-volume details are plausible after detailed information is estimated.

In the linear example a narrow wedge of profits begins at 7796 units and continues indefinitely. The defect of pure linear models is obvious: Reduced revenue or unit price as an additional quantity is produced can occur from a variety of economic happenings. Increased unit costs of production beyond that which is normal are typical as production increases indefinitely. Although linear methods can be used for production above or below normal capacity and for cost-cutting tactics, the simplest way is to examine the plots of the original data. This can be done by straightforward plotting of Table 7.5 and 7.6 or through a computer quadratic least-squares fit. This is shown by Figure 7.9 where two break-even points, a lower and upper, are indicated. C_v is an increasing function, while R_v is a decreasing function. With this dual intersection of the return and cost lines, additional analysis of the meaning of optimum profit is called for.

FIGURE 7.9

TOTAL RETURN AND TOTAL COST FUNCTIONS

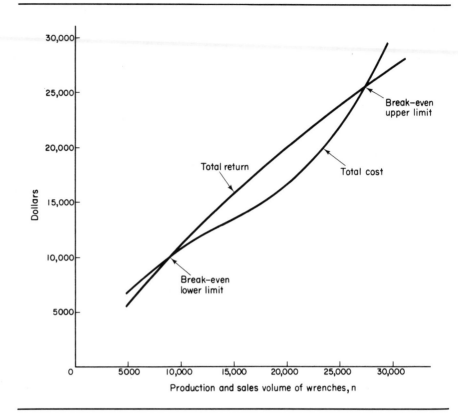

The intersection of the slopes of the total cost and total return functions is given as

$$\frac{dR_T}{dn} = \frac{dC_T}{dn} \qquad (7\text{-}13)$$

The finding of this intersection is the point at which marginal profit equals zero or $dZ/dn = dR_T/dn - dC_T/dn = 0$. If n increases beyond this point, the cost of each unit exceeds the revenue from each unit and total profits begin to decline as they do beyond the upper break-even point of Figure 7.9. The profits do not become zero until the profit-limit point, which is the beginning of a loss area. When the marginal profit is zero we have the critical production rate and maximum profit. This critical production does not necessarily occur at the rate corresponding to the minimum average unit cost, nor does it necessarily occur at the point of maximum profit per unit.

A plot of $dR_T/dn - dC_T/dn$ is given in Figure 7.6. The point at which $dR_T/dn = dC_T/dn$ is also zero for marginal profit or dZ/dn. The critical production rate may be determined directly from differentiation of the profit function Z. This critical production rate is zero at $n = 20,800$ units, which is the point of maximum gross profit. But it is not the rate corresponding to minimum average unit cost or the maximum profit per unit. The maximum profit per unit may be estimated from the level dome of the marginal profit curve, or approximately 12,200 units. Using Figure 7.6, minimum marginal cost is estimated as 14,500 units, and maximum marginal return is at the minimum sales projection of 5000 units.

Although we have shown that several points of operation can be obtained from minimizing total or unit cost, maximizing total or unit return, or maximizing gross or unit profit, one should not conclude that all is lost if these exact objectives are never precisely reached. There is a redeeming feature for the estimator if the actual n and the ideal point n do not coincide. Fortunately, most optimums are relatively flat near the optimum. For these domes operation on either side is in all likelihood insensitive. Naturally a converse example can be found that demonstrates a sharp peak as optimum. It is in this case that operation at this point becomes important.

7.6 SUMMARY

Frequently a preliminary estimate leads to the management decision of further consideration for the design. Before detailed methods of estimating are attempted, information that has been deliberately collected, structured, and verified to suit the method of estimating must be available. Most estimates are made using detailed methods.

The factor method and its many variations is a popular method. Using a wide assortment of functional models, the estimate is made for plants, processes, operations, and systems. In preliminary methods the approximate cost of a plant, for example, is estimated by multiplying the floor area by an appropriate dollars-per-square-foot factor. By increasing the amount of detail, greater accuracy is achieved by using separate factors for different cost items. Individual cost-per-unit-area figures can be used for heating, lighting, or painting. The power law and sizing model is another functional model used to estimate equipment.

Standard time data tabulated according to regression rules defining variables and constants are used to estimate labor. One set of data is required for each production method. In the process industries dominated by automation, labor may be estimated by a factor proportional to investment or materials or product cost, for instance. In construction, tables of time are available. In manufacturing, the application of standard time data is common. Both standard data and factor methods are used with each other, not separately.

Extensions of the factor method lead to cost-estimating relationships which are statistical and subject to various correlation tests for a best-model selection. Using CERs, detailed methods of estimating permit marginal analysis. Marginal analysis considers future design and cost under differing circumstances. The problem becomes one of how many additional resources are needed to acquire some specified additional capability, or conversely, how much additional effectiveness would result from some additional expenditure. The decision is made on relevant future estimated costs, not on costs which are sunk or average.

In the next four chapters the methods of detail estimating are examined further. The concern becomes one of using various cost-estimating relationships, ranging from simple cost factors to functional models, to estimate operations, products, projects, and systems.

SELECTED REFERENCES

For another viewpoint on estimating methods, consult

JELEN, F. C., ed.: *Cost and Optimization Engineering*, McGraw-Hill Book Company, New York, 1970.

VERNON, I. R., ed.: *Realistic Cost Estimating for Manufacturing*, Society of Manufacturing Engineers, Dearborn, Mich., 1968.

Various exponents for the power law and sizing model are in

PETERS, MAX S., and K. D. TIMMERHAUS: *Plant Design and Economics for Chemical Engineers*, 2nd ed., McGraw-Hill Book Company, New York, 1968.

The economics of marginal analysis are discussed in

> BAUMOL, WILLIAM J.: *Economic Theory and Operations Analysis*, 2nd ed., Prentice-Hall, Inc., Englewood Cliffs, N.J., 1965.

QUESTIONS

1. Describe the kinds of applications that are suitable for the factor method. What kinds are suitable for standard time data?
2. What is meant by the law of diminishing returns? Apply it to the factor method. Is the unit method of the previous chapter diminishing?
3. Outline the sequence of using factor data when the data are a benchmark 20 years ago, the evaluation is now, the contract-letting date is 2 years hence, and the project build date is 3 years hence.
4. Elaborate on the effects of fixed, semi-fixed, and variable costs on the collection of information, analysis, and construction of factor data.
5. Consider several potential applications for the power law and sizing model. What sort of information would be required for your examples? What range on the design would you think necessary?
6. Outline the method to be followed in marginal analysis. What distinguishes linear from nonlinear margins?
7. Define mathematically the point of indifference, the upper and lower break-even points, the maximum profit point, and the profit-limiting point.

PROBLEMS

7-1. A residential builder constructed a 2000-square-foot, two-story home on sandy loam soil for $31,000. This cost was exclusive of lot, taxes, and utility hookup costs. What is the unit estimate for his next home? An additional breakout of his costs revealed the following percentages for this same job:

Item	%	Item	%
Rough lumber	13.0	Earthwork	2.6
Rough carpentry	9.2	Flooring	6.4
Plumbing	13.6	Hardware	2.1
Finish lumber	1.2	Heating	3.0
Finish carpentry	2.8	Insulation	1.5
Cabinets	6.4	Lighting	0.5
Concrete	5.8	Painting	7.1
Wallboard	6.0	Roofing	2.2
Electrical wiring	5.5	All other	11.1

A 3200-square-foot, two-level home is to be built. Estimate his total and item costs. What elements would you adjust if this were a single-story home? Up or down?

7-2. Using a factor method, find the part cost for the following casting. A foundry is to cast a motor cylinder shown in Figure P7-2. Compute volume and cost of casting based on volume and the following factors: the shop yield is 54.5% (from prior experience), the metal loss is 10% or 5.4% using shop yield, furnace labor and overhead are $0.03 per pound of poured metal, the cost of metal charged is $0.06 per pound and the amount of remelted metal is 40% and is valued at $0.04 per pound. For cast iron, the density is 0.26 lb/in³. Allow $\frac{1}{8}$ inch of stock for all machined surfaces; the volume of cylinder is $\pi/4(\text{O.D.}^2 - \text{I.D.}^2) \times$ length. The pouring weight is finished weight/shop yield. The remelted metal weight is pouring weight $\times$ remelt factor. The cost of the metal is finished casting weight $\times$ metal loss factor. Consider what estimating factors are necessary. What items

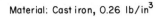

Material: Cast iron, 0.26 lb/in³

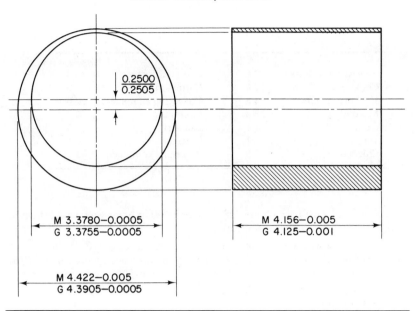

M 3.3780–0.0005
G 3.3755–0.0005

M 4.156–0.005
G 4.125–0.001

M 4.422–0.005
G 4.3905–0.0005

have been omitted in this estimate? Compute casting cost using your estimating factor. Note Problem 4-10 for similarity.

7-3. The hydrobromination bid received for another rising film and ozonation reactor is $175,000 and the current index is 121.5. Use the method discussed in Section 7.1. What is the benchmark cost? What are the factors for engineering, erection, and direct materials? An overall index of 123 is assumed for costs during construction. The regional factor is 7% more than the Gulf Coast for all factors. Items independent of the estimating process are found to cost $200,000. What is the total factor estimate for the chemical plant?

7-4. A processing plant is sufficiently designed for detailed estimating. Basic equipment is internally estimated as $249,000, while an outside consultant has suggested $325,000. The current Marshall and Stevens equipment index is 129.5. The index at the time of construction is forecast as 135.0 for equipment, erection, and direct materials and 148 for engineering costs and fees. The plant is to be constructed on the West Coast, and materials, erection, and engineering are 5, 17, and 0% more expensive, respectively. Site, utilities, and transportation roads and railroad spurs are estimated to cost $225,000 more. Find the total factor estimate for this plant.

7-5. A food factory is planned near a large area of truck farms. These farms sell to stores, but a surplus is generally available. A small general-purpose production plant is planned, and the following equipment is engineered and priced out by bidding:

Process	Cost	Process	Cost
Roller grader	$10,000	Extractor	$36,000
Peeling and coring	18,000	Two cookers	7,000
Pulping	30,000	Can seaming	54,000
Instrumentation	57,000	Packaging	22,000
Conveying	37,000	Six motors	35,000
Filler press	14,000	Heat exchanger	44,000
Tanks	10,000		$374,000

Factors to obtain installed cost, including cost of site development, buildings, electrical installation, carpentry, painting, contractor's fee, foundation, structures, piping, engineering overhead, and supervision, are listed for various equipment as

Process centrifuges	2.0	Can machines	0.8
Compressors	2.0	Cutting machines	7.0
Heat exchangers	4.8	Conveying equipment	2.0
Motors	8.5	Ejectors	2.5
Graders	1.6	Blenders	2.0
Tanks	2.1	Instruments	4.1
Fillers	0.8	Packaging equipment	1.5

Estimate a plant cost based on this information. What other way would there be to improve on the quality of this estimate?

7-6. An 80-kilowatt diesel electric set, naturally aspirated, cost $16,000 in 1964. A similar design, but 140 kilowatts, is planned for an isolated installation. The exponent $m = 0.6$ and index $I = 187$. Now the index is 207. A precompressor is estimated separately at $1900. Using the power law and sizing model, find the estimated equipment cost.

7-7. A firm that manufactures totally enclosed capacitor motors, fan-cooled, 1725 revolutions per minute, $\frac{5}{8}$-in.-O.D. keyway shaft has known costs on $\frac{1}{4}, \frac{1}{3}, \frac{3}{4}$, and 1 horsepower as $22.50, $25.00, $29.50, and $35.00. A 3450-revolutions-per-minute, 1-horsepower motor costs $33.50. Determine m for the 1725-horsepower motor series, and estimate the cost for a $1\frac{1}{2}$-horsepower motor. Should the 3450-revolutions-per-minute motor be included in the sample to estimate the 1725-revolutions-per-minute motor?

7-8. A small food-processing line has been designed and costs are known. Another similar line is to be estimated using the power law and sizing model. Major equipment, costs, and exponents are

Equipment	Reference Size	Unit Reference Cost	Exponent	Design
Dryer	27 ft²	$1800	0.20	38 ft²
Kettle	415 gal	2015	0.24	510 gal
Two motors	6 hp	625	0.53	7 hp
Two pumps	6 hp	1400	0.21	7 hp
Tank	800 gal	725	0.66	915 gal
Tank	500 gal	615	0.66	610 gal

If ancillary equipment for this anticipated line will cost $20,000, find the cost for the proposed equipment design.

7-9. Decide whether the following operations should be classified as constant or time variable. If the operation is constant, compute an average value for the time. If the operation is time variable, derive the equation relating time and the dependent variable. Use guide lines given in Section 7.3

(a) A foundry has data which indicate that it takes 82 seconds to deburr a 2-pound casting and 88 seconds to deburr a 4-pound casting.

(b) A pharmaceutical company has time-study records which show that for a particular product production runs of 30 and 55 hours produced outputs of 400 and 725 ounces, respectively.

7-10. Develop a set of standard data for an operation of "sawing and slitting" $\frac{1}{4}$-inch-O.D. maple dowels in a furniture factory. Several time studies are available from which normal times are removed for a recap sheet. It is assumed that methods engineering has standardized work-place layout.

Element	1	2	3	4	5
1. Pickup dowel	0.023	0.023	0.035	0.041	0.051
Reach distance	8 in.	8 in.	13 in.	16 in.	17 in.
2. Position for slotting	0.031	0.035	0.029	0.030	0.032
3. Move 14 in. and piece aside in tote box	0.024	0.021	0.031	0.030	0.030
4. Saw slot to length	0.093	0.126	0.103	0.210	0.193
Length	$\frac{1}{4}$ in.	$\frac{1}{4}$ in.	$\frac{1}{4}$ in.	1 in.	$\frac{7}{8}$ in.

Show the standard data in a tabular format.

7-11. A broadcast radio chassis weighs 4 pounds and has overall dimensions of $3 \times 6 \times 7\frac{1}{2}$ inches, box size. The methods engineer visualizes the elements of a paint operation as

Load part
Spray enamel inside surface less 2×4 inch opening
Spray enamel outside surface less 2×4 inch opening and bottom
Hang piece on hook
Spray bottom outside
Inspect
Unload

Using Table 7.3 determine the setup hours and standard minutes per piece. In this table setup hours are standard while the allowance for run time is 12%. What are the standard hours per 100 units? If the operator is paid $5.25 per hour, what is the total cost for a shop order of 60 parts? If the spray booth and materials cost $3.15 per hour, what are total and unit costs for the operation?

7-12. An eight-man drafting department concerned with size A, B, and C drawings is work-sampled by a management consultant over a standard 4-week period. A stick chart was summarized for the categories as

Item	
Drafting and tracing	778
Calculating	458
Checking prints	110
Classroom	125
Professional time off	172
Personal time, idle	270

During this period 55 drawings ($A = 20$, $B = 25$, $C = 10$) were produced with a total payroll of $8800. If the policy is to accept professional and personal time as necessary to the drafting of prints, what is the job factor? If personal time and idle time are prejudged at 10% only, what is the job factor?

7-13. Engineering aides conduct self-surveys of their daily work, and after a sufficient period of data collection, classification of the information reveals the following:

Description of Work	Minutes Per Occurrence	Frequency Per Print
1. Post drawing numbers	7.0	1/1
2. Duplicate drawings	19.0	1/1
3. Correct computer cards	38.0	2/1
4. Phone calls	1.5	1/4
5. Update drawing changes	27.0	1/2

A personal allowance of 15% is used for this category of work. Determine the job factor per engineering drawing. A new product design is anticipated and a separate staff will be collected. This product design will probably result in about 2500 prints over a 1-year period. How many aides will be required, and at $5 per hour, what amount would you estimate for this activity in a product expense budget?

7-14. A linear statistical cost-estimating relationship for public hospitals is given as $C_{ph} = 110,000 + 15\text{SFC}$, where SFC = square feet of all construction (including walls, floors, internal partitions, overhangs, but counted from one side). The correlation coefficient is $r^2 = 0.92$. The analyst concluded that SFC was a greater contributing feature than the common square feet for which $r^2 = 0.82$. What causal conclusions should the estimators who use this CER draw? Hospital construction requires a greater quantity of in-wall services, X-ray shielding, and sound-barrier installations.

7-15. Find the maximum and minimum points for the following curves using methods of classic calculus. Confirm the nature of the point by a second derivative test.
(a) $c = x^2 - 2x + 3$.
(b) $c = 2x^3 - 3x^2 - 36x + 20$.
(c) $c = 2x^{5/2} - 40x$.
(d) $c = (2x + 1)^{3/2} - 6x + 5$.
(e) $c = x^4 - 2x^2 + 12$.
(f) $c = x^2/(x - 1)$.
(g) If $c = (x + 1)^3(x - 3)^2$ and $c'(x) = (x + 1)^2(x - 3)(5x - 7)$, find the values of max and min points.
(h) $c = 2x_1^2 + 4x_1x_2 + 5x_2 - 2x_2^2$.

7-16. (a) An opportunity sale of an additional 200 units is received above an existing production level. The buyer is prepared to pay $350 for these 200 units. Cost for the current 500 units is $950. Total cost for producing 700 units is $1277. What is the marginal total cost? The unit cost? What is the profit or loss? What is the minimum opportunity sale price to break even?
(b) The total cost of producing a given output varies according to the following estimated data:

Output	Total Cost	Output	Total Cost
0	0	5	192
1	100	6	205
2	150	7	230
3	175	8	280
4	187	9	380

Plot the estimated total cost curve, average cost curve, and marginal cost curve.

7-17. Given the marginal return function $= 500 + 0.10n$, marginal cost $= \$1000$, point of maximum profit $= \$5000$, and fixed cost $= \$10,000$, find the demand function for price, the quantity per period, and the total cost function for the same period.

7-18. A motorcycle manufacturer is considering producing a new model of motorcycle designed for a specific market. Management has determined the total fixed cost for the plant and design to be $\$10^5$. The cost of producing the cycle after the initial investment has been made is estimated as a function $V(n)$ of the total number n of cycles produced.

(a) If $V(n) = 200n - 3 \times 10^{-3}n^2 + 10^{-7}n^3$, determine the marginal cost.

(b) What is the minimum marginal cost? How many cycles does this correspond to, and what is the average cost per cycle at this level of production?

(c) If the total return from sales $S(n)$ is given by $S(n) = 250n$, find the marginal return, the total profit, and the marginal profit.

(d) How many cycles should be produced to maximize profit?

7-19. The marginal cost function is $dC_T/dn = 5\beta + 10\alpha^2n^4 - (0.5n/\gamma)$ and gross profit is $0.0250n^{-2}\theta + (6\phi - 5\beta)n - (7n^4 + 2n^5)\alpha^2 + (0.25n^2/\gamma) - C_f$. Find the total cost, the variable cost, the average sales price, and the maximum profit per unit where α, β, γ, ϕ, and θ are constants.

Marginal Labor and Tool Cost.[3] The Don Boyle Co. makes zinc die castings. It has one 300-ton machine with a trim press located at the end of the quench conveyor. The machine is a hot chamber type and is capable of running automatically.   **CASE PROBLEM**

Andy James, Inc., manufactures leather goods. The purchasing agent for Andersons, Inc., calls Andy James (and several other manufacturers) for a quote on a 3-inch-wide "mod" belt. The belt is to be made with an adjustable buckle. The James engineers design a zinc die cast chrome-plated buckle, and call Don Boyle (and several other die casters) to quote on the buckle. Andersons, Inc., has estimated a volume of 75,000 belts annually for the next 3 years, at which time it is thought the item will be phased out.

Don Boyle must quote Andy James a unit price and a tooling price. He has a 300-ton machine capable of casting up to 16 buckles at one shot. His trim press is also capable of handling a shot this size. How does he quote this part to stay competitive? A 16-cavity die would minimize the unit price, but the tooling cost would be enormous. A single-cavity die maximizes the unit price but minimizes the tooling price. Which alternative should Don choose?

A single-cavity casting die, Don estimates, would cost about $4500. A trim die would cost an additional $900. Direct labor and overhead to cast one shot, no matter how many parts it has, is 6 cents. In other words, to cast the required 225,000 buckles in the next 3 years, the lowest cost for tooling is $5400. However, the unit cost (excluding tooling) is 6 cents per part, or $13,500 for the total of 225,000 units (no allowance is made for scrap).

Now, the marginal costs and marginal revenues must be examined. Marginal revenue is defined as the increment of total revenue (plus or minus) that results when the number of cavities is increased by one unit (or fraction thereof). Marginal cost likewise is defined as the increment of tooling cost that results from a similar increase in cavities. (Total revenue does not refer to the amount of money to be paid by Andy to Don. That would be a pricing problem, whereas this discussion concerns the problem of an efficient design.) Total revenue here is the total dollars saved by using a two-cavity die compared with a one-cavity die, or three cavities against two, and so forth. This total revenue may be called unit savings.

To cast the buckle for Andersons, Inc., the minimum starting point is single-cavity tooling. Total revenue at this point is zero. Don has not yet experienced a saving as a result of his tooling choice.

[3] Adapted from a paper by T. S. Platzer, Samsonite Corp., Denver, Colo.

A two-cavity design, Don estimates, would cost an additional $600 for the die-casting die and $200 for the trimming die. The marginal cost is $800, and this does not include unit cost. Unit costs are considered only in respect to changes that create a savings or loss. A two-cavity die will produce two parts per shot, reducing the unit cost to 3 cents per part, or $6750 for the 3-year requirement. This is a reduction from $13,500 to $6750, or a unit saving of $6750.

For each additional cavity added to the tooling, Don estimates that the tool cost will increase by an additional $800. With 16 buckles per shot, the maximum capacity of the machine has been reached. Anything over 16 buckles per shot would require a larger die-casting machine.

Where does Don maximize the profits from a tooling standpoint? How many cavities does this decision call for? What is the full cost for this decision? How does raw material affect your decision?

chapter 8

*For which of you, intending to build a tower,
sitteth not down first, and counteth the cost,
whether he have sufficient to finish it?*
Luke 14 : 28
The Holy Bible
King James Version

OPERATION ESTIMATING

This chapter stresses operation estimating. The heart of operation estimating is, in a substantive measure, a talent for breaking a task into essential elements. A design formulated sufficiently for this analysis is available, and the cost estimator solves functional cost equations that model the details of the design.

In manufacturing the design is a part print and a processing plan; in chemical industries a flow chart and layout with engineering calculations are at hand; in engineering construction a set of drawings and specifications is a part of the given information. These designs are used for purposes such as estimating, planning, scheduling, methods improvement, and production. They communicate the description and sequence of operations and are the hub on which individually and collectively many decisions, both large and small, are made.

Outmoded practices of operation estimating take the total cost of operating the plant for a period of time and divide by a total production quantity. Other practices determine a labor and material cost for an operation from a test run or prototype. Historical records of like operations are used. These methods are unsuitable for predictive cost estimates.

The operation estimator begins by subdividing the design task into large portions of labor and material. Progressively finer detail is determined until a description of labor and materials is very broad. At this point dollar extensions of labor and material are made to reflect the cost of the design.

There are many thousand kinds of labor and material operations. Regrettably, our choice of explanation extends only to few. Trade books, handbooks, internal sources of data, etc., must be consulted for data for any real-life estimating. Some sources are listed in the references.

8.1
OPERATION COST

An operation estimate is a forecast of labor and material expenses. While the labor and material may differ from design to design, techniques for evaluation are based on the same principles and practices. Operations are necessary to produce a change in value, and as a way of working the basic combination of a man and tool is the primary ingredient. Through man and tool activities the economic value of material is altered. The measure of the change in economic value is cost.

Cost in this instance implies a consumption of labor, materials, and tools in order to increase the value of some object. The use of expensive fixed assets involving capital cost in operations, while a consumption activity of wealth, is deferred until Chapter 10 where methods are introduced. In some cases automatic equipment produces units of output and may not require labor. These operations, however, consume materials and utilities and constitute the operation cost estimate.

Operation evaluations are limited as to the time horizon. The immediate future period rather than an extended time period is intended. The nature of labor is for a period of time such as "units of labor time per piece" or "units of labor per month or year."

The estimate may be undertaken to establish the operation cost for components, subassemblies, and intermediates of a product; to initiate the means of cost reduction; to provide a standard for production and control; and to compare different design ideas. It may be used to verify operation quotations submitted by vendors and to help determine the most economical method, process, or material for manufacturing a product. Whenever an operation is material- or labor-intensive, rather than capital-intensive, the methods of this chapter are primary.

8.2
ESTIMATING MATERIAL REQUIREMENTS AND COSTS

Direct materials consist of raw materials (such as flat plate, bar stock, and crude oil), castings, forgings, intermediate chemical products, and other standard parts normally not produced by the manufacturer. Materials have the characteristic of having been purchased, not manufactured, by the plant to which it is material. Thus sheet steel is a product from a rolling mill but a material in a sheet-metal forming plant. Direct materials become a part of the final product or are involved in the operation in a way that the material can be estimated. Indirect materials are those materials that are critical to the operation but do not become a part of the final product. This may include cutting lubricants, welding rods, clerical supplies, and so forth, but because their cost is difficult to assess, it is charged to operation cost by overhead distribution rather than through computational methods by estimating. As a consequence estimating evaluates the cost of direct materials and standard purchased parts contributed to the operation.

8.2-1
Standard Purchased Materials

Standard purchased materials are a class of materials normally not converted; rather they are accepted in a manufactured state, for instance, tires used by an automobile assembly plant. This category may be a significant proportion of total material cost, and it is considered separate from fabricated raw materials. Standard purchased parts and materials may carry a specific purchasing overhead rather than a larger general overhead rate, although the practice is not uniform. It may not be competitive to add a large markup on the same items that can be purchased by customers. Purchased parts and standard materials may be charged with out-of-pocket expenses for procurement, freight, receiving, handling, inspection, installation, and testing.

Standard purchased parts are costed either by estimating or purchasing. Estimating or engineering may provide a bill of material listing of the standard parts to purchasing, who will price them from catalogs or from quotations and return the information to estimating. Sometimes when a short lead time for sales prevails, estimating may compile the standard part costs. But with frequent price changes and negotiated price-volume breaks between the company's purchasing agents and seller, information from catalogs can be faulty. For some operations and products, the standard purchased parts such as nuts, bolts, and washers are estimated by a multiplying factor correlated to the direct materials. This happens whenever the standard purchased parts are insignificant.

8.2-2
Direct Materials

Direct material cost may be determined by the following function:

$$C_r = W(1 + L_1 + L_2 + L_3)P_m - R \qquad (8\text{-}1)$$

where C_r = material cost for a unit, dollars per unit
W = weight in pounds for a unit, or in compatible dimensions to price P_m

L_1 = losses due to scrap, decimal
L_2 = losses due to waste, decimal
L_3 = losses due to shrinkage, decimal
P_m = price per pound of material, or price per linear foot, or price per volume
R = unit price of anticipated material salvage, dollars per unit

Historical evidence is collected for prediction of material losses resulting from other than ideal material utilization. More than the desired final quantity should be started due to spoilage and rejection for defects. The rejection for defects results in scrap consisting of unusable raw materials or stock. A number of things are responsible for defects such as improper material layout, pin holes from an incorrectly poured casting, and fracture of formed sheet metal parts due to improper heat treating and aging. Scrap may include chips, short ends, and spoiled parts which cannot be recovered. Waste is illustrated by that which is left after parts have been blanked out. The deterioration of certain plastics within their processing pots is another loss. An overall shrinkage allowance compensates for physical shrinkage. The shrinkage allowance for lumber may be as high as 25 or 30%. Metal work may be 1% or less and plastics may amount to 10%. Each operation is analyzed for direct materials to produce the desired quantity. This computation is used to buy the exact amount of raw material, as well as cost the materials for the part or operation. In machining, the cost of the raw material for a piece is determined by multiplying the unit cost of the material by the weight of the rough stock used per piece. Rough stock includes overburden material to allow processing losses, gripping losses, scrap, and waste. Equation (8-1) requires for a machined part that the amount of stock overburden removed by machining be added to finished dimensions. For a component irregular in shape it is divided into simple geometric shapes, and the volume of the components is added to give total volume. The multiplier for volume would be density. Nonconvertible losses may include the tail-ends and short-ends of bar stock that can not be processed on bar machines.

Consider as a first example a $1\frac{7}{16}$-inch-O.D. carbon steel bar that is $1\frac{1}{8}$ inches long. The lathe cutoff tool will waste a $\frac{1}{8}$-inch width, and the machine selected for this task is limited to 6-foot bar lengths. Then the length (1.125 inch) plus parting tool width of 0.125 inches per unit gives the number of pieces from the bar stock as $72/1.250 = 57.6$. Reducing this quantity by 1 to permit collet gripping and with L_1, estimated as 2%, the anticipated unit material cost is

$$C_r = 0.42\left(1 + 0.02 + \frac{0.125}{1.125} + \frac{1.8}{72}\right)0.25 = \$0.121 \text{ per unit}$$

Losses for this type of material, including scrapped pieces, butt-ends, and overburden vary from 1 to 12% with 5% often added to material estimates to prorate the losses over the pieces produced. If the material is reclaimed as scrap, then reductions can be assessed against the estimate. This is illustrated by one of the exercises involving a plating operation. Competition within screw-machine shops is sometimes so severe that the profit may be limited to the reclaimed amount of sold scrap.

For stamping fabrication from coil stock or a blank and trim operation from sheets, modifications of Equation (8-1) are necessary. Raw material costs of sheet-metal stampings may be found as follows:

$$C_r = WLtP_m \tag{8-2}$$

where W = stock width, inches
L = blank length, inches

t = stock thickness, inches

P_m = price of material, dollars per cubic inch

Other models can be used depending on the engineering requirements, such as

$$\text{pieces per coil} = \frac{\text{coil length}}{\text{length of design}}$$

$$\text{weight of coil} = \text{gauge} \times \text{width} \times \text{length} \times \text{density}$$

(8-3)

$$\text{weight per piece} = \frac{\text{weight of coil}}{\text{pieces per coil}}$$

$$C_r = \text{piece weight} \times \text{cost of material per pound}$$

Processing losses for sheet-metal work are guided by experience. Note the work piece illustrated in Figure 8.1. In stamping fabrication, the distance

FIGURE 8.1

SHEET-METAL COMPONENT

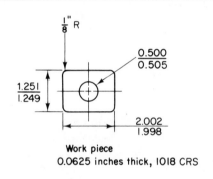

Work piece
0.0625 inches thick, 1018 CRS

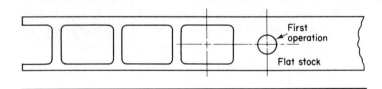

between blanks is restricted to a minimum of $0.75t$, while margins between the edge of the strip and blank are restricted to $0.90t$ for each side. Using Equation (8-2), the part width is 1.25 inches, margins are $2(0.90 \times 0.0625)$, and $D = 1.25 + 0.1125 = 1.3625$. The length is 2 inch $+ 0.75 \times 0.0625 = 2.047$. Our material, cold-rolled steel, has a density of 0.28 pounds per cubic inch and $0.15 per pound is used:

$$C_r = (1.3625)(2.047)(0.0625)(0.28)(0.15) = \$0.0073 \text{ per unit}$$

The sheet-metal part is considered a rectangle for purposes of material costs even though the slug in the center is not used for the part.

In construction cost estimating the labor cost and material cost are often calculated together, as this is the way the cost data are arranged. Consider the foundation wall given in Figure 8.2. Table 8.1 is the calculation of the unit cost per linear foot of the wall. The labor rate is related to the quantity and would be found through man-hour reporting systems, as shown in Chapter 3. The $16.40 cost per linear foot for a foundation wall is the factor, and when multiplied by the design length, the operation estimate is determined.

FIGURE 8.2

TYPICAL FOUNDATION WALL.

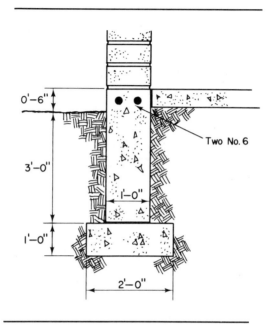

Consider an electrical equipment control chassis as another example. A material estimate is required before any of the electrical components have been mounted. Figure 8.3 is the isometric where a partition designated as P/N 673 is spotwelded to P/N 672 to become subassembly P/N 671. In this design, sheet stock 48 × 96 inches is used rather than coil stock. A waste allowance of 5% is added to account for losses in shearing of the sheet to blank size. The material cost for 5052-H34 Aluminum of 0.040 inches is $0.0030 per square inch and of 0.062 inches is $0.0041 per square inch. For P/N 673 and using Equation (8-2),

$$C_r = 12.50 \times 4.25 \times 1.05 \times 0.0041 = \$0.229$$

and for P/N 672

$$C_r = 7.75 \times 3.5 \times 1.05 \times 0.0030 = \$0.086$$

The information is presented in Table 8.2. Differences due to material costs that are due to rising prices are introduced. Inasmuch as the material quantity is too low for any price break, this consideration is unnecesary.

It is sometimes possible to achieve refunds for returned materials. In precious metal plating operations recovery techniques of noble metals in collection tanks are possible. Scrap is sold to a scrap dealer at prevailing prices. If

TABLE 8.1

COST DATA FOR FOUNDATION WALL THAT
INCLUDES LABOR AND MATERIAL, BUT
COSTING BASED ON MATERIAL 1 FOOT LONG[a]

Description	Quantity	Total Quantity	Labor Rate	Material Rate	Labor Amount	Material Amount
Excavation and disposal:						
$\quad$ 1-0[b] × 4-0 × 4-0	16.0	0.59 yard³	$2.00		$1.18	
Forms:						
$\quad$ 2 @ 1-0 × 4-6		9.0 ft²	0.60	0.15	5.40	$1.35
Reinforcing steel:						
$\quad$ 2-#6 (1.5 #/lf)		3.3 lb	0.05	0.10	0.16	0.33
$\quad$ × 1-0 + 10%						
Concrete:						
$\quad$ 1-0 × 2-0 × 1-0	2.0					
$\quad$ 1-0 × 1-0 × 3-6	3.5					
	5.5	0.20 yard³	3.00	13.00	0.60	2.60
Finish exposed concrete:						
$\quad$ 1-0 × 1-0		1.0 ft²	0.10		0.10	
Backfill and compaction:						
$\quad$ Excavation	16.0					
$\quad$ Less concrete						
$\quad\quad$ 1-0 × 2-0 × 1-0	(2.0)					
$\quad\quad$ 1-0 × 1-0 × 3-0	(3.0)					
	11.0	0.41 yard³	1.00	2.00	0.41	0.82
					$7.85	$ 5.10
						7.85
Insurance, taxes, union benefits (25% labor)						1.96
						14.91
Overhead (10%)						1.49
						$16.40

[a] From *Plant Engineering*, Sept. 1967, p. 125.
[b] One foot, 0 inches.

segregation of special materials into containers is encouraged, a better price becomes possible.

The selection of P_m can be complicated. P_m may be chosen from an inventory value, market cost, or a contract price for raw material, as the three may differ. Additionally, a scale-of-unit cost, decreasing with quantity, is realistic for purchased materials. The selection of a price for material depends on the circumstances of the operation.

The importance of material varies with respect to the operation cost. In some cases material is a small fraction of the cost, for example, the radio chassis. For automated production, material is a major item.

A procedure for direct material estimating may be summarized as

1. Determine the necessary additions to the final design dimensions to allow for processing losses from a rough to a final size.
2. Calculate the unit or lot size for a rough article.
3. Evaluate other nonprocessing losses and increase the bulk quantity of the unit or lot size by this loss proportion.
4. Multiply the quantity of material by the unit cost of the material considering price breaks for large orders and other fluctuations in base price.

FIGURE 8.3

SHEET-METAL PART

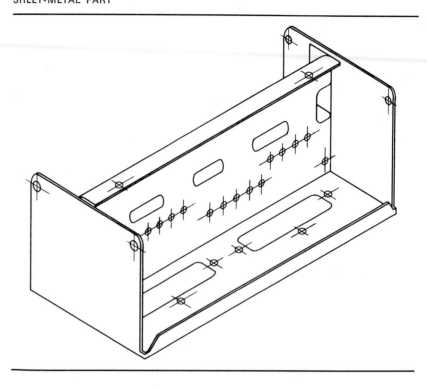

TABLE 8.2

COMPILED MATERIAL COST DATA
FOR ELECTRICAL CHASSIS

Manufacturing Estimate for Direct Materials

Product _____ Radio _____

Based upon quantity of ____ 100 units in year 1, 1000 units/yr. for years 2 and 3 ____

Date ____ March 19 ____

Main drawing number ____ 761—06734 ____

Line	Detail Drawing Number	Name	Material Specification and Thickness	Units	Blank Size	$/in² Cost Factor	Cost
1.	P/N 672	Chassis	5052-H34-0.040	in.²	7.75 × 3.50	$0.0030	$0.086
2.	P/N 673	Partition	5052-H34-0.062	in.²	12.50 × 4.25	0.0041	0.229
							$0.315
4.	P/N 672	Chassis	5052-H34-0.040	in.²	7.75 × 3.50	0.0032	$0.091
5.	P/N 673	Partition	5052-H34-0.062	in.²	12.50 × 4.25	0.0043	0.240
							$0.331
7.	P/N 672	Chassis	5052-H34-0.040	in.²	7.75 × 3.50	0.0034	$0.100
8.	P/N 673	Partition	5052-H34-0.062	in.²	12.50 × 4.25	0.0045	0.251
							$0.351

Year	Unit Cost
1	$0.315
2	0.331
3	0.351

Indirect costs cover those goods and services necessary for operation but are not traceable directly to specific operations or products. Nonassignable materials and tooling expenses are prominent sources of expenses that are indirect. Some materials can be classified as either indirect or direct depending on convenience. Fuel, such as natural gas, is a bonafide raw material, e.g., cracking refinery gases for the production of ethylene or heat-treating furnaces. They can be treated as a direct raw material cost or an indirect utility expense. Operating supplies include such things as lubricating oil and brooms and are too diverse and unimportant to be directly considered by the estimator. Company records should be used for this cost when they are available. Operating supplies have been found to be about 6 % of operating labor, but this is highly variable. Correlation of operating cost as a percentage of investment is often done for the chemical process industries.

In this section we shall demonstrate an example of a tool estimate. This is an indirect material from the company's tool room or an outside tool shop. Tool cost can be considered as an item of the total cost of an operation. In view of its importance to production a tooling estimate calls for special attention and is an exception to the practice of ignoring indirect materials. The cost of tooling can range greatly. Sometimes it is charged directly to a machine center, product, or proposal, and in other cases it is capitalized. To perform cost analysis for tooling there should be information available to compare the tooling required by the preliminary plan against that which is already on hand. Standard time data for the construction of tools are necessary.

After the preparation of the preliminary manufacturing plan is started, the cost engineer is able to evaluate perishable, special, and general-purpose tooling, inspection devices, and equipment. As equipment is normally capitalized and depreciated, it is considered in a subsequent chapter. The estimator normally does not concern himself with ordinary perishable tooling such as drill bits and cutters, which are commonplace. He does consider special tooling that pertains to the job and general-purpose tooling when it is uncommon for the operation being estimated.

In terms of test equipment and inspection devices, it too is weighted on equal footing with ordinary tooling. The estimator should be aware of what exists and is available for production and that which is special, requiring design or purchase. Special processing tools may be required for general-purpose or special-purpose equipment. In operation and postoperation inspection tooling before assembly could include gauges, holding devices, cutting tools, universal jigs, specially designed and constructed fixtures, and other purchased tooling. Inspection requirements for the estimate include such necessary tools for the receiving department, in-process inspection, postfabrication inspection requirements, and postassembly requirements. Environmental needs where engineering specifications call for peculiar ambient, weather, or shock tests cannot be overlooked.

Many types of jigs, fixtures, tools, gauges, and dies have fundamental details in common. This fact permits simplification in methods of estimating. The factor method is usually employed; the factors are not minute as found in labor estimating for mass production. That is too time consuming. In some cases the tool estimator may resort to the comparison method, which is the quickest but not the most accurate way.

Table 8.3 is an example of standard time data for tools. For Figure 8.4 we consider two manufacturing plans. It will be possible to compare their tool cost and see if any additional cost is made up by a cost reduction in labor. The first method employs a progressive die, which completes the part. The second method needs three dies. Note Table 8.4, which is a tool estimating form. The selected

TABLE 8.3

STANDARD TIME DATA FOR THE CONSTRUCTION
OF DIES MADE OF OIL-HARDENING TOOL STEEL
(TIME IN HOURS)[a]

1. Basic Piece Hole Dies

Hole Diameter (in.)	Hours for Basic Die (One Hole Only)	Hours for Each Additional Hole
Small holes:		
0.0156–0.1245	21	3.5
Pierce holes:		
0.125–0.500	16	2.5
Bored holes:		
0.501–1.000	21	9.0
1.001–1.500	31	10.0
1.501–2.000	33	12.0
2.001–2.500	36	15.0
2.501–3.000	39	18.0

2. Basic Blank Die

Periphery (in.)	Contour of Blank			
	Straight	Angular	Curved	Irregular
Up to 3	24	27	30	37.5
3–4	27	32	40	50.0
4–5	29	38	50	62.5
5–6*	32	45	60	75.0

*If periphery is greater than 6 in., use these values for every inch:

Straight line 5.0
Angular line 7.5
Curved line 10.0
Irregular line 12.5

3. Basic Form Die

Length of Bend	Width of Bend	90° Bends			Angle Bend
		L	U	V	
Up to 1	Up to 1	30	30	30	40
Over 1 to 2	Up to 1	32.5	32.5	32.5	43.5
Up to 1	Over 1 to 2	32.5	32.5	32.5	43.5
Over 1 to 2	Over 1 to 2	35	35	35	47
Over 2 to 3	Over 2 to 3	40	40	40	54

4. Pierce Holes and Pilots in Combination Dies

Type of Hole	Hole Diameter	Pierce Holes, Estimated Times (hr)	Pilots
Small	0.0156–0.1245	7 hr/hole for the first five holes. For each additional hole thereafter add 3.5 hr/hole.	3.5
Pierce	0.125–0.500	5 hr/hole for the first five holes. For each additional hole thereafter add 2.5 hr/hole	2.5
Bored	0.500–1.000	10 hr/hole for the first five holes. For each additional hole thereafter add 5.0 hr/hole.	5.0

TABLE 8.3 (Cont.)

5. Blanking Stations in Combination Dies

	Contour of Blank			
Periphery (in.)	Straight	Angular	Curved	Irregular
Up to 3	For odd-shaped blanks regardless of contour within blank, use a minimum of 15 hr/hole; if there are more than three similar holes, use 10 hr/hole thereafter.			
Over 3—add per inch of periphery	5.0	7.5	10.0	12.5

6. Bend or Forms in Combination Dies

Length of Bend (in.)	Width (in.)	90° Bends	Angle Bends
Up to 1	Up to 1	15 hr/bend	20 hr/bend
	Over 1 in. of length or width—add for every 1 in., either length or width, 2.5 hr for 90° bends and 3.5 hr for angle bends.		

*a*From Leonard Nelson, "How To Estimate Dies, Jigs, and Fixtures from a Part Print," *American Machinist/ Metalworking Manufacture*, Sept. 4, 1961.

TABLE 8.4

TOOL ESTIMATE FOR CLIP, FIGURE 8.4

Tool Estimate Sheet

1. Time Estimate

Method 1. Progressive Die

Item	Element No.	Calculation	Hours
3 Small holes	4	7×3	21
8 Small pilots	4	8×3.5	28
$1\frac{1}{2}$ Straight outline	5	2×5	10
2 Piece holes (tab)	4	2×5	10
$\frac{1}{2}$ In. of angle	5	$\frac{1}{2} \times 7.5$	4
4–90° Bends	6	4×15	60
		Total	133

Method 2. Three Dies

Item	Element No.	Calculation	Hours
Pierce and blank die			
3 Small holes	4	7×3	21
1 Small pilot	4	1×3.5	3.5
$1\frac{1}{2}$ In. of straight	5	2×5	10
2 Pierce holes (tabs)	4	2×5	10
$\frac{1}{2}$ In. of angle	5	$\frac{1}{2} \times 7.5$	4
			48.5
Form die (2 ears and end) 3–90° bends	6	3×15	45
Final form die Basic form die	3	1×30	30
		Total	123.5

TABLE 8.4 (cont.)

2. Materials

Method 1	Method 2
1 Die shoe	3 Die shoes
1 Die block	3 Die blocks
12 Pins	24 Pins
2 Strippers	6 Strippers
Misc.	Misc.
Est. $60	Est. $175

3. Adopted method: Progressive die

133 Total mach. hr @ cost $6:	$798
Total material & heat treat. cost:	60
Assem. & handling hr:	
Misc. cost, knobs, screws:	5
Try-out cost:	25
Total cost	$888

FIGURE 8.4

MANUFACTURED PART, 303 STAINLESS STEEL,
0.031 × 9/16 × 1.0 INCHES DEVELOPED SIZE

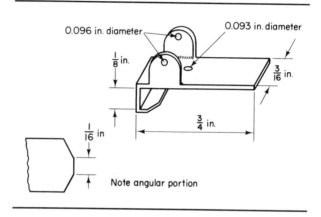

method uses the one-tool progressive die method ratherht an the three tools. The cost per hour of $6 includes expense chargeable to the tool room, space, power, tool drafting, and supervision. Some companies that maintain tool shops to manufacture their own tools insist there be no hidden costs, and when circumstances dictate, their tool shop competes for tool orders.

Why is a tool such as a progressive die labeled an "indirect materials expense" when the major amount of cost is indirect labor? The answer lies in the procedure used to recover this cost from customers. The recovery procedures visualize the output as a tangible material supplemental to direct labor and direct material expenses.

8.3
LABOR ESTIMATING

Labor comprises one of the most important items of operational cost. It has received intensive study, and many recording, measuring, and controlling schemes exist in an effort to manage labor. It can be classified a number of ways; for instance, direct-indirect, recurring-nonrecurring, designated-nondesignated, exempt-nonexempt, wage-salary, blue collar-management, and union-nonunion are choices.

Social, political, educational, and type of work are other divisions which classify labor. For purposes of this chapter we select direct-indirect as most appropriate because, simply, it is common in estimating circles.

Our examples are taken from a variety of work situations and deal with metal fabricating and construction. References are provided for additional information on other occupational fields.

The total direct labor required for an operation may be estimated by comparison with one or more similar jobs or it may be ascertained by detailed methods. For direct labor in production-type industries the detailed methods of estimating are encountered. This involves the breaking down of an operation into its basic elements, applying time factors to these elements, and finally arriving at a total time for the operation.

Fabrication unit operations, which involve processes on separate and distinguishable parts, are divided into a setup time and cycle time. Practice dictates that setup time standards for ordinary operations be determined from standard time data which have been constructed from any of the several methods of time measurement. The setup time is applied once to each lot of pieces and for cost estimating is prorated among the number of units in the production lot. Setup time is given in standard hours.

It has been previously stated that nonrecurring costs are one-time costs which do not accrue with production. Machine preparation or setup is one of the principal fixed costs that are treated differently. It is the custom that the cost of machine changeover include both the cost of setting up the machine from an inoperative condition and dismantling the machine or tools to another condition preparatory for the next lot. This setup and tear-down, jointly called setup, is considered to be a variable cost and is handled as a pro-rata cost element. Thus, setup time and its cost are distributed on a direct labor cost basis.

For continuous production or productions with extremely long batches, setup time may be negligible and setup is costed via overhead distribution. When setup is costed by overhead, cycle time is the only one considered by the estimator.

Cycle time, or, as it is sometimes called, run time, is expended on each piece after the setup has been made and consists of man or handling time, machine time, and delay time.

Machine time includes one or several elements of processing. As a machine operates at a uniform set rate, the time it takes for a cut can be accurately calculated from gear settings which control the speed of the cutter or the work piece and the movement or feed of the work piece past the tool. This calculation is handled by formulas, tables, or special slide rules that are widely available.

Machine cutting or processing by the equipment is a calculation where cutting speed in surface feet per minute is given. We then determine the revolutions per minute at which the material will be machined (say, by revolutions per minute = recommended cutting speed divided by π diameter) and, after this is known, take the part length and compute the time required to machine the travel length with a specified feed. A simpler formula relating feed to time is given by

$$t_m = \frac{L}{f} \tag{8-4}$$

where t_m = time to machine surface, minutes
L = length of cut, inches
f = feed or travel of cutter or work piece relative to each other, inches per minute

For a bar length of 20 inches, the feed of 2.750 inches per minute, $t_m = 20/2.750 = 7.27$ minutes. A variant would have f = speed in revolutions per minute $\times$ feed in inches per revolution for a lathe-turning operation.

Much more in the way of machine time calculations can be considered by the operation estimator. He may wish to indicate the tool life, the number of tools, and the down time required either to change or replace tools and resharpen if necessary. These exigencies add a tool allowance factor to the standard. Tool life is determined from the empirical function $VT^n f^m = k$, where V is in surface feet per minute, $T =$ tool life time in minutes, and n, m, and k are slopes and a constant determined from laboratory evaluation of a work piece and cutting material. These studies are extensive and the reader can contemplate the necessary calculations for the many combinations of materials. Each combination can be described by Figure 8.5. The minimum point of the operation cost can

FIGURE 8.5

STUDY OF THE VARIOUS EFFECTS OF CUTTING
SPEEDS SHOWS THAT AN OPTIMUM CUTTING
SPEED RESULTS FROM COMPARISON OF
FIXED AND VARIABLE COSTS

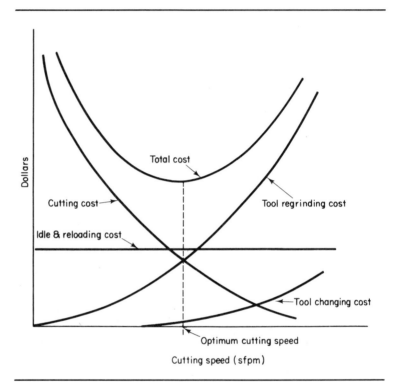

be determined using these graphical methods or, as is customary, formulas.

Equation (8-4) is the basis of processing machine cutting times. The length of cut has three parts which are the work-piece distance over which the tool takes a full cut, overtravel, and approach. As it is not always practical to start and stop a tool precisely at the edge of the surface that is to be cut, slight additions constitute overtravel and vary between $\frac{1}{32}$ to $\frac{1}{2}$ inch depending on the machine tool and part. Approach is excess travel caused by the geometry of the tool, such as drills and milling cutters. With a drill the common 118° point angle extends the approach length by about 0.3 times the drill diameter. Approaches for other types of cutters, such as milling cutters, vary but can be predetermined. Handbooks provide this information.

Methods for finding feed differ for all types of machine tools and cutters. Most methods begin with the speed of the cutter, linear rate in feet per minute, at which the cutter travels through the work material. Cutting speed and feed are affected by many variables.

Man or handling time covers the work of the operator whenever the machine is out of action. It may include such elements as starting and stopping, loading and unloading, clamping and unclamping, cleaning off chips, and replacing tools. This information is catalogued by standard time data and is given in normal minutes per element of work. Allowances expressed as a percentage for justifiable and re-occuring items such as personal, fatigue, and other delays are usually indicated by the standard time data. Man time is calculated using normal time data, and an allowance or decimal increases this value to the final standard time per unit for the operation. Sometimes the standard time data are reported with the allowances already added, as this reduces the arithmetic.

Consider now the standard single-point cutting equation on which the study of fundamental machining economics is based:

$$C_{sp} = C_0 t_m + C_t\left(\frac{t_m}{T}\right) + C_0 t_c\left(\frac{t_m}{T}\right) + C_0 t_h \qquad (8\text{-}5)$$

where C_{sp} = average unit cost of a single-point rough-turning operation
C_0 = cost of operating time, dollars per minute
t_m = time to machine the work piece, minutes per piece
C_t = tool cost, dollars per cutting edge
t_m/T = ratio of machining time to average tool life T, minutes
t_c = tool-changing time, minutes per operation
t_h = handling time, minutes per work piece

A cutter can be run at very high or low speeds. If operated at high speeds, cutter wearout is accelerated and vice versa. On the other hand, the concern is to not leave the cutter unchallenged for its performance.

The sum of four component costs are given by Equation (8-5) and are machining cost, tool cost, tool-changing cost, and handling cost. These costs are described by Figure 8.5. This basic tool life model and machining time t_m are related as

$$t_m = \frac{L\pi D}{12Vf} \qquad (8\text{-}6)$$

for which tables can be found and

$$VT^n f^m = k \qquad (8\text{-}7)$$

with Equation (8-6) expressing length L, diameter D, velocity V, and feed f in consistent units, and Equation (8-7) as the modified Taylor tool life relationship between the variables with exponents n and m and constant k as empirical parameters and T as tool life in minutes. The optimum cutting speed V^* is given by

$$V^* = \left\{\frac{C_0 n}{(f^{m/n}/k^{1/n})(C_0 t_c + C_t)(1 - n)}\right\}^n \qquad (8\text{-}8)$$

This optimum velocity can be determined by differential calculus. Instead it is derived in Chapter 12 by the method of geometric programming. This is the operating design for the machining-economics problem. A particular value can be determined by the appropriate substitution of empirical and cost data for Equations (8-5)–(8-8). For illustrative purposes, let $D = 3$ inches, $L = 6$ inches, $C_0 = \$0.08$ per minute, $C_t = \$0.50$ per tool, $t_c = 1$ minute per edge, $t_h = 1$ minute per piece, $m = 0.67$, $n = 0.125$, $k = 6.5$ for high-speed steel tool material, and $f = 0.010$ inches per revolution. We find $V^* = 87.04$ and the cost C_{sp}

$= \$0.49$. This is a calculation for a single-point tool. Other models more complicated than Equations (8-5) and (8-8) exist, but their inclusion is beyond the scope of this text.

The immediate discussion dealt with elements of an operation and the setup of an operation for metal working. While these are important to the estimate, the estimating procedure does not stop there. Following consolidation of elements into operations, the estimator determines operational cost by the following means: The direct-labor operational cost for a component may be found by either (1) multiplying summed operational standards for a cost center by the cost center average wage rate or (2) multiplying shop average wage times the total component standard. The operational cost for manufacturing direct labor is given by

$$C_{dl} = \sum_{i=1}^{n} LR_i\left(T_i + \frac{SU_i}{Q}\right) \tag{8-9}$$

where C_{dl} = direct labor cost on a per operation or unit or subassembly basis
$\quad LR$ = operator labor wage, dollars per hour
$\quad T$ = standard time estimated for operation, i.e., hours per unit
$\quad SU$ = setup standard for the operation, hours per lot
$\quad Q$ = lot quantity
$\quad i = 1, 2, \ldots, n$ operation

An operation to "drill complete" the 36 holes in a casting has been summarized using engineering standard time data as follows:

1. Setup radial drill, 7 drill diameters with fixture, 1.600 hours.
2. Cycle time per unit, 22.500 minutes.

The labor cost $LR = \$3.75$ per hour, $Q_1 = 200$, and $Q_2 = 10$ units. What is the pure labor cost for this operation? For 200 units,

$$C_{dl} = LR\left(T + \frac{SU}{Q}\right)$$

$$= 3.75\left(\frac{22.500}{60} + \frac{1.600}{200}\right) = \$1.44$$

For 10 units,

$$C_{dl} = 3.75\left(\frac{22.500}{60} + \frac{1.600}{10}\right) = \$2.01$$

A work station involving two operators working jointly on a subassembly of a tractor gear box requires the following estimated standards:

OPERATION 30, PART NUMBER 6682

Setup time.	24.000 hr
Assembler A, unit standard	2.650 hr
Assembler B, unit standard	2.500 hr
Assembler B, idle time	0.150 hr

The setup time is an estimated one-man standard where a setup operator $LR_{su} = \$4.14$ per hour is used. For assembler A, $LR_a = \$3.25$ and $LR_b = \$2.90$. Estimated quantities $Q_1 = 1000$, $Q_2 = 2000$, and $Q_3 = 5000$ units. It is evident for the subassembly operation that unit setup costs are insignificant when compared to cycle cost. In this example, idle time has been ignored; but usual practice recognizes the existence of this cost and includes it at some point in a cost calculation to assure full recovery of all costs.

| Quantity, Q_i | Setup Labor Cost | Unit Setup Cost | Cycle Cost | | Total Unit Operation Cost |
			Assembler A	Assembler B	
1. 1000	$99.36	$0.10	$8.61	$7.25	$15.96
2. 2000	99.36	0.05	8.61	7.25	15.91
3. 5000	99.36	0.02	8.61	7.25	15.88

The total labor cost for a component of a subassembly where work is performed on several or many operations in different cost centers is computed similarly. With average labor costs for a cost center and along with estimated standards for operations from standard time data, the component direct labor cost can be calculated.

The fabrication operations of the gear-box pinion gear are summarized in Table 8.5; 175 units have been predicted.

TABLE 8.5

COMPUTING DIRECT LABOR COST VIA WORK
CENTER USING OPERATION NUMBER AND
PART NUMBER OF DESIGN

Part Number: 443-3806 Pinion Gear

Operation Number	Work Center	$\sum SU$	$\dfrac{\sum SU}{Q}$	T_i	$T_i + \dfrac{\sum SU}{Q}$	LR_{ave}	Work Center Cost
1, 3, 9	Lathe	3.50	0.020	0.335	0.355	$2.80	$0.99
2, 4, 5	Gear	8.75	0.050	0.851	0.901	3.20	2.88
6, 7	Grind	10.25	0.059	1.265	1.324	3.45	4.57
8, 10	Bench	—	—	0.150	0.150	2.65	0.40
		Total C_{dl} for component:					$8.84

The previous example dealt with metal working. Other processes exist, and now we look at a sheet-metal subassembly operation. Two parts, 672 and 673, are processed separately and joined by spotwelding to form a radio chassis. We previously looked at this subassembly by considering its material cost.

Let the following ground rules be given. Marketing estimates that next year a total of 100 units will be sold and that 1000 units will be sold for the subsequent 2 years. The job will be built in half-size yearly lots so as to not build up inventories excessively. Setup cost, run-time rate, and other special charges are given as

8.3-2
Sheet-Metal Estimating

Year	Quantity	Lot Quantity	Setup Cost ($/hr)	Run-time Cost ($/hr)	Tooling Charges Per Hour	Fabrication Engineering Direct Charge Per Hour
1	100	50	$20	$10.00	$15	$5
2	1000	500	21	10.50	15	5
3	1000	500	22	11.00	15	5

Five hours are required to prepare a numerical control machine tool tape. Each operations chart requires 5 hours for preparation. Abbreviated preliminary-type engineering performance data are given by Table 8.6. The previous standard time data for tooling also applies here. Three process sheets are given by Tables 8.7, 8.8, and 8.9. These preliminary plans describe the method of manufacture and show the time estimates removed from Table 8.6. For P/N 672, a setup (irrespective of whether 50 or 500 parts are being run) requires 2.5 hours and 0.186 hours for run time. From experience the estimator knows that for operations 2, 5 and 6, and 9, Table 8.7, tools can be substituted for improved methods and an approximate savings of 0.7 hour of setup and 0.05 hour of run time will

TABLE 8.6

PRELIMINARY-TYPE STANDARD TIME DATA
FOR SHEET-METAL OPERATIONS
(INCLUDES ALLOWANCES)

Operation	Setup Hours	Run-time Hours per Unit
Shear	0.1	0.001
Punch press	0.4	0.0015
Tape-controlled punch press	0.2 + 0.03/station	0.008 + 0.0005/hit
Press brake	0.3	0.001 + 0.002/hit
Drill press	0.2 + 0.05/tool	0.015 + 0.003/tool + 0.001/hole
Degrease	0.1	0.001
Deburr	0.1	0.005
Spotweld	0.4	0.02 + 0.001/spot
Heliarc	0.2	0.003 + 0.008/corner
Chromate	0.1	0.003
Deoxidize	0.1	0.004
Silk screen	0.3	0.004

TABLE 8.7

PRELIMINARY PROCESS PLAN AND
ESTIMATING DATA

Process Sheet, P/N 672

Operation	Setup	Run Time
1. Shear	0.1	0.001
2. Tape-pierce & notch; assume 10 tools & 60 hits	0.5	0.038
3. Countersink (18) 0.120 holes (4) 0.125 holes	0.3	0.043
4. Degrease & deburr	0.2	0.006
5. Brake (2 lays)	0.3	0.005
6. Brake (2 lays)	0.3	0.005
7. Ream (2) 0.125 + 0.002 − 0.000 (4) 0.132 + 0.005 − 0.000	0.3	0.027
8. Degrease & deburr	0.2	0.006
9. Heliarc (4) corners	0.2	0.035
10. Grind welds	0.1	0.020
	2.5 hr	0.186 hr

occur if a compound pierce and blank die (150 hours × $15 = $2250) is purchased. Employing functional model (2-2), the break-even for replacing labor cost by tooling is given as

$$C = \frac{Na(1+t) - S}{I + T + D + M}$$

Notation was previously given in Chapter 2. We have 1000 units per year, $a = 0.05 \times 10 \times \$0.50$ savings per unit, $t = 10\%$ overhead savings, S = savings in setup or 0.6 hour × 20 = \$12, $I = 5\%$ allowance for interest, $T = 3\%$ for taxes, $D = 0$ for depreciation, and $M = 20\%$ annual allowance for maintenance. The permissible first cost for a tool is

$$C = \frac{1000 \times 0.50(1.10) - 12}{0.05 + 0.03 + 0.20} = \$1921$$

As our compound pierce and blank die is more costly than potential labor savings, we choose to process the part without the tool as given by Table 8.7.

A sheet-metal part cost can be computed at this point. We modify Equation (8-9) to

$$C_{dl} = T_{sm}C_{mcr} + \frac{SUC_{mcs}}{Q} + C_{rm} + \text{prorated special charges} \qquad (8\text{-}10)$$

where C_{dl} = sheet-metal cost center unit cost

$\quad T_{sm}$ = run-time standard hours per unit

$\quad C_{mcr}$ = sheet-metal run-time cost per hour

$\quad C_{mcs}$ = sheet-metal setup cost per hour

For the first year and from Table 8.10

$$C_{dl} = 0.340 \times 10 + \frac{6.15 \times 20}{50} + 0.315 + \frac{180}{50} = \$9.78$$

There is one discretionary policy in finding this cost. First, the prorated special charges have been distributed to first-year sales rather than to the total marketing quantity over 3 years. Subsequent yearly cost will be lower by this \$3 charge. In some cases it is possible for standard setup and run time to vary between various years, although it did not here. Changes in lot quantities would alter the production practices and therefore operation cost.

TABLE 8.8

PRELIMINARY PROCESS PLAN AND
ESTIMATING DATA

Process Sheet, P/N 673

Operation	Setup	Run Time
1. Shear	0.1	0.001
2. Tape-pierce & notch; assume 10 tools & 75 hits	0.5	0.047
3. Counterbore (2) 0.128 holes	0.25	0.020
4. Degrease & deburr	0.2	0.006
5. Brake	0.3	0.003
6. Brake	0.3	0.003
7. Brake	0.3	0.003
8. Brake	0.3	0.003
	2.25 hr	0.086 hr

TABLE 8.9

PRELIMINARY PROCESS PLAN AND
ESTIMATING DATA FOR THE ASSEMBLY

Process Sheet, P/N 674

Operation	Setup	Run Time
Material cost—Assembly (None)		
1. Deoxidize	0.1	0.004
2. Spotweld (10 spots) two parts	0.4	0.030
3. Drill through (1) hole B	0.3	0.021
Countersink (1) hole B		
4. Degrease & deburr hole	0.2	0.006
5. Chromate	0.1	0.003
6. Silk screen	0.3	0.004
	1.4 hr	0.068 hr

TABLE 8.10

ESTIMATING RECAP FOR
SHEET-METAL SUBASSEMBLY

	First Year			Second Year			Third Year			Special	Engrg.
P/N	Mat'l	SU	RT	Mat'l	SU	RT	Mat'l	SU	RT	Tooling	Charges
672	0.086	2.50	0.186	0.091	2.50	0.186	0.100	2.50	0.186	$15	$50
673	0.229	2.25	0.086	0.240	2.25	0.086	0.251	2.25	0.086	15	50
674	—	1.40	0.068	—	1.40	0.068	—	1.40	0.068	—	50
	$0.315	6.15	0.340	$0.331	6.15	0.340	$0.351	6.15	0.340	$30	$150

First year:

$$C_{sm} = 0.340 \times 10 + \frac{6.15 \times 20}{50} + 0.315 + \frac{180}{50} = \$9.78 \text{ per unit}$$

Second year:

$$C_{sm} = 0.340 \times 10.50 + \frac{6.15 \times 21}{500} + 0.331 = \$4.16 \text{ per unit}$$

Third year:

$$C_{sm} = 0.340 \times 11 + \frac{6.15 \times 22}{500} + 0.351 = \$4.36$$

8.3-3
Factors Involved in Estimating
Assembly Production

Assembly operations of a production process include the activities of manual labor, fastening and joining, packaging, and automatic or semiautomatic equipment assembly systems. This work is concerned with assembly categories of prototype assembly, lot, and mass production. Preliminary processing plans are, of course, linked with the volume and the methods of production. Prototype production is specified if there is but one final unit. Production is called batch or lot when the consecutive output of units is gathered in batches or lots for throughput in the production system. Examples include machine tools, computers, and steam boilers. Production is labeled mass whenever there is a large continuous output of articles of identical or nearly identical products. Cars, tractors, sewing machines, electrical appliances, and canned food provide everyday examples of mass production. Whether the flow line is zigzag or straight is immaterial to the needs of the estimator at the preliminary processing point.

The cycle time T is the sum of the assembly operations completed in sequence, i.e., $T = T_1 + T_2 + T_3 + \cdots + T_n$, where $1, 2, \ldots, n$ are the stations and $T_1, T_2, \ldots, T_n$ are the standard times of the individual assembly operations. As discussed previously, the operation estimator prepares detailed process or operation sheets that are likely to be used. These are like the detailed assembly instructions found in plastic model kits that are purchased in hobby shops. Exploded views, written instructions, and other visual or audio methods to minimize production effort are prepared to assist the worker.

Concurrent to the detailed assembly planning, the estimator evaluates the design, construction or procurement of special or standard tools, fixtures, and equipment that are effective for assembly requirements. Engineering performance time data are necessary for assembly operations in the same detail as that found for the fabrication sector of cost.

As an illustration, the estimator determines the preliminary processing plan for a light mechanical assembly operation involving the rolling of four clips to a radio chassis with eight eyelets using a manual foot-operated pneumatic riveting machine. Table 8.11 demonstrates this application.

TABLE 8.11

PRESS PARTS ON RADIO CHASSIS

Element No.	Description	Occurrence Per Cycle	Normal Minutes	Cycle
1	Get and aside primary part	1	0.025	0.025
2	Get hardware and move to tool	8	0.021	0.168
3	Position hardware to tool	8	0.008	0.064
4	Move and position primary part to tool	8	0.026	0.208
5	Get secondary part and move to tool	4	0.028	0.112
6	Position secondary part to tool	4	0.013	0.052
7	Align secondary part to primary part	4	0.028	0.112
8	Machine cycle	8	0.023	0.184
		Total minutes per cycle		0.925

8.3-4
Indirect Labor

Indirect labor is estimated differently from direct labor if it is estimated at all. The compelling reason is ease of costing. Direct costs are costs charged to a contract, job order, or machine center, for example; indirect costs, also known as overhead or burden, are apportioned to the contract or job order from an indirect cost account into which they were initially charged. The charges to these indirect accounts are liquidated by distributing to specific job orders and the like in proportion to a base such as direct labor dollars, direct labor hours, equipment hours, or other bases as the situation and operation requires. In the case of an estimate for an operation or product the indirect cost rates are usually acceptable for the pending future time period. If a new plant or a significant alteration of existing facilities is anticipated or a revised product cost is important, the analyst considers major expenses in addition to those computed for direct material and direct labor (certain fixed costs such as property taxes and interest on loans are overlooked). The estimate for indirect labor evaluates the essential auxiliary services and costs that are needed for operations, product research, engineering testing, selection and installation of capital equipment, tooling, and facilities engineering. The following example illustrates an estimate for indirect labor.

Cost centers are a common method of allocating indirect labor costs. The basic expression for the cost of indirect labor from a cost center would be

$$R_i = \frac{\text{total budgeted center cost}}{\text{number of available man-hours in center}} \qquad (8\text{-}11)$$

The expected or estimated cost of the service to the specific job would be

$$C_i = R_i T \qquad (8\text{-}12)$$

where C_i = estimated cost of the indirect service
$\quad R_i$ = rate of service per man-hour
$\quad T$ = total estimated man-hours
$\quad i$ = indirect service unit considered

Indirect labor time is handled on a specific job or service basis. One method to estimate the required man-hours of service from a support group is to use engineered standard data with a list of task elements and times. The allowed time to perform the specific job is estimated as

$$T = T_c + (T_c A_c) + N(T_t + T_j) + (T_j A_c) + T_p \qquad (8\text{-}13)$$

where T = allowed time, man-hours
$\quad T_c$ = craft time, man-hours
$\quad A_c$ = craft allowance, decimal
$\quad N$ = number of men on job
$\quad T_j$ = job preparation allowance, man-hours per job
$\quad T_t$ = travel time to job location, hours
$\quad T_p$ = partial-day influence, man-hours

T_c, A_c, and T_j are found from task time data tables.[1] The work elements visualized by the estimator are thought necessary for the job.

An experimental operation requires an enclosed area. The labor cost to build the required wall is to be charged to the operation. The plant maintenance group will build the wall and has 15 craftsmen, 3 shift foremen, and 1 manager. A quarter budget has been estimated:

FIRST-QUARTER BUDGET

Salaries and wages	$49,000
FICA	2,250
Workman's compensation	500
Group insurance	1,000
Pension fund	1,250
Fringe benefits	1,000
	$55,000

Cost center available man-hours are 8 hours per day × 66 days × 15 men = 7920 man-hours. The value of cost center service per man-hours is

$$R_i = \frac{\$55,000}{7920 \text{ man-hours}} = \$6.94 \text{ per man-hour}$$

[1]See the Navdocks bulletin listed last in the Selected References.

FIGURE 8.6

INDIRECT LABOR ESTIMATING FOR
MULTIPLE CRAFT-WORKER JOBS

Indirect labor calculation sheet

Shop order no: _____1_____

_____

Estimated by: _____ Date: _____

Job phase description: _Install 12-linear-ft high partition on 2×4 in. studs, 16 in. o.c.,_
gypsum wallboard on both sides, including shoe molding and baseboard both sides.
Install 100ft² of 12×12 in. acoustic tile without trim molding and using adhesive.

_____

Task description	Unit hours	Occur-rence	Craft time	
Install 12-lin.-ft high partition framework on 2×4 in studs, 16 in. o.c. gypsum wallboard on both sides, including shoe molding and baseboard both sides.	6.2	1	6.2	
Install 100 ft² of 12×12 in. acoustic tile without trim molding, using adhesive.	3.0	1	3.0	

	Totals	9
Travel zone _10_		
Multiplying factor _____ Allowance % _45_	Allowed time (craft time + general data)	23
Add factor per man _____ Preparation _0.5_	Job allowed (Total est. time + allowed time)	23
No. of craftsmen _3_		

The next step is to determine the required man-hours. This is found using craft time tables and is recorded in Figure 8.6. A nomograph considers diverse factors of travel time, craft allowances, preparation and setup, and the influence of crew work. Figure 8.7 shows an application with zone 10, 45% allowance, 0.5 hour of preparation, and three men giving an estimate of 23 hours for the job. With T determined, $C_i = 6.94(23) = \$160$, the labor estimate to build the enclosure.

FIGURE 8.7

NOMOGRAPH USED IN CALCULATING CRAFT TIME. (From U.S.N. Engineering Navdocks P-701, Aug. 1958.)

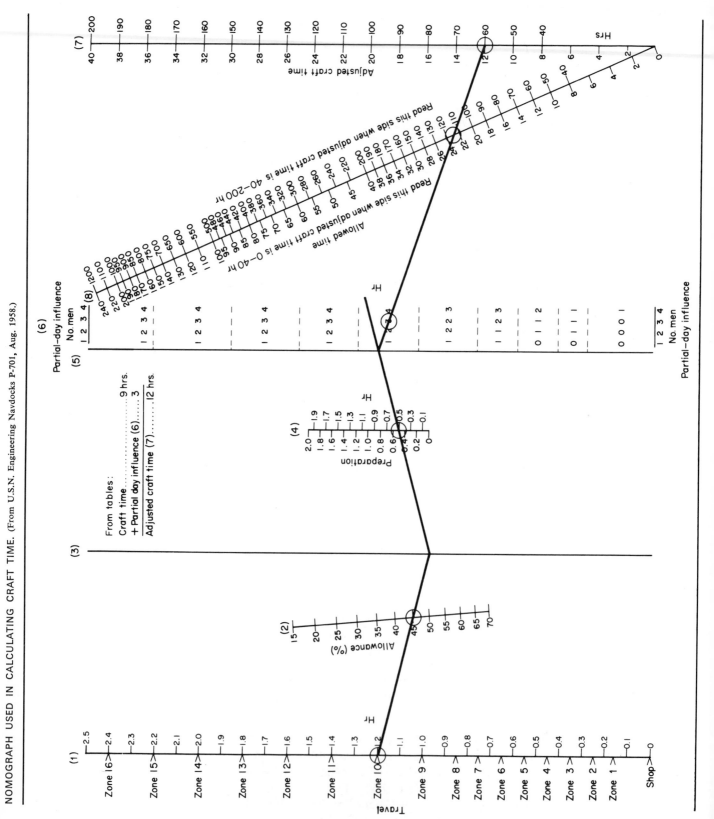

Detailed estimates of the engineering costs for operations are usually not undertaken. The magnitude of cost is insufficient to warrant special attention on an operation basis. But if it is estimated, normal average figures are applied, as they were for the sheet-metal example. When not estimated in this way, the cost of engineering service is distributed to the machine center or department via overhead practices. Larger-scaled designs, such as tool design, product, project, and system engineering, auxiliary services, and other special services, are estimated independently. Design estimating is considered in Chapter 9.

Cost is the major criterion for operation estimating. Material and labor are the principal elements of operation cost. The extent of labor breakdown for costing purposes is dependent on whether the work is repetitive-nonrepetitive, production-craft work, or direct-indirect. If the labor is designated and engineering performance data are available, estimates are found by multiplying the planned time by corresponding labor cost. Material is classified as direct and indirect. Scrap and spoilage are usually expected and increase the design amount by a historical factor. Tools can be estimated and appraised for their economic justification.

In the next chapter we shall consider product estimating. Operation estimating is a prerequisite to product estimating and provides information needed to cost and price out products.

For a discussion and details on chemical and processing industries, consult

ARIES, R. S., and R. N. NEWTON: *Chemical Engineering Cost Estimation*, McGraw-Hill Book Company, New York, 1958.

BAUMAN, H. C.: *Fundamentals of Cost Engineering in the Chemical Industry*, Van Nostrand Reinhold Company, New York, 1964.

JELEN, F. C., ed.: *Cost and Optimization Engineering*, McGraw-Hill Book Company, New York, 1970.

PETERS, MAX S., and KLAUS D. TIMMERHAUS: *Plant Design and Economics for Chemical Engineers*, 2nd ed., McGraw-Hill Book Company, New York, 1968.

POPPER, HERBERT, ed.: *Modern Cost-Engineering Techniques*, McGraw-Hill Book Company, New York, 1970.

RUDD, DALE F., and CHARLES C. WATSON: *Strategy of Process Engineering*, John Wiley & Sons, Inc., New York, 1968.

Information concerning electrical estimating can be found in

ASHLEY, RAY: *Electrical Estimating*, 3rd ed., McGraw-Hill Book Company, New York, 1961.

JOHNSON, RALPH E.: *Electrical Construction Cost Manual*, McGraw-Hill Book Company, New York, 1957.

For building trades, general and light construction, refer to

Building Cost File, Construction Publishing Company, Inc., New York, 1972.

The Building Estimator's Reference Book, 17th ed., Frank R. Walker Company, Chicago. 1969.

CE Cost Guide, Atlantic Highlands, N.J., 1970.

Construction Pricing and Scheduling Manual, McGraw-Hill Book Company, F. W. Dodge Division, New York, 1973.

DEATHERAGE, GEORGE E.: *Construction Estimating and Job Preplanning*, McGraw-Hill Book Company, New York, 1965.

Foster, Norman: *Construction Estimates from Take-off to Bid*, 2nd ed., McGraw-Hill Book Company, New York, 1972.

Godfrey, Robert Sturgis, ed.: *Building Construction Cost Data*, Robert Snow Means Company, Inc., Duxbery, Mass., 1973.

National Construction Estimator, annual revision, Craftsman Book Company of America, Los Angeles.

Page, John S.: *Estimator's General Construction Man-Hour Manual*, Gulf Publishing Company, Houston, Texas, 1959.

Parker, Albert D.: *Planning and Estimating Dam Construction*, McGraw-Hill Book Company, New York, 1971.

Richardson Publications, Downey, Calif.:

1. *Manual of Commercial-Industrial Construction Estimating and Engineering Standards.*
2. *Field Manual of Commercial Industrial Construction Costs.*
3. *Manual of Residential-Light Construction Estimating and Engineering Standards.*

Information for manufacturing and estimating data may be found in

Clugston, Richard, *Estimating Manufacturing Costs,* Cahners Books, Boston, Mass., 1971.

Hadden, Arthur A., and Victor K. George: *Handbook of Standard Time Data*, Ronald Press, New York, 1954.

Machining Data and Engineering Guidelines (*Revised*), Rock Island Arsenal Research Report SWERR TR 72-60, Publication No. AD 756365, National Technical Information Service, Springfield, Va.

Machining Data Handbook, 2nd ed., Metcut Research Associates, Cincinnati, Ohio, 1973.

Nordhoff, J. D.: *Machine-Shop Estimating*, 2nd ed., McGraw-Hill Book Company, New York, 1960.

Papas, Frank G., and Robert A. Dimberg, *Practical Work Standards*, McGraw-Hill Book Company, New York, 1962.

Parsons, C. W. S., *Estimating Machine Costs*, McGraw-Hill Book Company, New York, 1957.

Peat, A. P., *Cost Reduction Charts for Designers and Production Engineers*, The Machinery Publishing Co. Ltd., Brighton, England, 1968.

Vernon, Ivan R., ed.: *Realistic Cost Estimating for Manufacturing*, Society of Manufacturing Engineers, Dearborn, Mich., 1968.

Wilson, Frank W., ed.: *Manufacturing Planning and Estimating Handbook*, McGraw-Hill Book Company, New York, 1963.

A method for estimating craft work and engineering performance standards is found in

U.S.N. Engineering Performance Standards, Navdocks P-701, Bureau of Yards and Docks, Washington, D.C., Aug. 1958.

QUESTIONS

1. What is the purpose of an operation estimate? What restrictions usually govern an operation estimate?

2. Categorize materials. What structure is important to costing? Distinguish between standard purchased items and raw material.

3. List the kinds of losses that are encountered in materials. Which do you think most important?

4. Name some indirect materials. Why are they hard to estimate?

5. List the important types of labor for estimating. Which is easiest to estimate and why?

6. Why is setup, a fixed cost, treated as a variable cost? What is run time? What influences the cost of direct labor for fabrication?

7. How are engineering and special costs estimated?

8-1. Determine the operation cost for producing Figure 8.1. A quantity of 75,000 units per year is anticipated. Suppose two tooling methods are visualized: progressive die which completes the part in one operation, and two dies which punch out the hole and then blank out the work piece in two operations. Both methods use purchased strip stock material costing $0.0045 per square inch. Tool room cost is $14 per hour for the construction of tools. Find the total tool design and construction cost for the two tooling methods. Use the standard data given by Table 8.6 to determine operational time. Labor cost is $6.50 per hour. Engineering design costs for a progressive die or two dies are $200. Find operational cost including labor, material, and tooling. Which is the cheaper method? Is there a break-even point between the two tooling methods?

8-2. Estimate the unit cost of the material to produce the bayonet-clip half shown in Figure P8-2. The hot-rolled steel costs $0.22 per pound for 0.875-inch-diameter

FIGURE P8-2

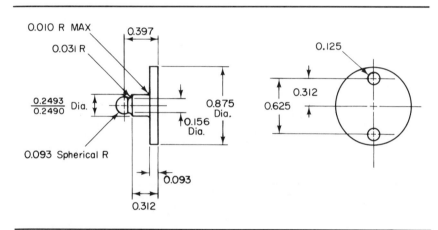

stock. The cutoff tool width is 0.125 inch and a 0.015-inch stock allowance for the 0.093-inch spherical end is sufficient. Allow 4% for spoilage and bar ends (0.875-inch-diameter bar stock weighs 2.05 pounds per foot).

8-3. Estimate the unit material cost for the two designs shown in Figure P8-3. The cold-rolled steel material costs $0.19 per pound. Raw material is available in strips $24\frac{3}{4}$-inches wide weighing 5 pounds per square foot. A slitting operation prepares suitable widths for the die at no loss in material. In addition to the blanking losses, add a 5% loss for scrap and ends. Find the losses for both designs. Determine the unit material cost. Estimate the tooling cost when engineering design costs are $125 and $190 and tool shop costs are $15 per hour. Assuming equal labor cost for the two designs, find the quantity break-even point for material and tooling. If minutes per piece is 0.0736 and 0.0529 for the 1.625-in. and 0.813 in. feed, and production costs equal $7 per hour, what is the break-even point?

8-4. A firm sells its plating operation, which is described by

plating line profit = revenue − cost

$$z = nS_a - (nV_c + F) + b_i - m_i$$

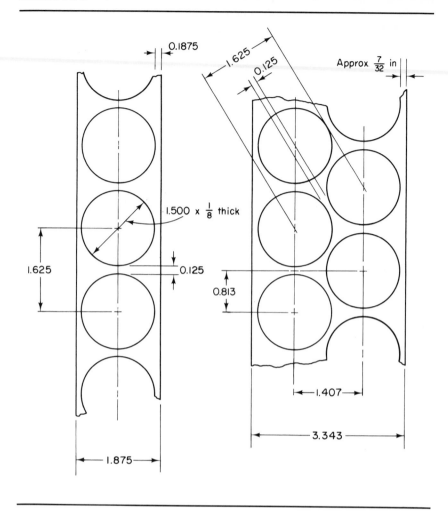

n = units of saleable product per period

S_a = average sale price, which can vary with both quality and quantity of product

V_c = variable conversion cost per unit of product n; includes consumable supplies

m_i = dollars of raw materials entering the process which appears as a part of the product

b_l = dollars per period of any by-product or waste materials which might be used for fuel or raw materials in any operation

F = fixed cost, dollars per period

A plating operation line uses raw materials equal to 1000 pounds per day and can operate at three conversion efficiencies of 65, 75, and 80. The output from any level is 250 tons of product. All raw material not appearing as plating on product is considered fully recoverable (although partial recovery can be considered, too) at a net $2 per pound following recovery procedures by the firm. The variable conversion costs differ with respect to the level of efficiency. The raw material costs are $7.50 per pound. Use 1 day as a basis of analysis. What are the net total costs and revenue for each conversion efficiency? Which one is the more profitable conversion?

	65%	75%	80%
Conversion efficiency	65%	75%	80%
Plating material usage, lb	1000	1000	1000
Revenues/ton for operation	$47	$48	$48.50
Total revenue	$11,750	$12,000	$12,125
Daily costs			
Variable	$1,825	$2,280	$2,320
Material	$7,500	$7,500	$7,500
Fixed (per efficiency)	$2,250	$2,350	$3,500
Total	$11,575	$12,130	$13,320
Credit for recoverable material	$700	$500	$400

8-5. The conversion of olefins to a soap intermediate product is partly a function of the percentage initiater concentration level and the percentage HBr in the gas recycle. While a high-purity product is desirable, it is not achieved without greater cost. Reaction completeness described by two variables is shown by Figure P8-5. A cost value associated with percent HBr in the gas recycle is given as

% HBr	Dollars Per Pound
60	$0.70
70	1.00
80	1.60
90	2.70

An intermediate product revenue value for purity can be established as

Product Purity	Dollars Per Pound
99	$18.50
97	18.35
90	17.00

FIGURE P8-5

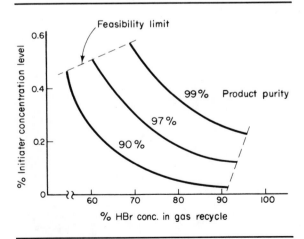

Considering only the values listed (no interpolation), determine the percentage concentration level for each feasible combination. Find the most favorable concentration for this operation.

8-6. Specially designed tools are not always necessary to produce sheet-metal parts. Figure 8.4 can be produced using general-purpose tooling. Starting with a 48 × 96 × 0.031 inch stainless steel sheet, the following preliminary processing plan indicates the operations:

Shear $\frac{9}{16}$-inch-wide strips 96 inches long
Shear $\frac{9}{16}$ × 1 inch blanks
Punch two 0.096-inch holes
Punch one 0.093-inch hole
Notch two corners for $\frac{1}{16}$-inch dimension
Tape punch press two ears
Brake
Brake
Deburr part
Degrease

Determine the setup and run-time hours using Table 8.6. A run quantity of 2500 is planned. Average setup and run-time labor will cost $6.30 and $5.15 per hour, respectively. The material cost factor is $0.0073 per square inch. What are the total direct labor and direct material costs? Unit cost? If a shearing operation wastes no metal, what is the yield from a 48 × 96 inch sheet for this part? How many 48 × 96 inch sheets are necessary to handle this volume?

8-7. (a) A cement contractor is asked to bid on a small concrete retaining wall along a sidewalk. The wall, on the average 2 feet high, will rest upon a 2 × 1 foot foundation. The length will be 75 feet. What is the expected cost? Adjust Table 8.1 to accomodate to the new wall height.

(b) An electrical contractor is required to install a conduit according to the sketch in Figure P8-7. His hours-per-foot standards, which do not conform exactly to the requirement, are as on page 255.

FIGURE P8-7

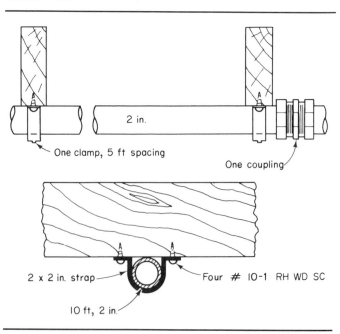

Labor costs $8.75 per electrical journeyman-hour, and material is $0.28 and $0.30 per pound for the $\frac{3}{4}$ and $1\frac{3}{4}$-inch sizes. He is required to install 7000 feet.

Size	Exposed	Slab Deck Wood Frame	Trench	Furred Ceiling	Notched Joists	Weight (lb/100 ft)
$\frac{3}{4}$	0.046	0.033	0.023	0.030	0.039	50
$1\frac{3}{4}$	0.080	0.062	0.040	0.054	0.063	100

What is his operational labor and material cost?

8-8. A large office with 10-foot ceilings by 250 foot sides is to be initially painted. Redecoration will be required every 4 years. Initially all holes will be filled and sanded, and one prime coat and one finish coat will be applied by spray gun. Initial protection will be nominal. The redecoration process calls for the walls to be washed and one finish coat to be applied with brush and rollers. A crew of three will be employed (one foreman and two journeymen). A three-section scaffold is 21 feet long × 5 feet wide × 5 feet high. Perform an operation estimate to forecast labor expenses for both the initial and redecorating processes and compare labor cost per square foot. Job elements are

1. Scaffolding time: 0.90 man-hours per scaffold, erection; 0.60 man-hour per scaffold, dismantle.
2. Sealing and sanding time: 195 square feet per hour, 0.50 man-hour per 100 square feet.
3. Spray paint concrete surfaces: 465 square feet per hour, 0.22 man-hour per 100 square feet.
4. Paint concrete surface by roller: 155 square feet per hour, 0.54 man-hour per 100 square feet.
5. Washing walls time: 0.87 man-hour per 100 square feet, 147 square feet per hour.

Assume that the scaffold will be moved 10 times per wall and that it will take 10 minutes to move for a new unpainted wall and 12 minutes for redecorating. The foreman has a job labor rate of $8 per hour, and each journeyman has a job rate of $7.45 per hour.

8-9. One operator controls four automatic screw machines. After these machines are set up, they work free of the operator. Even bar loading may be automatic. The cam time to make one piece is 4 seconds, or 900 pieces per hour at 100% efficiency, or 0.0667 minute per piece. The pro-rata setup is 0.002 minute per piece. If the actual efficiency is 85%, the operator controls four machines, and the labor rate is $0.0422 per minute, what is the operation cost per piece?

8-10. (a) Find the cutting time for a hard copper shaft 2 × 20 inches long. A surface velocity of 250 sfpm is suggested with a feed of 0.009 inches per revolution.
   (b) An end facing cut is required of a 10-inch-diameter work piece. The revolutions per minute of the lathe are controlled to maintain 400 surface feet per minute from the center out to the surface. Feed is 0.009. Find the time for the cut.
   (c) If the tool is K-3H carbide, the material is AISI 4140 steel, depth of cut is 0.050 inch, and the feed is 0.010 inch per revolution, what is the surface feet per minute for a 4-inch bar and a 6-minute life if the tool life equation $VT^{0.3723} = 1022$?
   (d) Consider the Taylor tool life model, $VT^n = k$ for the following tool materials and work materials:

Tool	Work	n	k
High-speed steel	Cast iron	0.14	75
High-speed steel	Steel	0.125	47
Cemented carbide	Steel	0.20	150
Cemented carbide	Cast iron	0.25	130

For a tool life of 10 minutes for each of these combinations, what is the cutting velocity?

8-11. Note the tool life curve of Figure P8-11 for SAE 3140, feed of 0.013, depth of cut

FIGURE P8-11

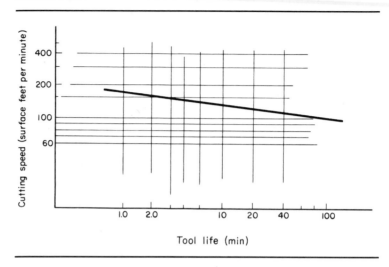

Tool life (min)

of 0.50, and a HSS tool material. Using the curve, find the parameters for $VT^n = k$. What is tool life for 60 feet per minute? What are the revolutions per minute for 2-inch bar stock and V_{150}?

8-12. Metal cylinders are to be turned on a lathe. The cylinders are 9 inches long and 4 inches in diameter. The cost of a single cutting tool edge is $1.50, the machine and operator cost per minute is $0.10, and tool changeover requires 1 minute. The maximum feed rate for an acceptable surface finish is 0.015 inches per minute. For the Taylor tool life equation ($VT^nF^m = k$) the constants are $n = 0.024$, $k = 23$, and $m = 0.45$. Determine the optimum spindle speed in revolutions per minute, the cutting time per cylinder, and the cost associated with each unit.

8-13. A cemented carbide tool is turning a 3-inch-diameter by 8-inch-long shaft of alloy steel with a depth of cut as $\frac{1}{8}$ inch, 0.015 feed, and $VT^{0.03}f^{0.5} = 25$. Tool-changing time is 2 minutes. The cost to grind a tool is $0.50, which includes the original cost of the tool. Labor and machine overhead is $0.12 per minute. Load and unloading time for the collet is 2 minutes. Find the minimum cost per piece.

* * * * *

Use the following standard time data for the next three problems. Assume a hypothetical machine capable of infinitely variable revolutions per minute (no fixed-gear revolutions per minute and direct-current direct drive) and feeds within its limits. Two tool holders are available, cross slide for turning, facing, threading, and so forth and tail stock for drilling and reaming. Engine lathe labor is machinist grade 3 and is paid $5.25 per hour.

(a) Setup: Includes allowances.

Operational Tools	Standard Hours
1	0.35
2	0.46
3	0.57
4	0.67

Add 0.07 for tolerance less than 0.010 inch. Add 0.06 for each multiple cut with the same tool.

(b) Cycle time, manual; normal minutes:

Load and unload collet	0.098
Turn end for end in collet	0.042
Load and unload three-jaw chuck	0.17
Turn on coolant	0.030
Slide splash shield on/off	0.078
Start and stop machine	0.050
Reverse spindle or adjust RPM	0.020
Adjust dial to cross slide	0.086
Blow piece with air	0.066
Center to work and lock tail stock	0.075
Adjust compound for additional cut	0.37
Move tail stock and lock: 1–5 inches	0.050
5–8 inches	0.070
Tail stock spindle to or from work: 1 inch	0.070
2 inches	0.10
Move saddle 1 inch	0.020
Each additional inch	0.010
Cross slide to work: $\frac{1}{2}$ inch	0.01
1 inch	0.02
3 inches	0.065
Hand file corner	0.121

Allowance for manual time = 20%.

(c) Machine time:

Work-piece Materials	High-Speed Steel Tool Surface Feet Per Minute	Turning (in./rev.)	Forming (in./rev.)	Drilling (in./rev.)
Aluminum	280–920	0.0015	0.0015	0.0025
Yellow brass	80–310	0.0015	0.0010	0.0025
Cast iron	40–130	0.0015	0.0005	0.0025
Steel:				
1020	60–90	0.0005	0.0005	0.0010
1112	75–160	0.0015	0.0005	0.0025
1040	45–85	0.0015	0.0005	0.0015
2015	25–70	0.0005	0.0005	0.0010

Allowance to cover tool wear = 10%.

8-14. Estimate labor and material costs for the pin in Figure P8-14. The stock is

FIGURE P8-14

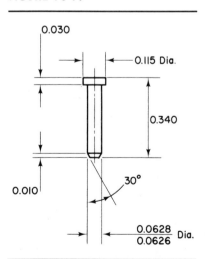

B-1112 steel with a *machinability rating* of 1.0, diameter = 0.125 inch, weight = 0.0417 pounds per foot with a cost of $0.37 per pound. Production quantity is 100. The elements of the lathe operation are

> Load ⅝-inch-long stock in collet
> Move cross slide to work
> Face end of stock, machine
> Position cross slide tool
> Adjust revolutions per minute
> Machine entire 0.340 length to 0.115 diameter
> Return tool to start
> Feed cross slide tool to preset 0.0628 diameter
> Adjust revolutions per minute
> Turn 0.0628 diameter less shoulder, machine
> Use hand file to chamfer 30° end
> Adjust revolutions per minute
> Cut off piece, machine
> Place piece aside and discard scrap

8-15. Estimate the material and engine lathe labor portion of the stand-off in Figure P8-15. Material is ground stock as 0.2502 diameter, SAE 2015 nickel steel, at

FIGURE P8-15

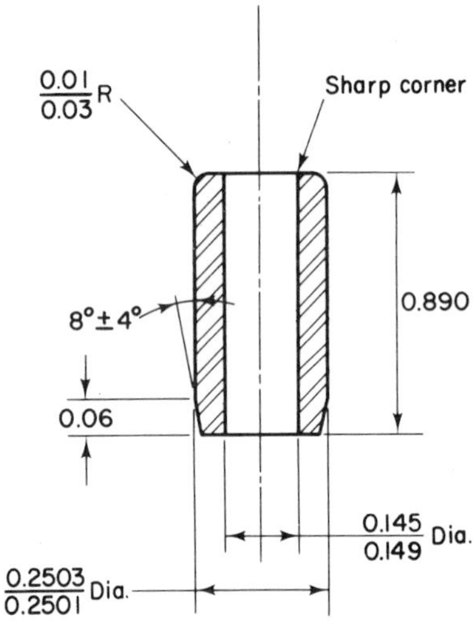

$1.07 per pound and 0.0139 pounds per inch, and 55 parts are required.

> Load 1-inch part in collet
> Move cross slide to work
> Adjust revolutions per minute
> Face end of stock, machine
> Position cross slide tool
> Change compound for taper-form cut
> Adjust revolutions per minute
> Plunge cut 0.06-inch taper, machine
> Pull back cross slide
> Move tail stock to position for drilling
> Adjust revolutions per minute

Drill hole, machine
Retract drill and reposition tail stock
Position compound
Cut off, machine
Part aside
Scrap away

8-16. Ninety brass spacer cups are required. Brass material that has an 0.625-inch diameter and a $\frac{1}{2}$-inch length is purchased that weighs 0.098 pounds each at a cost of $0.78 per pound. See Figure P8-16. Determine a preliminary processing

FIGURE P8-16

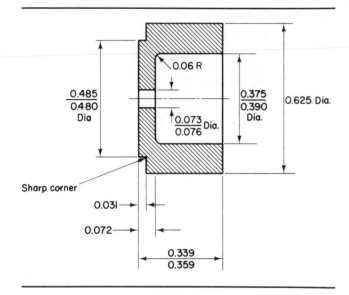

plan using the engine lathe to complete the part in one operation. Estimate the cost of labor and material.

* * * * *

The following standard time data are to be used for the following four problems:

Punch Press Standard Time Data

(a) All setups 0.15 hour (Includes allowances)

(b) Cycle time: Normal time:

(1)

Strip Feed Dimension	Minutes
$\frac{1}{2}$	0.043
1	0.046
$1\frac{1}{2}$	0.059
2	0.064
4	0.12
6	0.18

The element includes pickup strip, run machine, blow out slugs, and aside scrap.

(2) Coil stock with automatic feed, progressive die minutes: single die, 0.025 min. per piece; double die, 0.013 min. per piece.

Problems / Chap. 8 **259**

(3) Pickup, punch, or shear, and aside time for a single piece:

Length + Width	Minutes	Each Add'l Punch or Shear
3	0.12	0.03
4	0.10	0.03
6	0.09	0.03
15	0.08	0.03
32	0.11	0.03
44	0.15	0.04

For 180° flip or turn, add 0.02 per flip or turn for $L + W$ up to 25 and 0.03 per flip or turn for $L + W$ over 25.

(4) Personal and delay allowances for cycle time are 10%.

8-17. Using the punch press standard data, determine the punch press labor cost to complete Figure 8.4. The two methods are (1) progressive die and (2) three dies; pierce and blank, form two ears and an end, and form lip. Average labor costs $6.30 per hour. A quantity of 2500 is planned. Blank size $= 0.609 \times 1.028$ inches.

8-18. Find the unit and total operational cost for the design in Figure P8-18. The

FIGURE P8-18

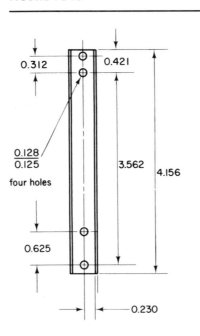

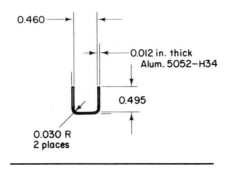

sheet-metal part is processed according to the following plan:

> Shear strips from 48 × 96 inch sheet
> Shear blanks from strips
> Punch four holes
> Form two lips using form die

Determine the unit material cost if the raw materials cost $0.004 per square inch. Estimated quantities are 12,000. Find the tooling cost using Table 8.3 and $14 per hour. The tool design cost is $90. Labor costs $4.80 per hour.

8-19. Find the unit and total operational cost for the design problem in Problem 8-18. A preliminary processing route sheet is

> Shear strips 1.450-inches wide from 48 × 96 inch sheet
> Punch and form complete with one progressive die

A progressive die design costs $410. Tooling construction labor and overhead costs $14 per hour and operational labor costs $4.80 per hour. Quantities are 12,000 units. Use Table 8.3 and sheet-metal standard time data.

8-20. Construct a preliminary processing plan to produce 40,000 of the soft brass springs shown in Figure P8-20, 0.005 inches thick, for which material costs

FIGURE P8-20

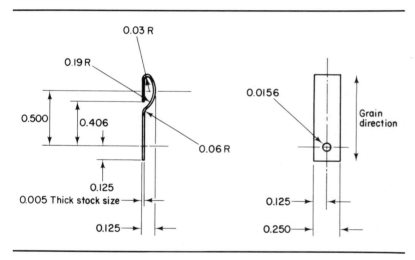

$0.0073 per square inch. Determine the tooling cost, standard labor cost (dollars of labor cost = $5 per hour and tool room labor cost = $14 per hour), and material cost. A hardening operation performed by an outside vendor costs $0.004 per spring.

8-21. A small mechanical assembly involving staking and riveting is planned with the following elements:

Element	Frequency
Get and aside primary part	1
Get hardware and move to tool	6
Position hardware to tool	6
Move and position primary part to tool	6
Get secondary part and move to tool	3
Position secondary part to tool	3
Align secondary part to primary part	3
Machine cycle	6

Using the data from Table 8.11, find the run time. If the setup is 0.10 standard hour and a small-machine assembler is paid $3.25 per hour, what is the unit labor cost for 500 components? Allowances are 18%.

8-22. An indirect labor department provides a service to other departments for which an estimate is required before the service department is given the job. Let $C_i =$ $7 per hour, craft time = 10 hours, the number of craftsmen is 2, Zone 5, allowance = 20%, preparation = 1 hour. Find the charge that the indirect labor time will bill for.

chapter 9

Our policy is to reduce the price, extend the operations, and improve the article. You will notice that the reduction of price comes first. We have never considered costs as fixed. Therefore we first reduce the price to the point where we believe more sales result. Then we go ahead and try to make the prices. We do not bother about the costs. The new price forces the costs down. The more usual way is to take the costs and then determine the price, and although that method may be scientific in the narrow sense; it is not scientific in the broad sense, because what earthly use is it to know the cost if it tells you that you cannot manufacture at a price at which the article can be sold? But more to the point is the fact that, although one may calculate what a cost is, and of course all of our costs are carefully calculated, no one knows what a cost ought to be. One of the ways of discovering is to name a price so low as to force everybody in the place to the highest point of efficiency. The low price makes everybody dig for profits. We make more discoveries concerning manufacturing and selling under this forced method than by any method of leisurely investigation.

Henry Ford, *My Life and Work*, Doubleday & Company, Inc., Garden City, N.Y., 1923.

PRODUCT ESTIMATING

With design details at hand the cost estimator begins the task of product estimating. While price is the object of the estimate, other documents are vital for management strategy. To permit the preparation of a product estimate, sales, marketing, and operation estimates must be concluded. Enroute to price setting, various procedures work on the task so as not to overlook the objectives of stockholders and consumers as well as management and estimating. These procedures focus on criteria for appraisal. The preliminary estimate provided early screening. Now a detailed product estimate is necessary in order to set a price and determine cash flow, rate of return, and profit and loss statements for a second look. The appraisal is made on new or continuing products and product lines to see if there is any deviation from the original target. As we shall see, Henry Ford's wisdom, while brilliant but eccentric in his time, must be broadened to include other basic objectives for product price.

9.1
PRODUCT PRICE

The product effort is on-going, all-inclusive, and basic to company survival. The care and stimulus for the product and its success rests on research, engineering, manufacturing, marketing, legal, and management. A product strategy calls for a wide assortment of decisions. All this is a large undertaking for any firm. The hazard is high as of some several hundred fresh ideas only one will congeal into a successful new product. In view of these complications we now direct our efforts toward cost-estimating aspects related to the product.

Even this portion of confined study is vast. In addition to price and cost of the product, there are problems concerning cash flow, rate of return, and meeting obligations to investors, owners, and shareholders.

Who is responsible for setting price? Practice varies, for are we talking about an old or new product? What precisely is the cash flow problem? As products provide income, the realization of revenue from the sale of newly introduced or established products must be offsetting of costs of these same products. Is the new or old product small in proportion to total income? If it is significant, individual products and product lines constitute an important factor to the cash flow problem. Some new products require a large outlay of cash for new plants and processing equipment in addition to engineering and construction costs. Investment analysis, called *profitability*, is one way to conduct an investigation. The capital obligations for this expansion may come from profit, current depreciation, loans, or new issues of stock. Expansion may be necessary to produce the product.

The conclusion of a product estimate is price. Along the way to this result are a number of analyses, not the least of which is cost. There is a school of thought that asserts that "price is not related to cost." An argument begins by assuming that house A and B are identical as to neighborhood, appearance, and the like. Price A is $20,000, while price B is $18,000. Now owner A argues that his cost for house A is $20,000. The buyer, however, considers this as irrelevant and so would choose B. But what about owner B? If his cost is under $18,000, he has made a profit, but if his cost is $18,000 or more, his prosperity is weakened. What the buyer is willing to pay is influenced by the lowest competitive price and is determined at the point of sale, not in the factory. If price is less than cost, the company must take steps to reduce cost or abandon the product. Of course, cost is not the only factor that sets a price, but it is a vital one. For long-term survival it is imperative to recover the full consumption of resources; a price strategy must accommodate this policy.

For a new product an initial screening analysis may be conducted by the exclusion chart method described in Chapter 6. The exclusion chart excludes an area which is free of competitive product price. To price a product inside that zone, however necessary because of unusual advantages and significant quality features, may be unwise.

Older products for which an existing market is well defined present a different situation. Here competition plays a bigger role. Custom products for one customer require different treatment than do products for which a large market exists.

A cost estimate for a product undergoing redesign requires less treatment than a new product. The components of the redesigned product are compared to the old design and classified as changed, added, or identical parts. The costs of the unchanged parts are found from records and may or may not be altered to reflect future conditions. Operation estimates are prepared for the altered and new parts. In each of these several product conditions, methods of product costing and pricing vary.

9.2 INFORMATION REQUIRED FOR PRODUCT ESTIMATING

Upon receiving a request for an estimate the estimator analyzes the design information, which includes

1. Due date for completion of estimate.
2. Quantity and rate of production.
3. Engineering bill of materials, drawings, and specifications.
4. Special test, inspection, and quality-control practices.
5. Packaging and shipping instructions.
6. Marketing information.

Other information, principally of a cost nature, has been discussed in Chapters 3, 4, and 5. Figure 3.1, which describes the traditional cost and price structure, has been recast in Figure 9.1 with price as the ultimate objective. The bottom two layers, direct material and direct labor, provide the grist for an operation estimate and was covered in Chapter 8. Addition of the upper blocks is achieved by the several methods of this chapter.

In some cases engineering costs for a product are not covered by overhead and must be estimated separately. These companion estimates would include research and development and engineering. For instance, high technology firms calculate engineering, development, and design as a separate line item cost.

The determination of administrative, marketing, distribution, and selling rates are found like overhead rates. However, the basic denominator for this calculation may be the cost of goods manufactured. Costs for administration and sales are structured and totaled, and finally the appropriate ratios are determined. Refinements such as separating divisional and corporate expenses or marketing from sales can be undertaken. These refinements are usually worth the trouble because these costs can be more accurately absorbed to a product for its eventual recovery.

Contingencies are another category in Figure 9.1. Uncertain costs may be

FIGURE 9.1

THE ELEMENTS IN A COST ESTIMATE

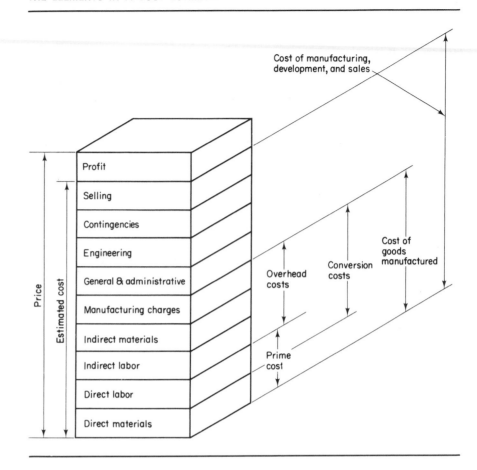

estimated here. Although this is not a desirable category, there are circumstances where it may be used. For instance, government imposed safety requirements may be required for product reliability. The firm, uncertain of this future behavior and unprepared to introduce this feature until the government has provided the specifications, uses this category for a visible cost. The contingency feature as an estimating practice has the advantage of providing special provisions for future legitimate costs. The use of the contingency, say, to cover careless detailed estimating practices, is not encouraged.

The elements of Figure 9.1 vary in importance depending upon the business sector. Some industries may be labor intensive and thus direct labor may be important; or material, like standard purchased parts, raw, (flat plate, bar stock), and direct utilities may dominate. In capital intensive industries, the recovery of capital money is important. The relative importance of these major elements may be seen by examining Figure 9.2. The example in this figure is an expensive product. The estimating emphasis given to the basic elements usually depends upon their percentage of total cost. The estimating department accommodates their information profile requirements, functional estimating models, and recap sheets as dictated by their individual product situation.

When is product estimating done? Figure 9.3 shows the location of this effort relative to design and production. This is a simple time scale defining the point at which information is processed.

FIGURE 9.2

PROPORTION OF MAJOR SEGMENTS OF COSTS
AS PRODUCTION VOLUME VARIES

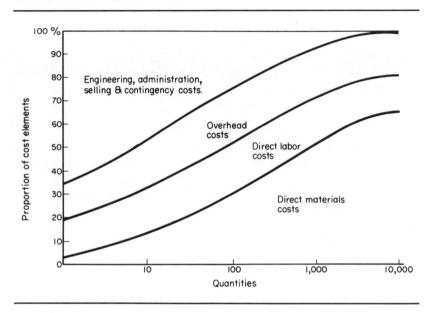

FIGURE 9.3

TIME SCALE LINE SHOWING CHRONOLOGY OF
PRODUCT ESTIMATE EVENTS

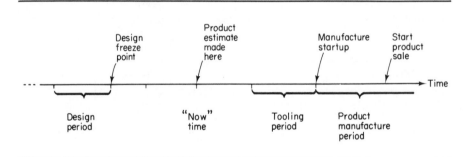

9.2-1
Estimating Research and
Engineering Costs

When products are comparable to existing products, or when there is a reasonable assurance about the state of art in terms of engineering, ordinary techniques useful in judging the costs of engineering may be adopted. For these conventional situations, engineering can be estimated on the basis of similar and past historical records using budgets and variance reports, such as shown in Chapter 4. This includes an appraisal of past performance on similar work and from it the determination of the probable efforts required for the proposed task.

Engineering provides services measured in terms of the design of the product, specifications, bill of materials, and maybe physical models. In addition, preliminary engineering may be done for a job for which orders are not received. The engineering task may embrace evaluation and production liaison, field and construction service, and maintenance engineering during the life of the product. For these specific functions, hours and a cost factor containing provisions for hourly rates for wages and salaries of the group, charges for supervision, housing, light, and heat are estimated. Engineering costs to develop proposals that do not materialize in orders are additionally collected into the conglomerate hourly rate and used as the multiplier for the estimated number of hours. The total of this cost is ammortized to the product by

$$\text{product charge} = \frac{\text{total engineering expenses}}{\text{projected product quantity}} \tag{9-1}$$

Model (9-1) is for exceptional charging that is not included in overhead practices. This model can be expanded into a recap sheet like Table 9.1. The design

TABLE 9.1

ESTIMATING FORM FOR DESIGN ENGINEERING

Design engineering estimate for _Speed Reducer_ products

Description _Design for Left hand speed reducer_

Customer _Not Known_ _____ Inquiry or quote no._____ Date _____
Based on quantity of _480 units_ _____ During period of _____

Type of labor	Hours per unit	Rate per hour	Extended labor
Scientist, research			
Engineer, senior design	40	$8.25	$330
Engineer, design	800	$7.50	$6000
Technician, electrical			
Designer / engr. aide	160	$3.75	$600
Tech. writer			
Illustrator			
Draftsman	80	$4.25	$340
Provisioning specialist			
Model shop			
Total design engineering labor	1080		$7270

engineering estimate is categorized by type of labor and rate per hour and is finally extended to total design engineering labor. The estimate can be associated with a custom design or for one lot of products. These estimates are based on comparisons to similar products and cost experiences.

There are cases where an engineering effort is large in proportion to a contract, or in special cases an engineering firm will be hired to do the engineering and, say, manage a large-scale construction job. Estimating these costs may become a competitive bid among engineering firms. If major equipment is installed, say, in the range of $500,000 to $40 million, engineering costs for complex pilot and chemical plants range from 21 to $7\frac{1}{2}\%$. In repetitive types of construction, engineering costs vary from $3\frac{1}{2}$ to 13% of total installed cost.

These engineering costs may be negotiated on a lump-sum turn-key basis where the cost of engineering is included in the erection package for an entire plant; more frequently, the engineering contracts are negotiated on some cost-plus basis. Variations would include

1. A straight cost plus a negotiated fee or profit for the engineering contractor.
2. A cost plus a fixed fee contract with a guaranteed maximum.
3. A contract for engineering design manpower to be supervised by the client's engineering staff.

No matter what the contract type, the functional model contains these elements as the total cost:

$$C_T = \sum S + \sum E + \sum OH + \sum F + P \qquad (9\text{-}2)$$

where C_T = sum of office and field expenses
$\quad S$ = salaries
$\quad E$ = variable expenses such as travel, living away from home, communication
OH = overhead, rent, depreciation, heat, light, clerical supplies, workman's compensation, and so forth
$\quad F$ = fees paid to other specialists and engineers
$\quad P$ = profit

In view of the proportion of salaries to the total cost, it is usual to find factors that multiply expected salaries to arrive at the estimate. These factors vary from 1.8 to 3.0 depending on complexity, novelty, or secrecy of the work. In chemical and architectural work, the ratio of design drafting is of the order of two or three times other types of engineering. In electronics this is reversed. Standards for engineering work have related size of drawing and design productivity. In tooling of dies, for instance, the size of the drawing can be associated to so many hours of design time. Other similar rules of thumb have been established. In any one industry there seems to be a tendency toward standardization in drawing size as well as the kind and quantity of information recorded on the print. To determine the cost of engineering, it must be estimated in order to be prorated to product charges like any other cost. Its impact may amount to very little, or it can be a major factor, particularly when a product, new or revised, calls for a new process or plant.

9.3 FINANCIAL DOCUMENTS REQUIRED FOR PRODUCT DECISION

The appraisal of products, old, newly introduced, or ones at the detailed stage, is handled by analysis. What money is necessary for the venture and what will come out of it? Management, stockholders, and money lenders are interested in this question. Operating capital and loans and notes for new equipment or plant enlargement may be needed before there is any revenue from the products. Thus it is mandatory to prepare several documents on products. Management takes action on the product on the basis of what these documents say. These documents are listed as

1. Cost estimate.
2. Cash flow statement.
3. Rate of return analysis.
4. Profit and loss statement.

The cost estimate is a document concerned with product details, and it indicates a probable selling price that the product will have to command to have a profitable future. Almost all of this chapter is concerned with this first document.

Cash Flow Statement. The cash flow document considers the value of trans-actions in and out of a firm. It may be likened to a reservoir receiving a stream of water. At certain times more water is received than at other times. Concurrently the demands placed on the reservoir fluctuate with a controlled quantity leaving. Money to a company behaves very much like this illustration and is frequently called a *stream.*

If the cost of the product venture is small in proportion to the inflow or accumulated surplus, the cash flow document may be unnecessary. If, on the other hand, the venture is a big one, the company must evaluate its cash position to meet obligations. If the product requires a tooling up period, long pilot runs, expensive equipment, and extensive engineering and pre-operation break-in, the construction of a cash flow document is vital. A small company may require operating capital before product revenue is received. They would find a cash flow document necessary.

A cost estimate involving price and quantity schedules, production rates, and marketing and sales rates, and capitilization costs for new equipment and plant enlargement is necessary in order to construct a cash flow statement. A sample is provided by Table 9.2. For this chemical plant operation the percent of capacity increases up to 100% at which point it is assumed that capacity would continue to grow due to learning effects. Obviously to have net profits after taxes, a market and price are known. Depreciation is a tax-sheltered fund while a tax credit (a product of the depreciation times the firm's tax rate) helps to provide the inflow of funds. Preoperating expenses and investment costs are first year cash out. Working capital is made up of several items such as accounts receivable, raw material inventory, work in process, and finished goods inventory. Increases in inventory require immediate cash outlays that delay cash flow from generating sales revenue. Ordinarily a higher requirement for cash on hand occurs during periods when operations are increasing. Determination of what constitutes working capital is usually meant to be incremental capital, i.e., differential capital between present and prior year. The arithmetic for net and cumulative cash flow is evident. A payout time calculation can be performed which measures the number of years required to regenerate, via profits, depreciation, and tax credits, the total investment of the fixed assets and pre-operating expenses that we required to launch the product. The payout time of 5 years illustrates that it will take this number of years to recoup the capital investment and pretax operation expenses. For a cash flow statement we define

$$F_c = (G - D_c - C)(1 - t) + D_c \qquad (9\text{-}3)$$

where F_c = total source of funds, dollars, year
$\quad\quad G$ = estimated annual gross product income, dollars
$\quad\quad D_c$ = depreciation charge, dollars
$\quad\quad C$ = annual costs not estimated elsewhere, dollars
$\quad\quad t$ = tax rate, decimal

The payout time and cash flow is based on a nondiscounted basis, i.e., the face value of the cash flows for each year. Methods taught in Chapter 10 show how these values are discounted.

Rate of Return Analysis. A broad view of rate of return analysis is one that considers the net effective profitability of a product over its life span. Sometimes this analysis is called *profitability.* This must evaluate the return for a price-volume situation over a period of time that will be maintained. A principle requires that a product, including equipment and plant enlargement capital costs and working capital, must, during the product's life, return to the firm a suitable fixed interest on the unpaid balance of outstanding cash flows. A com-

TABLE 9.2

CASH FLOW STATEMENT

	Year 1	Year 2	Year 3	Year 4	Year 5	Year 6
Percent of capacity	25	50	75	100	110	116
Production, 10,000 lb/yr	12	25	37	50	55	58
Net profit after taxes	$51,000	$148,500	$188,500	$248,500	$269,500	$282,500
Depreciation	70,000	70,000	70,000	70,000	70,000	70,000
Total inflow of funds[a]	$121,000	$218,500	$258,500	$318,500	$339,500	$352,500
Preoperating expenses after taxes	75,000					
Fixed assets (from project estimate)	700,000					
Working capital/yr (from operation estimates)	125,000	100,000	90,000	80,000	80,000	80,000
Total outflow of funds	$900,000	$100,000	$90,000	$80,000	$80,000	$80,000
Net cash flow	−779,000	118,500	168,500	238,500	259,500	279,000
Cumulative cash flow	−$779,000	−$660,500	−$492,000	−$253,000	+$6,500	$279,000

[a]Equation (9-3) can be equivalently $F_c = (G - C)(1 - t) + tD_e$

pany can value a product in terms of a constant annual rate of interest that will be produced on the unreturned balance of investment during a product's life. Thus profitability is an analysis technique that measures the desirability of risking money for new products, and the basic factors are capital, expense, revenue, and time. The end result of a rate of return analysis predicts the discounted net changes in the company's cash position. This topic is covered in the next chapter.

Profit and Loss Statement. The final document important to making decisions on whether to go ahead with commercialization of a product is the profit and loss statement. Chapter 3 introduced this statement along with several formats. Now we direct our attention to its role in the evaluation of a product. It is an important product that merits P & L attention. Uneventful products that do not impinge significantly on the profit and loss statement can be evaluated by other criteria.

The P & L takes into account sales volume, manufacturing costs, and research and design costs. For the analysis of a proposed product one must have an estimate of production level and an annual accrual of sales income. Direct manufacturing cost is a tabulation of the various cost components that are estimated. A P & L statement shows management the commitments that marketing, sales, and manufacturing operations must make in order to reach the predicted gross sales. P & L statements are made for several years ahead for significant products.

9.4
LEARNING CURVE

It is frequently recognized that repetition with the same operation results in less time or effort expended on that operation. In fact this improvement can be sufficiently regular to become predictive through ordinary estimating techniques. The observed characteristic of the improved performance is called *learning*. The first applications of learning were in airframe manufacture, which found that the number of man-hours spent in building a plane declined at a constant rate over a wide range of production. Figure 9.4 describes this phenomenon.

Other names abound for the learning curve including terms such as the manufacturing progress function and the experience or dynamic curve. They suggest that cost can be lowered with increasing quantity of production or experience. Knowing how much product cost can be lowered and at what point

FIGURE 9.4

LOG-LOG PLOT OF INDUSTRY AVERAGE UNIT
CURVE FOR CENTURY SERIES AIRCRAFT. (From
Aeronautical Material Planning Report Data)

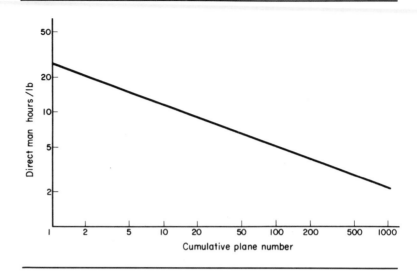

learning is applied in the estimating procedures are the reasons for studying learning prior to product estimating methods.

Applications of the learning curve are found in procurement, production, and the financial aspects of a manufacturing enterprise. In purchasing, a function can be used to negotiate purchase price or it may be used for the *make or buy* decision. Learning has been applied to equipment loading and manpower schedules. In cost estimating, decisions related to bidding, pricing, and capital requirements are based in part on the concept. Contract negotiation is sometimes re-opened after a first satisfactory model has been produced. With the experience of time and cost for the prototype unit known, the contract for later models is based on learning reductions. Aerospace firms do this with the Air Force and precision of $\pm 3\%$ has been reported.

There are explanations for this behavior and verification has been uncovered by independent researchers and companies. Principally, the reduction is due to direct-labor learning and the management process. The direct-labor learning process assumes that as a worker continues to produce, it is natural that he should require less time per unit with increasing production.

The management processes are those engineering programs which improve production, encourage quality, reduce design complexity, create technology progress, and foster product improvement. In short, they are those management programs that are considered extraordinary and exceptional to a particular product. These management programs inspire time and cost reduction and are credited with time and cost reduction of a product additionally beyond that which direct-labor learning would provide.

Some experience suggests that the operator is responsible for approximately 15% of the total reduction, while management and their programs contribute the remaining 85%. In the manufacturing industries the 85% has been broken down into 50% due to the product engineering endeavors, while manufacturing and industrial engineering activities are credited with the remaining 35%.

Ships, aircraft, computers, machine tools, and apartment and refinery construction have in common high cost, low volume, and discrete item production and can be treated by learning. Although the same principle applies to TV production, for instance, the effects may take years to uncover because of the large production volume. The learning curve is usually not applied to high-volume or low cost products.

The learning curve model rests on the following observations:

1. The amount of time and its cost required to complete a unit of product are less each time the task is undertaken.
2. The unit time will decrease at a decreasing rate.
3. The reduction in unit time follows a specific estimating model such as the negative exponential function.

To state the underlying hypothesis, the direct labor man-hours necessary to complete a unit of product will decrease by a constant percentage each time the production quantity is doubled. While the hypothesis stresses only time, practice has extended the concept to other types of measures. A frequent stated rate of improvement is 20% between doubled quantities. This establishes an 80% learning curve and means that the man-hours to build the second unit will be the product of 0.80 times that required for the first. The fourth unit (doubling 2) will require 0.80 times the man-hours for the second; the eighth unit (doubling 4) will require 0.80 times the fourth; and so forth. The rate of improvement (20% in this case) is constant with regard to doubled production quantities, but the absolute reduction between amounts is less. This is the reason why *follow-on* costs in low production are noticeably lower than original costs. The learning curve may be defined if the number of direct labor hours required to complete the first unit is established and if the subsequent rate of improvement is specified. Alternatively, the learning curve may be defined if direct-labor man-hours for a downstream unit and the learning curve rate are estimated. Other possibilities for defining the curve can be selected. The number of direct-labor hours required to complete the first production unit depends on these circumstances:

1. The previous experience of the company with the product. If it had little or no experience, the first unit time would be greater than for a product with considerable experience.
2. The amount of engineering, training, and general preparations that the organization expends in preparation for the product. In some cases the first several units are custom-made and tooling is not made until more sales can be assured. This "hard-way" production would inflate the first unit time.
3. The characteristics of the first unit. Large complex products would be expected to consume more direct-cost resources than something less complex.

The notion of constant reduction of time or effort between doubled quantities can be defined by a unit formula:

$$T_N = KN^s \qquad (9\text{-}4)$$

where T_N = effort per unit of production, such as man-hours or dollars required to produce the Nth unit
N = unit number

[1] Excerpts from W. J. Fabrycky, P. M. Ghare, and P. E. Torgersen, *Industrial Operations Research*, Prentice-Hall, Inc., Englewood Cliffs, N.J., 1972, pp. 172–200.

K = constant, or estimate, for unit 1 in units compatible to T_N
s = slope parameter or a function of the improvement rate

The slope parameter is negative because the effort decreases with increasing production. Figure 9.5 is the linear unit progress curve for 80% learning with cartesian coordinates, while Figure 9.6 represents a log-log curve.

FIGURE 9.5

80% MANUFACTURING PROGRESS FUNCTION WITH UNIT NUMBER AT 100 DIRECT-LABOR MAN-HOURS

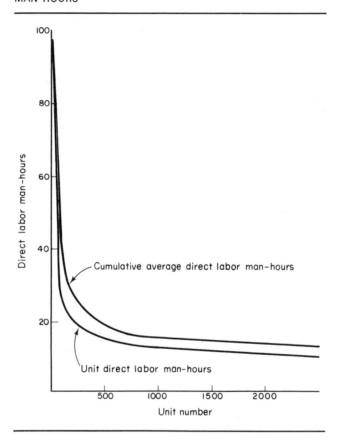

To understand the graphic presentation of the learning curve on logarithmic graph paper, first compare the characteristics of arithmetic and logarithmic graph paper. On arithmetic graph paper equal numerical differences are represented by equal distances. For example, the linear distance between 1 and 3 will be the same as from 8 to 10. On logarithmic graph paper the linear distance between any two quantities is dependent on the ratio of those two quantities. Two pairs of quantities having the same ratio will be equally spaced along the same axis. For example, the distance from 2 to 4 will be the same as from 30 to 60 or from 1000 to 2000.

The learning curve is usually plotted on double logarithmic paper, meaning that both the abscissa and the ordinate will be a logarithmic scale. For an exponential function $T_N = KN^s$ (the general class of learning curve functions)

the plot will result in a straight line on log-log paper. Because of this, the function can be plotted from either two points or one point and the slope, e.g., unit number 1 and the percentage improvement. Also, by using log-log paper the values for a large quantity of units can be presented on one graph, and these can be read relatively easily from the graph. Arithmetic graph paper, on the other hand, requires many values to sketch in the function.

FIGURE 9.6

LINEAR UNIT PROGRESS CURVE WHERE
$\phi = 80\%$ AND UNIT 1 = 100 HOURS

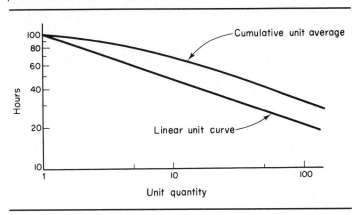

The slope constant s is negative because the time per unit diminishes with production. The units for T_N can be expressed in any of several compatible dimensions. Time or cost is frequently used, but when cost is used the underlying basis should be estimated with care.

The cartesian-coordinate curve representation indicates the range of major reduction. It is evident that any special management program for Figure 9.5 may be unnecessary beyond the five-hundredth unit or so, which is approximately where the slope begins to level out. This judgment is an evaluation of input resources to output success and points up that the savings occur early in the life of some products. As the product becomes older it becomes less susceptible to improvements.

The exponent s is defined (recalling the double quantity concept) as

$$s = \frac{\log \phi}{\log 2} \tag{9-5}$$

where $\phi =$ the slope parameter of the learning curve function.

Application of Equations (9-4) and (9-5) can be illustrated by an 80% learning curve with unit 1 at 1800 direct-labor man-hours. Solving for T_8, the number of direct-labor man-hours required to build the eighth unit gives

$$T_8 = 1800(8)^{\log 0.8/\log 2}$$

$$= 1800(8)^{-0.322} = \frac{1800}{1.9535} = 921 \text{ hours}$$

A table of decimal learning ratios to slope constant is given as

Slope Parameter, ϕ	Exponent, s
1.0 (no learning)	0
0.95	−0.074
0.90	−0.152
0.85	−0.234
0.80	−0.322
0.75	−0.415
0.70	−0.515
0.65	−0.621
0.60	−0.737
0.55	−0.861
0.50	−1.000

For unit 1 requiring 1500 hours and a projected learning of 75%, the time for unit 90 is

$$T_{90} = 1500(90)^{-0.415} = \frac{1500}{6.471} = 232 \text{ hours}$$

Tables shorten the calculations. Appendixes III and IV are unit and cumulative values, respectively.

9.4-2
Finding the Learning Curve from Two Points

It may be desired to find out if learning has materialized, and if it has, to determine the function for which other unit estimates can be found. Where only two points are specified it may be desirable to find the learning curve that extends through them. Let the two points be specified as (N_i, T_i) and (N_j, T_j). At each point

$$T_i = KN_i^s \qquad \text{and} \qquad T_j = KN_j^s$$

Dividing the second into the first gives

$$\frac{T_i}{T_j} = \left(\frac{N_i}{N_j}\right)^s \tag{9-6}$$

and taking the log of both sides we have

$$\log \frac{T_i}{T_j} = s \log \frac{N_i}{N_j}$$

$$s = \frac{\log T_i - \log T_j}{\log N_i - \log N_j} \tag{9-7}$$

K may be found by substituting s into $T_i = KN_i^s$ and solving for K:

$$\log T_i = \log K - s \log N_i \tag{9-8}$$

This has a linear form like $y = a + bx$, one of the linear models described in Chapter 5. It is seen that on log-log paper the intercept is K while the slope of the line is equal to $-s$.

An example will now illustrate this concept. A company audited two production units, the twentieth and fortieth, and found that about 700 hours and 635 hours were used, respectively. Now at the seventy-ninth unit, they want to estimate the time for the eightieth unit:

$$s = \frac{\log 700 - \log 635}{\log 20 - \log 40}$$

$$= \frac{2.8451 - 2.8028}{1.3010 - 1.6021} = \frac{0.0423}{-0.3011} = -0.1405$$

and

$$\log \phi = s \log 2$$

$$= (-0.1405)(0.3010) = 9.9577 - 10$$

Taking the antilog of both sides gives $\phi = 0.907$. The percentage learning ratio is 90.7%. Using the data for the twentieth unit,

$$700 = K(20)^{-0.1405}$$

$$\log 700 = \log K - 0.1405 \log 20$$

$$\log K = 2.8451 + 0.1405(1.3010) = 3.0279$$

Taking the antilog, $K = 1066$ hours. The learning curve function is

$$T_N = 1066N^{-0.1405}$$

The unit time for any unit can now be calculated directly. For the eightieth unit

$$T_{80} = 1066(80)^{-0.1405} = 576 \text{ hours}$$

This can be confirmed by using Appendix III and interpolating. It should be observed that

$$\frac{T_{40}}{T_{20}} = \frac{T_{80}}{T_{40}} = 0.907$$

9.4-3
Cumulative Curve

The unit formulation can be extended to other types of functional models. It may be necessary to determine a cumulative average number of direct-labor man-hours. This can be found by the cumulative total and is

$$T_T = T_1 + T_2 + \cdots + T_n = \sum_{i=1}^{n} T_i \qquad (9\text{-}9)$$

and its average is

$$\sum_{i=1}^{n} \frac{T_i}{n} \qquad (9\text{-}10)$$

A good approximation of the cumulative average number of direct-labor man-hours for 20 or more units is given by

$$V_n \doteq \frac{1}{(1 + s)} KN^s \qquad (9\text{-}11)$$

where $V_n \doteq$ cumulative average of number of direct-labor man-hours.

The cumulative average curve is above the linear unit curve. See Figure 9.6.

Unit 1 is estimated to require 10,000 hours and production is assumed to have an 80% learning curve. What will be the unit direct-labor man-hours, the cumulative direct-labor man-hours, and the cumulative average direct-labor man-hours for unit 4?

$$T_1 = 10,000$$

$$T_2 = 10,000(2)^{-0.322} = 8000$$

$$T_3 = 10,000(3)^{-0.322} = 7021$$
$$T_4 = 10,000(4)^{-0.322} = 6400$$

The cumulative direct-labor man-hours are

$$\sum_{i=1}^{n=4} T_i = 31,421$$

while the average using Equation (9-10) is

$$V_n = \frac{1}{4}(31,421) = 7855$$

$$= \frac{6400}{(1 - 0.322)} = 9440$$

Formula (9-11) is not accurate in this quantity range since it is on the upper left-hand hump of the curve. This can be seen by examining Figure 9.6.

9.4-4
Least-Squares Fit

No estimating device despite its attractiveness can be used indiscriminately. For this reason it is desireable that management insist on occasional audit exercises that measure actual learning as a function of production. By using the results from a least-squares fit study, estimating is able to compare actual learning against predictions. These actual plots have a two fold purpose. They tend to prevent abuses and encourage additional effort as a need is shown, or they reduce effort as various programs are brought under control.

One means of past data evaluation is to plot the data on log-log graph paper and sketch a straight line through the data. However, it is better to adopt a least-squares fit as first described in Chapter 5. The general learning curve can be expressed as a logarithmic straight line of the form $\log T_n = \log K + s \log N$. The method of least squares will yield

$$s = \frac{M \sum \log N \log T - \sum \log N \sum \log T}{M \sum (\log N)^2 - (\sum \log N)^2} \tag{9-12}$$

$$\log K = \frac{\sum \log T \sum (\log N)^2 - \sum \log N \sum \log N \log T}{M \sum (\log N)^2 - (\sum \log N)^2} \tag{9-13}$$

where M = sample number. An example illustrates the procedure. Five data points have been collected as

Unit, N	Time, T
10	510
30	210
100	190
150	125
300	71

The method is described by Table 9.3. For the data the learning slope is 70% and the initial derived first unit time is 1556 hours. An estimating model is then constructed as

TABLE 9.3

LEAST SQUARES ANALYSIS OF ACTUAL COST DATA
TO FIND INITIAL VALUE AND LEARNING SLOPE

Given $T = KN^s$

or $\log T = \log K + s \log N$ which is of the form $y = a + bx$

where $y = \log T$, $a = \log K$ intercept, $b = s$ slope, $x = \log N$, and $M =$ sample size

Unit, N	Man-Hours, T	$x = \log N$	$y = \log T$	$(\log N)^2$	$\log N \log T$
10	510	1.0000	2.7076	1.0000	2.7076
30	210	1.4771	2.3222	2.1818	3.4301
100	190	2.0000	2.2788	4.0000	4.5576
150	125	2.1761	2.0969	4.7354	4.5631
300	71	2.4771	1.8513	6.1360	4.5859
		9.1303	11.2568	18.0532	19.8443

$$s = \frac{M \sum \log N \log T - \sum \log N \sum \log T}{M \sum (\log N)^2 - (\sum \log N)^2}$$

$$= \frac{5(19.8443) - (9.1303)(11.2568)}{5(18.0532) - (9.1303)^2} = -0.515,$$

$$\log K = \frac{\sum \log T \sum (\log N)^2 - \sum \log N \sum \log N \log T}{M \sum (\log N)^2 - (\sum \log N)^2}$$

$$= \frac{11.2568(18.0532) - (9.1303)(19.8443)}{5(18.0532) - (9.1303)^2} = 3.19207,$$

antilog $= 1556$

$$T = 1556 N^{-0.515}$$

$$\log \phi = s \log 2 = -0.515(0.30102) = -0.15503$$

antilog $\phi = 0.6697$ or $\phi = 69.97\%$

$$T_N = 1556 \ N^{-0.515}$$

This is a customary analysis used to determine past learning performance. Too often estimators are prone to use learning slopes without an adequate test of the firm's experience.

9.4-5
Specific Cost Applications

In predicting costs it is useful for estimators to know the recurring costs of production and the learning slopes for the several aspects of production, e.g., manufacturing direct labor, raw material, manufacturing engineering, tooling, quality control, and other indirect charges. In our next application we consider learning for direct labor and factory overhead. Materials are now assumed to be insensitive to the learning curve.

$$C_i = \frac{KN^s}{s+1}(C_{dl}) + \frac{KN^s}{s+1}(C_{dl})(OH) + C_{rm}$$

$$= \frac{KN^s}{s+1}(C_{dl})(1 + OH) + C_{rm} \tag{9-14}$$

where $C_i =$ product cost per unit for product i

$C_{dl} =$ direct-labor hourly rate

$OH =$ overhead rate including engineering, tooling, quality control, and other indirect charges expressed as a decimal of the direct-labor hourly rate

Our example considers a situation in which 200 units are to be produced. $C_{dl} = \$4$ per hour, raw material cost is $250, and the overhead rate is 50%. The first unit is estimated to require 350 direct-labor man-hours, and a 90% learning curve slope is thought applicable. The average cost per unit is

$$C_t = \frac{350(200)^{-0.152}}{0.848}(4)(1 + 0.50) + 250 = \$1357$$

For a lot of 200, the total cost is $200(1357) = \$271,359$.

Now we examine a problem having several diverse items that can be estimated by a learning curve approach. Some factors can be associated as recurring and identified to production quantity. Portions of these same items are also insensitive to quantity. Engineering can be isolated in this fashion. The first unit will require the bulk of design, but subsequent units, if a customer and or specifications necessitate, also require additional design on a per unit basis such as *engineering-change orders*. The latter portion can be subjected to a learning curve philosophy.

Table 9.4 is an illustration of a design for 150 units. Selected cost items have been broken out and divided into the categories of fixed and variable. The learning slope is estimated on the basis of a similar past performance. Extensions to a total line cost are straightforward. Other major cost categories which were not considered via learning curves are then added as a percentage of total estimated costs. Finally a total price is determined for the 150 units.

9.5 METHODS OF PRODUCT ESTIMATING

Three methods of product estimating are studied. They are

1. Operation.
2. Department or cost center.
3. Variable cost.

World-wide firms that produce a product number in the millions. In the United States there are over 300,000 firms. Variations in these methods can be expected; but generally, these methods represent typical industrial practices.

All are widely practiced, and each has advantages that makes it superior to others at certain times. They depend on disclosure of detailed operational estimates, engineering data, and marketing information. The marketing information, although it may be vague and tentative for new products, must be available.

The bill of material is a fundamental design document, and the following data are obtainable from it:

1. Part or identifying number.
2. Description or title of part, subassembly.
3. Quantity required per unit of product.
4. Material specification, raw material stock number.
5. Component code classification such as vendor control number if purchased, standard or nonstandard purchased part.
6. Other products in which components may be used.
7. Cross reference to similar part numbers.

The bill of material is an explosion of information leading to other existing documents, products, and cost estimates. New product designs incorporate some parts, subassemblies, or major subassembly units of other manufactured

TABLE 9.4

LEARNING CURVE APPROACH
TO COST ESTIMATING

Selected Costs	First Unit Cost Estimate		Estimated Learning Slope	150-Unit Cumulative Factor	150-Unit Cumulative Cost	Total Line Cost
	Fixed	Variable				
Manufacturing	—	$4700	75%	30.93	$145,371	$145,371
Raw material	—	900	90	82.15	73,935	73,935
Engineering	$43,000	380	85	59.89	22,758	65,758
Tooling	5,100	240	70	21.97	5,272	10,372
Quality control	1,600	230	75	30.93	7,113	8,713
Equipment	8,400	330	90	82.15	27,109	35,509
Other direct costs	2,500	45	80	43.23	1,945	4,445
			1. Total selected costs			344,103
			2. Other indirect costs			28,000
					Subtotal	370,103
			3. Fixed and miscellaneous charges @ 0.28			103,628
			4. Distribution and administrative costs @ 0.21			77,722
			5. Contingencies @ 0.08			29,608
			6. Selling costs @ 0.02			7,402
					Subtotal	588,463
			7. Profit @ 0.18			105,923
					Total price	$694,386

products, and the product estimator uses the bill to find previously prepared estimates. These older estimates are updated using indexes or new wage rates and the like. For new parts he prepares new estimates. Thus the bill is a complete listing of parts, including fabricated and assembled parts, standard purchased parts (washers, resistors, and so forth) and raw material. Using a bill, the product estimator can be assured that all materials, both purchased and fabricated, will not be inadvertently overlooked.

9.5-1
The Operation Method

This is the oldest of the product estimating methods. Manufacturing process sheets are prepared for each component, subassembly, and assembly. These process sheets are either a preliminary or final document, depending upon the firm. They provide a manufacturing procedure, identify the operation setup and run time standards, labor grade and wage, and the operational overhead rate.

The estimating method calls for a standard labor cost estimate for each operation which is calculated using

$$\text{operation cost} = \left\{\left(\frac{SU_i}{Q} + T_i\right)LR_i\right\}(1 + OH_{wc}) \qquad (9\text{-}15)$$

where SU_i = setup for operation i in standard hours
Q = production quantity
T_i = standard time per unit(s)
LR_i = labor rate dollars per hour
OH_{wc} = overhead rate for work center

Practice dictates that setup time standards for operations be determined from any of the several methods of time measurement. For Q very large, setup time is negligible on a unit basis, and under circumstances as these cycle time is the only one considered. Setup is then costed via overhead distribution.

It is practice now for T_i to include normal elements and allowances for personal and delay. Composition of these items has been discussed in Chapter 4. On the other hand, the performance against these standards (either incentive or day-rate) invariably calls for adjustment to reflect actual rates of work. The means to reflect performance is usually handled by one of two ways: adjustments (either up or down) by dividing operational cost by efficiency, or adjustments in the operation overhead rate by the addition or subtraction of a predetermined standard performance variance.

The users of this method believe that as operational cost data (labor cost and individual machine or work station overheads) are available, precision in cost is greater than by other methods. Labor and machine rates can vary dramatically within a department.

A running total of direct material, labor, and overhead is carried along from the part to the minor, major, and final assemblies. Figure 9.7 is an indication of how this method is pursued. On completion of the product estimate with the product ready to be stored in finished-goods inventory, general and administrative overhead charges are added by multiplying the total of material, direct labor, and overhead by a factor. Engineering charges, tool costs, or other contingency charges could either be buried within prior overhead costs, or they could be isolated for individual line treatment. Should a lump sum cost like engineering and tooling be added it is required that they be converted to a unit charge.

The unit estimate column depends on the formula $SU/Q + T$, and for minor subassembly operation 20, Figure 9.7, "spotweld 10 spots . .", $SU/Q + T = 0.4/100 + 0.030 = 0.034$. The operation is multiplied by labor rate and by the overhead rate corresponding to the operation. The total of \$0.340 for material and \$5.887 per unit is an input to the final assembly shown as the top entry in Figure 9.7. Total unit costs for components are similarly added to the subassembly. The sums of the subassemblies are in turn merged at the final assembly. Only at the bottom of the last assembly operation sheet are other indirect costs, lines 2, 3, 4, and 5 added to arrive at a total unit cost of manufacturing, development, and sales. Some of that shown on Figure 9.7 can be handled through computer routines.

9.5-2
Department or Cost Center Method

This method differs from the preceding one in several ways. The emphasis is more upon cascading significant groups of costs by proper multipliers. In a sense Figure 9.1 is descriptive of the process we are about to describe. This figure could have a functional model of the form

$$C_T(Q) = \{(C_r + C_{dl}(1 + OH_M)\}(1 + OH_{SGA}) \qquad (9\text{-}16)$$

where $C_T(Q)$ = total cost for quantity Q units
$\quad\quad C_r$ = raw material cost, total dollars
$\quad C_{dl}$ = direct labor cost, total dollars
OH_M = factory overhead, estimated as a ratio of budgeted indirect cost items divided by anticipated direct labor costs for the same organizational unit, decimal
OH_{SGA} = cost of selling, general and administrative estimated as a ratio of budgeted items divided by cost of goods manufactured for the same organizational unit, decimal

If a unit cost C_T is desired, then $C_T(Q)$ would be divided by the number of units Q. Although this is a formal functional model approach, tabular representations are employed for simplicity. This is described as follows.

FIGURE 9.7

FLOW OF ESTIMATED DATA FROM COMPONENT
TO SUBASSEMBLY TO FINAL ASSEMBLY
USING OPERATION METHOD

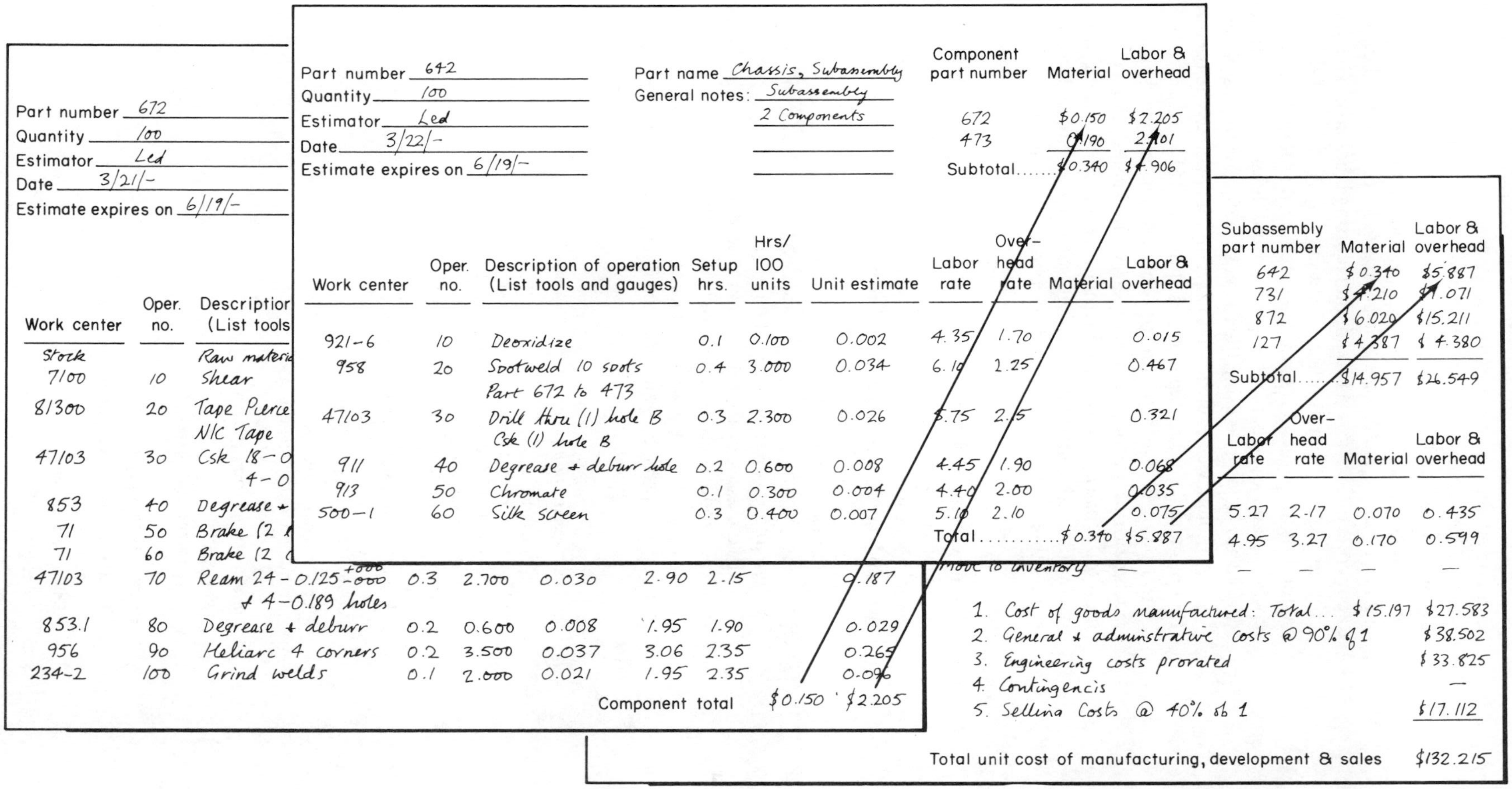

Initially an operation estimate is conducted for each part. The various component estimates are collected by department, and the time estimates are summed for each department. These sums are entered as in Table 9.5. The average wage rate is indicated for the department. Determination for the departmental labor cost is the product of the departmental time and wage rate. In this case a learning curve factor of 8% is subtracted from the total cost. Average fringe labor costs of 28% give a total manufacturing labor cost. The design engineering costs are found from another estimating form (such as Table 9.1) and entered. A total of all parts and materials is removed from operation estimates. Other costs are added as percentages of various lines. Both unit and lot costs are carried along. A price using the full cost method is the last determination.

Although we have chosen to identify the plant as the basis for the distribution of manufacturing expenses, as shown in Table 9.5, variations in the application point of overhead are possible. When full overhead is applied at the machine or production center, as shown in Tables 3.8 and 4.12, the method is more of a *machine hour rate* method. Techniques dealing with the finding of overhead for this method were discussed in Chapter 4 and Problem 9-13 demonstrates this uncommon method of product estimating.

9.5-3 Variable Cost Method

The final method depends upon segregation of the variable and fixed costs to be incurred. Variable cost first identifies and determines all variable costs associated with the product. These are considered to be the "direct" costs of the product. Direct labor and direct material are obviously included in this category. The variable costs included in manufacturing, overhead, administration, and sales must also be determined by an analysis of the respective cost behavior. A variable cost is one that changes directly with the volume of production and/or sales of the particular product.

Two classes of fixed costs must be determined. The first class are those that are product-related and would not be incurred without production and sale of the product. The second class are the standby fixed costs that are incurred regardless of producing the product. The following summarizes this separation.

Variable Manufacturing

Direct material
Direct labor (fringe benefits, premiums)
Direct overhead costs incurred specifically
  on manufacturing such as maintenance
Direct overhead costs incurred specifically
  on direct materials
Freight to warehouses and customers

Variable Marketing Costs

Cash discounts allowed
Promotion and selling expenses

Variable Administrative Costs

Bonuses
Various operating expenses

Standby (Fixed Costs)

Rent, real estate taxes, interest on mortgage,
  depreciation
Selling, advertising
Administrative

TABLE 9.5

EXAMPLE OF PRODUCT COST ESTIMATE
SUMMARY USING DEPARTMENT METHOD

Manufacturing estimate for _____ products

Description _Speed decreaser – LH_ _____

Customer_____ Quote no._____ Date_____
Base on quantity of _____480 units_____ During period of _____

Dept.	Description	Hr/Lot	Rate	Labor
103	Finishing	33.60	$5.60	$188.16
105	Machine shop	143.20	$6.00	$859.20
106	Miscellaneous machines	118.20	$5.75	$679.65
108	Precision assembly	28.80	$7.00	$201.60
131	Electronic assembly	68.05	$7.10	$483.16
125	Cable assembly	—	—	
233	Inspection	28.80	$7.20	$207.36
241	Stock room	48.00	$5.10	$244.80
	Subtotal	468.65		$2863.93
	Learning curve factor –8%			$229.11
	Subtotal-manufacturing labor			$2634.82
	Fringe labor cost + 28%			$737.75
	Total manufacturing labor			$3372.57

Item	Description	Unit	Lot
1.	Manufacturing labor (per above)	$7.026	$3372.57
2.	Manufacturing overhead @ 75% of item 1	$5.270	$2529.43
3.		$12.296	$5902.00
4.	General & Administrative @ 20% of item 3	$2.459	$1180.40
5.	Parts + materials per estimate	$2.321	$1113.80
6.	Overhead on parts and materials @ 15%	$0.348	$167.08
7.	Custom engineering labor per estimate	$8.330	$3998.50
8.	Custom engineering overhead @ 40%	$3.332	$1599.40
9.	Research and development per estimate	—	—
10.	Contingencies per estimate	—	—
11.		$29.086	$13,961.27
12.	Selling @ 25% of item 11	$7.272	$3,490.32
13.	Total cost	$36.358	$17,451.59
14.	Profit @ 13%	$4.727	$2,268.63
15.	Selling Price	$41.085	$19,720.22
16.			
17.			
18.			
19.			
20.			
21.			
22.			
23.			

Cost approvals		Price approvals	
Estimated by	_____	Accounting	_____
Approved by	_____	Product manager	_____
Division head	_____	Dir. of service engrg.	_____
	_____	General manager	_____

Product (Fixed Costs)

Selling, manufacturing, marketing, advertising
associated with product

The problem in structuring the cost elements is basically to select that cost which is variable and that which is fixed. One attitude in determining the amount of fixed cost is have the "bare-bones" or a turn-key cost as fixed and everything else as variable. This is at a zero operating level. Another viewpoint has the fixed amount include the present operating fixed expense with all executive, administrative, clerical and sales expenses considered fixed with the balance variable. These are extremes; the best policy is to set the level of fixed expense needed to support a minimum operating level below which the company would choose to quit.

A typical model for variable cost estimating is given as

$$C_t = \left(\sum P_m + \sum P_0 \right) + \left(\sum T_i \times LR \right)(1 + OH_{VM} + OH_f) + OH_{VA} + OH_P$$

(9-17)

where C_t = total cost per unit

P_m = purchased material costs including relevant materials overhead per unit summed over all components

P_0 = purchased vendor labor on materials/unit summed over all components, including relevant overhead

T_i = standard hours per unit summed over all operations

LR = average labor cost, dollars/hour

OH_{VM} is a variable manufacturing overhead rate based on direct labor dollars. Examples include indirect supplies and utilities.

OH_{VA} is a variable administrative unit cost and exists only if the product is manufactured and sold. It includes contract advertising, market and sales overhead related to the number of units sold, and certain administrative expenses. A lump sum for these expenses is estimated, allowing for variations in output and divided by the number sold. It is not strictly a constant but may vary as economics of scale affect the ingredients.

OH_f is a fixed overhead rate which represents the "bare-bones" or standby type of cost. This is the cost which is independent of current managerial policy or whether any particular products are produced. Interest costs on loans, property taxes, depreciation on equipment, plant protection costs, rentals, and certain salary expenses are typical.

OH_P are unit product costs and are fixed in the sense that small variation in output of products cause no change in the given budget of a planning period. They can vary from period to period given a change in the product mix. These costs are related to the decision of producing the product and include engineering costs for design and manufacture. If the product is made, OH_P exists, and vice-versa. Therefore, they are marginal costs for producing the product. Supervisory salary and plant management expenses are typical. Large variations in output of product would affect OH_P.

Although the formula is one means to describe this method, a product cost form as indicated in Table 9.6 is usually used, and, although a cost per unit is determined by Equation (9-17), a total cost model for a lot or batch can be found just as easily.

Note the original estimate for the wrench problem given in Chapter 7 and again in Table 9.7. For Table 9.7 it is seen that the total unit cost decreases and then eventually begins to increase. The estimator judged that plant limitations would impede lower costs and production penalty factors outweighed the

TABLE 9.6

EXAMPLE OF PRODUCT COST ESTIMATE
SUMMARY USING VARIABLE-COST METHOD

PRODUCT COST ESTIMATE SUMMARY

Preliminary or detail_____
Product line_____ Retail price $0.50 ___ Retail mark-up 44.5%

ITEM	AMOUNT	PER CENT
LIST PRICE	$0.2775	100.00
VARIABLE MANUFACTURING COSTS:		
Direct material	0.0741	26.70
Direct labor	0.0243	8.76
Variable manufacturing overhead	0.0550	19.82
Freight to warehouses.	0.0042	1.51
Irregular merchandise allowance	0.0022	0.80
Freight to customer	0.0021	0.77
TOTAL VARIABLE MANUFACTURING COSTS	0.1619	58.36
VARIABLE MARKETING COSTS:		
Trade discount allowed.	0.0028	1.01
Field allowances	0.0001	0.04
Promoting allowances.	0.0035	1.26
Royalties	—	—
Variable selling expenses.	0.0031	1.11
TOTAL VARIABLE MARKETING COSTS	0.0095	3.42
VARIABLE ADMINISTRATIVE COSTS:		
Cash discounts allowed.	0.0050	1.80
Variable administrative expenses.	0.0040	1.44
TOTAL VARIABLE ADMINISTRATIVE COSTS	0.0090	3.24
TOTAL VARIABLE COSTS	0.1804	65.00
STANDARD PROFIT CONTRIBUTION	0.0971	35.00
STANDBY FIXED COSTS:		
Manufacturing	0.0208	7.50
Selling and marketing	0.0077	2.77
Advertising	0.0002	0.07
Administrative.	0.0069	2.48
TOTAL STANDBY	0.0356	12.82
PRODUCT FIXED COSTS:		
Manufacturing	0.0066	2.38
Selling & marketing.	0.0039	1.40
Advertising.	0.0066	2.38
Administrative.	0.0200	7.24
TOTAL PRODUCT	0.0371	13.37
STANDARD EARNINGS	0.0244	8.79

advantages of increased volume. Both labor and material decline for reasons of learning efficiency and material price breaks. In both cases learning reaches a *floor* and then levels off. The variable overhead rates decline until midway in the production quantity range and then begin to increase. A similar pattern is

TABLE 9.7

DIRECT COSTING OF THE SPECIAL WRENCH
PRODUCT

Manufacturing Estimate Summary

Cost estimate no. _____ Description of product _Special Wrench_ _____ Date _____

Estimator _____ Plant _____ Estimate expires _____

Quantity	5000	10,000	15,000	20,000	25,000	30,000
Variable manufacturing costs:						
Direct labor	$0.251	$0.229	$0.201	$0.200	$0.200	$0.200
Direct material	0.600	0.500	0.440	0.384	0.380	0.380
Variable overhead rate on labor	(0.20)	(0.12)	(0.10)	(0.10)	(0.13)	(0.25)
Variable overhead rate on material	(0.10)	(0.02)	(0.01)	(0.01)	(0.02)	(0.03)
Variable manufacturing expenses	0.053	0.027	0.020	0.020	0.026	0.050
Variable material related expenses	0.060	0.010	0.008	0.007	0.010	0.011
Total	$0.964	$0.766	$0.669	$0.611	$0.616	$0.641
Variable marketing costs:						
Trade discounts						
Allowances						
Variable selling expenses						
Total	0.109	0.001	0.002	0.004	0.006	0.110
Variable administrative costs	0.100	0.077	0.011	0.013	0.021	0.022
Total variable costs	$1.173	$0.844	$0.682	$0.628	$0.643	$0.773
Standby (fixed) costs:						
Manufacturing						
Selling and marketing						
Administrative						
Total	$0.120	$0.120	$0.120	$0.120	$0.120	$0.120
Product (fixed):						
Manufacturing						
Selling and marketing						
Administrative						
Total	0.082	0.082	0.082	0.082	0.082	0.082
Total variable and fixed costs	$1.375	$1.046	$0.884	$0.820	$0.845	$0.975

seen for variable administrative and marketing costs. To reach these conclusions it is required to have concrete historical comparisons or several variable overhead rate calculations for various production quantities.

9.6 METHODS OF PRODUCT PRICING

Four methods of product pricing are studied:

1. Conference and comparison.
2. Investment.
3. Full cost.
4. Direct costing or contribution.

There is seldom a price solution that is without complexities of all sorts. The problems that appear first are those of competition and the consumer. Setting a price causes reactions by competition as well as a knotty evaluation of the kind of response to be expected from the consumer. Additionally, the law impinges on some price decisions. Prosecution is not unknown for prices established through industry-wide collusion. In decentralized companies price setting is a part of the integral management control system because the amount of profit is partially determined by the price that is paid to other elements of the company for products and services. In times of fierce competition salesmen exert pressure to reduce prices. A number of practicies exist and are closely related to price setting and add to the complexities. For instance, promotional pricing, premiums, coupons, trade-ins, extras, fire-sale gimmicks, volume discounts, repeat discounts, geographic price differentials, lease-buy arrangements, and reciprocal agreements are deals that squarely affect the price decision.

With all these ramifications, we can justify a few basic ways to give a price decision. The following have been suggested[2]:

1. Prices that are proportional to full cost, i.e., that produce the same percentage net profit for all products.
2. Prices that are proportional to incremental costs, i.e., that produce the same percentage contribution margin over incremental costs for all products.
3. Prices with profit margins that are proportional to conversion cost, i.e., that do not consider purchase material costs. Conversion cost corresponds to the value-added concept.
4. Prices that produce contribution margins that depend on elasticity of demand.
5. Prices that are systematically related to the stage of market and competitive development of individual members of the product line.

Costs are given stress in the first three rules. In the fourth rule, the theoretical desire to provide elasticity information in time and quality for use in product pricing meets with frequent failure. This method is seldom used.

The pricing situations presented to the price setter are diverse. They may range from a price for a one-of-a-kind product to one where there are identical units offered to many buyers. The firm may have only a single product or it may have a multiple-product line. The product may be brand new to the firm, or on rare occasions may be a product of research or invention. The product may have been manufactured for decades in the same form, or minor modifications may be introduced every so often. In all these situations one pricing method is superior and others are not. To make a rational price decision, we attempt to foresee the effects on the objectives of the company and on the several groups of people affected by the price.

9.6-1
Conference and Comparison

This nonanalytic method usually involves the people who know the product's market and its cost and understand the technical factors. They meet as a consequence of competitor's actions, or someone may notice that the price of an item may be out of line. Perhaps costs have gone up or down and adjustments are felt necessary. Discussion regarding the volume effect that each of the alternative prices would have and the product's average gross margin per unit as a result of alternative prices would be undertaken. Data on competitor's prices will usually be available, as well as comparisons of strengths and shortcomings of competing products. Past sales figures and the history of price changes may be at hand.

[2]Joel Dean, *Managerial Economics*, Prentice-Hall, Inc., Englewood Cliffs, N.J., 1951, pp. 473–476.

To assess the number of units that might be sold and their margin, we try to imagine the consumer response to each alternative. Furthermore, the response of distributors and salesmen, changes in the sales of other products, and probable response of competitors, and possibly the judgment on the share of the market we secure are evaluated. The predictions will be shaky, and when using them we must recall their shortcomings. While we discern the future actions of our consumers and competitors we can be certain that their reactions in the future will differ from those in the past.

96-2
Method of Investment

It is basic that if a new venture or a major improvement is desirable, then the net return from it must exceed the cost of capital required. This philosophy provides a pricing method that depends on invested cost. This requires that the cost of operation and the consumption of fixed assets and an acceptable rate of return be estimated. The acceptable rate of return on investment varies with numerous economic factors, but overall cumulative values have emerged. They range from a low of 5% to over 50%. A simplified pricing model based on investment is

$$S = \frac{iI}{(1-t)} + C_T \tag{9-18}$$

where S = sales dollars
i = rate of return, decimal
I = investment, dollars
t = tax rate, decimal
C_T = total cost excluding investment

This model is the total margin based on allocated investment.
Alternatively, the average markup in dollars per unit is

$$\frac{S - C_T}{N} = \frac{iI}{N(1-t)} \tag{9-19}$$

where N = units sold. In this relationship $(1/N)(S - C_T)(1 - t)$ is net profit per unit.

Where the total cost C_T is estimated and an acceptable rate of return i can be established, the markup and the final price can be made for a given management policy. Pricing for investment consideration is a long-term method. In estimating C_T it is necessary to consider long-run averages to allow for variation in production which occurs from time to time. This method is useful for capital-intensive type of industries.

Item	
C_T	$42,000
Apportioned investment	20,000
Production quantity	5000 units
Corporate tax rate	55%
Rate of return	25%

$$S = \frac{0.25(20,000)}{(1 - 0.55)} + 42,000 = \$53,111$$

and

$$SP = \frac{53,111}{5000} = \$10.62$$

where SP = sales price per unit.

The method emphasizes that it is fixed capital that provides the profit, as everything else is at cost.

9.6-3
Full Cost

This method, variously called the *cost plus* or the *markup*, sets prices that are proportional to full cost. It is probably the most popular method.

For custom design jobs the order is unique. The firm sells the product to only one customer, and it is unnecessary to estimate sales of this item at each of several prices.

The basic model is

$$SP = \frac{C_T}{(1 - R)N} \qquad (9\text{-}20)$$

where R = cost-plus rate, decimal. For Table 9.5, the selling price is found as

$$SP = \frac{17451.59}{(1 - 0.13)480} = \$41.79 \text{ per unit}$$

Simple as this appears, there are literally dozens of variations. For instance, the C_T term can be broken down to material, labor, burden, and engineering with each term having its own markup. This is found in government contracting where limitations on the markup of the cost components are sometimes required. Some companies use the same add-on percentage year after year. Others use markups which reflect the preceding year's actual percentages. For the most part these percentages vary with business conditions, and this feature along with its ease of understanding are its best features. As business falls the add-on is reduced and vice versa. When large companies use these procedures, it is usually as a starting place for a price decision. In some companies the sales managers in the territory eventually decide what price they will actually quote and the markup percentage serves as background. As you may notice in Table 9.5 the arithmetic varied from Equation 9–20. In Table 9.5 the 13% profit is a percentage of total cost leading to a price $41.085 indicating the variations in this pricing method.

9.6-4
Direct Costing or Contribution

As we have said, some costs vary closely with changes in volume while others do not. These previously developed thoughts are important for direct costing. Also called contribution pricing, this method is based on marginal profit motives. This method of pricing is used with variable-cost cost estimating. Assume that fixed and variable overhead have been separated and that the prospect of an additional opportunity sale is received.

In Table 9.8, a cost-volume-profit analysis, the product mix and price are assumed to be constant. The factory overhead, marketing, and administrative expenses have been divided into fixed and variable components. For the fixed portion they are constant over the volume increment and time period. This assumes that the firm has sufficient capacity to produce this opportunity sale. Only the variable costs are shown to move with volume changes. While profit is 10% at the original sales level, it increases to 15.7% because of a greater base for the fixed costs. The only cost incurred specifically to produce the additional 40% is variable cost; the profits on the increment are 30%. This is contrary to the cost-plus method that assumes that the same profit is earned on all revenue dollars.

TABLE 9.8

COST-VOLUME-PROFIT ANALYSIS

Item	Original Sales Level, $10,000,000	Increase in Sales, 40%, $14,000,000	Incremental Sales, $4,000,000
Variable costs:			
Labor	$2,500,000	$3,500,000	$1,000,000
Materials	3,000,000	4,200,000	1,200,000
Overhead	1,250,000	1,750,000	500,000
Marketing	50,000	70,000	20,000
Administrative	200,000	280,000	80,000
Total variable cost	$7,000,000	$9,800,000	$2,800,000
Fixed costs:			
Standby	750,000	750,000	0
Product	1,250,000	1,250,000	0
Total fixed costs	$2,000,000	$2,000,000	0
Total costs	$9,000,000	$1,180,000	$2,800,000
Pretax profit	$1,000,000	$2,200,000	$1,200,000
Profit %	10	15.7	30

What happens if a 10% price reduction on this 40% volume increment becomes necessary? Notice that Table 9.9 has the various fixed and variable classifications identified. Only the price on the 40% opportunity sale has been reduced.

TABLE 9.9

COST-VOLUME-PROFIT ANALYSIS WITH
PRICE REDUCTION

Item	Original Sales Level, $10,000,000	Increase in Sales, 40%, $13,600,000	Incremental Sales, $3,600,000
Variable costs	$7,000,000	$9,800,000	$2,800,000
Fixed costs	2,000,000	2,000,000	0
Total	$9,000,000	$11,800,000	$2,800,000
Pretax profit	$1,000,000	$1,800,000	$ 800,000
Profit %	10	13.2	22.2
Contribution	$3,000,000	$3,800,000	$ 800,000

Both the incremental number of units and the variable costs are the same in Tables 9.8 and 9.9. As fixed costs are not discontinued or adjustable with revenue, they remain the same. The data show that for a given reduction in price, profits for that portion are 22.2%. At the bottom of Table 9.9 is the amount for the contribution. This is obtained by subtracting the variable costs from the revenue. The contribution amount is the sum left over from the revenue after paying the variable costs and is used to pay the fixed expenses. Any overage is profit. This particular example shows that a 40% increase in unit volume with a price decrease of 10% on this increment (increase of 36% in revenue) produces an 80% increase in profit.

For the wrench problem we have a unitized example of contribution. The sales volume is estimated for each price. The arithmetic is straightforward.

TABLE 9.10

DIRECT COSTING ON THE BASIS OF UNIT STANDBY
AND FIXED OVERHEAD

a. Estimated sales at each price, units	5000	10,000	15,000	20,000	25,000	30,000
b. Retail price	$2.00	$1.86	$1.75	$1.61	$1.58	$1.50
c. Less middle men's discount of 40%	0.800	0.744	0.745	0.645	0.640	0.600
d. Mfg. net selling price†	1.200	1.116	1.005	0.965	0.940	0.900
e. Less variable mfg. cost/unit‡	1.173	0.854	0.682	0.628	0.643	0.773
f. Contribution to overhead/unit	+$0.027	$0.272	$0.323	$0.337	$0.297	$0.127
g. Total contribution to overhead	+$135	$2720	$4840	$6740	$7425	$3810
h. Less standby and product overhead/unit‡	$0.202	$0.202	$0.202	$0.202	$0.202	$0.202
i. Net contribution to overhead/unit	−$0.175	$0.070	$0.121	$0.135	$0.095	−$0.075

† From Table 7.6, total revenue ÷ quantity
‡ From Table 9.7

Reductions from the retail price to allow for distribution discounts lead to the manufacturer's net selling price. The variable manufacturing cost was previously given by Table 9.7. From these two a unit contribution to fixed overhead can be found, and total contribution is sales quantity times unit contribution. In this unitized method it is necessary to allocate a portion of the fixed costs to each unit. These standby costs are obligations that must be met—if they are not, foreclosure and legal repercussions ensue. Product costs will result if the decision is made to manufacture and market the product. The analysis can be converted to total values by using a total standby and product fixed cost. This total value was necessary before the unit value was determined because a portion of all fixed expenses were dedicated to this product.

It is possible to measure the relative profitability of products as a calculation of the difference between price by volume and total variable costs. This knowledge of the relative rates of contribution between the products helps the company to decide which products need emphasizing (or dropping) and which ones yield a low return on capital. Note Table 9.11. In this Table product costs have been identified with products, as they are incurred specifically for the benefit of the product. There are fixed costs which are shared in common with other products and cannot be validly apportioned because of the necessity of using arbitrary allocation methods. Conversely, this method provides no un-

TABLE 9.11

ANALYSIS OF PRODUCT-LINE PRICING
AND PROFITABILITY

Item	Monthly Total	Product A	Product B	Product C
Price × Volume	$6000	$5000	$200	$800
Total variable costs	4050	3450	160	440
Contribution	$1950	$1550	$ 40	$360
Contribution/sales revenue, %	32.5	31	20	45
Less specific product costs	$ 500	100	100	300
Adjusted contribution	$1450	$1450	$(60)	$ 60
Adjusted contribution/sales revenue, %	24.2	29	(30)	7.5
Common period costs	800			
Pretax profits	$ 650			
% of sales	10.8			

assailable trick to find the net profit for each product because of the arbitrary reduction of common period costs.

Table 9.11 estimates a single price by volume, although each product can be exploded separately by another estimate. Considering the ratio contribution-sales revenue, product C ranks highest. Note that this ranking is not in proportion to sales revenue. After specific product costs are assigned to products the ranking is reversed, with product A becoming first. Product B shows a loss, and if an attempt were made to arbitrarily allocate the common fixed costs, product B could not improve its position.

How is price determined using direct costing? Equation 9.20 can be used if we let R equal the contribution rate. If the total variable cost constitutes 65% of the list price, as shown in Table 9.6, and the contribution rate is prespecified as 35%, then the fraction $0.1804/(1 - 0.35) = \$0.2775$, which is the list price. The retail price is $\$0.50 = 0.2775/(1 - .445)$ where 44.5% is the retail markup.

9.7
LIFE CYCLE COSTING

Life cycle costing (LCC) of any item is simply the summation of all funds expended from its conception and fabrication through its operation to the end of its useful life. It is also defined as estimated total costs between the points in time when the item becomes a recognized object to its phase-out from inventory. Now the item is most generally a product, but a project or system would qualify as well. In an intuitive manner individuals have used LCC principles for decisions about cars when they concern themselves not only with initial cost (sticker price) but with operating expenses (gas mileage, worn parts) and residual value (resale price).

LCC analysis attempts to estimate all relevant costs, both present and future, in the decision-making process for the selection among various choices. Figure 9.8 illustrates engineering, total product cost and operation costs as

FIGURE 9.8

STAGES OF LIFE CYCLE COST

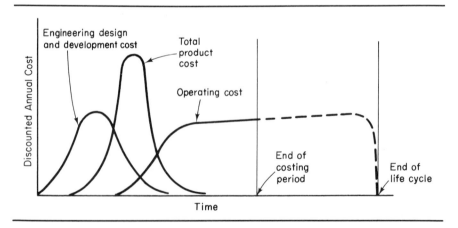

separate funds for a specified life cycle. Operating funds necessary to undertake a program can be easily greater than the original R & D or investment money required. Moreover, a product with higher development and investment costs but lower operating costs may, depending on service life, be a least-cost product. LCC encourages the trade-off analysis between one-time costs and recurring costs. This is described by Figure 9.9, where products A, B, or C are selected on

FIGURE 9.9

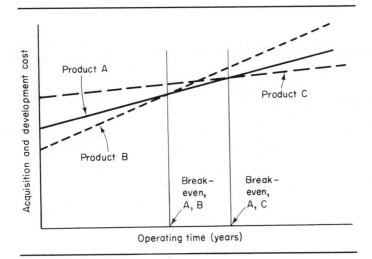

the basis of operating time over other choices simply because the operating expenses caused a lower net-discounted amount.

The spending of costs occurs throughout the cycle. Figure 9.10 is descriptive of the three major funding categories. Spending continues until some termination date, which may precede the mortality life of the product.

It has been shown that for military hardware systems that approximately two-thirds of life cycle costs are unalterably fixed during the definition and development phases of a product. Consequently, many costs and alternatives are limited and constrained by preliminary designs. We shall now consider a case study of LCC applied to a vacuum chamber.

Assume that an aerospace firm has requested a vacuum chamber design to simulate high-altitude pressures for electronic equipment. Figure 9.11 is a preliminary design of the product. Two firms have submitted designs and LCC analyses. The aerospace firm, having more than one technically qualified candidate, performs an LCC trade-off study so as to ascertain the vacuum chamber with lowest LCC.

In preparation for the design bid the following data are required for a LCC analysis:

1. Cost elements.
2. Operating profile.
3. Utilization factors.
4. Costs at current prices.
5. Current labor and material costs projected with inflation-deflation indexes.
6. Costs discounted to base period and discount factors.
7. Sums of discounted and undiscounted costs, summary.

Cost elements for the vendor are broken down into three major categories, engineering, product, and operation. Since they have been discussed before, we shall not concern ourselves with the vendor's problem; rather we assume that the vendor has provided a product price which includes these expenses along with a profit. Additional to the price the vendor provides other data such

FIGURE 9.10

ELEMENTS OF LIFE CYCLE COSTS

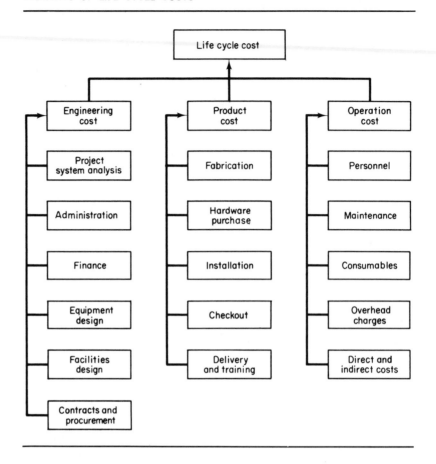

as Table 9.12. Most of these elements are self-explanatory. However, the operating profile describes the periodic cycle the equipment experiences. The operating profile has a repetition time and contains all the operating and nonoperating modes of the equipment. It is sometimes possible to have operating profiles internal to other operating profiles. For trade-off studies candidates must be evaluated with the same operating profile. In our example this means that we must know what use we are going to make of the vacuum chamber. The operating profile tells when the equipment will be operated; the utilization factors tell us how or in what way the equipment will be operating during each mode. In our environmental chamber, the diffusion pump may run 60% of the time, while in the shutdown mode it may not run. Instead, the roughing pump may hold a nominal vacuum.

Some factors indicate operating costs incurred during the life of the equipment. Two unusual cost parameters are *mean time between failure* (MTBF) and *mean time to repair* (MTTR). Time between overhaul, power consumption rate, and preventive maintenance routines such as the cycle and the preventive maintenance rate are additional information.

The aerospace firm estimates its own operating profile and determines that a continuous three-shift operation for about one-third of a year or 2920 hours per year is necessary. This is applicable to both vacuum pumps. Using the manning and a labor rate (accounting for time and a half and fringe costs),

A: 2920 hours per year × $8 per hour × 1 man per machine = $23,360

B: 2920 hours per year × $8 per hour × 2 men per machine = $46,720

FIGURE 9.11

VACUUM CHAMBER ENVIRONMENTAL
CONCEPTUAL TEST SYSTEM

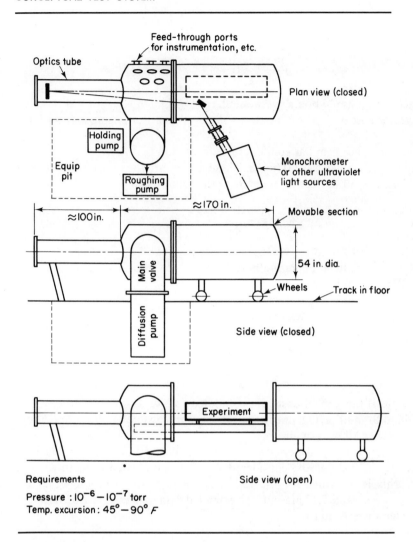

Requirements

Pressure : $10^{-6} - 10^{-7}$ torr
Temp. excursion : $45° - 90°$ *F*

TABLE 9.12

COST DATA FOR LCC ANALYSIS OF
ENVIRONMENTAL VACUUM SYSTEM

Cost Parameter	Data Source	Estimate, Vendor A	Estimate, Vendor B
Product price	Manufacturer	$200,000	$170,000
Equipment life	Customer's specification	2 yr	2 yr
Initial engineering	Manufacturer	$20,000	$30,000
Installation	Owner	$3,000	$4,000
Manning	Manufacturer	1 man	2 men
Mean time between failures	Manufacturer	500 hr	300 hr
Mean time to repair	Manufacturer	1 wk	2 wk
Preventive maintenance cycle	Manufacturer	160 hr	180 hr
Preventive maintenance down time	Manufacturer	4 hr	8 hr
Parts cost (% of product price)	Manufacturer	1%	2%
Input power	Manufacturer	8.0 kw	9.0 kw

Labor for preventive maintenance (PM) actions is calculated from the formula

$$\text{number of PM actions} = \frac{\text{scheduled operating hours}}{\text{PM cycle time}} \qquad (9\text{-}21)$$

For A the number of preventive maintenance actions is

$$\frac{2920}{160} = 19 \text{ actions}$$

Since each action uses 4 hours, we would have 19(4) = 76 hours of PM time for a total yearly cost of $455 (76 hours × $6 per hour). For chamber B we have $820.

The aerospace firm has studies and information that A will fail every 500 hours and B will fail every 200 hours. These figures are the mean time between failures. The cost of corrective maintenance is found from

$$C_{CM} = \frac{\text{SOH}}{\text{MTBF}}(\text{MTTR})C_{me} \qquad (9\text{-}22)$$

where C_{CM} = cost for corrective maintenance per year
  SOH = scheduled operating hours
  MTBF = mean hours between failures
  MTTR = mean hours to repair
  C_{me} = cost of maintenance labor

For example A we have

$$C_{CM} = \left(\frac{2920}{500}\right)(40)(6) = \$1400$$

For example B the annual costs are computed as $4672. Power consumption cost is found by multiplying input power in kilowatts by total hours of operation and the cost per kilowatt-hour.

Inasmuch as a 2-year period is concerned, effects of inflation or deflation on recurring costs should be considered and then discounted back to present time. As this is the topic for the following chapter, its inclusion serves no present purpose. Table 9.13 presents the extended data for a comparison between vacuum chambers A and B.

TABLE 9.13

LCC TRADE-OFF ANALYSIS OF TWO COMPETING
PRODUCTS FOR TWO YEARS

Cost	Vendor A	Vendor B
Product price	$200,000	$170,000
Engineering	20,000	30,000
Installation	3,000	4,000
Manning labor (2 yr)	46,720	93,440
Preventive maintenance costs (2 yr)	912	1,632
Corrective maintenance (2 yr)	2,800	6,220
Power requirements @ 0.025/kwh	1,163	1,312
Parts and supplies	2,000	3,400
Total	$276,600	$313,128

The results of an LCC study can now be seen. The higher first cost of A with its better maintenance performance and lower operating cost did return the economy in terms of operating performance and has the lowest LCC.

One of the lessons of LCC analysis is that the logistics of supplying spares or components of products in the field or repair of these same products is not uncommon. Certainly one of the real tests of a product occurs when it enters service in the uncontrolled environment of the user. If problems occur, the feedback is loud, and the only acceptable response time is short.

Usually, spares are quoted after a product has been sold, although the product may not have been delivered. Actual cost may be available to allow a basis for making the new estimate. When using actual costs, material and labor costs must be updated. If changes in setup quantity are noticeably different, adjustments in this, too, may be necessary. For companies who deal in selling spares, either on a single or lot basis, many of the same techniques as described in this chapter are used.

Repairs are more difficult to estimate. Until the product is examined, the work needed to restore the product to an acceptable condition cannot be fully identified. A number of practices are found with repairs. The product may be stripped at the site and estimated by a person experienced in identifying the work necessary. Or a time, material, and profit contract between the original supplier or repair company and the owner may be specified. The product may be shipped back to the factory and a repair inspection undertaken there. The time is estimated and a unit cost per hour is applied.

For the most part, product estimators "buildup" an estimate starting with operation estimates and concluding with either product cost or price. Sometimes a reverse procedure is required. Beginning with a competitor's market price for a given product, the estimator works backward to find total cost and the cost for various design elements. This practice is called *design-to-cost*.

Major and minor cost elements are uncovered by estimating or by the use of factors. The level of detail may even extend to components and hardware. In the design-to-cost procedure, and when detail exists, learning curve theory, tooling philosophies, quantities, rates of manufacture, and escalation or deescalation of costs can be introduced in the process.

The apportioning of cost is in accord with a logical design structure for the product. Designers may consult with the estimator in developing the logical structure. Each design is given a *target*; thus, the designer knows that he controls his design and hopefully is able to design to match the goal. The purpose is to insure that the product will meet a designated price to allow product competition.

Note Figure 9.12. Estimating ascertains that a product is sold for $555 per unit. The estimator reduces the price by the profit markup (about 10%) leaving a cost of manufacturing, development and sales of $500. In turn, general and administrative, selling, engineering, and cost of goods manufactured are determined. The same factors that were used to buildup a cost are used in a reciprocal way. With a overhead of $195 the process changes. At this point the division of remaining value is divided into direct materials and direct labor along the principal design lines.

This breakdown is identified and, in conference with the designer, the components and costs are estimated. The designer now has a cost goal. Figure 9.13 shows what these goals are for one section of a design.

Naturally, it is important that the total goal be distributed fairly and that no subassembly or component be favored at the expense of another. As the design effort progresses, the usual pattern of preliminary and detailed product cost estimates is undertaken. A cost and price would ultimately be found using

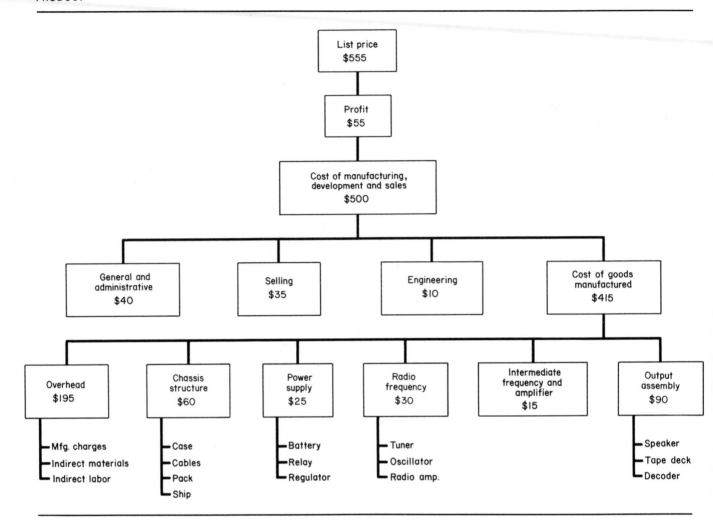

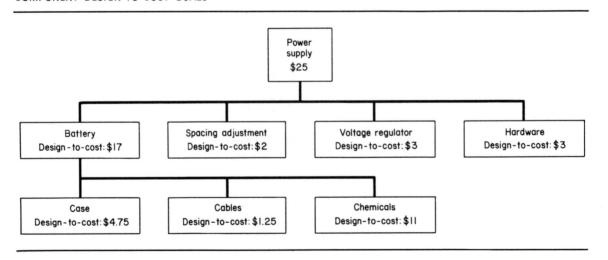

methods given in this chapter. Obviously, it would be desirable that the final price harmonize with design-to-cost goals.

9.10 SUMMARY

In this chapter we considered product-estimating procedures. Important as other motives may be in price setting, the cost of a product must be fully recovered with profit to assure that the firm will ultimately survive. Financial documents are important in the appraisal of the worth of a product to a firm. Learning curve theory, useful for large-scaled production goods, impinges on the product estimate in important ways. Methods of product costing and strategies of pricing lead to the information required for various financial documents and help to improve the product's success.

SELECTED REFERENCES

Information on product life costing may be found in

KLICK, ARNOLD: "Whither Life-Cycle Cost", *Economic Analysis and Military Source Allocation*, edited by T. Arthur Smith, Department of Army, Comptroller of the Army, Washington, D.C., 1968, pp. 79–99.

McCULLOUGH, J. D.: *Cost Effectiveness: Estimating Systems Costs*, The Rand Corporation, Bethesda, Md., Sept. 1965, pp 11–13, Vol 3229.

A general background in marketing is given by

HOWARD, JOHN A.: *Marketing Management*, Richard D. Irwin, Inc., Homewood, Ill., 1963.

A book with many methods on pricing is

TUCKER, SPENCER A.: *Pricing for Higher Profits*, McGraw-Hill Book Company, New York, 1966.

Learning curve theory, applications, and tables can be found in

COCHRAN, E. B.: *Production Planning Costs: Using the Improvement Curve*, Chandler Publishing Co., San Franciso, 1968.

FABRYCKY W. J., P. M. GHARE, and P. E. TORGERSEN: *Industrial Operations Research*, Prentice-Hall, Inc., Englewood Cliffs, N.J. 1972, pp. 172–193.

QUESTIONS

1. What are the strong and weak points of Henry Ford's pricing policy? Why is pricing an important part of the executive's responsibility?
2. What are the ways a product estimate will be used? List others not included in the chapter.
3. What kinds of information are required for a product estimate? Which is internally determined by the product estimator?
4. Relate the flow of information between the various financial documents. At what point would these documents be required? When would they be unnecessary?
5. Distinguish between a cash flow statement and a rate of return analysis. What is operating capital? Is it taxable? What is a tax credit?
6. Describe the purpose of learning for a cost estimate. In addition to production quantity, what other things show learning correlation? Where in the cost estimating process is the learning curve data applied?
7. Give the empirical observations that undergird learning behavior. Why is a log-log coordinate graph used in place of cartesian coordinates?
8. When is time estimated via learning? Cost?
9. How may the shape of the learning curve be defined? What factors contribute to this shape?

10. Segregate the methods of product costing. Outline these differences.
11. What are the advantages of variable cost estimating and direct costing for setting cost and price?
12. Indicate the complexities of pricing. Rank several objectives of a pricing policy.
13. When is the full cost method of pricing appropriate? What are its failings? Devise a new model that uses the full cost and investment method of pricing.
14. Is contribution pricing a marginal method? How important is it to have a contribution? What are the problems involved in determining fixed costs and associating them with various product lines?
15. How does life cycle costing differ from cost estimating? Would a consumer or manufacturer be inclined to LCC? In LCC analysis, what is given as fact and what must be estimated? Describe MTBF and MTTR.
16. What is design-to-cost? How does this procedure aid the control of costs during the design process? What features of this management procedure bear watching?

PROBLEMS

9-1. An investor is considering an invention, and receipts and disbursements have been estimated. Calculate the net cash flow representing this investment. Disregard the investor's tax and discounting.

End of Year	Disbursements	Receipts
0	$50,000	0
1	5,000	2,500
2	5,000	10,000
3	2,500	25,000
4	0	25,000
5	0	20,000

9-2. The cash flow for a product entering production is given as

Year	Disbursements	Receipts	Amount for Depreciation
0	$60,000	0	0
1	5,000	2,500	9000
2	5,000	10,000	9000
3	2,500	25,000	9000
4	2,500	25,000	9000
5	2,500	20,000	9000

Fixed assets are $45,000, and initial operating expenses are $15,000. If the owner is in the 25% tax bracket, determine the net cash flow. When is the break-even point based on a nondiscounted cash flow?

9-3. Construct a cash flow document. Investment and pre-operating charges are 100 and 25×10^5 for the first year; production is 10 million pounds and is expected to increase at 2 million pounds per year for the next 5 years. Net profit after taxes amounts to 20×10^5 for the first year, depreciation is straight line for 10 years, tax is 50%, and working capital is 8×10^5 dollars per year for year 1. Costs are assumed constant over the period. What is the payback period disregarding discounting?

9-4. (a) The cost-engineering department has determined an estimate of 10,000 hours at unit 100. They believe an improvement of 10% can be expected from unit 1 onward. Find the unit time to build unit 1 and 2.

(b) Unit 1 = 10,000 hours and unit 8 = 5120 hours. What unit and cumulative times can be expected at 16 and 32 units?

9-5. (a) For management proposal A, cost at unit 1 is $8000, and the learning rate is 80%. For management proposal B, unit 1 cost is $6000 and the learning rate is 90%. Which is the best proposal at unit 15? Assume that proposal B was adopted. Cost engineering discovers that at unit 15 the actual cost was really $3975, while for unit 1 it was as initially estimated. What was learning performance?

(b) An audit of learning curve performance revealed that at the eighteenth product unit the direct product cost was $181,000, while at the twenty-fourth product unit it was $169,000. What is the estimated direct product cost for the twenty-seventh product unit?

9-6. A quantity of 100 has cumulative average direct-labor man-hours of 1000 hours. Management has promised that it will apply concentrated efforts to achieve a 70% manufacturing progress function constant on this computer tape drive. Find the unit time at 100 using the conversion formula $\dfrac{1}{1+s}$.

9-7. Unit 30 has a unit time of 7 hours and an improvement anticipation of 20%. Find the cumulative average unit time using the formula $1/(1+s)$ as a multiplier and the Tables in the Appendix.

9-8. A company's experience in building planocentric gear transmissions indicates that the first unit will consume 100 hours of fabrication and assembly time. A learning rate of 75% is anticipated. A contract is being estimated in which 40 units will be supplied. Labor cost of $8 per hour covers labor time and indirect manufacturing expense and material expenses should be $25 per unit. If a 20% add-on for profit is historically applied, what should the bid price be? How much must the time estimates be off to consume the profit?

9-9. A prototype unit has been constructed with 45,000 hours of work recorded on job tickets. The direct labor hourly average rate is $6, and the overhead rate is 100%. Raw material costs charged to the job were $20,000. The company, after a period of time for design changes, will build units 2, . . . , 10 for sale. Customizing for specific customers is negligible. They anticipate a learning rate of 90% for direct labor and 95% for material. What are the total estimated costs for this product, the average, and the tenth-unit cost?

9-10. A company uses the learning curve approach to estimate its costs and prices out on this basis:

Cost Item	Prototype Cost Estimate		Estimated Learning Slope
	Fixed	Variable	
Direct labor	—	45,000 hr	90%
Direct material	—	$20,000	95
Manufacturing support	25,000	$5,000	95

Labor costs $6 per hour, fixed and miscellaneous overhead is 100%, distribution and administrative overhead is 20%, and selling costs are at 2%. The company expects to produce units 2–10, and it uses the full price method with profit at 10%. Find the price for the 10th unit. Find total price for 10 units.

9-11. (a) A product is priced at three levels:

Price	$4	$4	$5
Projected sales	10,000	30,000	15,000

(b) If the direct cost per unit is $3, what is the contribution for these three trials? A 40% contribution from each price takes care of manufacturing overhead, non-manufacturing overhead, and profit. If direct material and labor is $0.27 and $0.10 per unit, what is the selling price?

9-12. Two quantities of a product, 5000 and 10,000, are to be estimated using the formula for variable cost estimating. These data are given as

	5000	10,000
P_m	$0.60	$0.50
T_i	$0.062	$0.060
LR	$5.25	$5.25
OH_{VM}	10%	10%
OH_f	20%	20%
OH_{VA}	$0.14	$0.13
OH_P	$0.17	$0.16

Determine total cost for the product. In some instances OH_{VA} can be computed as a fraction of total variable labor and materials. If $OH_{VA} = 15\%$, find the total cost. Using the full markup of 20%, what price may be expected?

9-13. You are given the following data:

Item	Estimated Hours	Production Center Machine-Hour Rate
Light machining	5	$7.50
Finishing	10	6.00
Bench	20	4.50
Purchased materials	$70	
Purchased materials overhead rate (0.40)		
Fabricated materials	$16.20	
Forecasted production quantity	400	

What is the unit manufacturing cost? This firm uses the full cost method of pricing and has a markup of 30%. What price may be expected? (*Note:* General overhead is a part of the machine-hour rate.)

9-14. A new plant for processing uranium ore, U_3O_8, is planned. A consultant retained by the owner designs a scheme for the production of U_3O_8, and the following list summarizes his findings:

Item	Remarks
Capital cost	$7,000,000
Tons/yr processed	600,000
Pounds of U_3O_8 produced yearly	1,500,000
Operating cost	$3/ton processed
Hauling cost	$1.25/ton processed
Interest on investment	8%
Loan commitment	10 yr
Allowable depreciation time	10 yr

Find the cost of producing U_3O_8. Determine the price to charge per pound of U_3O_8 using any method. Make your own assumptions regarding profit. Do not consider depletion or effects of the equity-loan ratio for capital.

9-15. Estimating and accounting have submitted the following data to executive management:

Cost Item	Product A	Product B
Proportional investment	$60,000	$80,000
Direct labor	20,000	10,000
Direct materials	12,000	30,000
Other fixed costs	30,000	10,000

Find the prices for products A and B basing price on a return of investment of 10%. Using this price, what is the markup on total costs, marginal costs, and value added in production? The firm's income tax rate is 50% and 10,000 units of each product are produced yearly.

9-16. (a) A multi-national company designs and manufactures products, both complete and semi-finished, for sale to international divisions scattered throughout the world. While specific cost estimating practices depend upon which two trading countries are dealing, consider the following: A non-U.S. firm wants to buy a product for use as a standard purchased material for its own finished product. The American international division buys the product from the manufacturing division using a variable pricing formula, or variable cost plus 35% for the firm's contribution (price = variable cost ÷ (1 − .35)). The international division adds 20% for its own overhead based upon the manufacturing division's price. This transaction is F.O.B. (*freight on board* at the U.S.A. plant, and requires the consignee to pay all transportation costs from factory to destination). Export and import duty at the border is 0 and 20% of the *ad valorem* (invoice price at the port of shipment). Freight cost from the U.S.A. plant is $0.18 per unit. If the manufacturing variable cost = $1 per unit, what is the price that the importer ultimately pays? Recall that during production of this product, the importer will add labor, material, overhead, and contribution to the standard purchased material. Discuss what happens to the trade of this exported product if the rate and efficiency of production are approximately equal between these partners.

(b) A multi-national company is considering a "twin-plant" project where the two plants are adjacent in terms of transportation cost (separated by the border, or the material is non-bulky and transportation costs both ways are negligible on a per unit basis). On the U.S. side, labor rates and productivity conform to typical standards, while in the foreign country labor rates vary from 1/10 to 3/4 as much. The U.S. plant will process the material, transport the semi-finished material to the twin plant, whereupon the foreign plant adds labor value to the product, and returns the product to the U.S. plant. The U.S. manufacturing unit cost for the semi-processed material is $1. At the border, *custom fees* amount to 5% and return custom fees are 5% and an ad valorem of 20% of value added is assessed. In the foreign plant the labor work value is $.25 for an equivalent $1.25 per unit of U.S. work. Once back in the U.S.A., a 35% contribution is added to give final price. Find the product price for a twin-plant and a single-plant operation. Discuss the implications of a policy that sends goods for intermediate processing to other countries. What must the labor cost of the non-U.S. labor be to make the decision indifferent between the twin plants?

9-17. The following is the operation chart for the upper shaft in Figure P9-17. Material, H36A, $2\frac{1}{4}$ inches round by $14\frac{3}{8}$ inches long; weight, 16.5 pounds, $0.70 per pound.

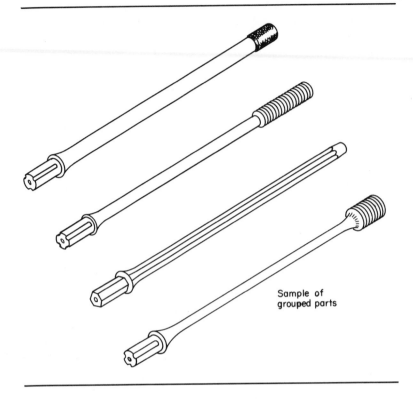

Sample of
grouped parts

Oper.	Dept.	Mach.	Unit Time	Description	Labor Rate	Machine Overhead Rate
5	Cutoff	B13	0.01	Cutoff to length	3.25	0.85
8	Lathe	B05	0.08	Face both ends to length center, turn thread end	3.50	0.75
22	Ship	FRM1		Send out to roll thread	—	—
28	Lathe	B06	0.03	Turn small dia. on end; finish turn behind thread	3.60	0.80
30	Lathe	B04	0.05	Drill large hole to depth	3.60	1.05
35	Lathe	B04	0.14	Drill small hole to depth	3.60	1.05
40	Copy lathe	B02	0.12	Face second end, turn, form relief, radii, recess	3.50	1.00
45	Lathe	B06	0.02	Center thread end	3.50	0.50
48	Bench	B12	0.02	Polish to 75 finish	2.75	0.10
50	Broach	B07	0.04	Broach splines	4.05	2.25
55	Bench			Wash	2.75	
60	Bench	B12	0.03	Burr and radius splines	2.75	0.10
65	Heat-treat			Heat-treat to spec.		
70	Inspection			Straighten		
75	Inspection			Check to part print, tape mark pack		

The operation to roll the thread costs $0.415 per unit. Heat treatment and inspection are indirect departments but levy a cost to the unit as 0.08 per pound and 5% of direct labor cost, respectively.

(a) Find the cost and price using the operation method of costing and the full price method. Fixed and miscellaneous and distribution overhead rates are

190 and 70% of direct material and labor, respectively. Price is at 30% of full cost.

(b) Find the cost and price using the departmental method of costing and the full price method. Average departmental wage rates are cutoff, $3.30; lathe, $3.50; bench, $2.75; and broach, $4.00. A general overhead rate is 85%, and the rate for cost of goods is 75%. Profit is at 25% of full cost.

(c) Find the cost and price using the method of variable cost and direct costing. Follow the lead given by Table 9.6 and use the estimates for direct material and direct labor from part (a) above. The breakdown of major categories is

Item	Per Cent
List price	100
1. Variable manufacturing costs	58
2. Variable marketing costs	4
3. Variable administrative costs	3
Total variable costs	65
1. Standby	13
2. Product	7
3. Standard earnings	15
Standard profit contribution	35

(d) Using any of the above three parts as a base, estimate by comparison the cost and price of the other three shafts.

9-18. A nonreparable electronic component is up for bid and an LCC approach is deemed mandatory. The purchasing agent advertises that an LCC model to select the winning bid will be based on the model

$$\text{cost per unit} = \frac{\text{unit price} + \text{unit stocking cost}}{\text{bid MTBF}}$$

The unit stocking cost of $110 per unit is the total cost for FOB, storage, installation, and dismantling. Each bidder supplies this information to the purchasing agent.

Bidder	Unit Price	Bid MTBF
1	1000	800
2	1250	615
3	1175	917

Which bidder wins the contract?

9-19. The government tells potential contractors that a product will be evaluated according to a LCC model:

$$\text{LCC} = \text{unit operating cost}$$
$$= (\text{unit price} + \text{logistic costs}) \div \text{service life}$$

The selected contractor demonstrates his service life in a post award reliability acceptance test. If the reliability test does not meet the level stated by the contractor, a penalty function deducts from the unit price as $(1 - \text{test value MTBF}/\text{quoted MTBF}) \times (\text{unit price} + \text{logistic costs})$. The logistic cost is $20.

Company	Unit Price	Hours MTBF
A	$35	1000
B	40	1200
C	70	2100

Find the winning company. Now suppose that the contractor failed to meet his quoted MTBF by 10%. What is the penalty and the final price?

9-20. The government establishes an LCC model to evaluate tires. The model is given as LCC = quantity (unit price + shipping cost + maintenance cost per unit). Three tire manufacturers were invited to bid. Each was asked to provide a sample for simulated landing tests to determine the number of landings per tire. The government then determined the number of tires for each company's quote based on a tire-landing index which is given by number of required landings ÷ best performance:

$$\text{number of required landings} = 1,200,000$$

Shipping costs were evaluated from the manufacturer to a central inventory. Maintenance cost is defined as the cost to change a tire ($4.75) × quantity quoted. Make a bid evaluation to determine an LCC price. Which company do you select?

Company	Landings Per Tire	Shipping Cost Per Tire	Bid Price Per Tire
A	110	$1.35	$138
B	105	1.30	128
C	95	1.26	136

9-21. Determine the factors that the estimator used in Figures 9.12 and 9.13 to uncover design-to-cost for the component. Future competition will probably drive price down as technology improves and more firms enter the market. What component costs do you expect for a $375 product price?

CASE PROBLEM *Unijunction Transistor Metronome.* Ray Enterprises has released a new design and hopes to market it through a well-established chain of catalog houses. President Art Ray says that the key to the quality of the product lies in its new electronic circuitry, given by Figure CP9-1. The bill of material is given by the parts list, and other materials cost $1.25 per unit which covers the case, plug, and vector boards. Marketing, accounting, finance, and sales have been asked for information and the following preliminary suggested P-V is received:

Annual Volume	Potential Market Price
200,000	$47.95
210,000	46.65
235,000	45.20
260,000	44.60

New investment: $20,000 in tooling; depreciation policy, 2 years. Current tax rate: 40%. Standards for production are determined from operation estimates as

FIGURE CP9.1

UNIJUNCTION TRANSISTOR METRONOME

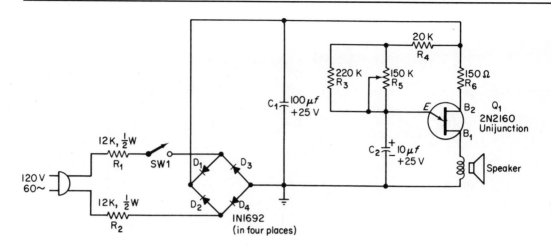

	Hours Per 100	Rate Per Hour
Finishing	18.00	$4.60
Machine shop	7.00	5.00
Assembly	6.25	5.00
Inspection	2.80	6.20

The learning curve for finishing and inspection is estimated as 90%, while all else is 75%. The manufacturing burden is broken into variable and fixed with a rate of 75% and 25%. Administrative and marketing burden costs amount to 50% fixed only. Engineering development costs have amounted to $8,000. Manufacturing startup expenses will be about $5000. Product costs are estimated at $0.25 for each unit.

The distributor's charge for a product of this sort is usually 40% less than list price. A full cost practice of adding 25% to all costs has been practiced before but is now used as a guide. A 20% return on appropriate investment is a minimum desired level. Ray Enterprises has used quality assurance techniques, and for products like this one has empirically determined that about 1 failure in 50 is expected. Its warranty policy is adamant: "Replace with new model if the old unit fails during the first year." The company reimburses the distributors for a new unit.

Construct a cost estimate and determine a price based on direct costing, investment and the full method. Provide a cash flow statement over the next 2 years and a profit and loss statement for this product. Should this product be made and sold? In view of the preliminary marketing data on price and volume and estimating data on cost, what recommendations on price and cost can you make to President Art Ray?

chapter 10

How to have your cake and eat it too: lend it out at interest.

Author unknown

PROJECT ESTIMATING

Cost estimators are frequently the first to recognize the need for new equipment, processes, plants, or their replacement. As the preliminary engineering plan is originated, and as a matter of good practice, the estimator contributes information to the budget defining the resources that may be called on. Should the preliminary estimate call for additional planning, a detailed estimate is made. If the estimate looks encouraging, the question to spend money becomes an executive-coordinated decision, particularly if the capital money is large when compared to readily available resources. Thus there is a special responsibility that rests within the confines of the project-estimating function.

To clarify the investment discussion for the remainder of this chapter, it is assumed that the technical feasibility of equipment, plants, and other physical services has been determined but that the project cost analysis is not yet revealed. Although project designs calling for evaluation differ, the economic techniques are common despite seemingly large differences in the engineering design. A project estimate provides a measure of the desirability of risking capital for a design proposal. The major concern is for the economic result arising from the expenditure of capital. The basic considerations are the design, capital, expense, income, and time.

Of the decisions which executive management makes, few affect the financial stability and the future earnings of the firm more than those pertaining to capital investment. These decisions commit the firm to manufacture or distribute certain products, to construct plants at certain locations, and to utilize certain materials, processes, methods, machines, or groups of machines. They establish the structure within which the organization will operate for years to come. These decisions involve thousands if not millions of dollars to any one firm. The design decision has a substantial impact on cost. Although engineering is able to influence the efficiency of the transformation of the design into the actual product or service, the approximate level of cost is nominally fixed after the plans and functional engineering concepts have been finalized. This holds true because these concepts and designs determine the limits and cost of the processes, materials, and labor that are used. Each of the various original design concepts results in a different final cost. It is important that the design concept be initially chosen in the light of the cost of the processes and the methods that it dictates and the capital investment requirements that result from these processes.

Due to the interaction between design objectives and cost, procedures that perform detailed analysis of investment opportunities and strategies are required. In early stages it is not uncommon to establish investment costs on information from a flow chart or sketch where details, concepts, tolerances, and specification information are not included. Later, a complete set of drawings, specifications, and bill of materials are available to guide a critical examination of investment opportunities. In both these cases the estimator evaluates the equipment, processes, method, tools, and so forth that convert the materials into finished products in order to arrive at the economic return that justifies the profitability of the project. In this chapter, as in other textbooks dealing with investment analyses, emphasis is given to the final arithmetic of the analyses, yet the greatest bulk of a project estimator's time is tied to evaluating the schemes and coming up with the capital requirements. The computation of the time-value-of-money concept is an important final step in the estimate, but it is valid as long as the estimating data are valid.

The purpose of converting capital into plants and equipment is to return an amount of money that exceeds the investment. This statement assumes that capital is productive and earns a profit for the owner of this capital. In efficiency terms this is related to a ratio of output to input. Unlike physical processes, the economic efficiency of capital, assuming long-term success and a capitalistic society, must exceed 1. The productivity of capital comes from the fact that money purchases more efficient processes for making goods and supplying services than consumers could employ themselves. These products are then offered to the public at attractive prices which pay a profit to the manufacturer.

In earlier chapters stress was placed on cost and price as the measure of importance. For purposes of capital investment, the estimator uses a term called *return*. Like cost and price, return can be expressed in several ways. Among these are total dollars, percent of sales, ratio of annual sales to investment, or return on investment. The last one is favored by most project estimators. The methods for calculation of return on investment are (1) average annual rate of return, (2) payback period, and (3) engineering-economic rate of return. Naturally, the selection of a method must be consistent with management's goals of profitability. The goal of any project estimate is to predict the net change in the company's overall cash position. The estimator makes studies of alternatives, looking for the change in cost and revenue that must be considered. These studies should provide factual quantitative data to measure the interaction of future events.

In this text we favor the compound-interest-based investment computation that considers the *time value of money*. However, as a background let us look at methods which are sometimes used because of their simplicity.

In some economic studies return on investment is expressed on an annual percentage basis. The yearly profit divided by the total initial investment represents a fractional return or its related percent return. This method recognizes that a good investment not only pays for itself but also provides a satisfactory return on the funds committed by the firm. There are several variations of which the following is one:

$$\% \text{ return} = \frac{\text{earnings per year}}{\text{net investment}} \times 100 \qquad (10\text{-}1)$$

The earnings are after tax, and deductions for depreciation usually represent some average future expectation. For instance, consider the following example: Investment for new equipment is $175,000, salvage will provide $15,000, and an average earnings of $22,000 after taxes is expected.

$$\% \text{ return} = \frac{\$22,000}{160,000} = 13.75\%$$

Now assume that a private investment opportunity of $25,000 has been brought to your attention. This is broken down to $20,000 for the investment and $5000 for initial working capital (cash, accounts payable and so forth). Annual operating and other expenses are estimated at $10,000 and income at $15,000 per year. This investment is analyzed as follows:

$$\text{Income} \quad \dots\dots\dots\dots \quad \$15,000$$
$$\text{Expenses} \quad \dots\dots\dots\dots \quad \underline{10,000}$$
$$\text{Net income} \dots\dots\dots\dots \quad \underline{\$\ 5,000}$$

$$\% \text{ return} = \frac{\$5000}{\$25,000} = 20\%$$

Another variation is expressed as

$$\% \text{ return} = \frac{\text{average earnings} - (\text{total investment} \div \text{economic life})}{\text{average investment}} \quad (10\text{-}2)$$

The earnings in the formula are the average annual earnings after taxes plus appropriate depreciation charges. It is seen that the original investment is recovered over the economic life of the proposal by subtracting the factor of total investment ÷ economic life from average earnings. This difference denotes the average annual economic profit on the investment. The average investment is defined as the total investment times 0.5. These methods acknowledge that the life of an investment for tax purposes and its true economic life are not the same. The first is based on the normal physical life, while the second represents the profitable life of the investment, which is frequently a shorter period of time. If management desires, it may incorporate a risk element by further shortening economic life.

Example: An average after-tax earning of $22,000 is expected from an investment of $175,000 with an economic life of 10 years. Straight-line depreciation is assumed for 12 years and salvage is $15,000.

$$\% \text{ return} = \frac{22,000 + (160,000/12) - (175,000/10)}{160,000 \times \frac{1}{2}} = 22.29\%$$

10.1-2
Payback Period Method

The payback method is easy and widely adopted. Essentially, the method determines how many years it takes to get the invested capital return. The formula as normally given is

$$\text{years payback} = \frac{\text{net investment}}{\text{annual after-tax earnings}} \quad (10\text{-}3)$$

The payback method recognizes liquidity as the basis for the measure of economic worth of capital expenditures. There are several variations in applying this method in which the following are significant: Divide the total investment by the additional annual pretax earnings or savings expected; divide the total investment by the earnings or savings after taxes; add the depreciation to the earnings after taxes before the division is performed. The object of each of these variations is to find how many years it would take for the investment to pay for itself.

The method separates proposals of doubtful validity from those which call for additional economic analysis. Obviously the method signals the immediate cash return aspect of the investment which may be desirable for corporations where a high-profit investment opportunity and limited cash resources exist. In some situations the payback is used for those investment situations where it is felt that the risk does not warrant earnings beyond the payback period. Let us use an example to illustrate this: The installed cost for new equipment is $175,000 and old equipment will be sold for $15,000. Better productivity of the new equipment will return $40,000. For a composite 55% corporate tax rate earnings amount to $22,000.

$$\text{years payback} = \frac{175,000 - 15,000}{22,000} = 7.3$$

Consider now two investment opportunities:

	Equipment A	Equipment B
Total investment	$ 60,000	$60,000
Revenue:		
Year 1	20,000	30,000
Year 2	20,000	30,000
Year 3	20,000	30,000
Year 4	20,000	—
Year 5	20,000	—
Total annual after-tax earnings	$100,000	$90,000
Payback period	3.0 years	2.0 years

In this case equipment B is preferred over A because of the smaller payback period. If sufficient resources were available, a management fiat could allow any investment that was under an arbitrary level such as 5.

The average annual rate of return is acknowledged to have faults. It assumes even distribution of earnings throughout the economic life of the asset. Even if this were true, there is a significant difference between the value of the dollars earned in the first year and those earned in later years. This method ignores the time value of money. A project yielding savings in early years of its life is more beneficial, as these funds become available for additional investment or for alternative use and often are subject to less risk than savings projected many years ahead. This method overlooks the differences in salvage values and their relation to the time element. Nor is interest on borrowed money in any way reflected.

The payback method suffers similarly. It is noted that the life pattern of earnings is ignored in payback formulas. In the example equipment B had a shorter payback period than equipment A, yet A will return $10,000 more. New equipment may not be profitable during the early part of the payback period; on the other hand, it may be quite profitable in the future. This method does not provide for a system of ranking with other investment possibilities, nor does it take into account depreciation or obsolescence, nor does it consider the earnings beyond the payback period. For example, it does not recognize that one investment with an earning of $10,000 the first year and $2000 the third year is more desirable than another which earns $6000 in each of the 3 years. The situation for which payback is suited, and then only provisionally, is as a rough measure of evaluation.

The engineering-economic method of determining return overcomes these shortcomings. The rest of this chapter is devoted to explaining this method. This method is applicable to every possible type of a prospective investment and can yield answers that permit valid comparisons between competing projects.

Gathering information is the first part of the project estimate. This involves determining the cash flows and their timing. Some information, however, is not reducible to data and is subject to the whim and caprice of human reasoning. Despite this shortcoming the selection of quality cost data is mandatory.

Cash flows are determined by any of the three types of information such as historical, measured, and policy. The general guidelines are:

1. Future data are preferred.
2. The time value of money is of major importance.
3. Cash or other equivalent measures are the significant criteria.
4. Differential revenues and costs are to be treated differently than total revenue and cost.

For the historical type of information, accounting and estimating data are assembled from records to indicate profits, losses, operating costs, net worths, and so forth for past period or periods of time. As accounting data are gathered for a multiple of purposes such as cost control, compliance with federal, state, and local laws, and company policies, the timing may not be indicated.

Measured cost data are assembled from material balances and quantity take-offs, flow chart information, time and motion studies, and operational and product estimates. Product estimates give a projection of records and forecasts to indicate future income, profits, losses, and operating costs for a proposed product or a modification of a product. The sales-volume forecast is an important source of information.

Operation estimates may provide information on the irregular stream of rentals avoided, or labor and material savings, in addition to the operation cost. They do, of course, supply labor and material costs for machine, process, or plant operation.

Policy data are necessary for project estimating. The value of money, expressed as an interest rate, is an important policy contribution. It may be a range or a single value. The source and quantity of money available for investment must be generally known.

The project estimator is concerned with the estimation of costs for the purchase of new assets or plants, or the enlargement of existing equipment and plants, or their refurbishment. This would involve estimating the items in Table 10.1.

The factor and power law and sizing models would be typical estimating methods used to determine the cost of equipment and plant addition. In forecasting the life of an asset there are two major concerns: annual deterioration and obsolescence. Where physical life or annual deterioration establishes the life, statistical data from past records become the basis for future prediction. If *life* is regarded as economic, statistical methods find the probability that the economic life of a proposal will terminate during each year of its service life. In either of these two cases, deterioration or obsolescence, the life of a particular proposal terminates because it is worn out or because the product or service which it produces is no longer profitable.

In the first instance, the physical condition has deteriorated and it does not produce the desired quality, or the cost of maintenance exceeds the cost of replacement. In the absence of statistical data, reliance must be placed on the judgment of people having experience with this equipment such as engineers, operators, and the people producing the equipment. In the case of obsolescence, competition may introduce substitute products, processes, or machines with better prices, qualities, or services. Even the corporation may make its own

TABLE 10.1

INFORMATION REQUIRED FOR
PROJECT ESTIMATE

First cost of asset constructed or delivered and installed and ready to run
Life period of the asset, both economic and functional
Hazards and losses relative to plant, equipment, material, and labor time
Detailed operational data:
    Insurance: property, inventory
    Efficiency levels and patterns of utilization
    Maintenance and repair costs
    Direct operating costs including labor, fuel, power, scrap material
    Indirect costs
    Changes in the inner cost of labor, power
Versatility to handle fluctuations in sales demand
Safety considerations
Production stoppages
Quality
Operator acceptance
Changing technologies
Operation and product estimates

---

product or services obsolete. When obsolescence is a factor, judgment is required, as the life of the equipment may suddenly be terminated by a change in company policy, buying habits, government legislation, competitive pressures, or new designs.

Another of the besetting predictions that can be bewildering is that of salvage value. An experienced appraiser may give opinions of future land and factory values. If the life is not expected to be great, the engineer is in a position to trust this source of information. For the longer-lived equipment, information may be unavailable. Despite the fear of distant predictions, errors in evaluating salvage value are not normally serious. Error in the present or the near future should be given more concern or study because the effects of these conclusions are greater. Another prediction for equipment and plants is that of the efficiency of the utilization. This efficiency is important because it controls cost to some extent. Errors in the degree of efficiency of equipment are considered serious, and predictions of this nature should be studied closely.

10.2-2
Working Capital

Net working capital is the difference between current assets and current liabilities. Sometimes we calculate working capital as the allowance in a capital estimate for necessary operating inventory (raw, in-process, and finished materials), cash, and net receivables. Working capital requires immediate cash outlays that are not realized until the sales revenue is generated. Thus *timing* is involved. The following should be tabulated and totaled to obtain total working capital: changes in current payables and accounts receivables, change in inventory of raw materials and supplies, changes in cash-on-hand balances, and changes in current liabilities and nondepreciable assets. Increases in working cash are required to support the operations resulting from the investment. A higher requirement exists during periods of expansion, and a lower requirement exists during contracting periods. Working capital is important because of the time delay in its recovery. At the termination date of the project, a credit for working capital is estimated to offset, although not equally, the initial requirements for working capital at the start of the project.

These costs have been previously identified, but special considerations are required for project estimating. Raw materials fall into two classifications. Materials from outside sources are charged at current market price delivered at the consuming point. Materials produced as a by-product within the organization are charged with out-of-pocket costs if there is no market for these materials. If there is no excess capacity of by-product materials and the by-products are commercially sold, then the net value is based on market price. This approach reflects the actual cost which counterbalances reduced sales revenue. If the raw material is an existing waste material, then its cost of disposal should be credited to the new product.

In addition to direct costs, distribution and administrative and selling costs have an effect on a project estimate. Should an investment create a change in these costs, then it is necessary to estimate that amount. Much of these costs would be unaffected. Areas such as legal, corporate public relations, and the like are more affected by changes in policy than by limited changes of a company's operation. Those overhead costs that could be affected are employee benefits, plant administration, employee training programs, insurance, and property taxes.

Taxation plays a complicated role in any estimating method whether the taxes are property, federal and state income, or sales. Income tax and to a lesser extent property and sales tax are crucial to the success or failure of a project. Income tax laws are the result of man-made legislation over a period of time. They incorporate diverse ideas, and some appear to be in conflict.

In most cases the desirability of a venture is equivalent to differences between income and cost. Since income taxes are levied on profit, they result in the reduction of its magnitude. A simplified basic tax equation can be constructed as follows:

$$NP = [G - (C + D + I)] - [G - (C + D + I)]t \qquad (10\text{-}4)$$

where NP = net profit after income taxes
$\quad G$ = estimated gross income for the venture considered
$\quad C$ = estimated accumulated annual costs not estimated elsewhere
$\quad D$ = annual depreciation charge acceptable to tax laws
$\quad I$ = interest paid on borrowed funds concerned with venture
$\quad t$ = applicable income tax rate

We can simplify Equation (10-4) to

$$NP = [G - (C + D + I)](1 - t) \qquad (10\text{-}5)$$

On examining the equation it should be noted that net profit is directly related to the applicable income tax rate. After the depreciation and other operating charges have been deducted from gross income, the remainder is subjected to the product of the difference between $1 - t$.

Depreciation (discussed in Chapter 4) is important since it affects the income tax liability. Tax liability is determined after depreciation, interest charges on investment (if a loan or note was used for its purchase), and annual costs are deducted. It should be pointed out that the depreciation charge itself is not a cash outlay. The cash outflow takes place at the time the asset is acquired. The depreciation charges are the assignment of that outflow over the economic or physical life and in no way involve a disbursement of cash. Increasing the rate at which the asset is depreciated does not increase the outflow of cash; it has the opposite effect since the depreciation charges reduce taxable income and consequently reduce the outflow of cash for taxes. The government recognizes several methods of calculating depreciation. The project return is vulnerable to the choice of method if the time value of money is considered. Although the ultimate recovery of the value of the asset is generally identical for the several methods, the timing of this recovery is different for each case. After-tax earnings are the only tangible basis on which to judge the relative attractiveness of competing projects.

There are many methods of capital investment analysis, and, unfortunately, only a few will be studied in detail. It is the analysis that leads to a strategy which in turn leads to the decision. The time value of money is the application of compound interest formulas to the additional cash flow produced by the investment. This concept enables management to place a value on the money which will become available for productive use in the future as well as for the money which is available today. Fundamentally, the time value of money begins with simple interest, or

$$I = Pni \qquad (10\text{-}6)$$

where I = interest earned, dollars
P = principal sum, dollars
n = number of compounding periods
i = interest rate, decimal

This formula can be restated as the amount including principal and simple interest which must be repaid eventually, or

$$F = P + I = P(1 + ni) \qquad (10\text{-}7)$$

where F = principal and interest sum collected at some future time. In the payment of simple interest, the interest is paid at the end of each time period or the sum total amount of money is paid after a given length of time. Under the latter condition there is no incentive to pay the interest until the end of the contract time.

   If interest were paid at the end of each time unit, the lender could put his money to use for earning additional profits. Compound interest considers this point and requires that interest be paid regularly at the end of each interest period. If the payment is not made, the amount due is added to the principal and interest is charged on this converted principal during the following time unit. An initial loan of $10,000 at an annual interest rate of 5% would require payment of $500 as interest at the end of the first year. If this payment were deferred, the interest for the second year would be ($10,000 + $500)(0.05) = $525, and the total compound amount due after 2 years would be $10,000 + $500 + $525 = $11,025. When interest is permitted to compound, as in the following computation, the interest earned during each interest period is permitted to accumulate with the principal sum at the beginning of the next interest period. This compounding is shown in Table 10.2. The resulting factor, $(1 + i)^n$, is referred to as the single-payment compound-amount factor. Values for this and other factors are found in the Appendix. The total amount of principal plus compound interest due after n periods is

$$F = P(1 + i)^n \qquad (10\text{-}8)$$

The single-payment compound-amount factor may be used to solve for a future

10.3
ENGINEERING ECONOMY METHODS FOR PROJECT ANALYSIS

TABLE 10.2

DERIVATION OF BASIC COMPOUND
INTEREST FORMULA

Year	Principal at Start of Period	Interest Earned During Period	Compound Amount F at the End of Period
1	P	Pi	$P + Pi = P(1 + i)$
2	$P(1 + i)$	$P(1 + i)i$	$P(1 + i) + P(1 + i)i = P(1 + i)^2$
3	$P(1 + i)^2$	$P(1 + i)^2 i$	$P(1 + i)^2 + P(1 + i)^2 i = P(1 + i)^3$
n	$P(1 + i)^{n-1}$	$P(1 + i)^{n-1} i$	$P(1 + i)^{n-1} + P(1 + i)^{n-1} i = P(1 + i)^n$

sum of money F, the interest rate i, the number of interest periods n or a present sum of money P when given the other quantities.

It has been stated that the engineering economic methods were preferred because they depended on time-value-of-money concepts. One should not conclude that all methods employing interest computations are useful for all occasions. Some have limited applicability. We present here four distinct variations; when these methods are given correct information and properly understood their answers are equally valid. They are:

1. Net present worth.
2. Net future worth.
3. Equivalent annual cost.
4. Rate of return.

Each of these methods measures a different factor of the investment; they can give quite different evaluations. Nonetheless, they will all give the same estimate for consistent decision making.

Each method is demonstrated with the same standard problem. Cents are dropped from calculations for ease of understanding.

Year	Disbursement	Revenue
0	$1025	$ 0
1	0	450
2	0	425
3	0	400

The annual compounding and end-of-year revenue convention is used to simplify understanding. It is assumed that the nonuniform revenue is instantaneously received at the end of year.

10.3-1
Net Present Worth Method

This method is also known as net present value or venture worth. The present worth of all future returned revenue is compared with the initial capital investment. It assumes a continuing stream of opportunities for investment at a pre-assigned interest rate. The procedure followed is to compare the magnitude of present worth of all revenues with the investment at the datum time 0. One way of defining the method is as the added amount that will be required at the start of a proposed project using a pre-assigned interest rate in order to produce receipts equal to, and at the same time as, the prospective investment. For a given interest rate of 10%, the net present worth of the previously given problem is computed by discounting all revenues to year 0 at this rate and subtracting the proposed investment, or

Period	$\dfrac{1}{(1+i)^n}$ Present Worth Factor @ 10%		
Year 1 to zero	450×0.9091	=	$ 409
Year 2 to zero	425×0.8264	=	351
Year 3 to zero	400×0.7513	=	301
		Total	$1061
Less proposed investment			1025
		Net present worth	$36

The $36 is the amount that must be added to the $1025 to set up the amount that would have to be invested at 10% to achieve receipts equal to and at the same time as those predicted for the recommended investment, or

$$(\$1025 + \$36) \times 1.1 = (\$1061) \times 1.1 = \$1167$$

$$\text{Less payment} \qquad \underline{\qquad 450}$$
$$\$ 717$$

$$717 \times 1.1 \qquad\qquad = \$ 789$$
$$\text{Less payment} \qquad \underline{\qquad 425}$$
$$\$ 364$$

$$\$364 \times 1.1 \qquad\qquad = \$ 400$$
$$\text{Less payment} \qquad \underline{\qquad 400}$$
$$0$$

10.3-2
Net Future Worth Method

The standard application of compound interest using $F = P(1 + i)^n$ is highlighted in this example. It uses the notion that assets and revenues can be invested at the pre-assigned interest rate where there is a continuous exposure of investment opportunities. A comparison of investment of the original sum plus reinvestment of revenues both at the pre-assigned interest is made against the standard alternative of investing only the original asset value. The calculation results in that added amount that is obtained at the end of the project economic life if the project's anticipated revenues were invested instead of the proposed investment. A common comparison uses the same stipulated interest rate.

For the sample problem and 10%, the net future worth is computed by compounding future revenues to the terminal year and then subtracting from this the amount that would have resulted from the other alternative of investing the original asset at the same pre-assigned interest rate to the terminal year:

Period	$(1 + i)^n$ Compound Amount		
Year 1 to 3	450×1.10^2	=	$ 545
Year 2 to 3	425×1.10	=	468
Year 3	400×1.00	=	400
			$1413
Less disbursement compounded to terminal year @ 10%			
Year 0 to 3	$1025 \times (1.10)^3$	=	1364
		Net future worth	$ 49

The calculations point out that if the project is funded and if the revenues materialize as estimated, then a surplus of $49 will be expected over the simple alternative of investing only the asset of $1025. The same period of time and equal interest rates are parts of this comparison.

10.3-3
Net Equivalent Annual
Worth Method

Management often wants a comparison of annual costs instead of, say, present worth of the costs. Here we refer to net costs, that is, the net difference between any cost or revenues or credits. This method considers a supply of opportunities for investment of both assets and receipts at the predetermined interest rate plus a supply of capital at the same interest rate. Now the sample problem does not have uniform annual receipts, and they first must be converted to total present worth and then to annual equivalents. The total present worth at time zero is $1061 (from Section 10.3-1). The annual equivalent is found by dividing by the sum of the present worth factors, or

$$\text{annual equivalent} = \frac{1061}{(0.9091 + 0.8264 + 0.7513)} = \frac{1061}{2.4868} = \$427$$

$$\text{net annual equivalent worth} = 427 - \frac{1025}{2.4868} = \$14$$

This $14 is the amount by which the anticipated revenues from the proposed investment, rated at 10% interest, exceed the annual equivalent of the proposed investment:

Anticipated equal annual receipts		$ 427
less equal annual equivalent worth @ 10%		14
Equal annual receipts to be generated by investing		$ 413
1025×1.10		= $1128
	Less payment	413
		715
715×1.10		= $ 787
	Less payment	413
		374
374×1.10		= $ 413
	Less payment	413
		0

Starting out with the investment which earns interest and subsequently subtracting payments leads to a balance of zero dollars at the end.

This procedure evaluates the rate of interest for discounted values of the net revenues from a project in order to have these present worths of the discounted values exactly equal to the present value of the investment. The method thus solves for an interest rate to bring about this equality. Other titles such as *true rate of return*, *profitability index*, and *internal rate of return* also exist. The adjective *true* distinguishes it from other less valid methods that have been labeled rate of return, for instance, Equations (10-1) and (10-2). For this method there is no assumption of an alternative investment and no predetermined interest rate. We define this interest rate at which a sum of money, equal to that invested in the proposed project, would have to be invested in an annuity fund in order for that fund to be able to make payments equal to, and at the same time as, the receipts from the proposed investment.

The solution for the interest rate is by repeated trials or by graphical or linear interpolation. For the sample problem

Year n	Receipts		$\dfrac{1}{(1+i)^n}$ PW Factor @ 15%		Discounted Value
1	$450	×	0.8696	=	$391
2	425	×	0.7561	=	321
3	400	×	0.6575	=	263
					$975

Year n	Receipts		$\dfrac{1}{(1+i)^n}$ PW Factor @ 5%		Discounted Value
1	$450	×	0.9524	=	$ 429
2	425	×	0.9070	=	385
3	400	×	0.8638	=	346
					$1160

The two trial values bound the initial asset value of $1025. Either graphical or interpolating methods can be used to locate the rate of return, or from which

$$\frac{0.05 - i}{1160 - 1025} = \frac{0.05 - 0.15}{1160 - 975}$$

it follows that $i = 12.3\%$. This rate of return is the interest rate at which the original sum of $1025 could be invested to provide revenues equal to, and at the same time as, the receipts of the prospective investment:

$$1025 \times 1.123 = \$1151$$
$$\text{Less payment} \quad \underline{450}$$
$$\$ 700$$

$$700 \times 1.123 = \$ 787$$
$$\text{Less payment} \quad \underline{425}$$
$$\$ 362$$

$$362 \times 1.123 = \$ 400$$
$$\text{Less payment} \quad \underline{400}$$
$$0$$

This shows that the earning rate is true and is the actual return of the invested money. It has the important advantage of being directly comparable to the cost of capital.

10.3-5
Comparison of Methods

A summary of the four different methods provides:

1. Net present worth at 10% $36
2. Net future worth at 10% $49
3. Net annual equivalent worth at 10% $14
4. Rate of return   12.3%

We can demonstrate that the first three are commensurable answers. For instance, the net present worth of $36 can be compounded to the terminal year using Equation (10-8) or

$$F = 36(1.10)^3 = \$49$$

The pre-assigned interest rate, 10%, gives the net future worth or $49. The net annual equivalent worth of $14 can be uncovered by dividing the net present worth value of $36 by a sum of present worth values, of 10% or

$$\frac{36}{2.4867} = \$14$$

On the other hand, the rate of return cannot be calculated from the foregoing answers. It is found directly from the data. Equivalency among the first three can be found for any arbitrary interest rate, and if these are equivalent to the rate of return, then it is equivalent at only one pre-assigned interest rate. The first three methods are only differences based on the choice of an interest rate. This arbitrary selection of an interest rate makes the three methods little more than decision tools for comparing projects. The rate of return is the only one that can be depended on to provide a consistent measure of the extent of the economic productivity of prospective investments. The answer can be compared directly with the cost of capital. However, no one single method or criterion of profitability analysis is preferred for all situations.

10.3-6
General Engineering Economy Models

Without any further ado, the following general engineering economic model[1] is proposed:

$$P_x = (1 - t)\left[\sum_{n=1}^{N} \frac{(S_n - C_n)}{(1 + i)^n}\right] + t\sum_{n=1}^{N} \frac{A(d)_n}{(1 + i)^n} + \frac{F_s}{(1 + i)^N} \qquad (10\text{-}9)$$

where P_x = present worth for reference year x; total discounted cash flow, dollars

S_n = sales or revenue in nth year, dollars

C_n = total cost (except depreciation charge) required to obtain S_n sales for nth year, dollars

$A(d)_n$ = annual depreciation in nth year; form allows any depreciation procedure to be substituted, dollars

F_s = future value of salvageable items (land, working capital, physical salvage), dollars

i = effective interest rate, decimal

[1] Following Herbert E. Schweyer's *Analytic Models for Managerial and Engineering Economics*, Van Nostrand Reinhold Company, New York, 1964, p. 265.

$t =$ tax rate, decimal

$n =$ end-of-year age for which computation is made

$N =$ life of asset, years

This model is able to commensurably measure (1) the rate of return or interest rate of the net cash inflows for a pre-established number of periods, (2) the net present worth of the excesses of revenues over expenses at a predetermined interest rate on capital for a specified number of future periods, or (3) the number of time periods required to pay off the initial investment given the interest rate. The model is sufficiently flexible to account for profits, tax, and variations in annual revenues, expenses, depreciation, deposits, and timing. Because of these factors, the fundamental relationship is a valid project-estimating model. When given the interest rate the first bracketed term represents the net present worth of sales over costs excluding depreciation; the second represents the tax credits for depreciation; while the third term represents the present worth of the asset's salvage value or land or working capital. The second term is called a tax credit and does not exist unless there is a profit t being zero. The datum period, x, requires an explanation. An investment, say, for plant construction, could be paid out over a few years as the construction progressed. From the commitment date of beginning construction to the time of the plant start-up date where revenue is received is classified as a negative time and the future worth method of Section 10.3-2 updates the amount of money to the datum time point, say 0, where revenues begin.

The final term of model (10-9) adjusts terminal values to the datum time. Tax is not a factor of salvage value inasmuch as the exact amount of return value is considered to have the same value as assumed originally. At the time of sale for the asset if a smaller amount of money is realized, then a tax credit applies to the portion of the difference in realized salvage from estimated salvage; conversely, if there is a gain in salvage dollars between estimated and actual values, the gain is treated as additional revenue subject to tax.

Simplifications of model (10-9) are possible. The variables S, C, $A(d)$, and F_s may have differing yearly values; however, average values are frequently used. Several illustrations will now describe Equation (10-9).

What is the net profit and the constant annual cash flow for a project with an initial total investment of \$175,000 and an economic life of 10 years with no salvage? Assume straight-line depreciation. After-tax constant earnings of \$22,000 are expected. Before-tax earnings are \$40,000. A tax rate of 0.55 is applicable. A summary table would provide

$S_n - C_n \,(= A)$	\$40,000
Depreciation 175,000/10, $A(d)$	17,500
Taxable profit	\$22,500
Profits tax @ 55%	12,375
Net profit	\$10,125

Constant annual cash flow $= \$10,125 + 17,500 = \$27,625$

The same result is obtained for constant year n using (10-9) as

$$(1 - t)(S_n - C_n) + tA(d) = 0.45(40,000) + 0.55(17,500) = 27,625$$

The following example considers the discounted cash flow method where nonuniform income is expected. It is seen that it is a trial and error approach to determine the nominal interest which makes the cash flow receipts equivalent to the initial disbursements for an investment. A machine will cost \$175,000 installed and will be capitalized prior to the installation. The earnings as anticipated are not uniform since the market will not be prepared for the product until several years have passed. Additionally, it is presumed that technology will improve on the product and that the income peak will decline. The economic life of the investment is forecasted to be 12 years, while the depreciable investment will be recovered over 10 years by an accelerated method such as sum-of-the-years digits (SYD). Salvage is estimated to return no value over disposal costs at the end of 12 years. The composite tax rate, including all relevant taxes, is assumed to be 55%.

Using

$$A(d)_n = \frac{2(N - n + 1)(P - F_s)}{N(N + 1)} \tag{10-10}$$

where $A(d)_n$ = annual depreciation amount varying with year n
$\quad\quad N$ = useful life
$\quad\quad P$ = investment costs
$\quad\quad F_s$ = salvage value
$\quad\quad n$ = current year number from start of investment, $n = 1, \ldots, N$

Depreciation charges can be calculated. Using various nominal values for yearly interest, the present worth can be computed. In the iterative plan, values of 0, 10, and 15% are used.

Analysis is given in Table 10.3. Using a linear graph of Figure 10.1 the rate

FIGURE 10.1

INTERPOLATION FOR RATE OF RETURN USING DISCOUNTED CASH FLOW MODEL

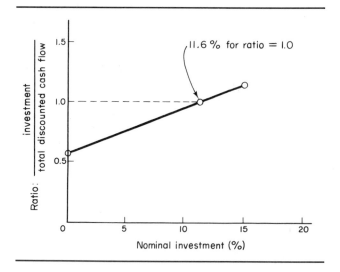

of return is found to be 11.6%. The summation procedure of Equation (10-9) must be followed year by year for each term from zero time to the last year of the project. The several forms of depreciation from Table 4.7 can be substituted for $A(d)_n$. Straight-line depreciation, sum-of-the-year digits, or other accelerated procedures can be considered. The refinements of nonuniform depreciation (other than straight line) are shown to be an advantage, particularly with larger interest values.

Model (10-9) may be further broadened by having it describe the unrecovered balance or net cash flow position at any time. The cash flow positions are

TABLE 10.3

ITERATIVE METHOD FOR EQUATION (10-9)

Year, n	Cash Earnings, $S_n - C_n$	Depreciation, $A(d)_n$	Taxable Income	Profit After Taxes	Net Cash Flow	$\frac{1}{(1+i)^n}$ 10%	Discounted Cash Flow	$\frac{1}{(1+i)^n}$ 15%	Discounted Cash Flow
0	(175,000)				(175,000)	1.000	(175,000)	1.000	(175,000)
1	35,000	31,818	3,182	1,432	33,250	0.909	30,224	0.869	28,894
2	35,000	28,636	6,364	2,864	31,500	0.826	26,019	0.756	23,814
3	35,000	25,454	9,546	4,296	29,750	0.751	22,342	0.658	19,575
4	40,000	22,273	18,727	8,427	30,700	0.683	20,968	0.572	17,560
5	45,000	19,091	25,909	11,659	30,750	0.621	19,096	0.497	15,283
6	45,000	15,909	29,091	13,091	29,000	0.565	16,385	0.432	12,528
7	45,000	12,727	32,273	14,523	27,250	0.513	13,979	0.376	10,246
8	50,000	9,545	40,455	18,204	27,749	0.467	12,959	0.326	9,046
9	40,000	6,364	32,636	14,686	21,050	0.424	8,925	0.284	5,978
10	35,000	3,183	31,817	14,317	17,500	0.386	6,755	0.247	4,322
11	25,000	—	25,000	11,250	11,250	0.351	3,949	0.215	2,419
12	25,000	—	25,000	11,250	11,250	0.319	3,589	0.187	2,104
					$300,999		$185,190		$151,769

discounted to time 0 with a predetermined interest rate. This model shows the number of earning periods to have a zero unrecovered balance between discounted revenues and the initial asset value. A modification of this type of analysis is given later where a lease versus buy comparison is made. In this case no revenues are assumed.

Figure 10.2 shows a type of the unrecovered balance discounted cash flow chart. It considers the patterns of cash flow during early and mature years. This one is particularly useful for projects where the investment in land purchase, R & D, engineering design, and construction is made over an extended time interval. The early stage *ab* represents negative expenditure on land. It is followed by the larger expenditure *bc* on design and investment, while *cd* is operating capital. At point *d* income begins and *de* is the start-up period. This is net income following the tax deduction. The curve reaches break-even at point *f* where the initial expenditure is balanced by the accumulated income. The zero time in this case is pushed back to the point of original spending. While interest need not be a factor in this approach, it is more attractive as a decision tool if it is.

Compound interest formulas are either discrete or continuous compounding. In the first case payments are considered *lump sum* and are concentrated at either the beginning or the end of the period. First cost for investments, land acquisition, working capital, and salvage value are lump-sum examples. There are other discrete payments that flow more or less continuously throughout a year as is typical with operating disbursements such as labor and salaries paid weekly and semimonthly, power monthly, and purchased materials at random intervals and bills due in 30 days or sooner. One of the principal reasons for the idea of continuous compounding is the fact that these various cash flows are actually scattered items and do not occur at precise intervals, which is usually considered to be the case. Certainly it is a safe approximation to say that much of a business' income is received in a steady stream. The concept of continuous interest rests on the assumption that earnings are created every day, hour, and minute of operation. The more frequent the compounding, the greater the future sum amount. For instance, the effective rate of 10% for semi-annual compounding is 10.2500; for monthly, 10.4713; and for daily and continuously, 10.5126 and 10.5171. It is seen that the number of periods of compounding has little effect when one goes from daily to continuous compounding. The differences between

FIGURE 10.2

CASH FLOWS FOR VARIOUS COMMITMENTS

a – b	Land expenditure
b – c	Design and investment
c – d	Operating capital
d – e	Startup
e – f	Income
g	Termination point

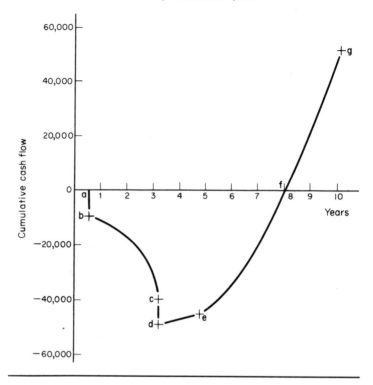

the continuous interest factor, both effective and nominal, are not large when restricted to low values such as below 10%. Most firms consider rates of return well above this region, especially on a before-tax base.

Previously, compound interest was presented with interest being computed annually, or semi-annually, or monthly as required by the dimensions of n. Continuous compounding is the limiting case of discrete compounding in that the frequency of compounding is taken to be infinitesimally small. In discrete compounding for m compounding periods per year we have the formula $F = P[1 + (r/m)]^{mn}$, which can be modified to

$$F \text{ after } n \text{ years} = P\left[\lim_{m \to \infty} \left(1 + \frac{r}{m}\right)^{mn}\right]$$

and by rearranging terms

$$F = P\left\{\lim_{m \to \infty} \left[\left(1 + \frac{r}{m}\right)^{m/r}\right]^{(rn)}\right\}$$

where $P =$ present sum of money
$F =$ future sum of money
$r =$ nominal annual interest rate
$m =$ number of compounding periods per year

Since the fundamental definition of e, the base of the natural logarithms, is

$$e = \lim_{m \to \infty} \left(1 + \frac{r}{m}\right)^{m/r} = 2.71828$$

one can now substitute and get the continuous compounding equivalent to $F = P(1 + i)^n$, which is

$$F = Pe^{rn} \qquad (10\text{-}11)$$

The equation for transforming nominal to effective rates of interest is now developed. Since

$$i = \left(1 + \frac{r}{m}\right)^m - 1 \qquad (10\text{-}12)$$

and

$$e = \lim_{m \to \infty} \left(1 + \frac{r}{m}\right)^{m/r}$$

we can substitute and get

$$i = e^r - 1 \qquad (10\text{-}13)$$

and

$$r = \ln(i + 1) \qquad (10\text{-}14)$$

This identity relating nominal and effective interest is tabulated in Appendix VII. With this, we say

$$e^{rn} = (i + 1)^n$$

and

$$F = Pe^{rn} = P(1 + i)^n \qquad (10\text{-}15)$$

which is consistent with our earlier findings. Equation (10-15) says where there is no flow of payments as in the situation with equal yearly payments, the compound amount factors are identical.

If the discrete and continuous systems have equal effective interest, each produces the same annual interest payment despite their differences in nominal interest. It is important to realize that tables of interest for continuous compounding are tabulated in two different ways: One uses effective rates of interest, and the other uses nominal rates of interest. Figure 10.3 shows these differences for simple, discrete compounding, and continuous compounding.

The continuous interest model equivalent to Equation (10-9) is

$$P_x = (1 - t)\left[\int_0^N [S(n) - C(n)]e^{-rn}\,dn\right] + t\int_0^N A(d)_n^{-rn}\,dn + F_s(N)^{-rN} \qquad (10\text{-}16)$$

where $S(n)$ = continuous function of the instantaneous value of sales, dollars per unit of time

$C(n)$ = continuous function of the instantaneous value of costs, dollars per unit of time; excludes depreciation and has negative sign

r = interest rate expressed as a decimal, nominal

n = any period, generally $n = N$

N = total life or total periods

$A(d)_n$ = continuous function of depreciation, annual charge

$F_s(N)$ = future salvage value expressed as a function of terminal year N

This model uses continous functions such as $S(n)$, $C(n)$, and $A(d)_n$. If these flow functions can be set up, the algebraic manipulations that can be used in economic

FIGURE 10.3

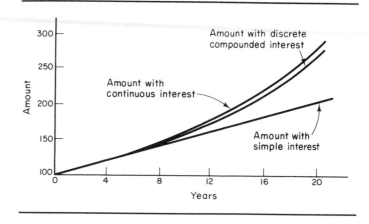

analyses become extended as to form and type. Models that reflect rising wage rates, learning for a new or existing plant, decaying prices and profit margins, and inflationary or deflationary pressures become possible. Along with electronic computation, dynamic application of these models allows for nonstationary estimates. Over the long run, wages (since the early 1930s), most materials, labor cost, and the cost of borrowing have increasing trends. Appendix VIII provides tables for continuous-compounding interest.

In Equation (10-16) the present worth values P_x discounted to zero time continuously by the factor e^{-rn} and then summed over N total periods are given by the integral $\int_0^N [S(n) - C(n)]e^{-rn}\, dn$. The net revenues corrected for taxes thus provide a real-dollar present worth. Additions for depreciation and salvage, a zero-time present worth, $P_{x=0}$, results for a life N. The various terms would appear as given in Figure 10.4.

A depreciation credit does not exist if there is no profit tax as $t = 0$. The depreciation charge is then a part of $[S(n) - C(n)]$. The terminal year, N, for a

FIGURE 10.4

PRESENT WORTH VALUES WITH CONTINUOUS
FUNCTIONS AFTER N YEARS

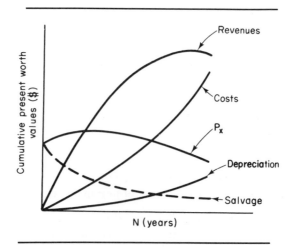

depreciating asset is project life, and the instantaneous depreciation function $A(d)_n$ concludes whenever $n = N$. The depreciation model can be straight line as $A(d) = (P - F_s)/n$, or sum-of-the-year digits $A(d)_n = [2(N - n + 1)/N(N + 1)](P - F_s)$. After the terminal year is reached, the salvage value of the asset and recovery of working capital $F_s(N)$ is shown as a function of N rather than n. $F_s(N)$ is not an integrated function.

The nature of the continuous functions $S(n)$ and $C(n)$ can be constant, linear increasing, or nonlinear, for example. In the case where sales and costs are linear with n, we have $S_0 + g_s n$, where S_0 is an initial value (but a pseudo-value, too, because it is at zero time) and g_s is the annual gradient for sales dollars of income. Costs can be handled in a similar way. For linear increasing sales alone we have $\int_0^N (S_0 + g_s n)e^{-rn}\, dn$ and $S(N) = S_0 + g_s N$. Excluding taxes and depreciation we have

$$S(N) - C(N) = S_0 + g_s N - (C_0 + g_c N)$$
$$= S_0 - C_0 + N(g_s - g_c) \qquad (10\text{-}17)$$

For the simplification of level sales and costs, Equation (10-17) further reduces to

$$S(N) - C(N) = S_0 - C_0 = A \qquad (10\text{-}18)$$

where A = end-of-period uniform cash flow.

In addition to approximations (10-17) and (10-18), Equation (10-16) may be modified by replacing the integrals with summations ($\sum$) as the following example illustrates. A company is considering the funding of R & D, design, construction, and operation of a plant to produce a new product.

The following cost functions are appropriate:

1. Sales = $2500 - 50n$.
2. Operating costs = $1000 + 150n$.
3. Depreciation method = straight line for 25 years.
4. Salvage value = $1500e^{-0.25N}$.
5. Investment cost = $1500.
6. Tax rate = 55%.
7. Earning rate = 12% after taxes.

The management desires to evaluate this proposition and compare it to other alternatives, which are not shown here. The main interest to the company is knowing the useful economic life of the product and operation of the plant. We want to know that optimum period of time that will give a maximum present worth. The problem can be programmed on a computer. A sample FORTRAN program is given by Table 10.4. The value of n is incremented and each term is evaluated. The life of the plant plus equipment is considered fixed at 25 years. The output of the program will give the present worth of the project for each year. See Table 10.5.

There will be two areas of thought that management must consider: (1) the total project dollars in terms of present worth, and (2) the economic or useful life of the project. The total project cost will be that total cost at the end. The economic life is different from the depreciable life.

The depreciation life of 25 years is longer than the economic life of approximately 16 years. The discounted tax credits are sustaining the projects thereafter. Notice also that revenues and costs are about equal at 8 years, obviously not a profitable project, and accelerated write off techniques for depreciation would not obviously improve the picture very much.

TABLE 10.4

FORTRAN STATEMENTS FOR NET PRESENT WORTH

```
*LIST SOURCE PROGRAM
       WRITE(5,13)
   13  FORMAT('RESULTS OF THE PROFITABILITY MODEL')
       WRITE(5,14)
   14  FORMAT('N',5X,'V',8X,'SN',6X,'CN',6X,'DN',5X,'SALN',5X,'BE')
       P=1500.
       T=.55
       SN=.0
       CN=.0
       DN=.0
       SALN=.0
       D03N=1,25
       SN=(1-T)*(SN+(2500.-50.*N)*1.12**(-N))
       CN=(1.-T)*(CN+(1000.+150.*N)*1.12**(-N))
       DN=(P/25.)*((1.12**N-1.)/(.12*(1.12**N)))*T
       SALN=1500.*EXP(-.25*25.)
       V=-P+SN-CN+DN+SALN
       BE=SN+DN+SALN-CN
    3  WRITE(5.15)N,V,SN,CN,DN,SALN,BE
   15  FORMAT(14,F10.2,F8.2,F8.2,F8.2,F8.2,F8.2)
       CALL EXIT
       END
```

TABLE 10.5

SUMMATION OF NONUNIFORM CASH FLOWS

Life	Revenues, $0.45 \sum_{1}^{N} \dfrac{(2500 - 50n)}{1.12^n}$	Costs, $0.45 \sum_{1}^{N} \dfrac{(1000 + 150n)}{1.12^n}$	$\dfrac{0.55(1 - e^{-0.25N})}{N(1.12)^n}$	Total Present Value
1	984	462	29	551
2	1303	674	56	685
3	1339	767	79	651
4	1260	803	100	557
5	1141	808	118	451
6	1015	796	135	354
7	894	775	150	269
8	784	748	163	199
9	685	718	175	142
10	598	685	186	99
11	521	651	195	65
12	454	616	204	42
13	395	581	211	25
14	343	547	218	14
15	298	513	224	9
16	259	480	230	9
17	224	448	234	10
.	.	.	.	.
.	.	.	.	.
.	.	.	.	.
25	69	245	258	82

Inasmuch as $S(n)$, $C(n)$, and $A(d)_n$ are constructed as continuous and if they are also differentiable, we can apply standard calculus operations on Equation (10-16) to uncover the point of maximum present worth. Equation (7-13) or $dR_T/dn = dC_T/dn$ also applies to the evaluation of projects. Each term of (10-16) can be differentiated to give its slope. For instance, with the sales function the slope is $S(N)e^{-rN}$. The maximum present worth is obtained mathematically from setting the derivative of Equation (10-16) with respect to N equal to zero. A solution for N finds optimum and terminal project life. With N determined, it is substituted back into Equation (10-16) to give the maximum value of present worth. Using a $dR_T/dn = dC_T/dn$ operation where n now is years rather than units of production, we obtain

$$\frac{dP_x}{dN} = e^{-rn}\left\{(1-t)[S(N) - C(N)] + tA(d)_n - rF_s(N) + \frac{d[F_s(N)]}{dN}\right\}$$
$$= 0 \qquad\qquad (10\text{-}19)$$

After minor adjustments we have that marginal gains equal marginal losses, or

$$(1-t)[S(N) - C(N)] + tA(d)_n = rF_s(N) - \frac{d[F_s(N)]}{dN} \qquad (10\text{-}20)$$

Each term can be evaluated by classic calculus techniques or with tables, or computer numerical methods can be used. After the terms are evaluated by substitution of various values of N, the optimum N can be found by trial and error values using Equation (10-20). The optimum age for maximum present worth occurs when marginal annual return (after taxes) on the left-hand side with an assumed terminating year N just balances the marginal losses for that same year. The marginal losses amount to interest on salvage value and initial working capital plus the change (loss) in salvage value for that year. Algebraically, the change in salvage value is negative, but the negative sign in the derivative makes the quantity consistent with interest on the salvage.

A sales cash flow from a project endeavor is estimated to have a function $1000 - 20n$ dollars per year, where n is continuous time from start-up. Continuous cost flows are assumed to be $300 + 60n$ dollars per year. Fixed project capital necessary to start, including plant, equipment, and working capital, is \$600, and a recoverable value function of $600e^{-0.3N}$, where N is the anticipated project commercial life is anticipated. The firm has a tax rate of 55%, a policy of a 20% minimum attractive rate of return, and uses constant yearly depreciation. Dollar values are really \$10^6, but this does not affect our problem. Determine the optimum life for this project and maximum present worth. In this example we use continuous discounting with nominal interest. To determine optimum life we establish a gain and loss equation. The intersection of gain and loss gives optimum N. Using $(1-t)[S(N) - C(N)] + tA(d)_n = rF_s(N) - d[F_s(N)]/dN$,

$$0.45[1000 - 20N - 300 - 60N] + \frac{0.55(P - F_s)}{N} = 0.2[600e^{-0.3N}] + 180e^{-0.3N}$$

$$0.45(700 - 80N) + \frac{0.55}{N}(600 - 600e^{-0.3N}) = 300e^{-0.3N}$$

$$315 - 36N + \frac{330}{N}(1 - e^{-0.3N}) = 300^{-0.3N}$$

A summary of this final line would give the following for values of $N = 1, 2, 4, 6, 8,$ and 10:

Year	Net Gain	Net Loss
1	365	222
2	317	165
4	229	90
6	145	51
8	65	27
10	−12	15

FIGURE 10.5

MARGINAL GAINS AND LOSSES

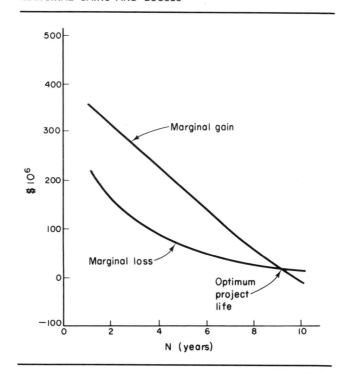

This plot is shown in Figure 10.5, and optimum intersection exists whenever net gains = net losses, or approximately 9 years. It is now possible to find the optimum and maximum amount of present worth using Equation (10-16). Ordinary exponential functions are additional tables that can be used to solve these forms.

$$P_{x=0} = (1 - 0.55)\left[\int_0^9 (700 - 80n)e^{-0.2n}\,dn\right]$$

$$+ (0.55)\frac{(330)}{9}\int_0^9 (1 - e^{-2.7}) + e^{-0.2n}\,dn = \$9.2 \times 10^6$$

This is a real flow of dollars inasmuch as taxation enters into the calculation.

10.3-8
Alternative Selection Models

The net present worth and rate of return models can be used for project estimates in making a decision among competing alternatives that provide identical services. However, the selection among alternatives for different duration lives or service calls for modifications of the basic models. The nature of the replacement may be like the existing asset, but the type of service is expected to linger beyond the economic life of the equipment. Consider now a discrete model that considers profit taxes and tax credits in replacement analysis or

$$P = P_I + (1 - t) \sum_{n=1}^{n=N} \frac{[C(n) - S(n)]}{(1 + i)^n} - t \sum_{n=1}^{n=N} \frac{A(d)_n}{(1 + i)^n} - \frac{F_s(N)}{(1 + i)^N} \qquad (10\text{-}21)$$

where P = present worth of all net costs for one project having a life N

P_I = initial investment sum

$C(n)$ = costs (excluding depreciation) for nth period, either nonuniform or average

$S(n)$ = end of period savings or credits for nth year, either nonuniform or average

N = total life, years

n = end-of-year age

$A(d)_n$ = depreciation charge for nth year, either nonuniform or average

This equation allows a tax credit for depreciation but a tax charge on savings or profits and makes P the present worth of net cost. The net cost of the service must be negative or zero. The terms $[C(n) - S(n)]/(1 + i)^n$ must be greater negatively than the absolute value of $A(d)_n/(1 + i)^n$ in any 1 year for the tax to apply. For actual problems the most negative of the choices is the best because the savings are negative in sign. The profitability model, Equation (10-16), determined the rate of return for $P_x = 0$. In model (10-21) an assumption regarding the interest rate is made, and for alternative comparisons equal N's are set. With two alternatives have differing lives, it is incorrect to compare the present worth involved. However, if a present worth analysis is desired, the analysis can be used if each asset is renewed at the end of its life until the elapsed time is the same for both assets. For instance, if $N_1 = 2$ years and $N_2 = 3$ years, the common points of renewals would be $3N_1 = 6 = 2N_2$. Thus N_1 has three renewals, while N_2 has two renewals for a common terminal life. If in Equation (10-21) we substitute integrals for summation signs and e^{-rn} for $(1 + i)^n$, the model becomes continuous compounding.

An analyst has two competing proposals for machine installation which will provide identical services. The Alpha machine costs $20,000 and uses $36,000 for direct labor. Other direct annual costs are 10% of the investment. Equipment of this sort is currently bringing $1200 for salvage, and this should not change 10 years hence, which is the anticipated project life. A second proposal, Beta, has a first cost of $35,000, annual labor cost requirements of $22,000, direct annual costs of $6000, and salvage value estimated as $2000. The project life is 10 years for Beta. This firm has a rate of return of 20% before taxes and 12% after taxes. The corporate tax rate is 40%. Depreciation will be considered straight line. Which is the preferable installation? Equation (10-21) can be simplified to

$$P = I + \sum_{n=1}^{n=N} \frac{C(n) - S(n)}{(1 + i)^n} - \frac{F_s(N)}{(1 + i)^N} \qquad (10\text{-}22)$$

to give a before-tax comparison. A comparison of present worth for two proposals is simple inasmuch as annual costs are constant at C_0 and the second term involving the summation of $(1 + i)^{-n}$ become $[(1 + i)^n - 1]/[i(1 + i)^n]$ for which tables are available in the Appendix.

For the Alpha installation,

$$P = 20,000 + \sum_{n=1}^{n=10} \frac{[36,000 + 2000]}{1.20^n} - \frac{1200}{1.20^{10}}$$

$$= 20,000 + 38,000(4.192) - \frac{1200}{6.192} = 20,000 + 159,296 - 194$$

$$= \$179,102$$

For the Beta installation,

$$P = 35,000 + \sum_{n=1}^{n=10} \frac{[22,000 + 6000]}{1.20^n} - \frac{2000}{1.20^{10}}$$

$$= 35,000 + 28,000(4.192) - \frac{2000}{6.192} = 35,000 + 117,376 - 323$$

$$= \$152,053$$

The present worth cost of the Beta installation is less by about $27,000. This is a cost evaluation and is a positive number. Continuous functions can be substituted, but the decision would not be altered, although different values would be found. It is seen that the salvage value really contributes very little to the decision. This is due in part to the relatively high rate of interest and the number of years. Future money at a distant time affects the decision only insignificantly. The precision for which these salvage sums must be estimated need not be as good as initial costs or early operating costs.

10.3-9
Criteria for Making Project Decisions

We have suggested present worth and the discounted rate of return as the methods most generally acceptable for profitability evaluation and the economic comparison of alternatives. Although other time-value methods give commensurable results, they were not discussed in view of many excellent texts which devote more space to engineering economics. Now we shall discuss the making of decisions after the project analysis is complete and calculations such as the rate of return or present worth are available.

As projects entail capital, we might belatedly ask: "What is capital?" From the standpoint of economic theory, capital is one of the broad classes of productive factors—a collection of stored services which are held for future rather than present consumption. It includes practically every accounting asset—land, buildings, machinery or equipment, inventories, receivables, and cash. We also include various kinds of intangibles—patents, franchises, and good will. The real assets can be handled in terms of their profitability as an alternative use of funds. Every kind of corporate liability or aspect of net worth (loans, equity funds, and so forth) can be handled in terms of its cost as a use or source of funds. This principle prevents the necessary detail about each type of financial instrument or source from obscuring the principles that are common to all sources.

Compensation for capital use is important for each financial decision involving a time period. Whether the compensation is called discount, interest, dividends, or premium, the difference between what is received (or invested) now and what is to be paid (or received) later is the cost of capital. There are two areas in which this concept may be applied: (1) the making of decisions within the firm as to how resources should be applied—which projects ought to be preferred to others, and (2) deciding what outside "sources" (whether loans, equity and amounts of contracts) should be used to provide the services, facilities, or money needed to operate the business. In project estimating we are more interested in the first one. Finance is more interested in the second one.

The application of the present worth method requires a value for the interest rate. This rate, sometimes called a minimum attractive rate of return (MARR), is crucial to the evaluation process. The lower this value, the more beneficial challengers appear and vice versa.

The policy that sets a return rate that serves as the floor level varies with the circumstances. An ambitious company has more demands for capital than it may have available. The supply of money for capital goods may come largely from the plowback of earnings, some from the liquidation and depreciation of equipment, or it may include money borrowed from the banks or new equity capital from issuance of stock. At any single time the supply of capital money available for investment is limited compared to the demands.

Some analysts prefer to understand a minimum attractive rate of return based on this rationale. If capital is placed in conservative investments where there is virtually no risk, an annual return of 3 to 5% might be obtained. Municipality bonds are capable of this rate. Investment in certain high-grade bonds might increase this yield to 4 to 6% for a somewhat greater risk and uncertainty, as the market price for bonds may vary. High-grade common stocks, on the other hand, may yield anywhere from 4 to 15% annually, with widely varying possibilities as to the stability of market price and future increase in value. Despite these types of investment businesses work under the principle of obtaining considerably higher rates of return than those already cited, knowing full well there are greater risks and uncertainties. Thus, business, in an effort to increase its profitability, follows a less conservative attitude in investments. In certain circumstances some companies anticipate a rate of return greater than 9 or 10% on an after-tax basis. If this is related to a before-tax basis, the required rates of return on ventures with typical risks are likely to be in the range of 20% or higher.

Assume that a new project promises to earn 22% on invested capital, which will assuredly increase the percentage of the firm's profit. On the other hand, another potential project promises to earn only 18%, which will reduce the firm's percentage profit. Is one considered more desirable than the other? This statement implies that the minimum rate of return is related to profit and loss statements that are established by accountancy. This logic suffers from the misunderstanding of the pursuits of the accountant and the project analyst. The accountant is interested in profit earned during past periods after income and expenses are known. Using his procedures he is able to calculate the results of operations and determine what the return on capital was. Nowhere in the accounting process is a return added to the costs or expenses. On the other hand, the project analyst is interested in the profitability of proposed operations. His practices lead to a minimum attractive rate of return to ensure profit. Hopefully, an optimized profit occurs. The project analyst is concerned with estimating the profitability of each dollar before an investment is made. To acknowledge that the minimum attractive rate of return is judged by the balance sheet or the income statement is inferior logic. Yet in total perspective, the criterion that judges the merits of new projects cannot overlook the profits previously earned. We need to be sure that the cost of the capital figure is one that does test proposals. This rate should be at least as high as the after-tax weighted-average cost of capital inflow to assure that proposals fit the general scheme of continuing operations and growth. Also, it should be as high as the internal rate currently earned by projects or activities with comparable risk. Management should therefore maintain some check to see how the various projects and risks tie into the desired overall weighted-average return. Table 10.6 is a collection of some discount rates but is by no means complete in the specific case.

In summary the calculation of the rate of return serves four essential functions. It provides

1. The price and cost of long-term capital.
2. The means of valuation of income and consumption at different points in time.
3. As a way of compensating for the return for taking risks.
4. As one criterion to evaluate engineering projects and designs.

TABLE 10.6

LIST OF DISCOUNT RATES USED

Classification	Various Discount Factors
Value analysis and engineering design	10–20%
Expansion of existing processes	20–40%
New processes, plants, inventions	30–100%+
Tennessee Valley Authority	$4\frac{1}{2}$%
Department of Defense:	
Shipyard projects	10%
Air stations	10%
Corps of Engineers	$3\frac{1}{4}$%
Department of the Interior	$2\frac{1}{2}$–12%
Public and politically inspired programs	-5–$5\frac{1}{2}$%
Public transportation program	5–15%

10.3-10
Tables and Formulas

Inasmuch as space does not allow full development of interest formulas, a summary of the standard engineering economy formulas along with factor notation are given in Appendix V. Interest rate factors are given for i and $r = 5$, 10, 20, and 30% in Appendixes VI and VIII. Additionally, a funds-flow table, Appendix IX, assuming a periodic amount of money flowing uniformly and continuously, or $\bar{A}$, is included. Instructions for its use should be clear from the table of formulas.

Each time a formula is used, it need not be calculated; rather table values expedite the solution. The convention for using these tables and factors is illustrated by this example. Find the future sum of a present investment of $1000 for 5 years and an interest rate of 10%. Using $F = P(1 + i)^n$, $F = 1000(1.10)^5 = 1000(1.611) = \$1611$. The Appendix interest factors are developed according to functional form and allow the single-payment compound-amount factor, which is designated $(^{F/P\,i,n})$, to be placed within the parenthesis or $F = P(^{F/P\,i,n})$ $= \$1000 \overset{F/P\,5\%,\,10}{\underset{(1.611}{}}{}_{)} = \$1611$. Several easy problems encourage this practice.

10.4
A PROJECT-ESTIMATING MODEL FOR PLANT INVESTMENT

So far methods that are germane to the analysis of future investments have been studied separately. Now several techniques will be composed into a model that is suitable for new or expanding plants where anticipated capital requirements are large. Details include the plant-sizing model and marginal strategies of Chapter 7, the forecasting linear demand model of Chapter 5, and the present worth of a sequence of investments.

The power law and sizing model may be rewritten as

$$F = I_r \left(\frac{Q}{Q_r}\right)^m \tag{10-23}$$

where F = future undiscounted sum of investment

I_r = known plant investment for known capacity Q_r

Q = future production capacity expressed as rate per time in units compatible to Q_r

m = empirical exponent, $0 < m \le 1$

The most elementary model for forecasting plant requirements is given by

$$D = bn_j \tag{10-24}$$

where D = yearly demand for product
$\quad b$ = slope
$\quad n$ = elapsed years
$\quad j = 0, 1, 2, \ldots$ expansion steps

Starting with zero, as now, presumes no existing demand inasmuch as a constant intercept for the vertical axis is overlooked. Slope b represents the increasing demand rate. If demand is linear-increasing and no end is in sight, the question arises how future investments to increase capacity can best satisfy these needs.

For the problem at hand the continuous-interest discrete-payment model for present worth best satisfies circumstances of this sort, so let

$$P = Fe^{-nr} \qquad (10\text{-}25)$$

where P = present amount of money
$\quad r$ = nominal continuous interest rate

We let this rate equal the minimum acceptable value, which discounts the future amount of money.

If there are several investments over a long period of time, the sum of these present values is

$$\sum_{j=1}^{\infty} F_j e^{-n_j r} \qquad (10\text{-}26)$$

This represents the stream of indefinite plant investments. This formula implies that plant expansions occur every so often or whenever $j = 1, 2, \ldots$, although they may not be on a regular yearly basis. Additional plant capacity is necessary whenever

$$n_j = \frac{1}{b} \sum_{K=1}^{j-1} Q_K \qquad (10\text{-}27)$$

where K = increasing steps of product throughput demand in the plant.

The increasing throughput $K = 1, 2, \ldots, j - 1$ continues until the time for plant capacity enlargement. This is described by Figure 10.6.

FIGURE 10.6

THE LINEAR INCREASING DEMAND REQUIRES
ADDITIONAL PRODUCTION. THE COST OF
$(Q_1 - Q_r) = I_r (Q/Q_r)^m$

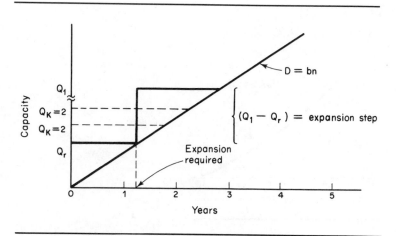

As demand is linear-increasing, the solution to expand is similar each time the problem arises. Therefore the time for the jth expansion is given by

$$n_j = (j - 1)\frac{Q}{b} \tag{10-28}$$

It is necessary, of course, that Q be an identical value because of linear demand.

Making a substitution, we now have

$$P_t = \sum_{j=1}^{\infty} I_r\left(\frac{Q}{Q_r}\right)^m e^{-r(j-1)Q/b} \tag{10-29}$$

where P_t = total present worth of a series of investment sums. As Equation (10-23) is a constant, we have

$$P_t = I_r\left(\frac{Q}{Q_r}\right)^m \sum_{j=1}^{\infty} e^{-r(j-1)Q/b} \tag{10-30}$$

and the infinite series $\sum_{j=1}^{\infty} e^{-r(j-1)Q/b}$ converges to the term $1/(1 - e^{-rQ/b})$. Next,

$$P_t = I_r\left(\frac{Q}{Q_r}\right)^m \frac{1}{1 - e^{-rQ/b}} \tag{10-31}$$

If an operation of $dP_t/dQ = 0$ is undertaken, the resulting solution for Q is that point where the marginal yield is neither rising nor falling, and the optimum value Q^* is reached, or

$$\frac{dP_t}{dQ} = \frac{-rQ^*e^{-rQ^*/b}}{b} + m(1 - e^{-rQ^*/b}) = 0 \tag{10-32}$$

This equation when plotted for values of rQ^*/b and m gives Figure 10.7. It is

FIGURE 10.7

OPTIMUM CAPACITY EXPANSION Q AS A
FUNCTION OF m

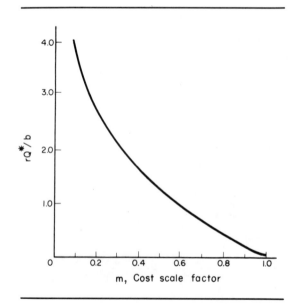

seen for higher values of m that the parameter rQ^*/b decreases. As $m \rightarrow 1$, smaller-capacity expansion steps are necessary, and if $m = 1$, there is no economic advantage to a larger system. Conversely, as $m \rightarrow 0$, expansion steps should be larger and more infrequent. There is a savings to expand on a larger scale rather than to defer the expansion until later. Capacity is found as

$$Q^* = \frac{b}{r} \qquad (10\text{-}33)$$

The previous discussion dealt with demand starting at some time $n_j = 0$, for which a hypothetical consumer demand was supposed to start. This situation is true where research and development has inaugurated a product, invention, or service that has not previously existed. This may also result from purchasing having procured the items. Now look at the problem of accommodating a forecast of an unsatisfied initial demand D_0 and a linear growth according to a time series such as $D = D_0 + bn_j$. In an original manufacturing operation with a capacity of $Q_1 > D_0$ the new investment problem is the anticipated enlargement at time n_2 which will occur whenever

$$n_2 = \frac{Q_1 - D_0}{b} \qquad (10\text{-}34)$$

where Q_1 = existing capacity.

This is demonstrated by Figure 10.8. Thus the problem reverts back to

FIGURE 10.8

EXPANSION PLAN FOR INCREASING LINEAR
DEMAND OF A PRODUCT

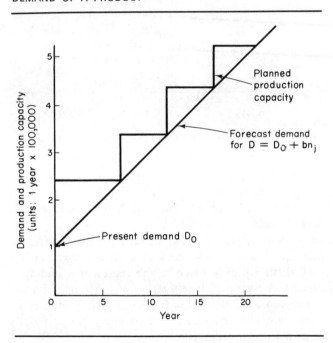

the earlier conclusions that after the initial system is no longer able to meet the demand, the enlargement capacity is given by Q^* using Figure 10.7 and the previous models. But what about investment for capacity Q_1? The present worth P_t of a sequence of investments is now given by

$$I_r\left(\frac{Q_1}{Q_r}\right)^m + e^{-(Q_1-D_0)r/b}\left[\frac{I_r(Q^*/Q_r)^m}{1-e^{-rQ^*/b}}\right] \qquad (10\text{-}35)$$

This first term on the left is the investment in the initial process of capacity Q_1 and the term on the right is the present worth of future expansions. Taking $dP_t/dQ_1 = 0$, the following expression for optimum initial capacity Q_1^* is found:

$$\left(Q_1^*\frac{r}{b}\right)^{m-1}e^{Q_1^*r/b} = \left(\frac{Q^*r}{b}\right)^{m-1}e^{(D_0+Q^*)r/b} \qquad (10\text{-}36)$$

In examining Equation (10-35), the factor $(Q_1^* - D_0)r/b$ can be defined as an *overdesign* factor which allows for future expansion. The dimensionless initial demand of D_0r/b and m are parameters. Figure 10.9 shows this relationship.

FIGURE 10.9

OVERDESIGN FOR THE INITIAL AND LINEARLY INCREASING DEMAND. (Following Dale F. Rudd, Charles C. Watson, *Strategy of Processing Engineering*, John Wiley & Sons, Inc., New York, 1968)

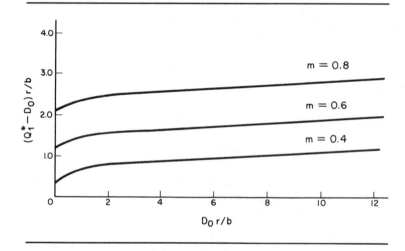

Your company has the requirement for 100,000 devices per year. A linear increase of 20,000 devices per year is forecast for the foreseeable future. It has been proposed that the company develop a facility which will produce these devices to meet all in-house needs.

Your task is to determine the proper initial capacity for the facility, the optimum incremental increase in capacity, and the time interval between expansions so that the sum of all the investment costs will represent a minimum present worth of investment. A historical investigation of other facilities has disclosed that the initial plant cost and the cost of future expansions follows the six-tenths rule. The return on capital invested in the company has previously shown a return of 20 cents per dollar on investments.

For all incremental increases in production after the initial construction, the problem is essentially that of the no initial demand situation. The present value of an expansion is

$$P_t = I_r\left(\frac{Q}{Q_r}\right)^m\left(\frac{1}{1-e^{rQ/b}}\right)$$

where the quantities are as previously defined. After the dP_t/dQ operation, the function

$$\frac{-rQ^*}{b}e^{-rQ^*/b} + m(1 - e^{-rQ^*/b}) = 0$$

is used. If rQ^*/b is solved for $m = 0.6$ and Figure 10.7 is used,

$$Q^* = \frac{0.95(20,000)}{0.20} = 95,000 \text{ units}$$

The present value of the initial investment plus the step increases is found by using Equation (10-35).

Using the result from the dP_t/dQ_1 operation, and when $m = 0.6$ and $D_0 r/b = 100,000(0.2)/20,000 = 1.0$,

$$(Q_1^* - D_0)\frac{r}{b} = 1.4$$

$$Q_1^* = 1.4\frac{b}{r} + D_0 = 1.4\left(\frac{20,000}{0.2}\right) + 100,000 = 240,000$$

The optimal arrangement would have:

1. Initial production capacity $Q_1 = 240,000$ units per year.
2. First expansion at $n_1 = (240,000 - 100,000)/20,000 = 7$ years.
3. Expansion increment $Q^* = 95,000$ units.
4. Period between expansions $Q^*/b = 95,000/20,000 = 4$ years, 9 months $\doteq 5$ years.

The results are shown graphically by Figure 10.8.

Some summary comments should be addressed to the procedure outlined here. It is admitted that the scheme is sharp-shooting the bull's-eye with a shotgun. This is not intended as an apology for the model; rather the reader is reminded that the technique is one of forecasting. Judgment is inescapable for the estimator, as he must submit project estimates for future plants which are dictated by the size. This model does not consider the effects of inflation on construction costs, nor does it anticipate the learning processes that would contribute to smarter manufacturing and increased production over the interval of time.

10.5 DECISION TREE MODEL

A technique,[3] here called decision tree, examines the effects of future alternatives and possible outcomes and decisions which result from an initial or present decision. This tool enables the project estimator to make an initial decision which includes consideration of risk. The title *decision tree* arises from the graphic appearance which shows branches for each possible alternative for a given decision and branches for the possible outcomes which result from the alternatives. Two examples will help in understanding decision tree analysis.

A fundamental form of the decision tree occurs whenever each alternative has a payoff income that is a certain outcome; i.e., the probability of expected income equals 1. As an illustration consider the replacement problem given by Figure 10.10. The problem is to replace an old machine with a new machine. This is not a one-time decision but is one that recurs periodically up to some time

---

[3]For more information about decision trees, see John F. Magee, "Decision Trees for Decision Making," *Harvard Business Review*, July, Aug. 1964, pp. 126–138.

FIGURE 10.10

REPLACEMENT EXAMPLE WITH INTEREST = 0%

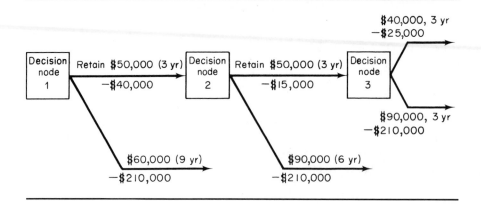

horizon. If the decision is to retain the old machine at decision node 1, then at a later decision node 2 the choice has to be made over again. Whatever the decision at node 2, the problem arises again at node 3. The graphic portrayal of the decision tree is now examined. Note that for each alternative a cash inflow is shown on top of the arrow and the cash investment required is shown below the arrow. For example, at decision node 3, if the decision is to retain the old machine, then an income of $40,000 for each of 3 years would be required. The renewal cost of $25,000 is necessary to permit continued production with the equipment.

In the usual situation the investment alternative calls for a decision at node 1, and little if any thought is given to downstream effects on this initial decision. To make a better choice for decision node 1 it is necessary to reflect on later alternatives and decisions which stem from this early one. Because of the deterioration and the passage of time, net revenue earned by a piece of equipment becomes less and less. To counteract this decay of productivity, a series of decisions of whether to keep an existing asset or purchase a replacement must be made from the set of future revenue and cost predictions for existing and subsequent replacements.

It is possible to derive a sequence of replacement decisions that maximizes total profit. The example problem considered here has a planning horizon of 9 years, and the decision sequence requires that the question of whether to purchase or to retain existing equipment be made every 3 years. The procedure to follow in decision tree analysis starts at the most distant decision node and determines the best alternative through a numerical study of that alternative and then "rolls back" to each earlier successive decision node until the initial or present decision node is determined. By following this procedure one is able to make a present time decision by considering the alternatives and expected decisions of the future. In actuality it is only decision 1 that is made. Presumably when 3 years pass, a new analysis will study the scene at that time.

The replacement problem posed by Figure 10.10 has interest equal to 0% and analysis is given by Table 10.7. When neglecting the time value of money the dollar has the same value regardless of the year in which it occurs. Note that the quest for one or the other alternative at decision node 3 leads to retaining the current machine as 3 × $40,000 = $120,000 minus the investment costs of $25,000 = $95,000. This exceeds the value of the purchase alternative of new equipment which is 3 × $90,000 − $210,000 = $60,000. Inasmuch as the retain option is greater than the purchase option the decision at decision node 3 is "retain." In a similar fashion the choice is made at decision node 2, and finally

Retain present investment

Purchase new investment

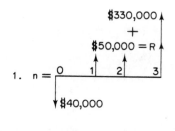

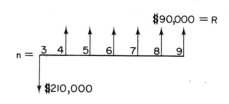

3.

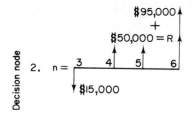

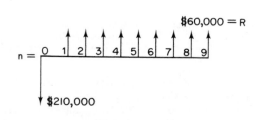

2.

Decision node

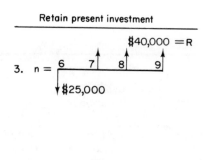

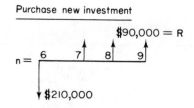

1.

the decision at decision node 1 is evaluated. Figure 10.11 is a graphic solution of the investment problem for three decision nodes. The upward arrow is designated as revenue, while the downward arrow is cost for the investment opportunity. Notice that the larger cash flow at decision node 3, $95,000, is used as a supplement for decision node 2 as the $95,000 is rolled back to decision node 2. Likewise at decision node 2 the purchase option and a cash flow of $330,000 is used as input to year 3 for decision node 1. The decision tree is analyzed on the tabular approach as shown in Table 10.7. The criterion used to choose between retaining or purchasing is the maximum cash flow at the decision node. The "a" denotes the chosen decision.

When one considers the time value of money, say 25%, a different set of decisions is uncovered. The decision tree approach is the same, which involves working from the most distant decision node to the earliest decision node. The simplest way to account for the timing of money is to use the present worth approach and discount all monetary income back to the decision node. Table 10.8 now shows the net cash flow and the decision table where interest is equal to 25%. When considering the effect of interest by calculating present worth at each decision, it is seen that the decision is to retain the present equipment at decisions 1, 2, and 3. This result is not surprising since high interest rates favor alternatives of low initial investments and give less weight to long-term returns.

TABLE 10.7

NET CASH FLOW AND DECISION TABLE
WITH INTEREST = 0%

Decision Node	Investment Alternative	Cash Inflow Per Year	Years	Cash Inflow	Maximum Previous Monetary Decision	Total Cash Inflow	Investment Opportunity Cost	Cash Flow
3	Retain	$40,000	3	$120,000	0	$120,000	$ 25,000	$ 95,000[a]
3	Purchase	90,000	3	270,000	0	270,000	210,000	60,000
2	Retain	50,000	3	150,000	$ 95,000	245,000	15,000	230,000
2	Purchase	90,000	6	540,000	0	540,000	210,000	330,000[a]
1	Retain	50,000	3	150,000	330,000	480,000	40,000	440,000[a]
1	Purchase	60,000	9	540,000	—	540,000	210,000	370,000

[a]Indicates the decision.

TABLE 10.8

NET CASH FLOW AND DECISION TABLE
WITH INTEREST = 25%

Decision Node	Investment Alternative	Cash Inflow Per Year	Years	Present Worth Factor[a]	Present Worth Cash Inflow	Maximum Previous Decision Present Worth Factor[a]	Cash Inflow	Present Worth Cash Inflow	Total Cash Inflow	Investment Opportunity Cost	Cash Flow
3	Retain	$40,000	3	P/A(1.95)	$ 78,000	—	$ 0	$ 0	$ 78,000	$ 25,000	$ 53,000[b]
3	Purchase	90,000	3	P/A(1.95)	175,500	—	0	0	175,500	210,000	−34,500
2	Retain	50,000	3	P/A(1.95)	97,500	P/F(0.512)	$53,000	$27,135	124,635	15,000	109,635[b]
2	Purchase	90,000	6	P/A(2.95)	265,500	—	0	0	265,000	210,000	55,000
1	Retain	50,000	3	P/A(1.95)	97,500	P/F(0.512)	109,635	51,700	149,200	40,000	109,200[b]
1	Purchase	60,000	9	P/A(3.47)	208,200	—	0	0	208,200	210,000	−2,000

[a]Key: $P/A = $ uniform series $= \dfrac{(1 + i)^n - 1}{i(1 + i)^n}$.

$P/F = $ single payment $= \dfrac{1}{(1 + i)^n}$.

[b]Indicates the decision.

10.6
LEASING VERSUS PURCHASING OF ASSETS

In recent years there has been an increase in the practice of leasing (renting) rather than purchasing many types of assets. Machine tools, automobiles, and buildings (almost any type of asset that can be purchased can be leased) are involved in lease arrangements. Generally the reasons given to lease rather than to buy are the following: (1) A lease may provide 100% financing and involve neither a down payment (customary with sales contracts) nor the compensating bank deposit usually required in the case of a loan; it therefore eliminates an initial drain on the company's working capital. (2) The lease term is relatively long. According to some authorities the lease averages 5 years in contrast to average loan terms of 3 years or less. With the payments being extended into the future, current expenditures are minimized. (3) There may be a savings in maintenance costs through leasing.

A number of studies have shown that there is no income tax advantage to leasing. This is true when accelerated methods of depreciation are used, for the owner can charge no more for depreciation than can the renter of the asset. If assets are leased, the annual lease payments are deducted in computing income tax. If the assets are purchased, the annual depreciation is deducted.

Inasmuch as leasing is only one of several ways of obtaining working capital, the decision to lease should consider the cost of obtaining capital by all methods. Borrowing capital does establish certain fixed obligations, however, and some leases cannot be canceled without costly penalties.

As an analysis of discounted cash flow for the leasing question, revisit a previous example. For the first cost of $180,000, a comparison is desired between three types of asset financing: purchase, bank loan, and leasing. From previous work it was discovered that the effect of the salvage value was minimal under present worth, and it will be ignored here. Additionally, a rate of return of 12% will be accepted as the minimum rate of return, which incidentally is near the average net profit on working capital among all types of business. The $5000 capital required to install can be financed by a leasing or bank loan arrangement. Therefore, the problem can be stated as follows:

Asset Cost: $180,000. Life is 10 years. The depreciation method is sum-of-the-year digits. There is no salvage value as the asset is retired to less productive service. The composite tax rate is 55%, and the minimum rate of return is chosen as 12%.

Lease Cost: $20 per month per $1000 of equipment cost, no advance payment. Leasing charges are on an annual basis for 10 years.

Bank Loan: Five-year loan at 6% interest; a 20% compensating bank balance is required. There are to be annual payments.

The solution in tabular form is given by Tables 10.9, 10.10, and 10.11. Figure 10.12 summarizes the competition and plots "cumulative cash flow discounted to present worth dollars" against years. For these data, it is seen that capital is conserved during the first $6\frac{1}{2}$ years by leasing over purchasing but that the differences to a bank loan balance at about 2 years. On a long-term basis the cash purchase option is more favorable than leasing for the way in which this problem has been formulated.

TABLE 10.9

LEASE CASH FLOW

Year A	Rental B	Tax Savings from Rental Expenses 45% of (B) C	Single Payment Present Worth Factor	Net Cash Flow (B − C)	Cumulative Net Discounted Cash Flow Cost (D × E)
0	$43,200	$19,440	1.000	$23,760	$ 23,760
1	43,200	19,440	0.893	23,760	44,977
2	43,200	19,440	0.797	23,760	63,914
3	43,200	19,440	0.712	23,760	80,831
4	43,200	19,440	0.636	23,760	95,942
5	43,200	19,440	0.567	23,760	109,414
6	43,200	19,440	0.507	23,760	121,461
7	43,200	19,440	0.452	23,760	132,200
8	43,200	19,440	0.404	23,760	141,799
9	43,200	19,440	0.361	23,760	150,377
10	—		0.322		

FIGURE 10.12

YEAR-BY-YEAR COMPARISON BETWEEN THREE
METHODS OF ASSET FINANCING

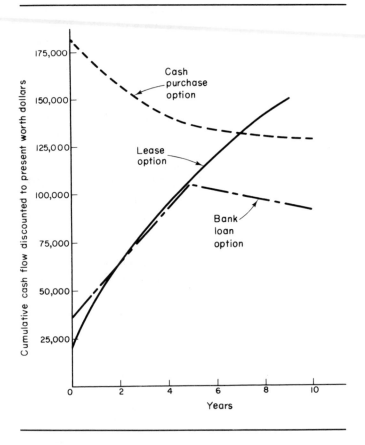

TABLE 10.10

PURCHASE CASH FLOW

Year	Depreciation (Expenses) G	Tax Savings from Depreciation (45% of G)	Cumulative Net Discounted Cash Flow Cost ($180,000 − ∑ DH)	Cash Flow Advantage: Lease Over Purchase
0	—	—	$180,000	$156,237
1	$31,818	$14,318	167,214	122,237
2	28,636	12,886	156,944	93,030
3	25,454	11,454	148,789	67,958
4	22,273	10,023	142,414	46,472
5	19,091	8,591	137,543	28,129
6	15,909	7,159	133,913	12,452
7	12,727	5,727	131,325	−875
8	9,545	4,295	129,590	−12,209
9	6,364	2,864	128,555	−21,822
10	3,183	1,432	128,094	−22,283

TABLE 10.11

BANK LOAN CASH FLOW

Year	Payments K	Depreciation L	Average Interest Charge M	Tax Savings $(45\%$ of $L+M)$	Undiscounted Cost $(K-N)$	Cumulative Net Cost Cash Flow $(D \times L)$	Cash Flow Advantage: Lease Over Bank Loan $(M-F)$
0	$36,000	—	—	—	$36,000	$ 36,000	$12,240
1	34,128	$31,818	$6,328	$17,165	16,963	51,147	6,170
2	34,128	28,636	6,328	15,733	18,395	65,808	1,894
3	34,128	25,454	6,328	14,301	19,827	79,925	−906
4	34,128	22,273	6,328	12,870	21,258	93,445	−2,497
5	34,128	19,091	6,328	11,438	22,690	106,310	−3,104
6	—	15,909	—	7,159	−7,159	102,681	−18,780
7	—	12,727	—	5,727	−5,727	100,092	−32,108
8	—	9,545	—	4,295	−4,295	98,357	−43,442
9	—	6,364	—	2,863	−2,863	97,323	−53,054
10	—	3,183	—	1,432	−1,432	96,862	−53,515

10.7 A PROJECT-ESTIMATING EXAMPLE

While a list of steps for the estimating of investment cost for engineered projects could be presented now, we prefer to narrate an example starting with the general problem statement and proceeding to the last and the smallest time-consuming step—the arithmetic of determining the rate of return, which will be left up to you. This example, admittedly a condensed version of the real thing, points up the importance of design, information, and knowledge in technique.

A uranium ore body has been discovered and has been assayed as to quality of U_3O_8. The firm chooses to evaluate potential designs, construction, start-up, and operation of a mill to refine the ore. Beginning with preliminary designs, several rough flow sheets are made, each being estimated along the way. The flow sheet establishes an investment and operating cost for milling the ore. While operating costs are affected by the grade, mill capacity, and reagent consumption, these, in turn, are originally specified by a preliminary flow sheet design.

One flow sheet is rated as follows:

Item	Flow-Sheet Design
Tons per year processed	612,000
Tons of ore processed daily	1,750
Pounds of U_3O_8 produced/year	1,513,000
Pounds of U_3O_8 produced/day	4,320

The company researches the literature and has experience in design, construction, and operation of uranium plants. A previous experience with a smaller plant having a throughput of 500 tons per day cost $5,500,000 for capital. The power law and sizing model, thought appropriate, is applied first. Using Equation (7-3), the preliminary investment cost is $C = \$5,500,000(1750/500)^{0.45}$ and $C = \$9,700,000$, where exponent $m = 0.45$ for this kind of plant. The operating expense, estimated to be 20% of the capital cost, results in $1,940,000 per year or $3.17 per ton. The sale price of U_3O_8 estimated at a time during normal operation 4 to 5 years hence is $6 per pound. Total revenue is about $9,000,000 per year.

If the processor does not own the ore pit, he must buy the ore. In our case we assume that the ore costs $1 per ton and that hauling 2 miles increases the cost to $1.25 per ton. As the plant does not own the mine, the plant cannot recover the depletion allowance. Plant salvage is assumed to be equal to zero inasmuch as the ore body will be exhausted on the last day of depreciation, which simplifies early calculations. The start-up working capital required will be ignored, even though a residual remains at end of life.

With the delivered cost of raw material as $1.25 per ton estimated or agreed to between the mine company and the processor, calculations can proceed.

Model (10-9) can be used to find the rate of return:

$$9,700,000 = (1 - 0.55)\left[\sum_{n=1}^{10} \frac{(9,000,000 - 2,760,000)}{(1 + i)^n}\right] + 0.55\left[\sum_{n=1}^{10} \frac{970,000}{(1 + i)^n}\right]$$

A trial and error approach shows that the rate of return is about 30 to 35% after taxes.

With this preliminary return, the firm chooses to proceed. The next step begins with the design, and a flow sheet as given by Figure 10.13 is developed.

FIGURE 10.13

U_3O_8 DESIGN FLOW SHEET

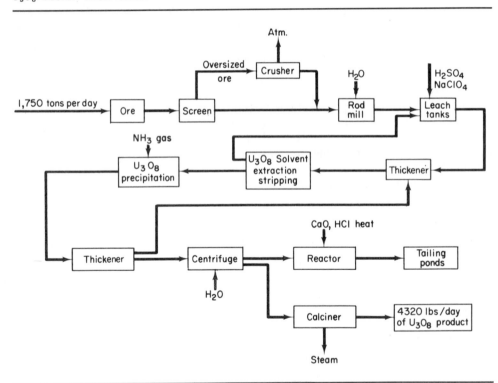

Next, major equipment is sized for material balances, horsepower requirements, flow rates, and so forth. Armed with this knowledge, an equipment list is determined and submitted for quotation externally or estimated internally. Table 10.12 shows the equipment list, quantity, size, and cost found through comparison pricing and external quotation. The total delivered and installed cost is found to be $C_e = \$1,700,000$. Next, factors are established that set other costs on the basis of the installed major equipment cost. These factors may be public

TABLE 10.12

SUMMARY OF ESTIMATING DATA
FOR MAJOR EQUIPMENT

Equipment	Quantity	Size	Major Equipment Cost
Conveyor	6	Miscellaneous length	$120,000
Ore crusher	1	735 tons/day	192,500
Leaching tanks	4	400 gallons per minute	270,000
Thickener tanks	7	1750 tons/day	230,000
Rod mill	1	1750 tons/day	128,525
Crushed ore calciner	1	0.2 tons/day	17,500
Steam generator	1		27,500
Tanks	20	Various	115,000
Water facilities and tailing pond	1	1750 tons/day	188,200
Strip storage tank	1	Various	17,260
U_3O_8 precipitation tanks	2	65 tons of solution	12,000
Charcoal column	1	17 tons/day	78,500
NH_3 reactor	3		38,100
U_3O_8 stripper	2	800 tons of solution	15,000
Bins	10	Various	76,000
Pumps	35	Various	172,000

$$C_e = \text{cost of major equipment} = \$1,698,585$$
$$\doteq \$1,700,000$$

Factors	%	Indirect Costs	%
f_1 Instrumentation	6	Contractor's overhead and profit	27
f_2 Electrical	35	Engineering fee	8
f_3 Piping	35	Contingency	1
f_4 Structural, building	90		$f_I = 36\%$
f_5 Painting	2		
f_6 Insulation	2	Working capital is at 20% of C_e	
f_7 Temporary construction	5	Property tax is at 2% of C_e	
f_8 Plant items	21	Insurance is at 1% of C_e	
f_9 Clean up	3		
f_{10} Yard improvements	1.5		
f_{11} Small buildings	3		
f_{12} Service facilities	4.5		
$\sum f_i = 208\%$			

knowledge or arrived at through the experience of the firm. Table 10.12 indicates the factor list relating to the plant capital cost. Additionally, it shows the development of the indirect costs of contractor's overhead and profit, engineering fees, and contingencies if appropriate. The summed value of the factors are 208%, while the sum of the indirect costs are 36%.

The value of the constructed plant, less land, is found using Equation (7-1), or

$$C = [C_e + \sum_i f_i C_e](f_I + 1)$$
$$= [1,700,000 + (2.08 \times 1,700,000)](1.36)$$
$$= \$7,120,000$$

This detailed estimate for capital investment is lower than the preliminary estimate but is within a 25% tolerance and is considered acceptable.

Operating costs for this plant are composed of the items listed in Table 10.13. Each item is determined differently and has a detailed estimate. The supervision list is based on personnel requirements of other plants in this

TABLE 10.13

SUMMARY OF OPERATING COST ESTIMATES

Item	Monthly Cost	Dollars Per Ton
Labor	$ 19,602	$0.287
Supervision	14,580	0.384
Reagents	52,337	1.027
Supplies	8,160	0.160
Maintenance	14,576	0.287
Laboratory	4,680	0.091
Power	15,300	0.300
Heat	5,950	0.117
Water	1,020	0.020
Other	14,520	0.285
	$150,725	$2.958
Insurance	5,937	
Property taxes	11,865	
Total	$168,527	$3.30

Job Classification	Number	Dollars Per Hour + 20% Fringe	Dollars Per Month
Front end loader operator	1	$4.20	$ 740
Crusher operator	1	4.50	792
Grind, leach operators	4	4.20	2,923
C.C.D. operators	4	4.50	3,132
U_3O_8 Sx ppt drying	4	4.50	3,132
Moly cct operators	4	4.20	2,923
Packing, chemical mix	1	4.20	740
Helpers	4	3.90	2,714
Laborers	4	3.60	2,506
			$19,602

Crushing plant: 5 days/week, 8 hr/day, 350 days/yr
Mill plant: 7 days/week, 24 hr/day, 350 days/yr

Operating Supplies	Dollars Per Ton
Grinding steel and liners, 0.8 lb/ton at 0.12/lb	$0.10
Crushing	0.02
Drums	0.02
Miscellaneous	0.02
	$0.16

Power is on the average 30 kwh/ton of ore processed at $0.01/kwh, or $0.30/ton
Hauling is $0.12/ton-mile

category. The mill labor requirements are composed from manning tables associated with equipment, the design, safety laws, and practices. Some details for the line items are also shown in Table 10.13. Reagents, power, operating supplies, and so forth are estimated from material balances from the flow chart. Annual property taxes and insurance of 3 and 1% are applied to the investment to determine these effects. Operating costs are finally set at $3.30 per ton up from $3.17 per ton. The purchased cost of ore is negotiated at $1 and hauling charges will be $0.24 per ton. Total cost for operation and material is finally $4.54 a ton.

Prices for the processed material are again checked and new information suggests that political forces will cause a drop to $5.90 to $6.00 per pound. New ore discoveries are an unknown force in the price. Finally, the estimate recap is given in Table 10.14.

At this point the calculation studies in this chapter are undertaken. Computer trials are run where variations in the data are introduced and tried. Continuous interest and continuous flow of receipts and disbursements could be assumed. Accelerated methods of depreciation could be used. The conclusion in finding the rate of return is left up to the student as Problem 10-26.

TABLE 10.14

SUMMARY OF ESTIMATING DATA FOR
ENGINEERING ECONOMIC RATE OF RETURN

Item	Flow-Sheet Design
Tons/yr processed	612,000
Tons/day processed	1,750
Pounds of U_3O_8 produced/yr	1,513,000
Pounds of U_3O_8 produced/day	4,320

Sales	Price
U_3O_8	$5.90–6.00/lb

Operating Costs	Cost
$/ton processed	$ 4.54
$/lb of U_3O_8 produced	1.83
$/yr	2,778,480

Capital Requirements	Cost
Equipment	$1,700,000
Instrumentation	100,200
Electrical	595,000
Piping	595,000
Structural, building	1,530,000
Painting	34,000
Insulation	34,000
Temporary construction	85,000
Plant items	357,000
Clean up	51,000
Yard improvements	25,500
Small buildings	51,000
Service facilities	77,000
Total construction	$5,234,700
Contractor overhead and profit	$1,413,369
Engineering fee	418,776
Contingency	52,347
Total plant cost (insurance and tax base)	$7,119,192
Interest during construction (plant)	569,535
Subtotal for depreciation	$7,688,727
Working capital	340,000
Land cost	50,000
Total investment	$8,078,727

The project estimator has the job of forecasting capital investment and operating expense to aid the management process in choosing and evaluating present and future equipment and plants. About 65% of all investment acquisitions are not to replace but to provide a potential where no potential existed before.

The estimating function eventually evaluates equipment, plants, and refurbishment of existing assets. The analysis stemmed originally from a preliminary design of some sort, and eventually plans are started to investigate investment possibilities. Examined in this chapter were traditional and crude rule-of-thumb procedures (which are not really recommended). The discounted cash flow and the present worth method can be modified to provide a wide-ranging assortment of models. Despite any choice of an analytical method, there remains the problem of predicting certain future events. Three models dealing with optimum plant sizing, equipment replacement, and lease-buy were discussed.

In the next chapter many of the engineering economic concepts are again considered for their value in system estimating.

SELECTED REFERENCES Texts dealing with engineering economics are

GRANT, EUGENE L., and W. GRANT IRESON: *Principles of Engineering Economy*, 5th ed., The Ronald Press Company, New York, 1968.

SCHWEYER, HERBERT E.: *Analytic Models for Managerial and Engineering Economics*, Van Nostrand Reinhold Company, New York, 1964.

THUESEN, H. G., W. J. FABRYCKY and G. J. THUESEN, *Engineering Economy*, 4th ed., Prentice-Hall, Inc., Englewood Cliffs, N.J., 1971.

Texts that stress engineering economics more from a cost engineering point of view are

JELEN, F. C., Ed., *Cost and Optimization Engineering,* McGraw-Hill Book Company, New York, 1970.

PARK, WILLIAM R., *Cost Engineering Analysis*, John Wiley & Sons, Inc., New York, 1972.

QUESTIONS
1. What is the purpose of a project estimate? Why are project estimates important?
2. Describe two psuedo-return methods and indicate why they are faulty. What advantages do they serve?
3. What kinds of information are necessary in order to undertake a project estimate?
4. Illustrate fixed capital, working capital, and direct and indirect costs.
5. Indicate the effects of taxes on profits, estimates, and calculations.
6. What is meant by "equivalent or equality" with the methods of net present worth, net future worth, and net annual equivalent worth?
7. What is a tax credit? Does the government actually refund this money to the firm? Where does it show up?
8. What criteria do you suggest for making project decisions? Will capital shortages influence your decision? Do you believe that a successful company is cash poor and bank-note rich?
9. Suppose that at a certain date an organization had a listing of engineered projects each showing the amount of capital required and the rate of return and that these can be plotted as shown in Figure Q9. Given a limited amount of equity capital, it asks the bank to cover these capital ventures, and you show them your curve. Now the bank, knowing your financial credit, offers its own curve with increasing interest because of their risk as more capital is loaned. What is the favorable aspect about this approach to the firm? Unfavorable? What does this method say about the marginal interest to be earned? At what point should one stop borrowing?

FIGURE 10-Q9

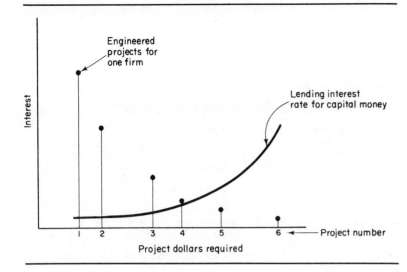

10-1. (a) What is the principal amount if the amount of interest at the end of $2\frac{1}{2}$ years **PROBLEMS**
is $450 for a simple interest rate of 5% per annum?

(b) If $1600 earns $48 in 9 months, what is the annual rate of interest?

(c) What is the present worth of $1000 6 years hence if money is worth 10%?

(d) For what period of time will $5000 have to be invested to amount to $6800 if it earns 10% simple interest per annum?

(e) Using the interest tables given in the text, interpolate and find the value of the single-payment compound-amount factor for 12 periods at 11% interest. For the equal-payment-series sinking-fund factor for 23 periods at 11% interest. The equal-payment-series compound-amount factor for 39 periods at 13% interest.

(f) What will be the amount accumulated by $6000 in 12 years at 10% compounded annually?

(g) What is the present value of the following series of prospective payments, $4000 for 15 years at 20% compounded annually?

(h) What is the present worth of a stream of income given as

Years	Amount
1	50
2	40
3	10
4	−5

when the cost of capital is 20%?

(i) What equal series of payments must be paid into a sinking fund to accumulate $5000 in 5 years at 10%?

(j) What equal series of payments are required to repay $6000 in 10 years at 15% compounded annually with annual payments?

(k) How many years will it take for an investment to double, triple itself if interest is 10%? 20%? Use tables.

(l) How would you determine an equal-payment present worth factor if you had only a table of single-payment present worth factors? Compute at 10% for five years.

10-2. (a) Investment for new equipment is $100,000 and salvage will be $10,000 eight years hence. An earning from this equipment will be $15,000 on the average after taxes. What is the non-time-value of money return? What is the payback?

(b) Two investment opportunities are proposed.

	Process A	Process B
Total investment	$60,000	$60,000
Revenue (after tax)		
Year 1	20,000	30,000
Year 2	20,000	30,000
Year 3	20,000	30,000
Year 4	20,000	
Year 5	20,000	

For an interest of 10%, which has the least period of time before capital recovery is complete?

10-3. (a) A project has the following estimated cash flows where disbursements and revenues are end of period.

Year	Disbursement	Revenue
0	$800	$ 0
1	0	450
2	0	425
3	0	400

If interest is 10%, find the net present worth, net future worth, and equivalent annual cost. Use the methods in Sections 10.3-1, 10.3-2, and 10.3-3 as a guide. Find the true rate of return. Present a summary of the four methods.

(b) Consider the cash flow statement given by Table 9.2. Use an interest rate of 10% and complete the profitability analysis.

10-4. A bridge designer is considering two designs to cross a small stream. A wooden design would cost $8000 and last 10 years. The steel design would cost $10,000 and last 15 years. Each structure has no salvage value and would require the same amount of maintenance. With 10% use the net annual worth method and compare the two methods. Also compare the two alternatives for a life of 30 years at an interest rate of 10% using the present worth method. Which is the better design?

10-5. A prospective venture is described by the following receipts and disbursements:

Year End	Receipts	Disbursements
0	$ 0	$800
1	200	0
2	1,000	200
3	600	100

For $i = 15\%$, describe the desirability on the basis of present value.

10-6. A drill press was purchased 5 years ago for $2400. It is being depreciated to no salvage value over a life of 10 years by the straight-line method of depreciation.

What would be the difference in current book value had the sum-of-the-year-digits method of depreciation been used? Discuss advantages and disadvantages of accelerated depreciation.

10-7. A 10-year-old machine has a $8000 book value but only a $4000 net realizable value. It is contending for its present job with a new machine costing $20,000 installed. Management insists that book value must be taken into account in the analysis and maintains that replacement at this instant will show a loss in profit equal to the excess of book value over net realizable value. They propose to avoid this by continuing the unit in the present service. Comment on this policy.

10-8. A construction job is to last 2 years. A belt conveyor costing $75,000 will have an operating cost of $20,000 and $15,000 salvage at the end of 2 years. Mobile equipment costing $30,000 could also be used having annual operating costs of $35,000 and $10,000 salvage at the end of 2 years. Both of these conveyors have longer physical lives than 2 years. The required rate of return is 10%. Which is the preferred investment?

10-9. An income producing property can be purchased for $60,000 and it has an annual gross income of $10,000 and operating cost of $3000. Estimated resale should be $20,000 in 10 years. If the required rate of return is 10%, should this investment be made?

10-10. The following are the facts for the choice between two sources of water: a water well sunk some 300 feet down, or the purchase of water from a water district, a utility created for the distribution of water to customers who may wish to subscribe. A well system would put a 1½-horsepower (using 2 kilowatts) pump 300 feet down. It would cost on a first-cost basis $2000. The pump would be expected to supply water at a rate of 10 gallons per minute. Pumping costs are $0.020 per kilowatt hour. The water district utility will sell "tap rights" for $400 which essentially provides for a stub at the property line. The monthly cost for water will be $6 for the first 6000 gallons and $0.50 for each 1000 gallons used thereafter. The family uses 20,000 gallons per month. This family can invest its money at 5% in commercial institutions. Find the break-even point in years between the two investments. Which is the best investment if the proposed owner will sell this house in ten years? 20 years?

10-11. A factory producing a special product can be constructed to meet present needs for $120,000. It would have an annual disbursement of $30,000 for operational requirements. In another 10 years the production requirements will be doubled, and another module, like the currently proposed design, will be necessary. Once two modules are constructed operating disbursements are $58,000 yearly. During the 10 years, inflationary pressures will increase the construction-cost index to 175. At the present moment, the cost index is 135. Another possibility is to construct a plant that will satisfy the capacity now as well as that which is predicted 10 years hence. The cost of constructing a two-module plant is $225,000, while its annual operating disbursements are estimated as $40,000 on half capacity during the first 10 years, and $55,000 on full capacity. These plants are predicted to have an indeterminately long life. This company is able to borrow the cash at 7%, it believes the risk to be 5%, and it desires a before-tax percentage profit on investment of 8%. Show the calculations to demonstrate whether the addition should be built now or later.

10-12. A proposal to substitute a laser interferometer inspection system in place of conventional gauge blocks for numerical control machine tools is advanced. Ten machines receive six inspections per year in a routine machine maintenance program to prevent performance degradation. The laser can do the job faster. Figure P10-12(a) is the annual conventional maintenance cost. Personnel training, gauge purchases, and refurbishment of equipment are annual costs. Figure P10-12(b) is the capital cost and the annual recurring costs of the laser interferometer project. If money is worth 30%, what is the present worth of savings over a 15-year horizon? When is the break-even point? Find the economic rate of return for a 15-year life. Consider discrete costs and savings. Estimate your information from the figures.

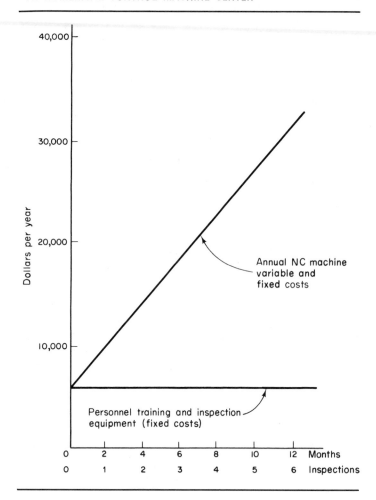

10-13. Evaluate the cash flow diagram on page 360 where arrows pointing down represent end-of-year disbursements and upward arrows are receipts. Determine present worth sum at time five, an equivalent annual payment, and a final future sum appropriate to the diagram. Use $i = 30\%$.

10-14. Two operation designs have been advanced. The first operation involves a general purpose machine costing $40,000 installed. Operational estimates for a composite design are 0.75 standard hours per 100 units. An operator costs $8 per hour including direct fringe benefits. A special purpose machine can be designed and will cost $55,000. Production for the bench-mark design is 0.25 standard hours per 100 units. This operator will cost $7.25 per hour including direct fringe benefits. Both machines will last 10 years under 4000 hours of use

Costs	General Purpose Machine	Special Purpose Machine
Power cost per hour	$ 0.30	$ 0.25
Maintenance per year, fixed	650.	700.
Maintenance per hour, variable	0.15	0.18
Other fixed-prorated overhead costs per year	2400.	2000.

LASER INTERFEROMETER FIRST-YEAR
INSPECTION COSTS FOR 10 NUMERICAL
CONTROL MACHINES

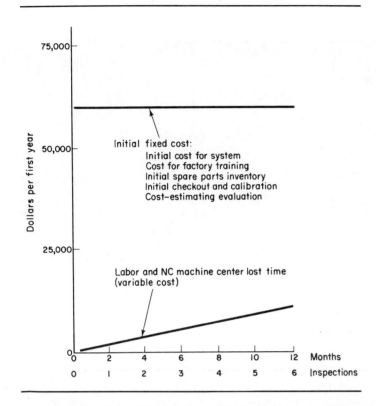

yearly and the machines will be retired to secondary service. Additional
estimated data are on page 358. For a before-tax rate of return of 20%, what
annual output rate is indifferent between the two designs? What is the decision
if production is above this point? What is the per unit cost if the special purpose
machine is used?

10-15. (a) What is the effective interest rate for a nominal 10% compounded semi-
annually, continuously? Confirm your answer with Appendix VII.

(b) What is the nominal interest rate for an effective rate of 10% compounded
continuously?

(c) What is the equivalent worth at the end of 12 years of a present expenditure
of $6000 if interest is 10% nominal compounded continuously? Compare
to Problem 10-1(f).

(d) What is the present worth of operating costs of $4000 per year for 15 years
incurred continuously if the nominal rate of interest is 20%? Compare to
Problem 10-1(g).

(e) Find the present amount of $1,000 per year flowing uniformly for a period of
6 years at an interest rate of 5% compounded continuously. Use the funds
flow conversion factor of Appendix IX.

(f) What is the present value of the following continuous funds-flow

Year	Continuous Flowing Amount
1	50
2	40
3	10
4	−5

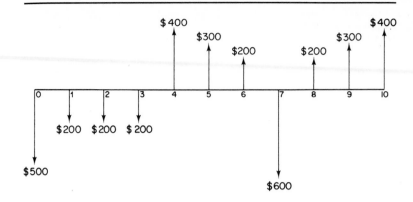

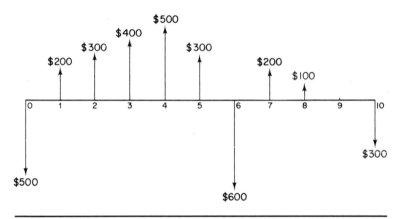

if the interest is 20% compounded continuously? Compare to Problem 10-1(h).

10-16. Two distinct refinery designs have been evaluated for a product manufacture. A summary of the economic data is given as:

	Plant A	Plant B
Turn-key cost	$12,000,000	$10,500,000
Annual costs, $C_o + K_n$	$1,100,000 + 200,000n$	$1,200,000 + 300,000n$
Salvage value at 20 years	900,000	300,000
Economic life of product	20 years	20 years
Internal interest rate for firm, %	20	20

The symbol n is for continuous time for payments and receipts. C_o is startup costs of operation including initial out of pocket charge. K is the slope of annual costs related to plant. With the evaluation to be made on the basis of present worth, which is the preferred plant?

10-17. Two plants have the following economic data where money is 10^6.

	Plant A	Plant B
First Cost	$12	$8
Annual costs, $C_o + K_n$	$1.1 + 0.2n$	$1.2 + 0.3n$
Salvage value at 10 years	1.0	1.1
Terminal life, years	10	10
Interests, %	20	20

The symbol n is continuous time. Which is the preferred installation?

10-18. Find the maximum present-worth amount for one-life cycle for Products 1 and 2 for the following data:

	Product 1	Product 2
(1) Life-cycle sales rate where n_2 is continuous years, units/year	2,500	$3,000 - 100n_2$
(2) Life-cycle selling price, \$/unit	10	$10 - 0.1n_2^{1.1}$
(3) Life-cycle costs, \$/unit	5	$3 + n_2^{1.1}$
(4) Project investment, \$	10,000	10,000
(5) Depreciation model	S.L.	S.L.
(6) Salvage value	1,000	$10,000e^{-0.25N_2}$
(7) Terminal life	N_2	N_2
(8) Profits tax	40%	40%
(9) Regeneration-of-capital rate		
Before taxes	30%	30%
After taxes	18%	18%

Notice that terminal project life is N_2 and is determined from Product 2 (*Hint*: Find optimum N_2 first using Product 2. Make assumption for Product 2 to avoid non-terminating integral.)

10-19. An engineer has estimated cash flow from a project design to have a function $800 - 15n$ dollars per year where n is continuous time from start up. Variable cost flows are assumed to be $200 + 40n$ dollars per year. Fixed project capital necessary to start is 500, and a salvage value function of $500e^{-0.3N}$ where N is the anticipated economic project life cycle. This firm has a tax rate of 55%, a policy of 20% nominal rate of return, and uses straight line depreciation. Dollar values as listed are \$10^6$. If a policy considers replacement of the same kind for this project, what is optimal project life? (Hint: Find the optimum present worth for this project. Is it a maximum or minimum, and how do you know? Use continuous interest.)

10-20. One of the alternatives to domestic natural gas is that of a liquid natural gas (LNG) which supplies sea-board regions with gas. The LNG is shipped aboard tankers at $-258°F$ from other countries in the world. A preliminary estimate of a LNG plant and special tanker is given as:

Capacity, 10^6 standard cubic feet per day	500
Shipping statute miles	2,500
Ship size MBBL	460
Capital cost:	
Liquefaction plant	\$210,000,000
Ships (4)	204,000,000
Receiving terminals	72,000,000
	\$486,000,000

	Dollars Per Year	Dollars Per 10^3 Standard cubic feet
Annual operating cost:		
Liquefaction Plant	\$ 12,100,000	\$0.071
Ships	10,400,000	0.052
Receiving terminals	2,700,000	0.015
	\$ 25,200,000	\$0.138

Fixed charges and profit @ 21%:

Plant	$ 44,100,000	$0.259
Ships	42,800,000	0.215
Receiving terminals	15,800,000	0.090
	$102,700,000	$0.564
Grand total	$127,900,000	$0.702
Gas costs at plant		0.130
Total value of gas into pipeline		$0.832

Fixed charges and profits include depreciation, interest, taxes, general and administrative overhead, and profit all evaluated at 21%. Asset life is assumed 25 years. Assume a price for gas and determine a rate of return. Assume discrete payments to simplify analysis. What is the cost per 1,000 British thermal units? A cubic foot of natural gas contains 1,100 Btu standard cubic feet per higher heating value.

10-21. The U. S. Bureau of Mines has studied the production of high-Btu gas from a coal gasification plant using Pittsburgh seam coal. The plant is sized and designed to produce 250×10^6 cfd gas using fluidized gasification followed by shift conversion, purification, and methanation.

	Cost
Capital Costs	
Coal preparation	$ 7,987,200
Gasification	24,459,100
Dust removal	7,376,000
Shift conversion	3,069,800
Waste heat recovery	4,912,100
Purification	23,837,800
Methanation	19,023,300
Pipeline compression	635,400
Oxygen plant	23,400,000
Sulfur recovery	944,000
Steam and power plant	22,020,000
Plant facilities	10,324,900
Plant utilities	14,799,000
Initial catalyst requirements	1,021,300
Total plant cost	$163,809,900
Interest during construction	$ 13,104,800
Subtotal for depreciation	176,914,700
Working capital	17,691,500
Total investment	$194,606,200

	Cost
Annual operating costs for $4 coal	
Raw water	$ 855,400
Catalyst and chemicals	552,400
Coal at $4	18,810,000
Direct labor	1,813,300
Maintenance	5,042,900
Payroll overhead	1,343,900
Operating supplies	1,008,600
Indirect costs	3,145,900
Taxes and insurance	3,276,200
Sulfur credit	− 677,200
Tax credit	− 1,920,200
Net operating cost	$ 33,250,200

The operating costs are highly dependent on the price of coal. What is the net adjustment of $6 and $8/ton coal supply? If the cost model in dollars/1,000 cf is (annual operating cost plus depreciation) divided by the production, what is the cost for $4, $6, and $8 coal? The life of this plant is 20 years. The anticipated sale price for $4, $6, and $8 coal is $0.63, $0.77, and $0.92 per 1,000 cf into the pipe at the plant. Calculate the rate of return for these three options. The tax rate $= 50\%$ and assume straight-line depreciation. Compare these costs to your own local gas supply prices and discuss the merits of coal gasification. What is a nominal cost of a Btu for your local natural gas? How does that compare to coal gas which will have about 925 Btu/scf. Use discrete payments to simplify analysis.

10-22. A process has been known to follow $F = I_r(Q/Q_r)^{0.4}$ where Q is in pound dimensions. For a process capacity of 15,000 pounds the cost is $180,000. Throughput time for the process is 3 hours. If we assume that the processing rate, pounds per hour is independent of size Q, what would be the best size process to design for initially where demand is 550,000 pounds per year? Let $r = 20\%$. The forecast is for a yearly linear increase of 30,000 pounds per year. What would be the size of future additions, and when would you schedule them?

10-23. A product has enjoyed national sales according to a trade journal:

Last year:	1,400,000 units
Two years ago:	1,200,000 units
Three years ago:	1,000,000 units

The cost of capital for a new venture in this company is 30% on a before-tax base. This company, which presently manufactures a related product, wants to enter this field. It assumes that through its existing markets it should capture 10% of future sales. Its history for plant construction is sketchy, but it does have relevant data for other production and the construction of a plant.

Annual Production	Plant and Internal Equipment Costs
40,000	$8,000,000

It also has knowledge of expected values of the exponent m, which is near 0.8. Find (a) the initial capacity for the facility, (b) the optimum marginal increase in capacity, (c) the time interval between expansions, (d) the present value of all future investments, and (e) the first expansion size.

10-24. What are the decisions for the following stage-wise problem for a planning period of 8 years at $i = 10\%$?

Years From Now	Investment Alternative	Cash Inflow Per Year	Cost
1	Old	$2	$4
	New	5	7
3	Old	3	5
	New	6	9
7	Old	2	6
	New	4	9

10-25. A chemical company that makes soap has a reagent pipeline that is of conventional 303 stainless steel. Due to the caustic that it transports, replacement of this 4-inch pipe is anticipated every 2 years for normal business conditions. A new stainless material, a larger diameter, and fewer elbows and pressure drops will increase the pipe life to 8 years. The following data are the economic input. The planning horizon is limited to 8 years.

Years From Now	Investment Alternative	Cash Inflow Per Year	Refurbish Buy Cost
0	Old pipe	$ 0	$ 8,000
	New pipe	3,000	14,000
2	Old pipe	0	10,000
	New pipe	4,000	20,000
4	Old pipe	0	12,000
	New pipe	5,000	26,000
6	Old pipe	0	14,000
	New pipe	6,000	26,000

The cash inflow represents the savings difference between the two investment opportunities. (a) Determine the decisions for each of the two-year choices with $i = 0\%$. (b) This chemical company requires a minimum attractive rate of return of 30% for investment decisions of this nature. What are the decisions now?

10-26. Complete the estimate started in section 10.9. Assume a salvage recovery of 5% of the depreciable amount and 75% of working capital at the conclusion of 14 years.

(a) Use an end-of-year convention and constant depreciation to find the rate of return.

(b) Assume an accelerated method for depreciation and complete part (a).

(c) Use constant depreciation and the continuous flow of funds convention.

(d) Plot a figure corresponding to Figure 10.2. Use an after-tax rate of return of 20% and assume a land purchase of $50,000 and constant depreciation.

Curve Segment	Monthly Time Phasing
Land	Immediate
Design and investment	36
Operating capital	42
Startup	48
Termination point	168

chapter 11

Observe how system into system runs,
What other planets circle other suns.
ALEXANDER POPE (1688–1744)

SYSTEM ESTIMATING

In this chapter we shall outline methods for applying system estimating to system problems. Unfortunately, a procedure cannot be set out beforehand in some recipe fashion. If that were possible, and after we have learned these procedures and by strict adherence to them, we would be guaranteed absolute confidence of result. The preparation of each estimate is unique and, to a large extent, the methods are adapted to the problem. Each system estimate is a product of the skill, experience, and resourcefulness of the engineer. Nevertheless, there is a methodology and a set of principles that aid the construction of system estimates.

We shall show that the measure of system estimating is effectiveness. Previously we had considered cost, price, and return as the appropriate measure for operation, product, and project types of designs. To make a cost effectiveness evaluation, it is necessary to contrive and then construct an information flow and system estimate structure. Given the information flow structure and the corresponding model it becomes possible to use various analytical aids to arrive at system effectiveness. As a sample of these kinds of analyses, several small-scale system studies of energy will be made.

11.1
SYSTEM EFFECTIVENESS AND SYSTEM ESTIMATING

The basic concepts inherent in system estimating have been applied successfully to a broad listing of problems such as water resources, health systems, social welfare programs, hospital planning, space systems, community relations interaction, and industrial growth. The techniques have been applied at levels ranging from preliminary to detailed.

What is meant by the word *system*? Three classifications are normally suggested, executive, operational, and physical. For the first we have in mind procedures and organizational forms that connect organized effort. In operational systems the central thought is about responsibilities, authority, and aspects of functional relationships as they deal with minor or major problems. Our concern is for the engineering and the social system, which we collectively term *physical*. Because this is a book about cost estimating our system shall deal with a configuration of operations, products, and projects; thus, scrutiny of these divisions is possible by the means of the previous three chapters.

11.1-1
Effectiveness

In system estimating the word *effectiveness* is the focus of a great deal of attention. Effectiveness is something like efficiency. In the traditional sense efficiency is nondimensional and as a quantity it approaches one. In engineering the object is to have as high efficiency as possible. But effectiveness means more than efficiency. For one thing it may have dimension.

The finding of system effectiveness measures is not an easy thing. Although many effectiveness measures are possible, only a few ever serve a practical purpose. For any meaningful evaluation, analysis cannot be conducted on lofty levels because of the lack of detail inherent in the general terms, and consequently more specific measures are desirable. To choose between two energy systems by listing the things we value, such as function, dependability, and lack of pollution, is not very helpful. This does not indicate that we should not form such a list early in the analysis. Effectiveness measures such as cost, heat loss, waste-heat recovery cost, and equal-marginal fuel cost are more specific in meaning.

With narrow measures of effectiveness, there is a greater chance of finding information to allow analysis.

While we could associate system effectiveness with a narrow meaning, this would restrict the actual character of the concept. There is no one best measure of effectiveness. With the present state of knowledge which obscures our perfection, we settle for something less. For cost engineering, that something less implies a dimension in monetary units such as dollars. This is our basic dimension of effectiveness.

The effectiveness measure is usually considered approximate. Precision, although it is not unknown in these matters, may be overshadowed by the intangible portions of the estimate. As a practical matter, to be able to make a mandatory choice between competing systems the tolerance need only be sufficient to indicate a selected alternative.

There are means to overcome the philosophical problem of effectiveness. The first of these is to make the effectiveness more specific. A major problem can be reduced to elemental problems. For each of the elemental problems, we assume that a special effectiveness measure can be constructed. This permits individual attention. A firm or governmental agency cannot have one system estimator examine all the problems of choice simultaneously and pick each course of action in light of all the decisions. The magnitude of the task requires that the problem be broken into its elements. The consequence of this action is to make some of the broader policy choices by high-level officials, while others are delegated to lower levels. The piecemeal system analysis makes it possible for more attention to detail. However, dangers are inherent in piecemeal analysis, as lower-level effectivenesses may be unrelated to the higher-level problem. If the chosen effectivenesses provide only approximate results for the smaller suboptimal problems, a hierarchy of crude effectiveness measures would be considered simultaneously, and potential inconsistencies are abundant.

11.1-2
Cultivation of Alternatives

A system estimator is expected to consider a broad range of alternatives in his study. He is encouraged to visualize most alternatives that are in the realm of practicality, denying acts such as perpetual energy, for instance. Although it may be appealing, it is really not worth the effort. The analyst rules out other less rewarding possibilities due to the shortage of time, effort, and money. The remainder are reasonably exhaustive. For instance, a system study of the electrical generating opportunities for a large central state may overlook hydroelectric power (however, in one central plains state there is hydro power from a river with very low head but a high volume rate of flow). For some situations it may be necessary to initially describe a list of possible alternatives that is at first quite large. Preliminary estimating methods will abridge the list to something more valuable. The converse of this is also true. As the analysis continues certain opportunities may present themselves which were not known at the start.

Selection of appropriate alternatives for further study may be guided by several rules. First, the scope of the system to be compared in conjunction with the selection of the effectiveness will tend to deny alternatives of doubtful value. When called on to narrow the selection of alternatives the estimator may want to logically state restrictions that heretofore had only been verbalized. This will prevent certain types of alternatives from creeping into the evaluation process. Naturally, a good rule is familiarity with overall system objectives.

11.1-3
Intangibles

An analyst may be unable to commit some aspect of a design to ordinary monetary units. Such things as function, beauty, safety, quality of life, and ease, are difficult to evaluate. These factors are referred to as intangibles or irreducibles. Although special and nonfundamental units for a scale of measurement or ranking might be arbitrarily forced on the intangibles, the analyst usually believes that it is not worth the effort and that it is truly imponderable in terms of ordinary units.

If intangibles are not to be treated in monetary units, there are other ways to consider them. If the stakes are not high, it may be convenient to ignore the intangibles. While the intangible may be advantageous to some, there may be a contrasting disadvantage to others. A mere listing of the intangibles, both pro and con, may be a sufficient examination. In some cases the intangibles may be classified and ranked by some ordinal scale method using the Delphi method (see Chapters 5 and 6).

Despite the ability to render, although superficially, various intangibles to a deterministic scale, the preponderance of practice chooses to accept most intangibles as closed to numerical estimating. Thus, it leaves the estimator in the position of considering what is tractable in dollars and that which is not.

11.1-4
Long-Term Uncertainty

The cost estimates mentioned so far can be called average or expected outcomes. When viewed as a random variable they may be off their mark. For this present discussion there are two types of uncertainty (disregarding now risk, certainty, and uncertainty). In the first case the uncertainty may be about the state of the world (or nature) in the future and is called long-term uncertainty. It includes factors like technology uncertainty or strategic uncertainty by competitors. Statistical uncertainty is different and it results from purely chance elements. This exists even if the uncertainties of the first type are zero. They are usually less troublesome to handle in system studies. Attention to detail, meticulous care, and other techniques deal with statistical fluctuations. Discussion about how to cope with random statistical variation is found in Chapter 5.

Probabilistic techniques and sensitivity analysis are tools to use with long-term uncertainties. Analysis to recognize these long-term uncertainties can be involved or simple. If the problem warrants attention for this reason, the analyst may choose to provide a low, medium and high value for critical input variables. These would determine how sensitive the results are to variations in certain parameters. The number of combinatorial variations rises at remarkable rate for a statistical evaluation of this kind. Policies that are dominant irrespective of minor or major variations in input data are desirable solutions. However, it is an unusual system design that is so simple.

11.1-5
Time Horizon

Undoubtedly, most, if not all, of decision making is a part of the prior history of decisions. Previous choices have affected the now time, and current decisions will influence future actions. Viewed this way all system-estimating problems can be classified as having no time limit. Sometimes dynamic models have an unbounded horizon while others have a terminating date. Project-estimating models are mostly terminating. Life cycle costs and the rate-of-return and time-value-of-money policy actions are examples. Obsolescence and depreciation are two factors that influence cost-engineering systems significantly. As a rule the particular system problem is posed in a dynamic context, and the estimator considers time explicitly.

With time-phased costs from the cost streams of different alternatives, the irregular amounts are discounted using an appropriate rate. Methods dealing with this idea were introduced in the last chapter, and they fit system-estimating problems. For sociopolitical problem environments the specific rate may be difficult to set. An upper and lower rate can be used to see if there is any difference in the final conclusions of the system design.

11.1-6
System Effectiveness Policies

Discussion has been leading up to the manner in which system studies are undertaken. The two approaches are either fixed effectiveness or fixed cost. For a fixed effectiveness system, various alternative means for achieving a prescribed capability or level of effectiveness are studied to determine how such effectiveness can be attained with the least amount of resources. In other words, there is a given objective and the question is to find that alternative or feasible combination of alternatives to achieve the objective. The second type of system analysis deals with a fixed budget or specified cost. Given a fixed level of resource, the estimator looks for the most efficient manner to achieve the highest level of effectiveness.

For either or these two approaches, there are two ways of conceiving the effectiveness measure. The analyst can consider either relative or absolute effectiveness. For relative effectiveness, the measure is viewed more as an index, and only differences are considered important. For comparisons between competing systems, it is simpler to use relative differences to make the decision. The second method is that of the absolute effectiveness or the total cost measure. In addition to the selection among alternatives, it indicates the magnitude and is an additional bit of wisdom for the decision maker.

11.2
SYSTEM ESTIMATE FORMULATION AND INFORMATION NETWORK

To a large degree the nature of a system estimate is determined by the information that can be obtained. In this respect the system estimate is like other types of estimates, as accuracy, precision, and its general goodness are no less dependent on the available package of information. There is one distinction, however. In system estimating the kind, quality, and amount of information are usually established after the start of the estimate. With an operation, product, and, to a lesser extent, project design, information is reused, and agencies outside of estimating are in the habit of supplying raw data, performance figures, and the like. This is not true for the system estimate. As this type of estimate is unique, the information is determined after the estimate is started. Usually it is gathered by the system estimator.

The purpose of this present discussion is to show an algorithm[1] for finding kinds of data. While there are others, this one centralizes attention on a *flow* of information that aids model development and visualization.

An engineering system may have a great number of physical stages. For instance, a boiler-turbine-generator can be visualized as a three-stage system. When the system has a distinguishable modular structure the components respond both in an individual and systematic way. Individual behavior is unlike the system response, but we single out the component for study and then impute a change in behavior which reflects the system alteration on the component. Mathematical modeling, experience, and logical statements are information ingredients that connect these modules. The way to represent a system is to begin

[1]D. V. Steward, "On an Approach to Techniques for the Analysis of the Structures of Large Systems of Equations," *SIAM Review 4*, No. 4 (1962), and "Partitioning and Tearing Systems of Equations," *Journal of SIAM, Numerical Analysis*, Series B, *2*, No. 2 (1965).

by "black-boxing" each major physical device and connecting the boxes by arrows indicating input-output between the boxes. This is followed by itemizing the relationships and variables for each box or node in this network.

A number of algebraic equations (linear or nonlinear and real) exist with these cost-engineering models. In formulating the system estimate model, the number of variables appearing in any equation is small when compared to the total number of variables in the system. A system estimator will conceive of many equations which reflect the physical, economic, and social constraints, and normally each equation has only a few variables. There are advantages to cost-engineering system approximations of this sort. Errors in a large variable-equation model are more possible because of the understanding necessary to know that which is extraneous from that which is not. When using a many-variable-equation model, debugging becomes troublesome. It may be desirable to find a means to isolate parts of the model, which can then be studied independently of other parts. Indeed, how to choose one piece of a design to study at a time such that there is minimum outside interference is the heart of the estimator's art.

Consider the following two sets of functional equations. There are a number of approaches to their solution.

		a	b	c
$F_1(x_a) = 0$	1.	x		
$F_2(x_a, x_b) = 0$	2.	x	x	
$F_3(x_b, x_c) = 0$	3.		x	x

(11-1a)

		a	b	c
$F_1(x_a, x_b) = 0$	1.	x	x	
$F_2(x_b, x_c) = 0$	2.		x	x
$F_3(x_c, x_a) = 0$	3.	x		x

(11-2b)

On the right are the same equations in the form of an array indicating those variables appearing in each equation. The first set [Equation (11-1a)] is solved as follows: Independent of other equations, the first equation is solved for the number represented by x_a. After a number for x_a is at hand, the second is solved for the number represented by x_b. With a number for x_b available, the third is solved for the number represented by x_c. This sequence of substitutions and solutions can be imagined as a flow of information.

For the second set [Equation (11-1b)] there is no equation that can be solved independently of the others. Methods of iteration or elimination can be used, and both processes imply a flow of information. For either iteration or elimination, the first consideration is for which variable each equation is to be solved. This variable is called the output variable (or dependent variable). The remaining variables of the equations are called input variables (or independent variables). For the second set it may be easier to solve the first equation for x_b, the second for x_c, and the third for x_a, which is described by an output set ($1b$, $2c$, $3a$).

Properties of an output set have the following characteristics:

1. Each equation has one output variable.
2. Each variable appears as the output variable of exactly one equation.

Difficulties appear in giving so few properties. An equation may contain a variable for which we are unable to solve uniquely. If the iteration set ($1b$, $2c$, $3a$) is to be used, then we initialize by arbitrarily guessing at a number for x_a. Using this number the first equation is solved for a number represented by x_b. Eventually the third equation is computed for a number x_a. For a converging system, x_a must eventually agree with the x_a used in the first equation. This is called a loop, and it implies that no equation appearing in the loop may be solved independently of the other equations in the loop.

In this discussion we ignore a set of equations like

$$x^2 = -y^2$$
$$z = 1$$
$$2z = 2$$

as this system cannot be solved uniquely. A social-engineering set like this is considered a poor approximation.

For a system design there are a certain number of variables which are not specified. They are called free system design variables. The number of these variables is called *degrees of freedom*.

The behavior of each black box or node can be described approximately by equations, computer programs, manufacturer's recommendations, tables and graphs, and judgment of experienced estimators. The design relations consist of M sources of information about the system including reference to N variables x_j (people statistics, operating conditions, physical factors). These design relations must be independent. In the case where $M > N$ the system problem is not well formulated and it is generally not possible to find values for all the variables which satisfy the design relations. In the case $M = N$ there exists no freedom in the selection of the values of the variables. True system problems are never as trivial as this. When $N > M$ there are more variables than system relations and a set of variables is open to selection by the estimator. The number of degrees of freedom equals $N - M$ and may be selected as system design variables. The remaining variables are called state or dependent variables. After the engineer has assigned particular values to the design variables, the values of the state variable are obtained by solution of the design relations. Now let us reconcile what has been said in this paragraph in mathematical form, or

$$f_i(d_j, s_k) = 0 \qquad \qquad (11\text{-}2)$$

where f_i = number of design relations, $i = 1, \ldots, M$
$\quad\ d_j$ = independent design variables, $j = 1, \ldots, N - M$
$\quad\ s_k$ = state or dependent variables, $k = 1, \ldots, M$

How do we go about specifying the system variables? Obviously the constants are fixed outside of the cost engineer's persuasion. But degrees of freedom are open to selection and are consumed in two ways:

1. Certain variables are assigned definite values to provide a connection between the nodes. For instance, a social problem may require a basic income per family unit for subsistence. For a physical system the coolant temperature required for a heat exchanger may be fixed at the temperature of the available supply of cooling water from a river. These are *external* factors.
2. The remaining degrees of freedom are used up in the selection of variables to be adjusted in optimizing system effectiveness. These variables are classified as the *economic degrees of freedom*.

For example, if $M = 6$ and $N = 12$, we have 6 total degrees of freedom. Assume that for a physical component of a system 4 of 6 degrees of freedom are used in integrating the component to its environment. The remaining 2 free variables can be selected by the estimator to achieve cost economy.

Now tie in what has been said to flow of information. Given that we have $M < N$ and M constraints and N variables where each i has only a few design variables, our objective is to model a network where one bit of information is flowed into relationships to achieve an output variable, which is in turn flowed into another equation. This is a precedence-succession type of calculation of networks. We so structure the information that after the flows of output and input variables are finally concluded, the network economic degrees of freedom are the significant system variables open to selection by the designer.

There are other ways to solve a set of equations with a few variables each. The set can be solved simultaneously (that is, theoretically in one operation) with the total degrees of freedom appearing either as independent variables or as parameters of the solution (as in a linear programming situation equal to zero). By solving sets of equations sequentially or in stages or by iteration, we simulate a flow of information within a network.

Consider the system outline for a boiler-turbine-generator heating plant given in Figure 11.1. The blocks represent major items of equipment. A sub-

FIGURE 11.1

BOILER-TURBINE-GENERATOR SYSTEM

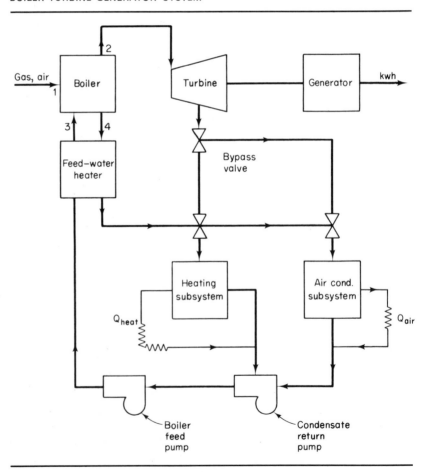

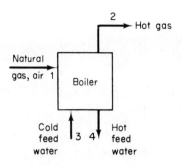

Design Relationships:

1. $M_1 = M_2$
2. $M_3 = M_4$
3. $Q = M_1 (h_2 - h_1)$
4. $Q = M_3 (h_4 - h_3)$
5. $Q = UA (h_2 - h_1)$
6. $U = f(M_1, ..., M_4, h_1, ..., h_4, ..., E)$

Major Variables:

E = efficiency of boiler
Q = heat transferred, Btuh
A = boiler area
U = overall heat–transfer coefficient

$\left.\begin{array}{l} M_1 \\ M_2 \\ M_3 \\ M_4 \end{array}\right\}$ = mass flow rate, lb–m/hr

$\left.\begin{array}{l} h_1 \\ h_2 \\ h_3 \\ h_4 \end{array}\right\}$ = enthalpy, Btu/lb

Degrees of freedom: $12 - 6 = 6$

---

system diagram of the boiler is given in Figure 11.2. For this subsystem 6 design relationships and 12 variables exist for 6 ($= 12 - 6$) degrees of freedom.

The boiler provides a special service to the larger system by heating a hot gas $M_2 = 1675$ pounds per hour with an enthalpy $h_2 = 1194.1$, while the hot feed water has properties of $M_4 = 295$ pounds per hour and enthalphy $h_4 = 1194.1$. Four of the 6 degrees of freedom are consumed in integrating the boiler into the system. The remaining 2 degrees of freedom are open to design selection and optimization for economy. Two variables, Q, the heat transferred, and E, the boiler efficiency, are desirable as design variables.

After the equation and the variables are gathered, as shown in Figure 11.2 for an isolated subsystem, we proceed to the next step. Note Figure 11.3. The equations are listed along the left-hand margin and the variables across the top,

FIGURE 11.3

TABULAR LISTING OF VARIABLES AND
EQUATIONS

Variables

Equations	E	Q	A	U	M_1	(M_2)	M_3	(M_4)	h_1	(h_2)	h_3	(h_4)
1					X	X						
2							X	X				
3		X			X				X	X		
4		X					X				X	X
5		X	X	X					X	X		
6	X			X	X	X	X	X	X	X	X	X

○ Fixed by outside system

□ Preferred as design variable

with the X associating the variable for each equation. Circles are used to indicate those variables that are fixed by the outside environment. As cost is closely associated with boiler efficiency, we have chosen to identify this as a preferred design variable. This is indicated by a square.

Next we eliminate from further consideration those variables beyond immediate control. They are predetermined, and these four columns do not appear in the first pass in Figure 11.4. Those variables which appear once in any equation are found, and the column and row are lined. This concludes the first pass.

The removal of the variable which appears in only one equation is a successive elimination. With A and the fifth equation dropped, U and the sixth equation are lined in the second pass. The third and final fourth pass leave Q. This is the input parameter.

The precedence order of equation calculation is shown in Figure 11.5. The last equation solved is 5, the first one removed, while the first equations jointly determined are 1 and 2, which were removed last. This is a reverse ordering of unknowns-equation solution. The design variable A is then paired to cost. Preferred values of E are used for the correlating function sixth equation. Cost is then finally established for a boiler system where E and A are cross-related.

Observe that there must be as many nonredundent equations as unknown values when dealing with physical systems. Any equation can normally be solved for an unknown quantity provided the other unknown values are supplied from other equations. Eventually the equations are successively solved for unknowns in such a way that each equation solves for one of its most significant quantities based on estimates of the physical aspects of the problem.

Not all the algebraic refinements for constructing an information flow algorithm have been considered. The general appearance of this algorithm is like

FIGURE 11.4

PASSES THROUGH DESIGN-VARIABLE
SELECTION ALGORITHM

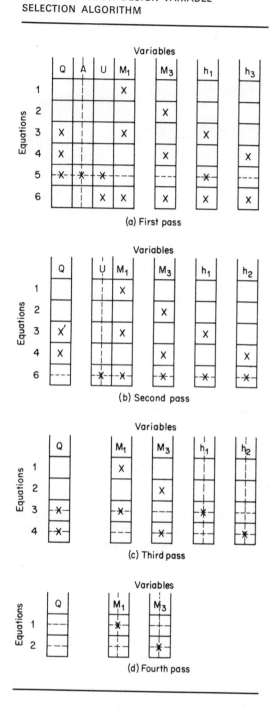

(a) First pass

(b) Second pass

(c) Third pass

(d) Fourth pass

the assignment case of linear programming, a subject mentioned in Chapter 12. Given a much larger system of equations than the small example presented and a computer with the ability to perform many arithmetic operations sequentially, the concept of information flow and system estimating can be used to determine system information requirements leading to the evaluation of system designs.

FIGURE 11.5

PRECEDENCE ORDER OF EQUATION
CALCULATION

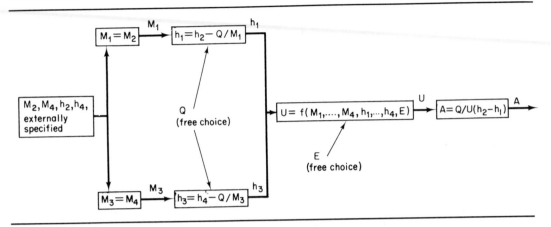

11.3
ANALYTICAL AIDS FOR SOLVING SYSTEM ESTIMATES

We shall now examine other popular devices used to analyze system proposals. These are brief encounters, and the reader is encouraged to examine the Selected References for greater detail on the mechanics of special methods. In the previous section we studied an algorithm which highlighted the flow of information as if the quantities and variables were the real substance itself. A network model, while it is a self-contained tool for visualization and computation, is often inappropriate and other analytical aids for system estimating are necessary. These aids may be subroutines to the network model, or they may be freestanding techniques useful for system estimating.

11.3-1
Budgets

While emphasis so far has suggested that system estimating is the formulation and solution to mathematical statements, budgeting is one aid that cannot be suppressed. Its importance is made clear by its popularity, for the budget is indispensable for planning and traceability.

We have previously studied fixed, variable, operating, and appropriation types of budgets and have shown their construction in Chapter 4. Minor modifications for concepts of opportunity cost, inherited (sunk or residual) cost, and marginal cost make these standard procedures suitable for system budgets.

From the viewpoint of system analysis, cost continues to be broadly interpreted and can be considered to be the amount paid or given for anything whether it is labor or self-denial to secure a benefit. In the minds of many system analysts, cost is only one element of value forgone in order to secure a benefit. In short, cost is a negative benefit. In these terms, cost includes money, time, performance, consumption of scarce resources, and ordinary human skills.

It is relatively easy to determine a value scale which relates the relative worth of one resource to another if an interchange is indexed by the dollar. The money value of inputs is not difficult to establish, as the market place provides a mechanism for assessing these costs. For a commercial system the dollar value measure system applies, and competition aids in establishing the price. In other situations the market mechanism may be unavailable, for example, weapons system or river basin development. Electrical power, one component of a river basin development, may have market value, but recreation or flood risk do not have this advantage. Whenever a competitive market action and reaction is nonexistent the analyst can determine an opportunity cost function for those components which require a value. This is equivalent to imputing a price which might be comparable to a market price.

Inasmuch as resources are always scarce, the options for their application are limited. The selection of one system design precludes another. Thus an opportunity cost arises from the fact that the expenditure of money on one design pre-empts its use for another. Failure to take advantage of the other opportunity may result in foregoing a profit or benefit that otherwise might have been obtained. This is a true cost which can be assessed against the alternative selected. In this context it is called opportunity cost. This indirect cost measurement approach evaluates the resource, which may not have been clearly identified and measured for the selected alternative.

Inherited and marginal costs cannot be separately treated in matters of system budgeting. Inherited or residual cost is a value of earlier resources committed to the system, while marginal costs are the additional costs resulting from a change in objectives or level of the decision. In earlier chapters we suggested that a policy can be made based on marginal analysis irrespective of inherited values. For instance, an optimum operating point n existed when $dC_T/dn = 0$. It is not really this simple. Not all decisions can be made separate of inherited costs and values such as marginal cost theory seems to imply. Certainly the logic to ascertain a marginal decision is based in part on inherited values. To remove or ignore that base and to deny its importance for future decision making overlooks its value as a base representing existing capability. In determining how many additional resources are needed to acquire some specific capability or, conversely, how much additional effectiveness will result from some additional cost is a marginal cost budgeting problem.

How are these concepts incorporated into the budget? The opportunity cost can be a line item in the accounts of the budget, as it is certainly germane to the design selection. A clarification must be made as to what it precisely means on the budget form to avoid misleading interpretations.

Two budgeting extremes for inherited and marginal costs are total of inherited values plus marginal cost, or *full*, and marginal costs only. Because of their extreme position, both can be faulty or correct at various times. In the full method, the overstatement of fixed values may cause insensitivity. One means to overcome a disproportionate statement is to identify that which is fixed inheritance from that which is variable or proportional to the decision. In the latter case, an allocation of a part of the inherited system value tends to reduce the invariant part of the fixed budget.

If one uses an absolute budget, then full inherited values and their amortization are indispensable methods. If a relative general budget is to be employed, marginal cost values can be used. Of course, there is the type of budget that is in between these two. Caution in the preparation and understanding must be exercised.

11.3-2
Discounting

The previous chapter considered aspects of discounting as they related to project estimating. The methods presented there are sufficient for our current need. We now want to point out that which is relevant insofar as system estimating is concerned. The central conclusion that the time-value-of-money concepts can be selected as an optimization criterion for time-dynamic system models is valid. It allows a systematic method of comparing streams of costs and returns.

Generally the discounting or present worth model is used. If the annual rate is 5%, a dollar received 1 year from now is really worth $(1 + 0.05)^{-1}$ and in n years $(1 + 0.05)^{-n}$ dollars right now. Or if you presently had $\$(1 + 0.05)^{-n}$ and lent it at 5% interest compounded, you would have back $\$1$ n years from today.

Regardless of the costing period chosen it is desirable to time-phase system cost throughout the total life period. Thus, the annual cost model is not as appropriate as life cycle costs or the present worth of system design and development, acquisition, and operation costs. This raises the point, What is the life cycle? In public works 30 to 100 years can be used. In weapons system the horizon may terminate within a period of months to 20 years. In commercial enterprises, a system may logically extend between a fad period to an enduring life of 20 to 40 years. All kinds of priorities, political and social, affect the length of time. As the system life cycle becomes longer the actual length becomes less crucial, as the discounting factor drops rapidly if either i or n or both increase.

The total system life cycle is used for another reason. A trade-off analysis between design, investment for construction and acquisition, or operating cost becomes possible if the birth-retirement or death horizon concept is used. In this event it is possible to time-phase all costs and see if, say, additional costs in design might be warranted over excessively large investment and operating costs. Life cycle cost notions were first discussed for product estimating in Chapter 9. Those same practices apply if you substitute the word *system* for *product*. Figure 11.6 is representative of system time phasing of costs.

FIGURE 11.6

COST TIME PHASING FOR A SYSTEM LIFE CYCLE

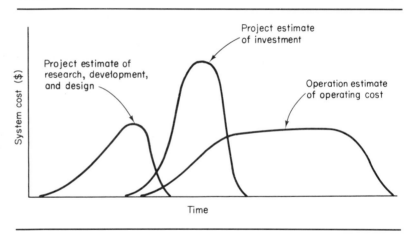

When considering present value, the higher the interest rate i, the smaller the discounted value. The interest rate pertinent for a system estimate is an important subject in its own right and is a lively topic among scholars and practitioners. An unrealistically low or high rate can give a distorted value to the proposed design. In addition, what are the rates for public and governmental units as compared to commercial enterprises? Inasmuch as the government is concerned with social good, there is an undeniable mixture of politics associated with economics. The general effect of an increase in the rate of discounting is not only to make it undesirable to engage in some investment but also to change the character of those system designs adopted, making them less capital-intensive and more labor-intensive.

11.3-3
Benefit-Cost Analysis

Benefit-cost analysis is a practical way of assessing the desirability of system alternatives where it is important to take a long-time and broad view. It is an effectiveness measure in its own right, as it implies the enumeration and evaluation of all or nearly all the selected costs and benefits.

The benefit-cost analysis has a long history, initially applied to the federal improvement of navigation back in 1902. It was used in France before that. For the most part benefit-cost is used by governmental agencies. B/C analysis is applied for water improvement and related land use and its application is controlled by federal statute. Throughout the world the majority of water resource systems are financed with public funds. Other governmental applications have sprouted from these basic two.

As with other effectiveness measures, the B/C ratio has limitations which we cannot cover fully. It is a ratio—simply benefits divided by costs. At that point, however, complications set in. Which costs and whose benefits are to be included? How are they to be valued? At what interest rate are they to be discounted? What are the relevant constraints? There is bound to be arbitrariness in answering these questions.

There are two methods of calculating the benefit-cost ratio. The first involves subtracting annual benefits from annual costs to establish a net annual benefit. Annual net benefits are discounted back to the date of the program's inception and summed to establish a present value of discounted net benefits. The benefit-cost ratio is then formed by relating this figure to the capital cost of the program. This approach is vaguely similar to a business calculation of the rate of profit that can be earned by capital. The second approach is to establish the gross benefits and costs for a typical year. The costs include annual operating costs and amortization of investment. No discounting is used.

In most cases the scope and nature of the system which is to be analyzed are clear. There is a wide class of costs and benefits which accrue to organizations other than the one sponsoring the system and an equally wide issue of how the parent agency should consider them. For instance, a hydroelectric dam can be costed and benefits determined, but what about the recreational amenities, water for farming, and improvements in scenery? The net rise in rents and land values is a result of the benefits of hydro power. These secondary benefits may be more important to one governmental agency than another, and calculations can impute these pecuniary spillovers in a more or less favorable way.

Some advocates suggest that a B/C analysis side-step the issue by requiring public agencies to operate on a commercial basis, leaving resource allocation to be resolved through an artifice of the pricing system. But welfare economics, the well-being of people, income redistribution, market imperfections, and the like make a reasonable demonstration of B/C in a commercial environment difficult. Thus a B/C philosophy has to have a comprehensive public viewpoint. When constrained by laws and appropriations, the B/C ratio is best used as a means of ranking various systems. Those which are higher are considered better from the B/C viewpoint.

Once the tangible benefits and costs are recognized, classes of reimbursable and nonreimbursable cost allocations are made. We have seen the problem of cost allocation before. In a B/C situation though, it is the proper distribution of the costs of the features that serve several purposes that are the problem. This problem does not arise in the cost of a single purpose project, nor when national policy has determined in advance that the purpose to be served outweighs all costs. The costs of a multiple-purpose project are composed of the costs of individual project features such as irrigation canals, power houses, or navigational works, which serve only a single purpose. A dam, of course, serves these several single purposes and its cost must be allocated to both reimbursable and nonreimbursable services.

Broad principles of cost allocation are possible. Each purpose should share equitably in the savings resulting from multiple-purpose construction within the limits of maximum and minimum allocations. The maximum allocation to each purpose is its benefits or alternative single purpose cost whichever is less. The minimum allocation to each purpose is its specific or its separable cost. Joint costs are apportioned without regard to the ability of any particular purpose to pay.

Of the several ways to estimate a benefit-cost ratio, the one chosen here uses discounting. Principally, the B/C ratio has in the numerator the present worth of all benefits, while the denominator is the present worth of all costs. Annual values can equivalently be chosen without affecting the B/C quantity, which is the customary practice for federal projects.

A present worth approach to benefit-cost ratios is determined by

$$B/C = \frac{P_b}{P_c} \qquad (11\text{-}4)$$

where $P_b = \sum_{n=1}^{N} \frac{B_n}{(1+i)^n}$

$P_c = \sum_{n=1}^{N} \frac{C_n}{(1+i)^n}$

B_n = future benefits in year n, dollars
C_n = future costs in year n, dollars
i = interest rate
n = year index, $n = 1, \ldots, N$

As an example, consider the following. A project is planned largely to provide water for industrial, municipal, and domestic use in connection with anticipated development of coal and oil shale reserves. It would increase irrigation supplies for production of livestock feeds and also would benefit recreation, fish, and wildlife, and flood control. Table 11.1 is a final summary of estimated costs of construction and annual operation, average annual benefits, a B/C calculation, and cost allocations and repayment for the project.

TABLE 11.1

BENEFIT COST SUMMARY OF DAMS FOR
OIL SHALE DEVELOPMENT

1. *Estimated costs of construction*

Dam #1	$11,850,000
Dam #2	6,400,000
Dam #3	5,100,000
Canals and diversion dams	23,870,000
Laterals and drains	5,350,000
Operation equipment	330,000
Fish and wild life equipment	610,000
Recreational facilities	1,246,000
Construction total	$54,756,000
Estimated annual operation, maintenance, and replacement costs (100 years)	147,200

2. *Estimated average annual benefits* (100 years)

Industrial, municipal, and domestic water use	$ 4,590,900
Irrigation	869,800
Recreation	245,800
Fish and wildlife	137,700
Flood control	11,000
Less benefits lost	9,900
Net benefits	$ 5,845,300

3. B/C Calculation

$$\sum_{n=1}^{100} \frac{C_n}{(1+i)^n} = 54,756,000 + 147,200 \overset{P/A\ 5^{3/8},\ 100}{(18.5056)}$$

$$= \$\ 57,480,024$$

$$\sum_{n=1}^{100} \frac{B_n}{(1+i)^n} = 5,845,300 \overset{P/A\ 5^{3/8},\ 100}{(18.5056)}$$

$$= \$108,170,783$$

$$B/C = \frac{108,170,783}{57,480,024} = 1.88$$

4. Cost allocations and repayment

	Construction Costs	Annual operation, maintenance, and replacement costs
Reimbursable costs[a]		
Industrial, municipal, and domestic	$39,330,000	$ 61,200
Irrigation	10,160,000	40,100
Recreation	166,500	17,800
Fish and wildlife	375,000	
Subtotal	$50,031,500	$119,100
Nonreimbursable costs		
Recreation	$ 2,765,500	$ 27,500
Fish and wildlife	1,823,000	400
Flood control	136,000	200
Subtotal	$ 4,724,500	$ 28,100
Total	$54,756,000	$147,200
Repayment		
Industrial, municipal, and domestic use[b]		
Prepayment[c]	$ 474,000	
Water conservancy district	38,856,000	$ 61,200
	$39,330,000	$ 61,200
Irrigation		
Prepayment	$ 126,000	
Water conservancy district	6,720,000	40,100
Apportioned to others	3,314,000	
	10,160,000	40,100
Recreation, fish, and wildlife[b]		
Non-federal interests	541,500	17,800
Total	$50,031,500	$119,100

a. Reimbursed over 50 years
b. Repayment at rate of 3.5% annually
c. Nonreimbursable expenses for project investigations

If there are several designs requiring a decision for one system, invariably the cost and benefits are different, and a marginal method must be used to find the best system. Several guidelines are necessary. First, the same interest rate should be used to figure costs and benefits. The same period of life, N, is necessary for the system alternatives. We initially calculate the B/C ratio for each system, such as that which is given by Table 11.2. In this table there are four mutually exclusive choices. System choices that were less than 1 would be rejected. The system choices are ranked in order of increasing cost.

Inspection of these B/C ratios would suggest that alternative B is chosen because its ratio is the greatest. This is an improper selection. The proper choice depends on the principles of marginal return.

Table 11.3 shows the required calculations. System D is used as the initial base since it requires the minimum present worth cost. The first row of this table

TABLE 11.2

BENEFIT-COST RATIOS FOR FOUR MUTUALLY
EXCLUSIVE SYSTEM CHOICES

System Design	Present Worth Benefits	Present Worth Costs	B/C
A	$120,000	$60,000	2.00
B	112,000	53,000	2.11
C	75,000	58,000	1.29
D	64,000	34,000	1.87

TABLE 11.3

MARGINAL BENEFIT-COST RATIOS

System Design	Marginal Present Worth Benefit	Marginal Present Worth Costs	Marginal B/C Rates	Decision
D	$64,000	$34,000	1.87	Accept D
C − D	11,000	24,000	0.46	Reject C
B − D	48,000	19,000	2.52	Accept B
A − B	8,000	7,000	1.14	Accept A

repeats row D from the previous table. Insofar as this first calculation is concerned, the decision is to accept D. Next, the marginal increase in cost and benefits is determined by using the next alternative above the least costly alternative. This would be C, and the C minus D values are indicated for the row. The marginal B/C ratio is 0.46, which is less than 1, and C is rejected. We proceed by considering the alternatives in order of increasing costs. Now choice B is compared to D and the B/C ratio exceeds 1 and design B is preferred to alternative D. System design B is the current best choice. Last, the marginal gain and loss of design A minus design B is computed. The ratio of marginal present worth benefits to marginal present worth costs is greater than 1, indicating that design A is preferable to B. The final choice is A, and it assures that the equivalent present worth benefits will be less than the equivalent present worth costs and the system return is maximized. Choice A is contrary to the initial selection B.

11.3-4
System Boundaries

While discussion has been devoted to the effectiveness issue, little has been said in this chapter about boundaries or constraints that limit the effectiveness. Constraints, whether implied or unimplied, are always existent. These bounds may be in the form of a budget limit in terms of a fixed amount of cash, or a nonlinear mathematical statement may be used. In the *variational* method of estimating, the constraints are explicit, while they are not written down for functional estimating.

The finding of well-expressed constraints is not an easy task. Usually the process starts by verbalizing a known constraint situation. Creating a notation and devising an algebra that translates the verbal statements may be the next step. Or the constraints may be well known, and empirical evidence may have substantiated its statistical behavior. The constraint may be theoretical, resting on mathematical hypothesis. However the constraint is derived, it is of major importance to system estimating.

Sometimes the constraints are obvious and thus easily overlooked. For instance, a constraint type of importance in optimization is that of nonnegativity, i.e., $x_i \geq 0$, $i = 1, 2, \ldots, n$, which says that the x variable cannot assume negative values. Negative production, meaning that the sunk cost of production could be recovered by a backward process such as disassembly of a product, is prevented by the statement that production $x_i \geq 0$.

A translation of a verbal description for a *go no-go* alternative could be

$$x_i = \begin{cases} 1, & \text{if system } j \text{ is accepted} \\ 0, & \text{otherwise} \end{cases} \tag{11-5}$$

This zero-one variable can be exploited to represent combinatorial kinds of restrictions that are often present in capital budgeting problems.

Constraints can be written in a linear way as

$$\begin{aligned} a_1 x_1 + a_2 x_2 + \cdots + a_n x_n &\geq a \\ b_1 x_1 + b_2 x_2 + \cdots + b_n x_n &= b \\ c_1 x_1 + c_2 x_2 + \cdots + c_n x_n &\leq c \end{aligned} \tag{11-6}$$

where an effectiveness function such as

$$p_1 x_1 + p_2 x_2 + \cdots + p_n x_n \tag{11-7}$$

is minimized or maximized depending on the context of the problem. This constrained optimization problem may have

1. No feasible solution, that is, no values of all the x_j's, $j = 1, 2, \ldots, n$, that satisfy every constraint.
2. A unique feasible solution that is optimal.
3. More than one optimal feasible solution.
4. A feasible solution such that the effectiveness function is unbounded; i.e., the value of the function can be made as large as wished in a maximization, or as small as wished in a minimization problem.

These and other matters are covered in Chapter 12, where boundary optimization problems are considered. It has been our purpose to mention some of these notions because of their close connection to system estimating.

11.3-5 Sensitivity

Analyses undertaken for system problems are rarely confined to single numerical values of the optimal solution. We want to know how much the input parameter values can vary without causing alterations in a computed optimal solution or the composition of some policy. An investigation of this sort is termed a sensitivity analysis. Shadow prices and dual variables are other terms used to describe sensitivity factors. In a sensitivity analysis study, fluctuations of the unit profit, or item cost, or time horizon, or product demand, or rainfall, or indirect secondary benefits are permitted. Sensitivity questions are sufficiently complicated to require electronic computation.

In the simplest case, say for straightforward computation of system data, each parameter is varied in turn to determine its effect on the model. Those parameters which are shown to have little or no effect may be treated as constant, or the analyst could be tolerant of variations in these data. Those parameters which when incremented show large variation in effectiveness and alter the optimal solution or reverse policy decisions are called sensitive and bear watching. These kinds of data should be examined in detail since they are important to system effectiveness.

In system problems where mathematical models are important, sensitivity has a more precise meaning. Sensitivity in these cases is used for postoptimality analysis. For instance, where the objective function and constraints conform to the linear or nonlinear models the right-hand constant of a constraint [equation or inequality of the form of Equation (11-6)] represents quantities of scarce resources. The shadow price as used in linear programming indicates the unit worth of each resource and is indicated within the optimal solution. These thoughts are expanded on by the discussion of linear programming and the Lagrange method of undetermined multipliers.

As used in this context, sensitivity means the effect on a system design based on a tolerance range of input variables or postoptimality of a mathematical program modeling the system. Figure 11.7 is a graph showing the first kind.

FIGURE 11.7

SENSITIVITY ANALYSIS WITH RESPECT TO
SYSTEM EFFECTIVENESS

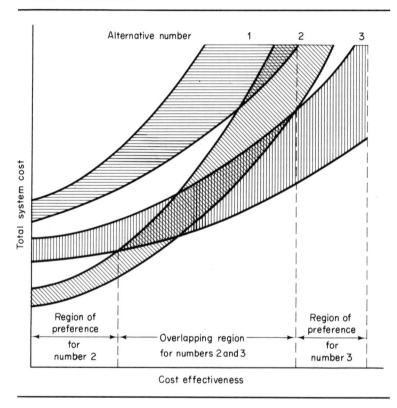

With respect to effectiveness, a total cost can vary for three alternatives to make a number of alternatives acceptable. The notion of sensitivity provides an overlapping region where either of two choices is acceptable in terms of effectiveness.

11.3-6
Probability
The application of the probability art and science is commonplace in system estimating. In making system, product, project, or other design decisions, risk and uncertainty cloud future events. We deal with the future about which we cannot know, but certain things are inferred from what we know about the past. Time-series models, discussed in Chapter 5, are typical. It is information gained from past events used to predict future events. This involves the study of probability. To a small extent, probability has been discussed piecemeal through-

out the previous chapters. (See especially Chapters 5 and 6.) But depth and rigor are necessary for system estimating. For that the student will want to check the Selected References.

As future events are random the insistence on single-valued determined data for input and output is misleading. It is frequently more meaningful to compare system alternatives in terms of a probability of being attained rather than by comparing mean values. As was shown in Figure 6.9, cost estimates can be shown as probability distributions. The mean and variance are important in selecting between two system alternatives. The dilemma of total system cost of an alternative with a lower mean cost but a higher standard deviation—how much higher is naturally germane to the selection.

Given that neither system cost nor effectiveness can be calculated precisely, the determination of the amount of uncertainty is accorded a most-probable cost-effectiveness value. This requires finding the probability distribution of the major units of information. A cost estimate using this reckoning would be given as \$250,000 $\pm$ \$75,000 for a probability of 0.95 of being in this range.

One way to handle probability system cost is via Monte Carlo analysis. A simulation of the system elements is an effective way to model these problems. Many conditional combinations of random costs can be contrived to simulate the system.

11.3-7
General Summation Model

In the previous three chapters on estimating, it is seen that the summing of cost elements of various kinds is a popular method. System estimating is no different. Aggregation of all modes of cost may be handled by a general summation model of the form

$$C = \sum_t e^{-rn_t}(\sum E + \sum M + \sum L + \sum OH + \cdots) \qquad (11\text{-}8)$$

where e^{-rn_t} = continuous discounting factor for n_t number of years for nominal interest r

E = sum of engineering cost [i.e., Equation (9-2)]
M = sum of direct materials [i.e., Equation (8-1)]
L = sum of direct labor [i.e., Equation (8-9)]
OH = sum of overhead [i.e., Table 9.6]

This description is for a product design. A system design formulation would be similar. For instance, a model employing utility[2] is given for system i as

$$U_i = \sum_t e^{-rn_t} \sum_s P_s \sum_j E(u_{jsrt}) \qquad (11\text{-}9)$$

where U_i = utility assigned to alternative A_i, dimensioned utiles

P_s = conditional probability $P(S_s | A_i)$ and S_s is state of nature s while alternative A_i is treated as if it were to be implemented

$E(u_{jsrt})$ = set of utility ensembles considering their probability and criteria function and j is the design criterion

Application of this equation is beyond the needs of this text. Those that are interested in a further digression can consult the Selected References. Terms such as recap sheet, tabular form matrix format, or cumulative total imply the same thing as the general summation model. These formats do not show any mathematical structure. Summation notation is an excellent way to grasp the comprehensiveness of a system design in a language that is understood by practitioners from many disciplines.

[2]From M. W. Lifson, "Value Theory," *Cost-Effectiveness. The Economic Evaluation of Engineered Systems*, edited by J. Morley English, John Wiley & Sons, Inc., New York, 1968, Chap. 6.

The preceding section dealt with several techniques for formulating and solving system-estimating problems. There is no unique way. Each problem must be studied, and then one or several particular means must be used to find system effectiveness. "Use the simplest procedures" is a good rule to follow. This section considers a system problem having national and world implications. Its purpose is to examine the energy conversion problem in general and the mathematical terms and to demonstrate several of the techniques. It is not intended as a thorough treatment; rather it shows the complexities that develop. Comparisons among alternative systems,[3] optimum operation of a grouping of energy converters, and a computational procedure for selection of components are developed.

One of the necessary foundations for a nation's economic growth is the production of energy, heavily manifested in electricity. The rate and need for energy has been growing faster than the population. Since 1900 in the United States the generation of electrical energy has been doubling every 8 to 10 years. When the existing and potential energy needs are considered in terms of energy sources, the problem of energy supply assumes a new mangnitude of importance. If this country is to maintain its standard of living, much less the problems of worldwide growth in living standards, then the production of energy and sources of energy can be viewed as a current fundamental problem. In the near future considerations will have to be given to more efficient forms of utilizing existing energy supplies, to new sources of energy, and to advanced technologies of converting energy to electrical form. Without research and engineering design in all the aspects of this area, the results to society could be severe for two key reasons: First, long-time research efforts will be required for solutions to the complex engineering problems; second, our society of high population density could not exist without energy in the usable forms. An example of new energy source utilization is evident in the construction of nuclear power plants instead of building conventional steam-turbine generation plants.

Looking ahead to the end of this century, growth in electrical energy generation in the United States to a level of 6000 billion kilowatt-hours is clearly indicated. The implications of sixfold expansion over present levels in electrical energy are capital resource requirements amounting to perhaps $300 billion, a buildup in annual primary energy needs to a level of 1600 billion tons of bituminous coal equivalent, and the numerous complex technological problems created in designing, building, and operating the generation, transmission, and distribution facilities and the prevention of an assortment of pollution problems. These are sobering prospects.

Potential sources of nonfossil "fuels" appear to be tidal, wind, and solar energy. Projections about electrical generation are unclear for all three energy sources despite their large theoretical potential. The use of these energy sources is retarded by high cost or inefficient conversion technologies. With practical methods of conversion the above energy sources could help meet the future demands for energy. Underlying the most widespread possibility of utilizing these energy sources is the inherent requirement for storage. The load-factor availability of these sources is random, and the availability of energy is independent of demand needs. It is this erratic and irregular output of energy that especially affects wind generator development, even though its design state of the art is advanced.

[3]One system is due to Arthur Bruckner, II, W. J. Fabrycky, and James E. Shamblin, "Economic Optimization of Energy Conversion with Storage," *IEEE Spectrum*, 5, No. 4 (April 1968), p. 101.

A broad spectrum of progress is being made in the electrical industry. Developments in power transmission are notable. Improvements in operating costs have advanced in nuclear power generation. In some areas nuclear plants with fuel subsidies are becoming competitive with conventional generation.

Behind these advances are the longer-range history and current status in power generation. Conventional conversion plants (steam-turbine generation plants) are nearly engineering-optimal in design and efficiency of energy source conversion as individual units. An area of research is caused by power-peaking energy requirements. The problem is one where the average yearly generation capacity requirements are often on the order of 50% of peak-demand capacity requirements with concomitant losses in efficiency and plant investment cost. Special-purpose gas turbines, pumped-hydro storage, exchange of power, long-range transmission for different time zones, and liquid natural gas for special turbines are solutions to overcoming peak-demand requirements.

To define the terms considered, simplified performance curves of a turbine-generator-boiler unit are shown in Figure 11.8. These three curves are labeled input-output and power output versus heat rate and marginal fuel rate. Corresponding vertical and horizontal scale units may be noted. In the top curve, the fuel input in British thermal units per hour is plotted as a function of output in megawatts. The corresponding heat rate, which is obtained by dividing input by the corresponding output, is shown next. A marginal (or sometimes called incremental) fuel rate is given as

$$\text{marginal fuel rate} = \frac{\Delta \text{ input}}{\Delta \text{ output}} \tag{11-10}$$

The marginal fuel rate is equal to a small change in the input divided by the associated small change in output. As Δ quantities become infinitesimal and progressively smaller, it may be inferred that

$$\text{marginal fuel rate} = \frac{d(\text{input})}{d(\text{output})} \tag{11-11}$$

The dimensional units associated with the marginal fuel rate are British thermal units and kilowatt-hours and are the same as the heat-rate units. The marginal fuel rate is converted to marginal fuel cost by multiplying the marginal fuel rate in Birtish thermal units per kilowatt-hour by the fuel cost in cents per million British thermal units. The marginal fuel cost is stated in mills per kilowatt-hour or dollars per megawatt-hour.

The marginal production cost of a given unit or output is composed of marginal fuel costs plus marginal costs of items such as labor, supplies, maintenance, and water. For a rigorous analytical analysis, it is necessary to express the costs of these items as a function of instantaneous output. In view of this dilemma, arbitrary methods of allocating the labor, supplies, and maintenance costs are resorted to. The most common of these is to assume that these costs are a fixed percentage of the marginal fuel costs.

Underlying the problems of energy supply, the storage of energy is the age-old quest of man. Earliest man wished that he could use winter's winds to pump water in summer and to grind grain in the fall; he wished to store summer's heat for the winter. But the sources of energy were often not available at time of need. Availability of energy and its demand were generally independent of each other. As civilization developed and populations grew, greater demands for energy arose. Coal and oil, which had been stored slowly by nature, became the world's major source of energy fuels. The problems of the future on a worldwide basis require consideration of unconventional sources of energy if the demand for

FIGURE 11.8

INPUT-OUTPUT, HEAT-RATE CHARACTERISTIC,
AND INCREMENTAL FUEL RATE CURVE. (From
Leon K. Kirchmayer, *Economic Operation of Power Systems*,
John Wiley & Sons, Inc. New York, 1958, p. 10)

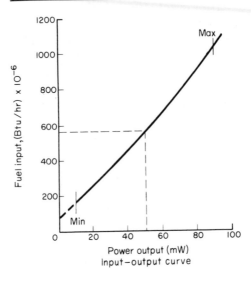

Power output (mW)
Input—output curve

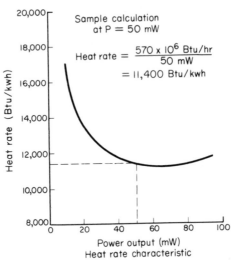

Sample calculation
at P = 50 mW

$$\text{Heat rate} = \frac{570 \times 10^6 \text{ Btu/hr}}{50 \text{ mW}}$$

$$= 11{,}400 \text{ Btu/kwh}$$

Power output (mW)
Heat rate characteristic

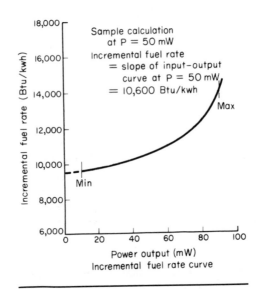

Sample calculation
at P = 50 mW
Incremental fuel rate
 = slope of input–output
    curve at P = 50 mW
 = 10,600 Btu/kwh

Power output (mW)
Incremental fuel rate curve

power is to be met. In addition, technological advances in these areas can also prove to be of value to the conventional generation problems of reduced plant investment and increased fuel efficiency.

Another fundamental approach is to make the generation of energy independent of the demand for energy. This is accomplished by the storage of energy. Conventional conversion systems can be visualized as a two-block model: a conversion block which generates at essentially the same rate as the demand function block. This relationship is portrayed in Figure 11.9, where *blocks* are

FIGURE 11.9

TWO-BLOCK MODEL OF CONVENTIONAL ENERGY
CONVERSION

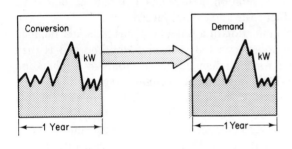

entities of detail having logical wholeness in system interrelationships. When energy conversion utilizes an uncontrolled energy input (or when combined with conventional generation) a three-block model is necessary to overcome the inequalities over time of input and output energy rates. See Figure 11.10. The first block can be visualized as directly supplying demand, or the storage subsystem, or both. The second block, storage, in turn holds supplied energy during some time periods and sometimes supplies energy to the demand block. The third block, demand, obtains its energy from either generation, or storage, or both.

FIGURE 11.10

THREE-BLOCK MODEL FOR CONTROLLED INPUT
GENERATION WITH BALANCED ENERGY
STORAGE

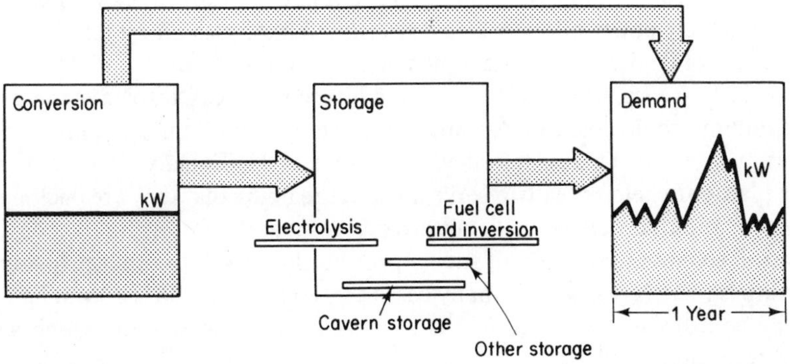

One system under research is the *electrolysis-fuel cell* storage subsystem of a complete energy conversion with storage system. Energy generation input is received by the electrolysis component, which breaks water into hydrogen and oxygen. The hydrogen, in turn, is stored in some container which acts as the time equalizer of input and output energy rates. When stored energy demand occurs, this hydrogen with oxygen or air is released to a fuel cell as the fuel for *combustion* of which electricity is a direct product. This electricity can be sent through an inverter to provide an alternating-current energy supply.

The closest existing parallel of conventional forms of generation to an energy storage system is hydroelectric generation. The purpose of the dam is to equalize the difference in time rates of input and demand output. Unfortunately not enough hydroelectricity is available to meet more than a small fraction of the needs for energy. However, dams with a small watershed supply have been built to act as an energy storage subsystem. These systems are called *pumped storage*. There are a number of power companies utilizing an integrated combination of hydroelectricity, steam-turbine generation, and pumped storage to gain some overall system economies in fuel costs and capital investment.

The *stored energy* is a combination of a watershed plus the water pumped into the reservoir during off-hours' demand such as night-time and is then utilized to meet peaking loads for high-demand output hours in combination with conventional generation. This is an example of engineering utilizing natural advantages to improve efficiency. Cost decisions for all equipment-loading operations are based on the utility industry practices of marginal heat rates. Choice of unit is the straightforward comparison of marginal steam fuel costs versus the cost of hydro power. It should be noted that at best this storage furnishes only a small percentage of peak load and that the pumping capacity is about one-fourth of the pumped storage generating capacity. This hydrogeneration cost is in the upper ranges of mills per kilowatt-hour when compared to the conventional generation units in this system. When demand is at lowest generation power levels, water is pumped into the reservoir. At such times the conventional generation cost for pumping is lowest because the most efficient equipment is used, and the fuel conversion efficiency is increased by an improved running load percentage. A major advantage of pumped storage is its ability to reach full-load generation in about 2 minutes. This rapid start-up capability offers a significant advantage to emergency loading problems, especially when independent of power failure. A desirable factor of feasibility for any energy storage block is the capability of rapidly reaching full load independent of any external power supply. This pumped water development was based on the new designs of combined unit pump-turbine generators for which capitalization economies made feasible such systems where site conditions were desirable. Economic relationships in this study are pertinent to trade-off investigations.

About 3 kilowatt-hours of pumping are required to store enough water to produce 2 kilowatt-hours of on-peak generation. This poses an almost insurmountable economic problem in the minds of many. Put in focus, the night-time pumping is at low cost, whereas the energy delivered on peak displaces energy which frequently would cost twice as much as the energy used for pumping, and sometimes even more. This is described by Figure 11.11. Equally important is that these plants are held as ready reserve most of the hours that they are backing up the system and that they actually cut peaks a comparatively few hours a day. In the overall economic equation of such a project, the cost of the energy lost in the pumping is relatively small. In many cases this could easily be less than half, and maybe as little as one-third, of the savings on manpower costs alone as compared to a thermal plant.

There is also the power system consideration of the mix of generation units. A system with a high percentage of older, inefficient steam-turbine generation equipment and low base loads has higher marginal fuel costs at peak loading for comparative economies. The rate of change of marginal heat-rate

FIGURE 11.11

PUMPED STORAGE INCREASES OFF-PEAK
LOADING OF BASE LOAD UNITS. (From "Pumped-
Hydro Peaking Unit Displaces Steam Units," by R. F. Walker
and R. U. Hugo, *Electrical World* July 26, 1965)

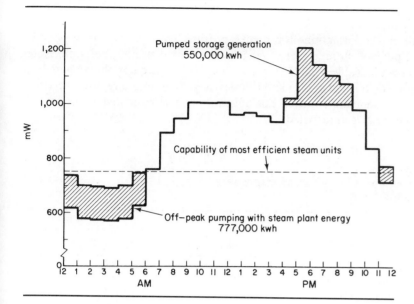

curves for major generation equipment is significant to the costs of pumping. The capacity rating of the most efficient generation units relative to the daily low demand loads also affects the possibilities for economic load factor gains by pumping.

The marginal cost curve is a nonlinear approximation of a step function. Presuming that the cheapest operating unit is on line first, and then in succession other systems come on line, in order of economy, a step function with increasing marginal fuel rates (British thermal units per kilowatt-hour) is formed. A nonlinear least-squares fit will give something like Figure 11.12, where the

FIGURE 11.12

INCREMENTAL COSTS FOR STANDARD AND
PUMPED STORAGE FUEL COSTS

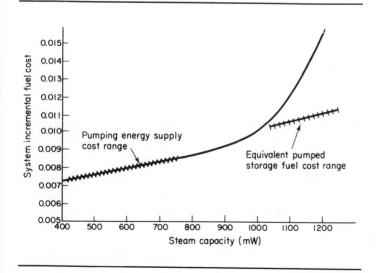

crosses indicate statistical fuzz. The curve indicates the two separate marginal heat rates. In addition to savings in marginal fuel costs, there can be a savings in fixed heat requirements for older units due to availability of the hydrogenerators. A pump storage plant has the opportunity of being self-modernizing since the cost of providing pumping energy improves as better fuel costs are realized with future base load units.

With this information it becomes possible to construct an economic balance of pumped-storage energy versus conventional steam generation. Note the one shown in Table 11.4. Direct-line comparisions are not always possible, and therefore equivalent credits and charges must be found to remove inconsistencies for a comparison. For this illustration it is seen that a savings of $750,000 per year is anticipated.

TABLE 11.4

SUMMARY ANALYSIS OF PUMP STORAGE OVER
STEAM EXPANSION

	Steam	*Pumped Storage*
Unit size, name plate (MW)	—	300
Average net capacity (MW)	265	290
Annual generation (kwh)	1300×10^6	100×10^6
Investment:		
Total including transmission	$\$34 \times 10^6$	$\$31 \times 10^6$
$/kW at max. capability	123	96
Annual charges:		
Fixed charge rate	12	11
Annual cost	$\$4.1 \times 10^6$	$\$3.4 \times 10^6$
Annual operating costs:		
Fuel	$\$2.86 \times 10^6$	$\$0.3 \times 10^6$
Fuel differential	—	$\$3.05 \times 10^6$
Operating and maintenance	0.69×10^6	0.1×10^6
Total operating costs	3.5×10^6	3.45×10^6
Total annual costs	$\$7.6 \times 10^6$	$\$6.85 \times 10^6$
Savings		$750,000

Now consider the economic scheduling of separate power-producing systems. For example, one unit may have continuous day and night operation, while a second is brought on line for increased load. This is a simple scheduling problem. Theoretically, the minimum input in dollars per hour for a given total load is obtained when all generating units are operated at the identical marginal production cost. A mathematical representation of this is given as

$$\text{minimize } F_t = \sum_n F_n \tag{11-12}$$

where F_n = input to unit n in dollars per hour subject to

$$\sum_n P_n = P_r \tag{11-13}$$

where P_n = output of unit n

P_r = received load

As shown in Chapter 12, Equations (11-12) and (11-13) are satisfied when the Lagrangian multiplier is found, or

$$\lambda = \frac{dF_n}{dP_n} \tag{11-14}$$

where λ = marginal production cost of unit n in dollars per megawatt-hour. The value of λ is chosen to have $\sum P_n = P_r$. An increasing λ results in an increase in total generation, while decreasing λ results in a decrease in total generation.

This same result could be obtained by intuitive reasoning. Assume that all units are not operating at the same marginal cost. Consequently, some sources would be operated at higher marginal costs than others. It would be possible to decrease the dollars-per-hour input to the system by increasing the generation at the lower marginal cost sources. In the limiting case, it can be seen that minimum cost is found at the equilibrium point of identical marginal costs.

Using these ideas, now consider a two-source system. For total load $F_t = F_1 + F_2$ and outputs $P_1 + P_2 = P_r$, or $P_2 = P_r - P_1$, find the values of P_1 and P_2 that will result in a minimum value of F_t for a given value of P_r. According to calculus, F_t will be a minimum when the first derivative of F_t with respect to P_1 is zero, or $dF_t/dP_1 = 0$, and

$$\frac{dF_t}{dP_1} = \frac{d(F_1 + F_2)}{dP_1} = \frac{dF_1}{dP_1} + \frac{dF_2}{dP_1} \tag{11-15}$$

As $dP_2 = -dP_1$ and $dP_r = 0$, $dP_1/dP_2 = -1$, we have, by combining equations,

$$\frac{dF_t}{dP_1} = \frac{dF_1}{dP_1} - \frac{dF_2}{dP_1}(-1)$$
$$= \frac{dF_1}{dP_1} - \frac{dF_2}{dP_1}\frac{dP_1}{dP_2} = \frac{dF_1}{dP_1} - \frac{dF_2}{dP_2} \tag{11-16}$$

As $dF_t/dP_1 = dF_1/dP_1 - dF_2/dP_2 = 0$, then finally $\tag{11-17}$

$$\frac{dF_1}{dP_1} = \frac{dF_2}{dP_2} \tag{11-18}$$

That is, the marginal costs of sources 1 and 2 are equal to each other.

The conversion of fossil, water, or nuclear energy into electrical energy is only one part of the energy picture. Electrical transmission from source of generation to user may involve 5% or more in transmission losses. This is a significant amount; furthermore, in the conventional central-steam plant involving coal-to-electrical energy thermodynamic efficiencies of 40% seem to be an upper limit, as technology has reached diminishing returns in economies of efficiency for conventional plants. Because of these two factors, a large institution such as a college, hospital, shopping center, or large apartment can install its own electric generating system, and in appropriate situations the fuel costs to power the generating system are less than the cost of electricity from a utility. Furthermore, the system captures the *waste heat* and converts it into steam or hot water and uses this by-product for heating, air conditioning, and domestic hot water. These self-contained systems can show significant savings in operating costs.

The idea of generating electricity from a privately owned power source is not new. At the turn of the century, many large commercial buildings generated their own electricity for elevators, call bells, fire alarms, arc, and incandescent lighting. In the early 1900s private on-site power generation was replaced by the purchase of power from one central source. Since then, purchased power has been the norm.

While on-site power is not new, the idea of reclaiming its waste-heat by-product and using it to power other building services certainly is. Self-sufficient building complexes using the energy source—usually gas, oil, or even coal—can be utilized to operate a completely integrated system of mechanical services.

A typical total energy system consists of several carefully combined components. An electric generator is powered either by a turbine or a reciprocating engine. A recovery system reclaims the heat left over from the turbine or engine in the form of either steam or hot water. The recovered heat is used for heating or, through the use of absorption-type refrigeration units, for cooling. The turbine and the reciprocating engine handle heat recovery differently. For the turbine, exhaust gases pass through waste-heat boilers which convert the energy into steam or hot water. In the reciprocating engine system, both the engine's jacket cooling water and the hot exhaust gases are passed through heat exchangers to obtain hot water or steam. The lure of this energy concept lies in this recovered heat. It is in a sense free energy since its equivalent would have to be purchased from other sources in the form of other fuels or electricity.

When we begin to plan for evaluation of equipment modules such as engine-switchgear-turbine and the like various systematic procedures can be used to structure the problem. One of these special structures is that of *dynamic programming*. This computational technique is based on a principle of optimality which states that an optimal policy has the property that no matter what the initial state and initial decision are, the remaining decisions must constitute an optimal policy with respect to the state resulting from the first decision. Assume that a system engineer has provided the formulation network given in Figure 11.13 for a high school. This resembles many aspects of Figure 11.1 but

FIGURE 11.13

NETWORK OF A THERMO-ELECTRICAL CYCLE

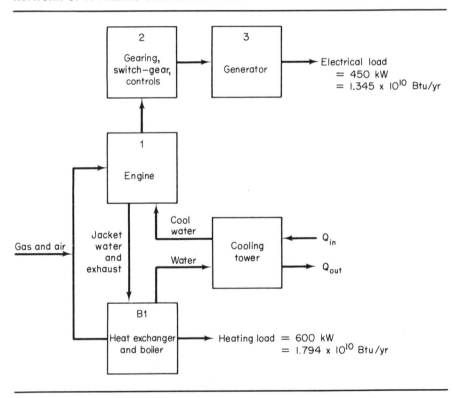

is an abbreviated version. The formulation of this network follows many of the practical steps outlined in Section 11.2.

Note that in Figure 11.13 the electrical load is determined as 450 kilowatts ($= 1.345 \times 10^{10}$ British thermal units per year), while the heating load in equivalent units is 600 kilowatts ($= 1.794 \times 10^{10}$ British thermal units per year).

For this physical system the rate of return is the policy figure of 5% and a planning horizon of 10 years is the life of equipment of this sort. Furthermore, separate analysis has concluded that a configuration using a natural-gas naturally aspirated reciprocating engine is preferred over a turbine. Other technical details have led to this preliminary network and are summarized in Table 11.5.

TABLE 11.5

SUMMARY OF TECHNICAL DETAILS FOR
SELF-CONTAINED ENERGY UNIT

Engine Design	Thermal Efficiency	Capital Cost	Maintenance Cost Per Hour
A	0.30	$18,000	$0.35
B	0.31	19,200	0.34
C	0.31	26,000	0.33

Gearing Switch-Gear Controls Design	Equivalent Thermal Efficiency	Capital Cost
D	0.99	$8000
E	0.97	7800
F	0.95	7700
G	0.90	7000

Generator Design	Equivalent Thermal Efficiency	Capital Cost
H	0.93	$38,250
I	0.90	37,350
J	0.88	35,000

Heat Exchanger & Boiler Design	Primary Fuel Efficiency	Engine Recovery Efficiency	Capital Cost	Operating Cost Per Hour
K	0.80	0.53	$8000	$5
L	0.83	0.43	7000	5
M	0.81	0.46	9000	5

Stage	Verbalized Cost Effectiveness Relationship
Engine	Capital fixed cost + maintenance + cost of unrecovered engine heat
Gearing, etc.	Capital fixed cost + cost of heat losses
Generator	Capital fixed cost + cost of heat losses
Heat exchanger	Capital fixed cost + cost of primary boiler heat losses + operating cost

The problem is to find the minimum capital cost plus costs due to unrecovered heat plus maintenance and operating costs. Other effectiveness measures can be constructed. This one uses vendor specifications certified as minimum operating levels. The system designer is to pick an engine, control, generator, heat exchanger, and boiler set. To select the set purely on fixed cost overlooks inefficiencies, operating cost, and maintenance cost. While there are only 108 combinations (= 3 × 4 × 3 × 3) and all these can be exhaustively compared, a better computational procedure is required for the typical real problem. This is the purpose of multistage analysis: to reduce the arithmetic burden. It offers an improvement over complete enumeration.

A formulation according to the principle of optimality can be written as

$$C_t = \underset{\text{(stage)}}{\text{minimim}}\{r_t \text{ (capital, heat losses, operating costs)}$$
$$+ C_{t-1} \text{ (capital, heat losses, operating costs)}\} \qquad (11\text{-}19)$$

where C_t = optimal system selection for stage t, expressed in effectiveness
dimensions of cost dollars

r_t = functional stage cost of stage t, function of capital, heat losses, costs

C_{t-1} = previous functional cost for $t-1$ stage where optimal policy has
been pursued

t = particular stage up to N

Table 11.5 is a summary of the cost and technical details for a self-contained energy unit. Various designs, A, B, C, and so forth, are indicated for the several modules. The object becomes one of finding the minimum cost of capital, heat losses, and maintenance recursively over the life cycle. Can you find a lower cost combination than the following?

Design Component	System Choice
Engine	B
Gearing	D
Generator	H
Heat exchanger & boiler	K

11.5 SUMMARY

As system problems are configurations of operations, products, and projects in any configuration, a standardized treatment for estimating is largely illusory. An estimating method is invented for the problem. A measure, called effectiveness and having a dollar dimension, is the usual one for cost-estimating activities. Sometimes in the approach a network model is determined. This may or may not have mathematical relationships; if not, experience and several analytical methods are employed. The task of determining the kinds and sources of information is next. As always, data are vital, and careful methods of analysis can be used to extend the usefulness of the information. If the system is long-range and complex, the effort to set a system estimate is formidable. Patronage and an element of forgiveness is a kindly trait to have for estimates of this sort.

SELECTED REFERENCES

For discussion about cost effectiveness and system estimating, the following are pertinent references:

ENGLISH, J. MORLEY, ed.: *Cost-Effectiveness the Economic Evaluation of Engineered Systems*, John Wiley & Sons, Inc., New York, 1968.

FISHER, GENE H.: *Cost Consideration in Systems Analysis*, American Elsevier Publishing Company, Inc., New York, 1971.

McKEAN, ROLAND N.: *Efficiency in Government Through Systems Analysis*, John Wiley & Sons, Inc., New York, 1958.

SEILER, CARL: *Introduction to Systems Cost-Effectiveness*, John Wiley & Sons, Inc. (Interscience Division), New York, 1969.

Electrical power systems and their related economics may be seen in

GALATIN, M.: *Economies of Scale and Technological Change in Thermal Power Generation*, North-Holland Publishing Company, Amsterdam, 1968.

KIRCHMAYER, LEON K.: *Economic Operation of Power Systems*, John Wiley & Sons, Inc., New York, 1958.

SPORN, PHILIP: *Technology, Engineering, and Economics*, The M.I.T. Press, Cambridge, Mass., 1969.

For excellent treatement of operations research, system analysis, and mathematical model building, consult

DE NEUFUILLE, RICHARD, and JOSEPH H. STAFFORD: *Systems Analysis for Engineers and Managers*, McGraw-Hill Book Company, New York, 1971.

FABRYCKY, W. J., P. M. GHARE, and P. E. TORGERSEN: *Industrial Operations Research*, Prentice-Hall, Inc., Englewood Cliffs, N.J., 1972.

GUE, RONALD L., and MICHAEL E. THOMAS: *Mathematical Methods in Operations Research*, The Macmillan Company, New York, 1968.

RUDD, DALE F., and CHARLES C. WATSON: *Strategy of Process Engineering*, John Wiley & Sons, Inc., New York, 1968.

WAGNER, HARVEY: *Principles of Operations Research with Applications to Managerial Decisions*, Prentice-Hall, Inc., Englewood Cliffs, N.J., 1969.

WILDE, DOUGLAS J., and CHARLES S. BEIGHTLER: *Foundations of Optimization*, Prentice-Hall, Inc., Englewood Cliffs, N.J., 1967.

For benefit-cost analysis, see

KENDALL, M. G., ed.: *Cost-Benefit Analyses*, American Elsevier Publishing Company, Inc., New York, 1971.

QUESTIONS

1. Define, in your own words, system effectiveness, cost effectiveness, and engineering efficiency. What are their characteristics?

2. What are the advantages and disadvantages of specific effectivenesses? Visualize several hypothetical systems and state appropriate effectivenesses.

3. When is a relative effectiveness better than absolute effectiveness? If accuracy of data is a problem, how do the two methods rank?

4. Cite the philosophical problems of system effectiveness. Can you counter these arguments?

5. For an imaginary system problem of your choosing, list several intangibles that pertain. Indicate the intangibles which have pros and cons. Using the Delphi methods of Chapter 5, determine an ordinal scale and rank your intangibles.

6. What are various distinctions used for uncertainty? How does risk-certainty-uncertainty differ from technological and strategic uncertainty? Give some methods that might make uncertainty visible.

7. There are several effectiveness policies given in the chapter. Point up their major differences.

8. Why is a network visualization of a system problem useful in analysis?

9. Why are budgets important in system estimating? Define opportunity cost, inherited cost, and marginal cost. How would you go about including these concepts as a line item on a budget?

10. What makes the benefit-cost ratio like an effectiveness measure? How would one go about determining secondary benefits?

11. Provide several definitions of sensitivity. Why is it an important concept? How can it be abused?

12. Safety is an element that is often selected for inclusion within the B/C ratio as tangible and measurable. Yet, one quickly gains the impression that while it is generally considered that there are too many accidents, as a nation we seem unwilling to pay the cost required for a modest reduction in accidents. Thus the B/C element overstates the benefits. Discuss.

13. An effectiveness measure for operating a private automobile is given as cents per mile. For the standard, compact, and subcompact size, measurements indicate 13.6, 10.8, and 9.4 cents per mile for a total effectiveness value. If a strict estimating

fundamentalist is aware of these costs and deliberately chooses to drive the standard car, can you measure the benefits foregone due to the intangibles of owning a standard-sized car?

11-1. There is no uniform method for the analysis of system problems. If one were available and after we had learned the procedures and by strict adherence to the method, we would be assured of a good chance for a successful result. Thus system evaluation is still dependent on the skill and resourcefulness of the engineer. Considering the system concepts of (a) effectiveness and system models, (b) cultivation of alternatives, (c) intangibles and tangibles, (d) time horizon, and (e) fixed, variable, inherited, and marginal cost streams, how do you propose to analyze a technical problem that you are or would be concerned with? State your own problem, give its ramifications, and provide the system procedure for its evaluation.

11-2. A system is conceived of in terms of three goals. Management considers each desirable and of equal importance. The available alternatives are mutually exclusive and are measured by a relative effectiveness consistent for the goals and alternatives. Higher effectiveness is considered more desirable.

	G_1	G_2	G_3
A_1	80	22	19
A_2	50	40	43
A_3	35	70	21
A_4	39	47	48
A_5	25	25	75

(a) Which is the preferred alternative for each goal?

(b) Assuming that goals G_2 and G_3 are equally desirable, what weighting would goal G_1 require to favor alternative A_1?

11-3. The system for producing soap is processed-controlled. A dynamic controller responds to sensors and measures physical states and instructs a selector to identify percent HBr in the recycle gas. See Figure P11-3. Controlling on a

FIGURE P11-3

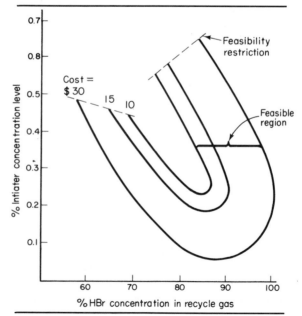

one-to-one basis, i.e., 0.60 to 1.00, the percent HBr is normalized to a 0–1 scale. This dynamic monitoring can be graphically plotted on the isocost map. During the previous period the normalized value is 0.10. Now the value is 0.40. What is the previous and current percentage initiator concentration level to obtain a minimum isocost level? There is to be no interpolation between lines. Draw a line between these two points. The next point fluctuates off the control line in a right-angle direction to another minimum isocost curve point. Where would you expect the next point? Discuss: What kind of a system problem is this?

11-4. A welfare program that is to be discounted at a 5% factor costs $350,000 with a $20,000 annual operating cost. Benefits are anticipated to be $40,000 in the first year and to increase by $30,000 each year for 4 years. Following the fifth year it will decline to no benefit in 2 years. What is the benefit-cost ratio with and without discounting?

11-5. A sociopolitical system has a first cost of $1 million and an annual maintenance cost of $25,000 each year over a 50-year life cycle. Benefits average out as $50,000 per year.
(a) At 5% interest, what is the annual system cost?
(b) What is the benefit-cost ratio?

11-6. See Figure P11-6. An existing highway ABC is 8.5 miles long and consists of a 20-foot × 6-inch concrete pavement with 6-foot oiled shoulders on each side. Constructed in 1924 on a 600-foot minimum radius alignment, it carries an

FIGURE P11-6

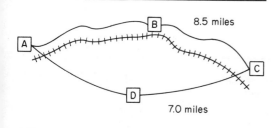

A
B
8.5 miles
C
D
7.0 miles

average daily traffic of 10,000 vehicles with 10% trucks. Now requiring reconstruction, unit estimated costs of widening and improvements to the existing highway on its present alignment is $200,000 per mile. Right of way will cost $55,000 per mile. An alternative location, ADC, can be constructed on a 2000-foot minimum curve. Unit estimating data for a four-lane divided highway is $300,000 per mile. Right of way costs $22,000 per mile. ADC requires two railroad viaducts with estimated costs of $250,000 each. Maintenance cost of either line is $600 per mile-year. Traffic will double in 20 years. Truck operating costs $0.25 per mile and cars cost $0.10 per mile. If ADC were constructed, it is estimated that 40% of the traffic between points AC would stay on ADC and that the remainder would use ABC. Based on a project estimate of capital cost only, which route would you advise? What other factors should be considered? What would be the savings in traffic in 20 years if ADC were built? Construct a discounted B/C ratio for 5% considering all cost effects. Does this change your recommendation?

11-7. The B/C test has been applied to welfare and job training for underprivileged youth. In this case an undiscounted B/C ratio is given as

$$\text{benefit-cost} = \frac{B_p - B}{C_a - T_n}$$

where B_p = graduate earnings, B = original earnings of student, C_a = annual amortization payment, and T_n = taxes on net increased earning of student. The following items are estimated:

Direct program training cost	$ 725,000
Allocation of center overhead based on planned enrollment	950,000
Subtotal	1,675,000
Capital investment cost at 5%	83,750
Job corps cost at 25%	418,750
Total cost	2,177,500
Number of graduates	400
Total cost/graduate	5,443
Five-yr amortization cost (C_a)	1,089
Average starting salary	4,222
Five-yr average salary (B_p)	5,026
Five-yr average taxes (T_n)	562
Original earning power of students (B)	1,040

Find the undiscounted and discounted benefit-cost ratio. Let the interest rate be 5%. What happens to the B/C ratio as the interest rate increases? Decreases?

11-8. Construct a personal B/C ratio for your own education along the lines of the one in the previous problem.

11-9. Four mutually exclusive designs have their benefits and costs estimated:

Design	Present Worth Benefits	Present Worth Costs
A	$48,000	$38,000
B	35,000	24,000
C	37,000	31,000
D	45,000	34,000

Determine the individual B/C ratios and analyze the four choices on the basis of marginal yield to find the best one.

11-10. Using Figure 11.8, what is the heat rate for 50 megawatts in British thermal units per kilowatt-hour? What is the marginal fuel rate for 50 megawatts?

11-11. Units 1 and 2 have fuel inputs given by

$$F_1 = (8P_1 + 0.024P_1^2 + 80)10^6$$
$$F_2 = (6P_2 + 0.04P_2^2 + 120)10^6$$

where P_1 and P_2 are unit outputs in megawatts. Plot the input-output characteristics for each unit expressing input in British thermal units per hour and output in megawatts. Assume that the minimum loading of each unit is 10 megawatts and that the maximum loading is 100 megawatts. Calculate the heat rate in British thermal units per kilowatt-hour and plot against output in megawatts. Assume that the cost of fuel is $0.20 per 10^6 British thermal units. Calculate the marginal production cost in dollars per megawatt-hour of each unit and plot against output in megawatts. Assume that the system load starts at 50 megawatts and increases to 200 megawatts. Which unit is placed in service first, and also what is the system load when the second unit comes on line?

11-12. Problems in providing energy to large cities are immense considering the logistics of supplying coal or gas to these centers. To avoid that problem of transporting huge supplies of coal to these centers, coal-fired plants are located at the mine mouth and the energy is delivered by 500-kilovolt alternating-current transmission lines. Four remote plants have been engineered, and data on the basis of delivered energy to a large city are estimated:

Power Plant Site	Bus Bar Cost (mills/kwh)	Transmission Fixed Cost (mills/kwh)
Green River	4.13	1.37
Black Mesa	4.09	0.99
San Juan	3.89	1.37
Victorville	5.19	0.17

Which power plant site offers the best potential for delivered energy? A gas-fired plant located in the city has a total cost of 5.70 mills per kilowatt-hour. Discuss the point that electrical energy can be delivered to a city from remote coal-fired plants via extra high-voltage transmission lines cheaper than from natural gas-fired plants located in the city. What are the benefits? Problems?

11-13. A five-potline aluminum reduction operation will have a demand of 400 to 485 megawatts and an annual use of 3.8 billion kilowatt-hours. If a $0.01-per-$10^6$-British-thermal-unit differential in energy cost can make a difference in cost of 0.09 mill per kilowatt-hour, what is the annual cost differential due to this fuel reduction? Discuss the importance of very small fuel economies to the economic welfare of a large aluminum reduction plant. Discuss the political question of charging a large user of energy a higher unit cost for energy and the smaller user a lower unit cost.

11-14. The production of synthetic natural gas (SNG) from a gasification plant of coal is offered as a system to overcome shortages of natural gas. A system to produce SNG consists of a coal gasification unit; oxygen plant; steam, power, and water supply plants; processes to handle the effluents; and coal- and ash-handling facilities. A system to produce 250×10^6 standard cubic feet per day of high-British-thermal-unit gas costs approximately $250 million. In the event of stringent local environmental air pollution requirements, these costs could be higher. On the other hand, a power plant consuming the same amount of coal as a 250×10^6 standard cubic feet per day high-British-thermal-unit gas plant, namely 800 short tons per hour with a lower heating value of 9500 British thermal units per pound with a thermal efficiency of 38%, produces 1700 megawatts. Two 850-megawatt power plants cost approximately $250 million. Evaluate the cost of a British thermal unit by either of these two routes. Determine your own local energy rates and compare. Are these proposed costs realistic? Discuss: Investment costs for both systems of processing coal to a high-grade energy are equal related to the coal input. The investment cost of a gas-producing plant is therefore competitive with an electrical generating plant, and if we consider the more favorable thermal efficiency of the coal-to-gas process (claimed to be 70%), the coal gasification system should have an important role in an overall system for the utilization of coal.

11-15. A gas energy system uses natural-gas reciprocating-engine generation sets to provide electrical power. The heat in the jacket-cooling water and exhaust gas is recovered to heat the building. The system block model is shown in Figure P11-15. The electrical loads establish the primary requirements in selecting the size and number of engines. Electrical loads must be met. If any additional heat is required for air conditioning or heating beyond that which is recovered from the engines, supplemental gas is purchased. But air conditioning and heating

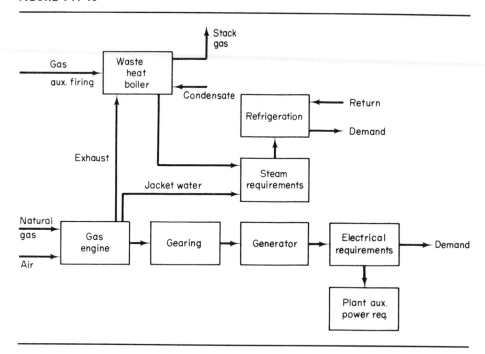

supplemental requirements are not sensitive factors for this problem. A frequency distribution of total kilowatt demand for the 8760 yearly hours is determined.

Condition	Kilowatt Plateau	Yearly Hours (%)
1. Low load	175	5
	225	10
	270	10
2. Intermediate load	175	5
	225	10
	400	10
	525	10
3. High load	225	5
	400	5
	525	10
	900	10
	950	10

Two engines are being considered and their performance is as follows:

175-Kilowatt Engine		225-Kilowatt Engine	
Efficiency (%)	MBtu/hr	Efficiency (%)	MBtu/hr
50	1250	49	1523
63	1453	63	1786
72	1580	74	2003
81	1695	80	2120
90	1825	89	2250
95	1890	98	2430
100	1960		

Gas costs $0.20 per thousand cubic feet for a gas quality of 840,000 British thermal units per thousand cubic feet. The installed cost per kilowatt is $188 and $169 for 175- and 225-kilowatt engines. It is usually desirable to load gas engines at 90% rated capacity. Heat rejected to the jacket water is 3400 British thermal units per kilowatt-hour and is 90% recoverable. Heat rejected to exhaust is 3400 British thermal units per kilowatt-hour but is 66% recoverable. About two-thirds of waste energy is in the jacket. Maintenance costs are the same for either engine at $0.20 per hour per operating engine. One engine is required as a standby unit. Ignore property taxes, insurance, and operating personnel costs. Determine which of several effectiveness measures would be suitable to determine the number of engines. Generally, most all-gas investment systems are optimized if a maximum number of engine-generator sets are of minimum power. The idea is to have the engines operating on their maximum ideal gas consumption point. Estimate the number of engines, their size, capital costs, gas costs, and credits for heat recovery. Try two different schedules of engines to attempt a minimization.

11-16. A one-component separator is being considered for a design variable selection

FIGURE P11-16(a)

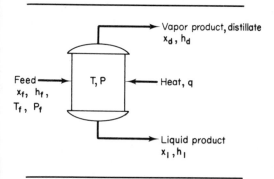

algorithm. The sketch is shown in Figure P11-16(a). The design equations are as follows:

1. Material balance: $x_f = x_d + x_l$

2. Energy balance: $x_f h_f = x_d h_d + x_l h_l + q$

3. Enthalpy of feed: $h_f = f(T_f, P_f, x_f)$

4. Enthalpy of vapor: $h_d = f(T, P, x_d)$

5. Enthalpy of liquid: $h_l = f(T, P, x_l)$

6. Equilibrium constant: $K = \dfrac{x_d}{x_l}$

How many equations, variables, and degrees of freedom are there? Inasmuch as the feed is known, x_f, P_f, h_f, and T_f are quantities that are circled in Figure P11-16(b). Additionally it is desired that K be the preferred design variable, which is boxed.

Complete the four passes to this problem. The final pass is given as

	x_d	x_e
1	x	x
6	x	x

	x_f	p_f	h_f	T_f	x_d	h_d	h_e	x_e	q	T	P	K
1	X				X			X				
2	X		X		X	X	X	X	X			
3	X	X	X	X								
4					X	X				X	X	
5							X	X		X	X	
6					X			X				X

Construct a precedence chart. Discuss: The flow of information as a consequence of the precedence chart.

11-17. A rotary dryer using countercurrent flow is used to dry wet salt. Figure P11-17(a) is a sketch giving pertinent symbols. Eight equations describe this system [see

FIGURE P11-17(a)

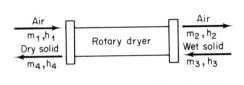

FIGURE P11-17(b)

	M_1	M_2	M_3	M_4	A	h_1	h_2	h_3	h_4	U	a	h_s	h_σ	E	Q
1	X	X													
2			X	X											
3		X				X	X								X
4				X				X	X						X
5					X					X	X	X	X		
6								X	X			X			
7						X	X						X		
8	X	X	X	X						X		X	X	X	

Figure P11-17(b)]. Exceptional notation:

a = surface of solid particles exposed to air

A = cross-sectional area of dryer

h_s = internal enthalpy of solid salt

h_a = internal enthalpy of air

Complete the passes. The precedence chart in Figure P11-17(c) is an aid. Discuss: What information is specified externally to the problem, and what information must the estimator uncover? What are the variables that are cross-related?

FIGURE P11-17(c)

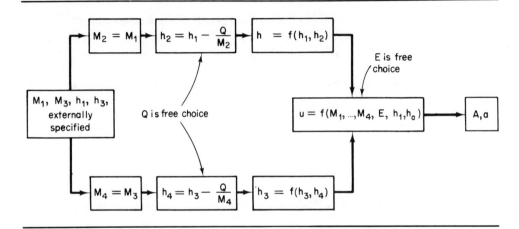

11-18. A system has 13 variables, $x_1, x_2, \ldots, x_{13}$ and 7 constraints. x_1 is considered an important variable and would be a desirable economic degree of freedom. Variables x_5, x_9, and x_{11} are specified externally of the subsystem. The design relationships are

1. $x_2 = x_3 x_4 x_{13}$

2. $x_{13} = \dfrac{(x_9 - x_{12}) - (x_{10} - x_{11})}{\ln[(x_9 - x_{12})/(x_{10} - x_{11})]}$

3. $x_5 = x_6$

4. $x_7 = x_8$

5. $x_2 = x_5(x_9 - x_{10})$

6. $x_2 = x_7(x_{12} - x_{11})$

7. $x_4 = f(x_1, x_5, \ldots, x_{12})$, where x_4 is a correlated function of the nine variables

Determine a network diagram showing the solution to the equations. Discuss: What are the variables cross-related to cost?

11-19. A communication satellite system is being designed to broadcast TV signals from space to n earth stations. See Figure P11-19. The total cost of the satellite C_s is a function of its transmitter power P_s, or

$$C_s = 3 \times 10^5 P_s \quad \text{dollars} \tag{1}$$

The cost of each station C_e depends on the size on the antenna it employs, or

$$C_e = \frac{20 \times 10^5}{D^{1/3}} e^{D/45} \quad \text{dollars} \tag{2}$$

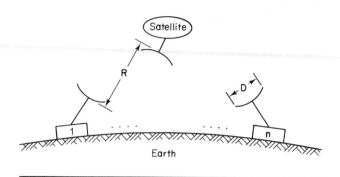

for $15 < D < 250$, where D is the diameter of the antenna in feet. The total system cost TSC is then

$$TSC = C_s + n \times C_e \tag{3}$$

The strength of the signal received at each earth station S_e depends on the gain of the satellite antenna G_s, the transmitter power P_s, the gain of the earth station antenna G_e, the wavelength of the signal λ, and the distance between the satellite and the earth station R. The relation is

$$S_e = \frac{P_s G_s G_e \lambda^2}{(4\pi R)^2} \tag{4}$$

The gain of the earth station antenna G_e depends on its diameter D:

$$G_e = \tfrac{1}{2}\left(\frac{\pi D}{\lambda}\right)^2 \tag{5}$$

This can be substituted to yield

$$S_e = \frac{P_s G_s (\pi D)^2 \lambda^2}{2\lambda^2 (4\pi R)^2} = \frac{P_s G_s D^2}{32 R^2} \tag{6}$$

For a satellite in geostationary orbit R is approximately 125×10^6 feet. For the purposes of this problem assume that G_s is a constant equal to 400. Thus

$$S_e = \frac{400}{32(125 \times 10^6)^2} P_s D^2 = 8 \times 10^{-16} P_s D^2 \tag{7}$$

For proper reception assume that S_e must be equal to or greater than 10^{-10}. Compute the values of P_s and D which minimize total system cost and meet the minimum signal requirement of $S_e = 10^{-10}$ for $n = 10$. [*Hint*: Substitute $S_e = 10^{-10}$ in equation (7).] Compute the values for $n = 100$ and $n = 1000$. Discuss what happens when n gets larger in a practical sense. In a trade-off sense.

CASE PROBLEM   *The Search for an Effectiveness Measure in Telecommunication Planning.* The next 20 years will see a tremendous growth in the telecommunication industry, and what is now a $30-billion business will experience a fivefold increase. This expected growth in the volume of communication will result from color TV, telemetry, graphical and computer links, defense needs, and commercial and industrial needs. Coincident to these pressures is the influence of the population and urbanization of large cities. It is suspected that these demands will congest modes of telecommunication traffic and that telecommunication planners will be forced to more costly and less efficient modes of communication.

It should be apparent that the title of this problem is appropriate since there is probably no one "ideal" effectiveness measure. Despite this shortcoming one may well imagine that the government and telecommunications industry will find the results of a suboptimum solution useful to guiding management in the control of literally millions of dollars (billions by the entire industry) to achieve a competitive market place.

Four major telecommunication systems will be studied in this case problem. They are line-of-sight (LOS) microwave, tropospheric scatter, direct-burial ground coaxial cable, and synchronous satellite systems.

The following summaries indicate today's cost for construction and operation of the four modes of communication. Their presentations vary since the compiler of these data did not follow uniform disclosure methods. Service life and the investment rate differ with respect to a particular mode. It has been ascertained from similar and earlier cost data that construction and operating costs are decreasing; for electronics construction and equipment costs are decreasing at a rate of 7% per year while annual operating expenses are decreasing by 4% per year. Conveniently, inflation effects can be ignored. A government-controlled satellite utility system is restricted to a rate of return of 4% on capitalized investment. Private telecommunications industries will be seeking 20% on investment before taxes.

LOS MICROWAVE SYSTEM COST BREAKDOWN

Link length between towers: 30 miles

Item	Constant	Variable
1. Terminal studio (2 required)	$270,000 each	—
2. Repeater stations (number depends on length)		$162,000 each
3. Operating expense, yearly	732,000	18,000/tower

Number of towers = repeaters + terminal studios
Life of repeater towers: 20 years
Life of terminal stations: 5 years

TROPOSPHERIC SCATTER SYSTEM COST BREAKDOWN

Link length between towers: 225 miles

Item	Constant	Variable
1. Terminal station (2 required)	$2,250,000 each	
2. Relay stations (number depends on distance)		$3,050,000 each
3. Operating expense, yearly	2,000,000	320,000/station

Number of towers = repeaters + terminal studios
Life of repeater towers: 20 years
Life of terminal stations: 5 years

DIRECT-BURIAL COAXIAL CABLE SYSTEM COST

Average cost: dollars per channel mile

Channel Capacity	Construction	Operating Cost
240	$ 61	$14
120	122	27
72	204	45

Life may be assumed to be 20 years.

SATELLITE COMMUNICATION SYSTEM COST

1. Construction cost per land channel mile is $46.
2. Annual operating cost per land channel mile is $15.

A 10-year life for ground equipment and a 7-year life for the satellite with an interest rate of 4% is assumed.

SUBSCRIBER COMMUNICATION CENTERS

The adoption of a particular system depends on the routing between subscribers. Five cities have been selected as typical, for they represent a variety of technical considerations and suggest that communication is geographically and population interdependent.

Subscriber Route	One-Way Miles	Feasibility LOS	Tropo	Ground	Satellite
New York to Paris	2625	No	No	No	Yes
New York to Washington	200	Yes	Yes	Yes	No
New York to Los Angeles	2500	Yes	Yes	Yes	Yes
New York to Denver	1400	Yes	Yes	Yes	Yes

Megatropolis	Population	Estimated Maximum Upper Limit for Channels of LOS, Tropo, Satellite Modes
New York	30,000,000	90,000
Paris	15,000,000	37,500
Los Angeles	15,000,000	37,500
Washington, D.C.	5,000,000	12,500
Denver	1,500,000	3,750

Urbanization and population growth of these centers are expected to continue at the rate of 3% per year compounded annually.

Usage of telecommunications systems has been correlated to urban population, and for the free world the estimate relates activity to channels as follows: One channel of voice communication of a space mode = 500 population. Cable channels may be assumed infinite; i.e., they can be buried over the entire earth, except as it becomes uneconomical.

The mode of activity depends on the technical attractiveness and economy of the device at any particular time. For example, LOS systems are popular now, but in the future other modes will become more competitive in engineering and cost. The current usage of the several systems is known and specialists have predicted trends at year 2000.

PREDICTED WORLD PROPORTION OF TELECOMMUNICATION BUSINESS

	Today	2000
LOS	50	30
Troposcatter	10	20
Overland cable	20	25
Satellite	5	10
Submarine cable	15	5

Mode	Bandwidth	Channel Width
Microwave	0.72–24 MC	4 KC
Troposcatter	0.072–2.4 MC	4 KC
Submarine	0.72–8 MC	4 KC
Overland	0.72–8 MC	4 KC
Satellite	0.72–24 MC	4 KC

Number of mode channels = net band width ÷ channel width.

You are to find an effectiveness measure for New York telecommunication. First, in this system case problem, list the apparent problems as discussed. Next, look at the solutions, and attempt to write basic effectiveness functions. Do not try to be exhaustive or accurate. Consider the specific route problem of New York to Washington, and determine the number of voice channels that population arithmetic will lead to. Satisfy these local needs by determining the cost of four systems to satisfy that requirement. Find which modes are filled, and determine and rank the cost of the communication. Continue with other cities.

A telecommunication planner will buy and operate a system that meets his technical requirements. If two or more systems are technical competitors, then the systems planner will select the lower-cost one. Any mode can be filled until the estimated upper limit is reached. The planner will then search for available open modes following the line of least cost resistance. If all systems are blocked, no additional space communications business is tolerated and that portion of the telecommunication spectrum is filled. Channel requirements will be scheduled on the basis of one channel per 500 population. Traffic will be considered one-way for a city. For example, New York has allotted a maximum of 90,000 channels for communication to the other four cities. Return messages are classed as a shadow problem and will be ignored. Messages are sent to population centers on the basis of population proportions of the other cities.

Improve on your effectiveness measure. Consider the number of messages per dollar, or the annual cost, that attempts to solve the original problem. Do not be discouraged because of the size of this case problem. Actually, telecommunications planners have no perfect solution to this problem either.

A more successful approach would result from a computer simulation model where the effects of increasing telecommunications activity resulting from population pressures can be programmed. If time permits, construct a computer model to solve this problem for all modes of communication leaving New York to the other cities.

chapter 12

*How is it possible to budget for solvency
in dealing with matters of chance?*
GEORGE BERNARD SHAW
Everybody's Political What's What

OPTIMIZATION WITH ENGINEERING-ECONOMIC BOUNDARIES

Cost estimators help in the economic optimization of engineering designs. The effort to achieve a lower cost or improved rate of return or effectiveness requires a broad base of support within the organization. Designs are too complicated to expect one individual to design, estimate, and optimize. Cross talk between many interested parties is habitual in optimization. In view of his special understanding of the design's economic facts, the estimator is a contributor to an optimization program, or just as often, he may be the chief individual in its management and followthrough. First, he is the central source of estimates for a design. Using his functional models, tabulations, and recap sheets, he is able to supply the coefficients, parameters, and exponents for a design. Labor? This information comes from a labor time standard or a factor percentage. Material? This information comes from a bill of material and design. Purchase price for materials or outside labor? This information is in his file. The essential informational requirements for optimization are accessible to him. Despite any mathematical refinements, the optimization is only as good as the estimates on which it is based. Before meaningful answers become possible, realistic information is necessary.

The previous chapter touched lightly on some of the optimization concepts. In this chapter we shall discuss those ideas again. Now we take a functional estimating model, or, as it is otherwise called, an objective function, and construct mathematical constraint equations. The objective function may be viewed as an extension of the popular summation models found in each of the estimating chapters. As the title of this chapter suggests, the constraints are engineering and economic in character. They may be either linear or nonlinear, but they relate to the purpose of the objective function. Constraints may result from limitations on the quantity produced due to production bottlenecks, faltering of price as quantity increases, or empirical evidence of physical laws, to name a few.

We study the methods of linear programming, the Lagrange method of undetermined multipliers, and geometric programming. There are many more, but texts in operations research and optimization should be examined for greater detail and depth.

12.1
LINEAR PROGRAMMING

Linear programming is a mathematical technique used to optimize systems of equations. To the extent that the system of equations approaches the real thing it pretends to be, linear programming can be exploited. There is a great variety of practical applications that have been solved by linear programming. This is understandable in view of the ability of the computer to handle complex problems and the ability of the analyst to resolve the relationships into straightforward linear equations.

Of all the programming methods (computer programming is not implied here) linear programming has seen more growth in theory and practice than its younger brother, nonlinear programming. Since 1947, when Dr. George B. Dantzig published his first paper on the simplex method, progress in the field has been rapid. Although the first applications were military, it was not long until there were important industrial applications. Linear programming problems are structured according to the following: (1) There is some optimal solution to be obtained such as maximum profit, minimum cost, or minimum elapsed time of the system being studied. (2) There are a very large number of variables that require consideration simultaneously. These may be outputs of the system (such as products and components), while others are classified as inputs to the system, such as cash flow dollars. The inputs are sometimes known as resources. (3) Some function to be optimized is restricted from the practical consideration of the problem at hand. These restrictions result from conflicting objectives of the firm, scarce resources, or critical interactions placed on the variables. Briefly then, linear programming is a procedure for determining an optimum program of interdependent activities when restricted by available resources.

The word *linear* implies that the problem conforms to the linear programming model with a simple straight-line algebraic relationship between the variables, and linearity can be closely approximated by first-order equations. If this condition is violated, methods of nonlinear programming are required. Now the term *programming* will henceforth refer to a process of solving for a particular set of solutions, or a program, or a plan of action which optimizes (either maximizes or minimizes) some mathematical objective.

The central characteristic of linear programming is to optimize a primary linear function having many linear restrictions. Programming is the mathematical method for the analysis and computation of optimum decisions which do not violate the restrictions as imposed by inequality side conditions with computation by a so-called iterative plan. *Iteration* is a term that denotes a systematic trial and error procedure and leads to an answer that will not ordinarily be arrived at directly. The reader should not presume that iteration is exhaustive enumeration or wild guessing. Consider this situation. If a problem involved 10 simultaneous equations with 30 variables, the number of solutions required, if they are all to be tested for optimality by exhaustive enumeration, would lead to better than 30 million solutions. Even with electronic computers this is an impossible task. The frustrating part about the 30 million solutions is that a large percentage would turn out to be impractical or impossible, but there would be no way to identify these in advance. It is clear that there is a need for an efficient procedure.

12.1-1
The General Linear Programming Problem

A linear programming problem requires that a linear function

$$y = c_1 x_1 + c_2 x_2 + \cdots + c_n x_n \qquad (12\text{-}1)$$

be maximized or minimized. Without any constraints that limit the selection of the x_n variables, the solution is absurd. In the quest for a maximum we would choose x_n's having positive coefficients and assign any large arbitrary value. If we were interested in minimizing, then decreasing the variables (for positive-sign coefficients) or increasing the variables (for negative-sign coefficients) would ultimately force our objective to a minimum value of zero, an obvious trivial example. Thus restrictions of the form

$$a_{11}x_1 + a_{12}x_2 + \cdots + a_{1n}x_n \leq b_1$$
$$a_{21}x_1 + a_{22}x_2 + \cdots + a_{2n}x_n \leq b_2$$
$$a_{i1}x_1 \qquad + \cdots + a_{ij}x_j \leq b_i \tag{12-2}$$
$$\vdots$$
$$a_{m1}x_1 + a_{m2}x_2 + \cdots + a_{mn}x_n \leq b_m$$

restrain a maximizing solution. Furthermore, the variables x_j must be controlled by

$$x_j \geq 0, \qquad j = 1, \ldots, n \tag{12-3}$$

This requirement, known as nonnegativity, stipulates that the variables must be zero or positive. Negative variables are ruled out. This nonnegativity feature excludes reversible processes, such as negative production having no physical counterpart. The a_{ij}'s are called technological coefficients while the b_i's are resource restrictions.

12.1-2
The Graphical Method

The linear programming problem can be solved by graphics, systematic and exhaustive enumeration, vectors, and a simplex method. Although it is the simplex method that is important, understanding the graphical method where easy problems are solved permits the reader to grasp relationships between the solution stages of the simplex method. The graphical solution is limited to problems having two or three activities. In the three-dimensional problem each restriction consists of a relationship between the three variables and provides a plane in the three-dimensional space. First, however, consider the two-dimensional case.

A manufacturing company has two products, A and B. Standard time data developed through consistent work study practices have provided reliable information. Product A requires 2 hours of sheet-metal work, 5 hours of machine-shop work, and 0.40 hour for painting operations, while product B requires 1 hour for sheet-metal work, 12 hours for machine-shop work, and 0.33 hour for a painting operation. The manufacturer is limited to 80 hours in the sheet-metal department, 600 hours of machine-shop time, and with one painter he has 40 hours available. He makes a profit of $150 on each product A and a profit of $200 for product B. He can sell all that he can make. The question to be considered is how the manufacturer should allocate his production capacity to models A and B. That is, how many of each model should he make in order to maximize his profit? The data for this problem are given in the following table and the objective is to determine that mix of products A and B which yields the maximum profit:

Department	Production Time Required Per Product		Capacity Restriction Per Week
	A	B	
Sheet-metal shop	2	1	80
Machine shop	5	12	600
Paint shop	0.40	0.33	40
Profit per unit	$150	$200	

In a graphical construction of the constraints one first assumes that the information is an equality rather than an inequality. For sheet metal, the equation would read $2A + 1B = 80$, which expresses that for each unit of A, 2 hours are required, and 1 hour for unit B. However, is this what the cost analyst really means? He presumes that the sheet-metal department could be working less than 80 hours if this proves more optimal and should have the restriction of $2A + 1B \leq 80$, which is more in line with what is generally desired. From this restriction we see from Figure 12.1 that the area above and to the right of this line is excluded, and the cross-hatching area satisfies, for the moment, acceptable production. The sheet-metal restriction $2A + 1B \leq 80$ defines a half-space on the graph denoted by the axes A and B. However, the linear programming rationale imposes other restrictions which have been previously stated to be the nonnegativity requirements with only positive value variables permitted. This restriction $A, B \geq 0$ limits the output of products A and B either to zero or positive values; thus the region of interest as represented by the shaded area in Figure 12.1 is misleading inasmuch as negative values for A and B, which mean

FIGURE 12.1

HALF-SPACE RESTRICTION OF SHEET-METAL
DEPARTMENT CONSTRAINT

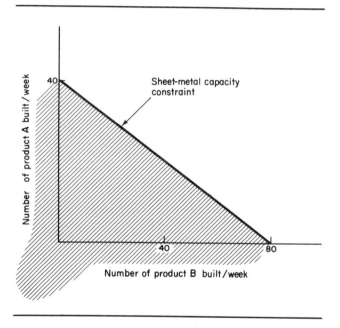

negative production, have no real physical counterpart. Thus, the introduction of the nonnegativity constraint restricts the zone of possible solutions to the first quadrant of the AB-plane. Similarly, the other technical specifications can be transformed into the following inequalities:

$$5A + 12B \leq 600$$
$$0.40A + 0.33B \leq 40$$
$$A, B \geq 0$$

If the technical constraints for the machine shop and the paint shop are treated in a similar fashion, Figure 12.2 may be constructed. It is seen that the cross-

FIGURE 12.2

THREE CONSTRAINTS WITH TWO ACTIVE AND
ONE INACTIVE

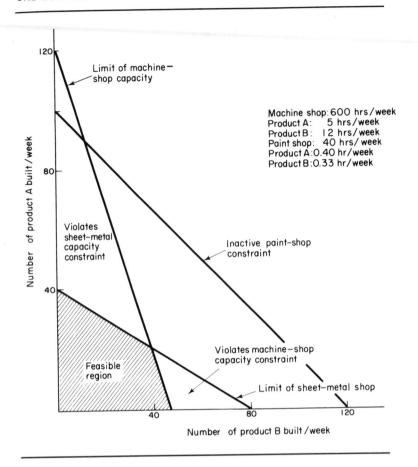

hatched area represents the region of all possible solutions. Any integer point in this area and/or on the boundary of the area is a possible solution. There are an infinite number of solutions if we assume divisibility of each of the production units. It will be noted that the paint shop constraint is inactive; that is, the machine shop and the sheet-metal shop are more restrictive in the production of products *A* and *B*. Another way of looking at this is that the paint shop has more capacity than is allowed by the production limitations of the machine and sheet-metal shops.

The objective function, in this case, to be maximized is found by converting the data from the table into the equation $y = 150A + 200B$. The next step is to map $150A + 200B$. This may be done by recognizing that at 40 units of *A*, 0 units of *B*, and similarly at 30 units *B* and 0 units *A*, a profit line of a $6000 value can be drawn. With a little reflection it is seen that an isoprofit line is the locus of all points (all possible combinations of *A* and *B*) which provide the same profit. Proceeding with this thought, other isoprofit lines yielding different levels of profit could be drawn. It is apparent that by going to the production set of (50*B*, 0*A*) there is a $4000 increase over the previous set (40*A*, 0*B*). This set yielding the $10,000 profit is parallel to the 6000 line but located farther away from the origin. This is to be expected since the contributions of the two products are fixed, and larger profit results as we move away from the origin. A little thought shows that we should keep drawing isoprofit lines for higher profits so long as we are within the area of feasible solution. Obviously we shall have to stop whenever we find that the isoprofit line lies either at a corner point of the convex polygon of Figure 12.3 or lies coincident to one of its boundaries. In either case we have found our optimum solution(s).

FIGURE 12.3

CONSTRAINTS WITH ISOPROFIT LINES

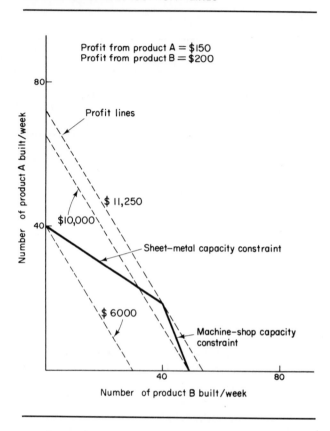

If it happens that the isoprofit line is coincident with one of the boundaries in the convex polygon, all points on that boundary line are feasible as well as optimum solutions. It is necessary that the isoprofit line have an equal slope with one of the boundary lines. In the more frequent case the isoprofit line does not coincide with one of the boundary lines, and one of the corner points provides the optimum and only solution. Thus, among all the infinite possible solutions that may be selected from the feasible region, that corner provides an optimum solution. This optimum point may be determined directly from the graph or by the simultaneous solution of the two intersecting boundaries at this point. This optimum set shows that $A = 18.9$ units and $B = 42.2$ units. Obviously, fractional production units are impossible, and the reader need merely enumerate locally about integer production quantities of A and B to find that point which gives the maximum profit. This particular point will be interior to the boundary lines.

Consider now the graphical maximization for three activities. While similar to that of two activities, this problem will later be solved by the simplex method. Suppose that three different types of fan belts are to be produced by The Rank Rubber Company. One unit of belt x_1 requires 5 minutes of belt-flipping time,

2 minutes of belt-skiving time, and 3 minutes of belt-curing time and produces a profit of $1.20. One unit of x_2 requires 4 minutes of flipping time, 3 minutes of skiving time, and 10 minutes of curing time and produces a profit of $1.60. One unit of belt x_3 requires 3 minutes of flipping time and 5 minutes of curing time and produces a profit of $1.45. The available capacities of the belt-flipping department, the belt-skiving department, and the belt-curing department are 8000 minutes, 4000 minutes, and 12000 minutes per day, respectively. These data may be summarized by the following table:

Department	Belt x_1	Belt x_2	Belt x_3	Capacity
Belt flipping	5 min	4 min	3 min	8000 min/day
Belt skiving	2 min	3 min	0 min	4000 min/day
Belt curing	3 min	10 min	5 min	12000 min/day
Profit per belt	$1.20	$1.60	$1.45	

The three products compete for available production capacities in each department. Converting these data into the linear programming format,

$$\text{maximize } y = \$1.20x_1 + \$1.60x_2 + \$1.45x_3$$

subject to

$$5x_1 + 4x_2 + 3x_3 \leq 8000$$
$$2x_1 + 3x_2 + 0x_3 \leq 4000$$
$$3x_1 + 10x_2 + 5x_3 \leq 12000$$

and $x_{1,2,3} \geq 0$.

The graphical equivalent of these algebraic statements is given by Figure 12.4. The solid space circumscribed by the numbers 0–7 represents a volume of feasible solutions. There is a family of profit planes parallel to each other and represented by the plane $\$1.20x_1 + \$1.60x_2 + \$1.45x_3$. The scheme for determining the optimum point in a three-dimensional case is similar to that of a two-dimensional case, although more cumbersome. If a series of parallel planes are consecutively moving away from the origin, eventually an extreme point will be reached where there is no other higher profit plane parallel to this plane without it being infeasible. Thus the reader can visualize that as we move away from the origin there is some particular plane which will just touch the extreme most corner of a convex polyhedral surface which is in the volume of all our feasible solutions. That point(s) of contact in the three-dimensional space is the optimum solution. In our case here, the solution is unique and will later be found as $x_1 = 242$, $x_2 = 0$, and $x_3 = 2254$.

A two-dimensional minimization approach is examined in light of the following problem. A company produces a king-size rubber drain stopper used for certain environmental applications controlling caustic fluids. This stopper consists of two mixed materials called rubber I and rubber II which cost $7 and $9 per pound. Design specifications require that the rubber stopper weigh 7.5 pounds and contain four ingredients x_1, x_2, x_3, and x_4. In addition, the design specifications require that the granules for each charge into the die weigh at least 1, 3, 1, and $1\frac{1}{2}$ pounds for the ingredients.

Rubber I and rubber II are constituted of x_1, x_2, x_3, and x_4 in the following proportions:

FIGURE 12.4

VOLUME OF FEASIBLE SOLUTION SPACE FOR
A LINEAR PROGRAMMING PROBLEM OF
BELT PRODUCTION

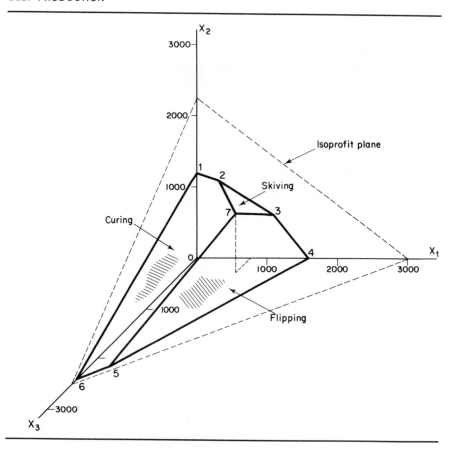

	x_1	x_2	x_3	x_4
Rubber I	0%	50%	35%	15%
Rubber II	45%	30%	0%	25%

Ingredient	Rubber I	Rubber II	Engineering Design Requirement, Weight
x_1	0	0.45	1
x_2	0.50	0.30	3
x_3	0.35	0	1
x_4	0.15	0.25	1.5
Cost	\$7.00	\$9.00	

The objective is to determine the quantity of rubbers I and II to minimize total cost. This requires minimizing

$$\text{total cost} = \$7\text{I} + \$9\text{II}$$

subject to

$$0I + 0.45II \geq 1.0$$

$$0.50I + 0.30II \geq 3.0$$

$$0.35I + 0.00II \geq 1.0$$

$$0.15I + 0.25II \geq 1.5$$

where I, II $\geq$ 0.

The graphical equivalent of the algebraic statements of this mixture problem is shown in Figure 12.5. The cross-hatched area represents the possible

FIGURE 12.5

MINIMIZING TOTAL COST FOR RUBBER
DRAIN STOPPERS

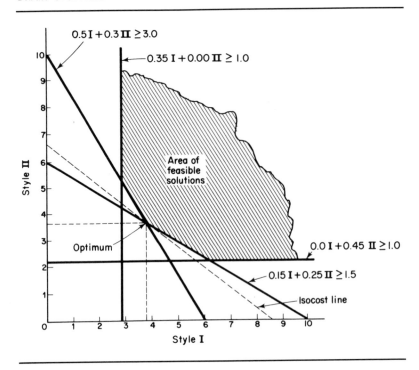

feasible solutions to the problem. There is a family of total cost lines beginning in the upper right-hand zone that are parallel to each other and represented by $7I + $9II. One of these lines will touch a corner of the feasible solution at its minimum point. The coordinates of this point represent the optimal quantities of rubber I and rubber II that minimize total cost. If these two products can be continuously separated into fractional weights, it is found that 3.8 and 3.8 pounds are the optimum values for rubber I and rubber II. Total cost is calculated as

$$\$7(3.8) + \$9(3.8) = \$60.80$$

Any other mix schedule for rubber I and rubber II that meets the engineering design requirements increases total cost.

The student correctly surmises that what has been proposed so far would be incorrect if the boundary restricting the feasible zone were not a convex set. For the extreme point to be the optimum point, the feasibility set must always be convex. A set of constraints is convex if all points on a line connecting any two points in that set are also in the set. In Figure 12.6, set *a* is convex while set *b*

FIGURE 12.6

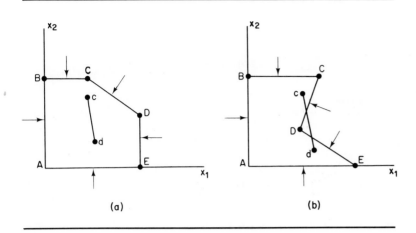

(a)   (b)

is not. In the latter, part of the line joining points *c* and *d* falls outside the constraint set. Although the illustration is in a two-dimensional space, the concept of a convex set is general and can be extended to any *N*-dimensional space.

There are other linear programming considerations that can be explained via graphical methods. For example, if a constraint is an equality, the optimal solution must lie on it rather than on either side of it. Considering products *A* and *B* again, if for some purpose the resource sheet metal time must be fully used (they have no other assigned work), the equality constraint, now $2A + B = 80$, restricts the feasible zone to a line having extreme points limited by the machine-shop and nonnegativity requirements. For Figure 12.2 it is evident that the maximum objective function still exists as in this illustration at a corner

FIGURE 12.7

SIMPLE PROBLEMS WITHOUT FEASIBLE
SOLUTIONS

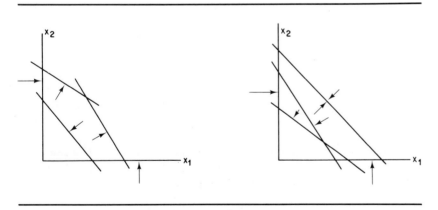

solution. In Figure 12.7 we see simple problems without feasible solutions. The system designer has improperly established his constraints. Finally, the last illustration, Figure 12.8, demonstrates an unbounded maximum objective function. Here again the problem has been improperly established.

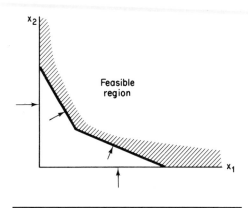

12.1-3
Conversion of Inequalities
to Equalities by the Addition
of Slack Variables

Constraint equations are generally of two forms: less-than-or-equal-to or more-than-or-equal-to. The less-than-or-equal-to type is of the general form

$$a_{11}x_1 + a_{12}x_2 + \cdots + a_{1n}x_n \leq b_1 \qquad (12\text{-}4)$$

where $a_{11}, a_{12}, \ldots, a_{1n}$ are technological constants and b is a scaler. This less-than-or-equal-to inequality can be transformed into an equation by the addition of a nonnegative variable. For products A and B, we can rewrite the equations in terms of x_1 and x_2 (formerly A and B) as

$$5x_1 + 12x_2 \leq 600$$
$$2x_1 + x_2 \leq 80$$

The addition of variables, say x_3 and x_4, convert the inequalities to

$$5x_1 + 12x_2 + x_3 = 600$$
$$2x_1 + x_2 + x_4 = 80$$

These variables x_3 and x_4 are known as slack variables, and they assume whatever value is necessary for the equation to be satisfied. If in the third equation $x_1 = x_2 = 0$, then $x_3 = 600$.

The more-than-or-equal-to inequality is similar except for the inequality notation:

$$a_{11}x_1 + a_{12}x_2 + \cdots + a_{1n}x_n \geq b_1 \qquad (12\text{-}5)$$

It is converted by the subtraction of a nonnegative variable. Consider the rubber drain stopper problem. If rubbers I and II are denoted x_1 and x_2 (ignoring the resulting confusion with ingredient proportions),

$$0.45x_2 \geq 1.0$$
$$0.50x_1 + 0.30x_2 \geq 3.0$$
$$0.35x_1 \geq 1.0$$
$$0.15x_1 + 0.25x_2 \geq 1.5$$

By the subtraction of surplus variables these become

$$0.45x_2 - x_3 = 1.0$$
$$0.50x_1 + 0.30x_2 - x_4 = 3.0$$
$$0.35x_1 \qquad - x_5 = 1.0$$
$$0.15x_1 + 0.25x_2 - x_6 = 1.5$$

If there are m constraints in n variables, the increase in variables is m, i.e., $x_1, x_2, \ldots, x_n, x_{n+1}, \ldots, x_{n+m}$. These slack variables $x_{n+1}, \ldots, x_{n+m}$ cannot violate the nonnegativity requirement and are one of the x_j variables, $j = 1, 2, \ldots, n, n+1, \ldots, n+m$. Additional variables are interposed to allow an initial solution. After the inequalities that exist in the restrictions are converted to equalities, the entire set of constraints make up m equations in $m + n$ unknown variables.

A physical interpretation can be attached to slack variables. They can be thought of as fictitious products each requiring for its production one unit from only one of the resources and zero units of resource from the others. Each fictitious slack product yields a profit of zero. Surplus variables represent over-fullment. The product profit function becomes

$$y = 150x_1 + 200x_2 + 0x_3 + 0x_4$$

A zero coefficient is assigned to output products x_3 and x_4. A similar logic gives the minimum cost equation for the rubber mix problem as $y = 7x_1 + 9x_2 + 0x_3 + 0x_4 + 0x_5 + 0x_6$.

12.1-4 Systems of Simultaneous Linear Equations

As linear programming is concerned with answers from systems of simultaneous linear equations, it is useful to review their algebra. Systems of simultaneous linear equations can be classified into three groups:

A system of n linear equations having n unknowns: A solution may be unique, inconsistent, and without a solution or have an infinite number of solutions. The unique solution provides little incentive to the operations analyst and is trivial. This category does not apply to the linear programming case.

The system of linear equations where there are more equations than unknowns or $m > n$: It is possible that no set of variables may simultaneously satisfy all equations. Frequently a number of equations may be discarded to create an interesting problem. This is generally not the linear programming problem.

The system of linear equations having more unknowns than equations: In this case $n - m$ of the variables assume arbitrary zero value. An infinite number of solutions are possible. Of these, one or more are selected to obtain an optimum policy. This is the conventional linear programming problem.

For the $n > m$ system of simultaneous linear equations, the task becomes almost infinitely reduced in magnitude whenever $n - m$ of the variables assume zero values. What has been graphically proved makes this reduction of effort possible. The idea that for a convex set of constraints one extreme point provides the optimum has enormous implications. For the two-product problems with $m = 2$ and $n = 4$ (x_3 and x_4 are slack variables), the number of corner combinations are $n!/[m!(m - n)!] = 6$. The clue to the enumerative method sets two variables equal to zero and solves two equations uniquely for the two unknowns, as is done in the following table:

Solution	x_1	x_2	x_3	x_4	Profit y
1	0	0	600	80	0
2	0	80	−360	0	Infeasible
3	120	0	0	−160	Infeasible
4	40	0	400	0	$6,000
5	0	50	0	30	$10,000
6	18.9	42.1	0	0	$11,255

Solutions 2 and 3 involve negative variables and are not evaluated for profit because the nonnegative restriction has been violated. This type of solution is called infeasible, as it fails to satisfy the conditions of the problem. A feasible solution is any solution containing more than m positive variables and satisfying the linear and nonnegativity constraints. Thus solutions 1, 4, 5, and 6 are feasible. These solutions are also referred to as a program.

Solution 6 is the optimum program or the maximum solution. Of course it is understood that production is limited to integer numbers and that the value of the objective function as conceived would become 19 and 42 units of product A and B for a profit of $11,250.

12.1-5
Maximizing by the Simplex Method

The simplex algorithm operates on systems of linear equations, both constraints and objective, in a manner which selects successive extreme point intersections between constraint equations, tests them for optimality against the objective function, and ultimately arrives at the optimum solution for the given system. It is necessary that the optimum fall on the plane boundaries of the region of feasible solutions. Thus, the simplex algorithm is a computational device with the characteristics that it proposes feasible solutions, that the proposed solutions fall at intersections of the boundaries of the feasible region, and that it starts with a feasible solution and works through successive improvements until it arrives at and indicates the optimum solution.

Now reconsider the fan belt problem given previously; rewriting into equations, we have

$$5x_1 + 4x_2 + 3x_3 + x_4 = 8000$$
$$2x_1 + 3x_2 + 0x_3 + x_5 = 4000$$
$$3x_1 + 10x_2 + 5x_3 + x_6 = 12000$$

where $n = 6$ and $m = 3$. The objective function becomes

$$\text{total profit} = y = 1.20x_1 + 1.60x_2 + 1.45x_3 + 0x_4 + 0x_5 + 0x_6$$

The computational procedure starts with the initial tableau where the coefficients of the objective function and constraint equations are arranged as follows:

Current Value c_i	Basis Variable	c_j	1.20	1.60	1.45	0	0	0
		b	x_1	x_2	x_3	x_4	x_5	x_6
0	x_4	8000	5	4	3	1	0	0
0	x_5	4000	2	3	0	0	1	0
0	x_6	12000	3	10	5	0	0	1

The coefficients of the objective function appear above their corresponding variables in the c_j row. Notice also that the coefficients of these same variables in the constraint equalities are entered in their appropriate positions in the body of the table. Last we enter the quantities which form the right-hand side of the equalities under the column head b.

The simplex process provides a means to start at the origin which is feasible. Thus at $x_1 = x_2 = x_3 = 0$ the slack variables are active and equal to their corresponding b-value. For the initial feasible solution the slack variables x_4, x_5, and x_6 are entered beneath the basis variable for the active constraint having a 1 in its row. This now leads to the solution improvement, testing for optimality, and iteration, as given by the following steps.

1. Within each column, $x_1, \ldots, x_6$, in the body of the tableau, multiply the coefficient times the corresponding coefficient c_i in its row and sum the products thus formed for the location y_j, or $y_j = \sum_{i=1}^{m} a_{ij}c_i$, $j = 1, \ldots, n + m$, where the a_{ij}'s are the matrix elements. For x_1, $5(0) + 2(0) + 3(0) = 0$, and similarly for each column. This appears at the foot of the column as now seen in Table 12.1 within the initial feasible solution block.

2. Subtract these y_j-sums from the c_j-coefficients of the objective function entered at the top of the table and show the $(c_j - y_j)$-difference below their x_j-entries. This is not done for the b column.

3. If all these differences are less than or equal to zero, then the optimum solution has been reached, and the values for the variables shown as solution variables will appear on their coresponding line under the heading b. If this condition does not exist, continue with the next step, for if any $c_j - y_j$ is positive for at least one j, a better program is possible.

    Referring now to the fan belt tableau, Table 12.1 shows the first element, under x_1 in the $(c_j - y_j)$-row, as $1.20 - 0 = 1.20$; the second is $1.60 - 0 = 1.60$; and the third $1.45 - 0 = 1.45$. As $c_j - y_j$ is positive for at least one j, the initial solution is not optimal. This is to be expected when the initial feasible solution is obtained by allocating all capacity to the slack variables. As this optimality test indicated the optimal program has not been found, the following iterative scheme must be employed.

4. Examine the $(c_j - y_j)$-values for the largest positive value and designate the column as the pivot column k. The variable at the head of this column will be an incoming basis variable in the next iteration.

5. Next find the pivot row by dividing the quantities in column b by the coefficients appearing at the intersection of the row in question and the pivot column k, or $b_i/a_{ik} = \theta_i$, where $i = 1, 2, \ldots, m$, and enter these new quantities in a column headed θ_i. Ignore denominator terms with negative or zero terms. The row showing the smallest positive θ_i will be the pivot row r and the element rk which then appears at the intersection of the pivot row and the pivot column is designated as the pivot element. Pivot elements are designated by an asterisk in Table 12.1 and in the following table:

b	x_1	x_2	x_3	x_4	x_5	x_6	θ_i	
8,000	5	4	3	1	0	0	2000	
4,000	2	3	0	0	1	0	1333	
12,000	3	10*	5	0	0	1	1200	r
	k							

TABLE 12.1

SIMPLEX LINEAR PROGRAMMING

Initial Feasible Solution

c_i	B.V.	c_j \ b	1.20 \ x_1	1.60 \ x_2	1.45 \ x_3	0 \ x_4	0 \ x_5	0 \ x_6	θ_i	
0	x_4	8,000	5	4	3	1	0	0	2000	
0	x_5	4,000	2	3	0	0	1	0	1333	
0	x_6	12,000	3	10*	5	0	0	1	1200	r
y_j		0	0	0	0	0	0	0		
$c_j - y_j$			1.20	1.60	1.45	0	0	0		
				k						

First Iteration

0	x_4	3,200	3.80	0	1	1	0	−0.40	842	
0	x_5	400	1.10*	0	−1.50	0	1	−0.30	364	r
1.60	x_2	1,200	0.30	1	0.50	0	0	0.10	4000	
y_j		1,920	0.48	1.60	0.80	0	0	0.16		
$c_j - y_j$			0.72	0	0.65	0	0	−0.16		
			k							

Second Iteration

0	x_4	1,817	0	0	6.17*	1	−3.46	0.63	294	r
1.20	x_1	364	1	0	−1.36	0	0.91	−0.27		
1.60	x_2	1,091	0	1	0.91	0	−0.27	0.18	1198	
y_j		2,182	1.20	1.60	−0.18	0	0.66	−0.04		
$c_j - y_j$			0	0	1.63	0	−0.66	+0.04		
					k					

Third Iteration

1.45	x_3	294	0	0	1	0.16	−0.56	0.10		
1.20	x_1	764	1	0	0	0.22	0.15	−0.13	5093	
1.60	x_2	823	0	1	0	−0.15	0.24*	0.09	3429	r
y_j		2,660	1.20	1.60	1.45	0.26	−0.25	0.13		
$c_j - y_j$			0	0	0	−0.26	0.25	−0.13		
							k			

Final Iteration

1.45	x_3	2,254	0	2.34	1	−0.19	0	0.31
1.20	x_1	242	1	−0.63	0	0.32	0	−0.19
0	x_5	500	0	4.17	0	−0.63	1	0.38
y_j		3,268	1.20	2.64	1.45	0.11	0	0.22
$c_j - y_j$			0	−1.04	0	−0.11	0	−0.22

6. A new matrix is formed with the pivot row calculated first. Calculate new elements a'_{ij} for the pivot row as

$$a'_{ij} = \frac{a_{rj}}{a_{rk}} \qquad \text{for } i = r$$

This reduces the pivot element to 1. At iteration 1, the operation for the pivot row would result as

$$\frac{12000}{10} = 1200; \quad \frac{3}{10} = 0.3; \quad \frac{10}{10} = 1; \quad \frac{5}{10} = 0.5; \quad \frac{0}{1} = 0, \quad \cdots$$

The new incoming variable x_2 is now a basis variable in the first iteration and replaces x_6. Likewise its associated cost element 1.60 is placed in the c_i-position in the same x_2-row. This new row may be seen by examining the third row of the first iteration. The departing nonbasis variable $x_6 = 0$ is in the next iteration.

7. With a row operation performed on the pivot row, the next step is to operate on remaining rows to remove the new solution variable from other constraint equations by making its coefficient in the nonpivot row constraint equation zero. New elements a_{ij} are calculated by

$$a'_{ij} = a_{ij} - a_{ik}a'_{rj} \qquad \text{for } i \neq r$$

As $r = 3$ and if $i = 1$, then

$$a'_{1j} = a_{1j} - a_{12}a'_{3j} \qquad \text{and} \qquad j = 1$$
$$a'_{11} = 5 - 4(0.3) = 3.8$$

For $j = 2$, $a'_{12} = 4 - 4(1) = 0$. For $j = 3$, $a'_{13} = 3 - 4(0.5) = 1$. For $j = 4$, $a'_{14} = 1 - 4(0) = 1$, and so forth. For the $(i = 1)$-row and b-column we shall have $b' = 8000 - 4(1200) = 3200$. Consider the computation of $i = 2$. $a'_{2j} = a_{2j} - a_{22}a'_{3j}$. If we let $r = 1$, then $a'_{21} = 2 - 3(0.3) = 1.1$.

8. After setting up the new table, return to step 1 and iterate until step 3 is satisfied.

Four iterations are required for the fan belt problem following the initial feasible tableau. The initial solution starting at the origin has slack variables x_4, x_5, and x_6 active. The solution at this point is $x_1 = 0$, $x_2 = 0$, $x_3 = 0$, $x_4 = 8000$, $x_5 = 4000$, and $x_6 = 12,000$ for a profit of $y_j = 0$. Step 4 identifies x_2 as the new variable, while step 5 indicates that x_6 will be absent in the next iteration.

In the initial feasible solution the 1.60 designates x_2 as the pivot column. $\theta_i = 12,000/10 = 1200$ is the smallest positive θ_i and determines the outgoing variable x_6. This interchange between the variables is called simplexing and is an important step in the algorithm.

In the first iteration y_j has improved to $1920, and the solution is $x_1 = 0$, $x_2 = 1200$, $x_3 = 0$, $x_4 = 3200$, $x_5 = 400$, and $x_6 = 0$. This corresponds to point 1 in Figure 12.4. The optimality steps, 1, 2, and 3, require additional programming. Steps 4 and 5 indicate a simplexing interchange between x_1 and x_5 where the pivot element $a_{rk} = 1.10$.

The second iteration forms the new matrix and employs steps 6 and 7. Optimality evaluation requires improvement even though $y = \$2182$ increases by an amount $\$2182 - \$1920 = \$262$.

The solution $x_1 = 364$, $x_2 = 1091$, and $x_3 = 0$ now resides at location 2 in

TABLE 12.2
MINIMIZING BY THE SIMPLEX LINEAR
PROGRAMMING TECHNIQUE

Initial Feasible Solution

c_i	B.V.	c_j:	-7	-9	0	0	0	0	$-M$	$-M$	$-M$	$-M$	
		b	x_1	x_2	x_3	x_4	x_5	x_6	x_7	x_8	x_9	x_{10}	θ_i
$-M$	x_7	1.0	0	0.45	-1	0	0	0	1	0	0	0	
$-M$	x_8	3.0	0.50	0.30	0	-1	0	0	0	1	0	0	6
$-M$	x_9	1.0	0.35*	0	0	0	-1	0	0	0	1	0	2.9 *r*
$-M$	x_{10}	1.5	0.15	0.25	0	0	0	-1	0	0	0	1	10
y_j		$-6.5M$	$-M$	$-M$	M	M	M	M	$-M$	$-M$	$-M$	$-M$	
$c_j - y_j$			$-7+M$	$-9+M$	$-M$	$-M$	$-M$	$-M$	0	0	0	0	
			k										

First Iteration

c_i	B.V.	b	x_1	x_2	x_3	x_4	x_5	x_6	x_7	x_8	x_9	x_{10}	θ_i
$-M$	x_7	1.0	0	0.45	-1	0	0	0	1	0	0	0	
$-M$	x_8	1.57	0	0.30	0	-1	1.43*	0	0	1	-1.43	0	1 *r*
-7	x_1	2.86	1	0	0	0	-2.86	0	0	0	2.86	0	
$-M$	x_{10}	1.07	0	0.25	0	0	0.43	-1	0	0	-0.43	1	2.5
y_j		$-20-3.6M$	-7	$-M$	M	M	$-1.86M+20$	M	$-M$	$-M$	$1.86M-20$	$-M$	
$c_j - y_j$			0	$-9+M$	$-M$	$-M$	$1.86M-20$	$-M$	0	0	$-2.86M+20$	0	
							k						

TABLE 12.2 (Cont.)

Second Iteration

												r
$-M$	x_7	1.0	0	0.45*	-1	0	0	1	-1	0	0	2.2
0	x_5	1.10	0	0.21	0	1	-0.70	0	0	-1	0.70	5.2
-7	x_1	6.0	1	0.60	0	0	-2.00	0	0	0	2.00	10
$-M$	x_{10}	0.60	0	0.16	0	0	0.30	0	-1	0	-0.30	3.8
	y_j	$-1.6M-42$	-7	$-0.61M-4.2$	M	0	$-0.3M+14$	M	$-M$	$-M$	$0.3M-14$	
	c_j-y_j	0	0	$0.61M+4.8$	$-M$	0	$+0.3M-14$	0	0	$-M$	$-1.3M+14$	

k

Third Iteration

												r
-9	x_2	2.22	0	1	-2.22	0	0	2.22	0	0	0	1.3
0	x_5	0.63	0	0	0.47	1	-0.70	-0.47	0	-1	0.70	3.5
-7	x_1	4.67	1	0	1.33	0	-2.00	-1.33	0	0	2.00	0.7
$-M$	x_{10}	0.25	0	0	0.35*	0	0.30	-0.35	0	0	-0.30	
	y_j	$-0.25M-53$	-7	-9	$-0.35M+10.7$	0	$-0.30M+14$	M	$0.35M-10.7$	$0.30M-14$		
	c_j-y_j	0	0	0	$+0.35M-10.7$	0	$+0.30M-14$	$-M$	$-0.35M-10.7$	$0.30M+14$		

k

Final Iteration

-9	x_2	3.75	0	1	0	0	1.90	0	-6.34	-1.90	6.34
0	x_5	1.00	0	0	0	1	-1.10	0	1.34	1.10	-1.34
-7	x_1	3.75	1	0	0	0	-3.14	0	-3.14	3.14	-3.80
0	x_3	0.71	0	0	1	0	0.86	1	30.46	-4.88	-30.46
	y_j	-60	-7	-9	0	0	5	0	30	-5	-30
	c_j-y_j	-30	0	0	0	0	-5	0	30	5	-30

Figure 12.4. The slack variables, not plotted of course, are $x_4 = 1817$, $x_5 = 0$, and $x_6 = 0$.

The third iteration now has x_1, x_2, and x_3 in solution, but optimality has not been obtained. This solution corresponds to point 7. The isoprofit plane reveals that profit $y = \$2660$ for this combination of fan belt production.

The fourth and final iteration finds the optimality test satisfied and shows that fan belt x_2 production should be discontinued. It is better to concentrate on producing x_1 and x_3. $c_j - y_j$ is either 0 or negative here, and there is no need to identify the incoming or departing variable. Point 4 of Figure 12.4 is the graphical counterpart.

12.1-6
Minimization by the
Simplex Method

The rubber drain stopper will be used to explain the variations in the simplex process for a minimizing algorithm. Repeating the constraints,

$$0.45x_2 - x_3 = 1.0$$
$$0.50x_1 + 0.30x_2 - x_4 = 3.0$$
$$0.35x_1 \qquad\qquad - x_5 = 1.0$$
$$0.15x_1 + 0.25x_2 - x_6 = 1.5$$

and the cost equation $y = 7x_1 + 9x_2 + 0x_3 + 0x_4 + 0x_5 + 0x_6$. The graphical equivalent of the inequality algebraic statements of this problem has already been described. The shaded area of Figure 12.5 represents the area of feasible solutions.

A maximizing algorithm as previously described will work for a minimizing function if the signs of the cost coefficient are multiplied by -1. The principle that maximizing the negative of a function is the same as minimizing the function applies here. Thus values for $-y$ are entered in the same locations of the initial solution.

In examining these constraint equalities, an initial feasible solution does not exist at $x_1 = x_2 = 0$, which was the device used with the maximizing scheme. At the origin $x_3 = -1.0$, $x_4 = -3.0$, $x_5 = -1.0$, and $x_6 = -1.5$ violate nonnegativity. An initial feasible solution having correct algebraic form can be invented by adding "artificial" or temporary variables to each constraint, resulting in

$$0.45x_2 - x_3 + x_7 = 1.0$$
$$0.50x_1 + 0.30x_2 - x_4 + x_8 = 3.0$$
$$0.35x_1 \qquad\qquad - x_5 + x_9 = 1.0$$
$$0.15x_1 + 0.25x_2 - x_6 + x_{10} = 1.5$$

The artificial variables, x_7, x_8, x_9, and x_{10}, with a value greater than zero destroy the equality required by the general linear programming model. To prevent the artificial variables from appearing in the final optimal solution, a large penalty is associated with each. This penalty, an extremely high cost, is larger than any of the objective function coefficients and is designated nonnumerically as $-M$. With these changes the total cost equation becomes

$$y = -7x_1 - 9x_2 - 0x_3 - 0x_4 - 0x_5 - 0x_6 - Mx_7$$
$$- Mx_8 - Mx_9 - Mx_{10}$$

The conversion of constraints and the objective function are now complete. The initial matrix required for the simplex algorithm is shown by Table 12.2. The initial solution variables are the artificial variables, and the feasible solution, although imaginary and prohibitively costly, does allow the computation scheme to start. As before, the incoming variable is selected as the most positive of the $(c_j - y_j)$-row. The outgoing variable controlled by the b_i/a_{ik}s will be the least positive. The algorithm first proceeds to force the artificial variables out of solution in successive stages and creates a variable interchange with real and slack variables. Outside of the -1 multiplication of the y and the creation of artificial variables with $-M$ costs, the computational procedure as developed for simplex maximizing may be used directly for a minimizing problem. This assures that the value of the program will decrease as computation continues. Table 12.2 now gives the simplex algorithm for a cost of $60 and x_1 and $x_2 = 3.75$ pounds after multiplying the final $y_j = -60$ by -1.

12.1-7
Equality Constraint Considerations

To understand linear programming more thoroughly, it is necessary to consider equality constraints. It sometimes happens that a resource must be fully used and an equality results. In the two-product example, A and B, suppose that the sheet-metal time must be completely filled as this department has no other duty. This or other reasons may require that certain constraints be strictly adhered to. If sheet metal is critical, the two-product profit-maximization constraints now appear this way:

$$5x_1 + 12x_2 \leq 600$$
$$2x_1 + x_2 = 80$$
$$x_{1,2} \geq 0$$

Using the artificial variable idea,

$$5x_1 + 12x_2 + x_3 = 600$$
$$2x_1 + x_2 + x_4 = 80$$

x_3 and x_4 are slack and artificial variables, respectively. Now the profit function to be maximized becomes

$$y = 150x_1 + 200x_2 - 0x_3 - Mx_4$$

The penalty profit $-M$ results in the removal of the artificial variable early in the iterative process. The simplex procedure can be applied, and computation proceeds as outlined before. Table 12.3 provides the initial feasible solution.

The reader may want to uncover the optimal iteration. There he will find that the optimal resides precisely as found where both constraints were inequations. The maximum isoprofit line corresponds to the intersection of these two boundaries in this case.

TABLE 12.3

INITIAL FEASIBLE TABLEAU FOR
AN EQUALITY CONSTRAINT PROBLEM

Initial Feasible Solution

c_i	B.V.	c_j	150	200	0	$-M$
		b	x_1	x_2	x_3	x_4
0	x_3	600	5	12	1	0
$-M$	x_4	80	2	1	0	1

12.1-8
Duality of Linear Programming Problems

The linear programming problems presented so far may be called *primal*. There is another closely related linear programming method that extends the versatility of the model. Called the *dual*, this method permits, in certain situations, refinements that reduce the computational work. When *primal* (the original) problems have their rows and columns transposed, the inverted problem is called the *dual*. After the dual is structured, computation proceeds as in the primal simplex format. Furthermore, after a dual solution is found, it may be converted to a primal solution.

Now consider a minimizing primal to be given by

$$a_{11}x_1 + a_{12}x_2 + \cdots + a_{1n}x_n \geq b_1$$
$$\vdots$$
$$a_{m1}x_1 + a_{m2}x_2 + \cdots + a_{mn}x_n \geq b_m$$

where variables $x_j \geq 0$ and which minimizes

$$y = c_1x_1 + \cdots + c_nx_n$$

After a matrix transpose operation, the dual would become

$$a_{11}w_1 + a_{21}w_2 + \cdots + a_{m1}w_m \leq c_1$$
$$\vdots$$
$$a_{1n}w_1 + a_{2n}w_2 + \cdots + a_{mn}w_m \leq c_n$$

where dual variable $w_i \geq 0$ and which maximizes

$$g = w_1b_1 + w_2b_2 + \cdots + w_mb_m$$

For the minimization primal previously given, the dual becomes maximize

$$g = 1.0w_1 + 3.0w_2 + 1.0w_3 + 1.5w_4$$

subject to

$$0.50w_2 + 0.35w_3 + 0.15w_4 \leq 7$$
$$0.45w_1 + 0.30w_2 + \qquad\quad 0.25w_4 \leq 9$$

As each of the two dual constraints is a less-than-or-equal-to inequality, we add nonnegative slack variables to get

$$0.50w_2 + 0.35w_3 + 0.15w_4 + w_5 = 7$$
$$0.45w_1 + 0.30w_2 + \qquad\quad 0.25w_4 + w_6 = 9$$

and the objective function now becomes a maximize or $g = 1.0w_1 + 3.0w_2 + 1.0w_3 + 1.5w_4 + 0w_5 + 0w_6$. When solved this yields an optimum of $w_2 = 5$, $w_4 = 30$ with shadow prices of $x_1 = 3.75$ and $x_2 = 3.75$ at the optimum, or $g = \$60$. This is shown as Table 12.3. Values for the activity variables of the primal appear in the columns of slack variables in the dual with negative signs. Also the value of y is in the same location for the primal and dual but with opposite sign. Compare the final Table 12.4 iteration with Table 12.2 to note the dissimilarities.

Before leaving the primal-dual relationships, consider another primal problem. Maximize $5x_1 + 4x_2 + 3x_3$ subject to

$$x_1 + 0.5x_2 + 0.7x_3 \leq 20$$
$$5x_1 \qquad\qquad + \quad x_3 \leq 10$$
$$0.2x_2 + 0.5x_3 \leq 14$$
$$0.3x_1 + 0.5x_2 + 0.5x_3 \leq 12$$

TABLE 12.4

DUAL SIMPLEX SOLUTION TO
PRIMAL PROBLEM

c_i	B.V.	c_j	1	3	1	1.5	0	0	
		b	w_1	w_2	w_3	w_4	w_5	w_6	θ_i
0	w_5	7	0	0.5*	0.35	0.15	1	0	14 r
0	w_6	9	0.45	0.3	0	0.25	0	1	30
	y_j	0	0	0	0	0	0	0	
	$c_j - y_j$		1	3	1	1.5	0	0	
				k					

First Iteration

3	w_2	14	0	1	0.7	0.3	2	0	
0	w_6	4.8	0.45*	0	-0.21	0.16	-0.6	1	11 r
	y_j	42	0	3	2.1	0.9	6	0	
	$c_j - y_j$		1	0	-1.1	0.6	-6	0	
				k					

Second Iteration

3	w_2	14	0	1	0.7	0.3	2	0	46
1	w_1	10.7	1	0	-0.47	0.36*	-1.33	2.22	30 r
	y_j	52.7	1	3	1.63	1.26	4.67	2.22	
	$c_j - y_j$		0	0	-0.63	0.25	-4.67	-2.22	
					k				

Final Iteration

3	w_2	5	-0.74	1	1.09	0	3.13	-1.88	
1.5	w_4	30	2.81	0	-1.31	1	-3.75	6.25	
	y_j	60	1.98	3	1.31	1.5	3.75	-3.75	
	$c_j - y_j$		-0.98	0	-0.31	0	-3.75	-3.75	

The dual is written as

$$w_1 + 5w_2 + \qquad\qquad 0.3w_4 \geq 5$$
$$0.5w_1 + \qquad 0.2w_3 + 0.5w_4 \geq 4$$
$$0.7w_1 + \ w_2 + 0.5w_3 + 0.5w_4 \geq 3$$

and minimize $g = 20w_1 + 10w_2 + 14w_3 + 12w_4$.

Slack variables, w_5, w_6, and w_7, along with artificial variables, w_8, w_9, and w_{10}, would ultimately be required. The objective function would appear as

$$g = 20w_1 + 10w_2 + 14w_3 + 12w_4 + 0w_5 + 0w_6$$
$$+ 0w_7 - Mw_8 - Mw_9 - Mw_{10}$$

It is evident from the increased dimensionality that certain types of problems prosper under duality while others do not. The last one would be best solved by primal simplex methods. Although a primal and a dual can be solved for any given problem, it does not follow that it is equally easy to do so. The choice of computation would favor solving the primal if it has more variables than constraints. From a practical point of view, operations within a row are fairly simple. Operation between rows are more difficult.

12.2
LAGRANGE METHOD OF UNDETERMINED MULTIPLIERS

We previously mentioned classic calculus. In that discussion we did not consider any boundary constraints. While the classic techniques are not classified as computational devices, they are important theoretical tools. These concepts have aided much of the standard theory of production and consumer behavior in economics and give insights to cost-estimating problem formulation and understanding.

Usually in the calculus method of maxima and minima one can, if possible, eliminate some of the variables by substituting in the side constraints and eventually reduce the problem to an ordinary maximum and minimum problem. This procedure is not always convenient, and the Lagrange[1] method of undetermined multipliers may be more tractable.

In the effectiveness function $y(x) = f(x_1, x_2, \ldots, x_n)$ it becomes necessary that the total differential

$$dy = \frac{\partial y}{\partial x_1} dx_1 + \frac{\partial y}{\partial x_2} dx_2 + \cdots + \frac{\partial y}{\partial x_n} dx_n = 0 \qquad (12\text{-}6)$$

$$dy = \sum_{j=1}^{n} \frac{\partial y}{\partial x_j} dx_j = 0 \qquad \text{for } j = 1, 2, \ldots, n \qquad (12\text{-}7)$$

The effectiveness function is restricted by the following constraints:

$$g_1 = g_1(x_1, x_2, \ldots, x_n) = 0 \qquad (12\text{-}8)$$
$$\vdots$$
$$g_m = g_m(x_1, x_2, \ldots, x_n) = 0 \qquad \text{and } i = 1, 2, \ldots, m \qquad (12\text{-}9)$$

Each of these functional constraints may be differentiated to give

$$dg_i = \sum_{j=1}^{n} \frac{\partial g_i}{\partial x_j} dx_j = 0, \qquad i = 1, 2, \ldots, m \qquad (12\text{-}10)$$

Each of the m equations is multiplied by an (at this point) undetermined parameter λ_i different for each equation, called a Lagrangian multiplier, and $\lambda_i \neq 0$ initially.

Usually we have $m < n$:

$$\lambda_1 dg_1 = \sum_{j=1}^{n} \lambda_1 \frac{\partial g_1}{\partial x_j} dx_j = 0 \qquad (12\text{-}11)$$
$$\vdots$$
$$\lambda_m dg_m = \sum_{j=1}^{n} \lambda_m \frac{\partial g_m}{\partial x_j} dx_j = 0 \qquad (12\text{-}12)$$

where the additional λ's remove the variational aspects of the problem, as there are $n - m$ open spaces. Adding these equations together along with dy, there results

$$dy + \lambda_1 dg_1 + \lambda_2 dg_2 + \cdots + \lambda_m dg_m = 0 \qquad (12\text{-}13)$$

$$\sum_{j=1}^{n} \left(\frac{\partial y}{\partial x_j} + \lambda_1 \frac{\partial g_1}{\partial x_j} + \lambda_2 \frac{\partial g_2}{\partial x_j} + \cdots + \lambda_m \frac{\partial g_m}{\partial x_j} \right) dx_j = 0 \qquad (12\text{-}14)$$

[1]Joseph Louis Lagrange (1763–1813), a great French mathematician, is known for his major work in the calculus of variations, celestial and general mechanics, differential equations, and algebra.

As each of the parameters x_j are independent, we require of the λ_m's that they be chosen to have the parenthesis zero for every dx_j. Therefore there are n equations of the type

$$\frac{\partial y}{\partial x_j} + \lambda_1 \frac{\partial g_1}{\partial x_j} + \lambda_2 \frac{\partial g_2}{\partial x_j} + \cdots + \lambda_m \frac{\partial g_m}{\partial x_j} = 0, \qquad j = 1, 2, \ldots, n \quad (12\text{-}15)$$

Additionally, one should recall that there are m constraint equations of the form $g_i = 0$, $i = 1, 2, \ldots, m$. There are $m + n$ equations and $m + n$ unknowns with the n parameters (x_j's) and m Lagrangian multipliers (λ_i's). The solution of this set of $m + n$ simultaneous equations gives the desired optimum.

It is desired to build a tank with as little material as possible. The effectiveness function is the surface area, $A = 2\pi r^2 + 2\pi r l$, and the constraint is on volume, $V = \pi r^2 l$, where V may be chosen to fit a specific case:

$$A = 2\pi r^2 + 2\pi r l$$

$$\frac{\partial A}{\partial r} = 4\pi r + 2\pi l$$

$$\frac{\partial A}{\partial l} = 2\pi r$$

The constraint equation is

$$g_1 = V - \pi r^2 l = 0$$

$$\frac{\partial g_1}{\partial r} = -2\pi r l$$

$$\frac{\partial g_1}{\partial l} = -\pi r^2$$

The three equations are

$$\frac{\partial A}{\partial r} + \lambda \frac{\partial g_1}{\partial r} = 4\pi r + 2\pi l + \lambda(-2\pi r l) = 0$$

$$\frac{\partial A}{\partial l} + \frac{\partial g_1}{\partial l} = 2\pi r - \lambda \pi r^2 = 0$$

$$V - \pi r^2 l = 0$$

Simultaneous solution gives $\lambda = 2/r$, $l = 2r$, and $r = (V/2\pi)^{1/3}$.

Here, then, we see the emergence of a mathematical tool which can solve optimization problems through the simple use of multipliers whereby the constraints become a part of the Lagrange equation. This equation is partially differentiated by each of the original independent variables and the Lagrange multipliers, yielding a set of simultaneous equations solvable by conventional methods. Their worth can be appreciated in analyzing and optimizing cost-associated functions in connection with business decisions. Not only can we determine the optimum operating conditions for the situation at hand, but also an analysis can be made of this optimum to evaluate the consequences of moving off this optimum. Marginal cost analysis is a natural extension of Lagrangian multipliers. See Chapter 7 for a differing approach to marginal analysis.

Now with this understanding of Lagrangian multipliers, marginal cost, marginal revenue, and differential calculus, we proceed to an application where all are used in explaining a simple problem: Maximize the function $y = 5X_1 + 4X_2$ subject to $X_1 \leq 2$ and $X_2 \leq 4$ using the method of Lagrange multipliers.

State the values of λ_1, λ_2, and y at the maximum point. Form the Lagrange expression for the function,

$$y = 5X_1 + 4X_2 + \lambda_1(X_1 - 2) + \lambda_2(X_2 - 4)$$

Take the partial derivatives with respect to X_1, X_2, λ_1, and λ_2:

$$\frac{\partial y}{\partial X_1} = 5 + \lambda_1 = 0$$

$$\frac{\partial y}{\partial X_2} = 4 + \lambda_2 = 0$$

$$\frac{\partial y}{\partial \lambda_1} = X_1 - 2 = 0$$

$$\frac{\partial y}{\partial \lambda_2} = X_2 - 4 = 0$$

For this easy problem the values are found to be $X_1 = 2$, $X_2 = 4$, $\lambda_1 = -5$, and $\lambda_2 = -4$. Suppose that we relax the second constraint by one unit, making it $X_2 \leq 5$, and find the new maximum point with the rest of the problem unchanged:

$$y = 5X_1 + 4X_2 + \lambda_1(X_1 - 2) + \lambda_2(X_2 - 5)$$

Taking the partial derivatives with respect to the X_1, X_2, and λs and setting them equal to zero as before, we find that $X_1 = 2$, $X_2 = 5$, $\lambda_1 = -5$, and $\lambda_2 = -4$. If we evaluate the original function y using the original values of X_1 and X_2, we find $y = 5(2) + 4(4) = 26$. Next evaluate the function using values obtained by relaxing the second constraint by one unit:

$$y = 5(2) + 4(5) = 30$$

The difference between these two values is numerically equal to the negative value of λ_2 or 4. A similar test can be made by tightening the constraint by one unit ($X_2 \leq 3$) whereby the value of y is 22. Here again the difference is 4.

We see that the value of the Lagrangian multipliers, as solved in the system of equations, is useful in predicting the change in the value of the function at the optimum or maximum point brought about by a marginal change in the constraint. With the added knowledge of the Lagrangian multipliers, this solution for an optimum operating level gives cost or revenue for related marginal changes in any of the constraints.

The λ's are sometimes called *sensitivity factors*, *dual variables*, or *shadow prices*. This procedure works when effectiveness functions and constraints are nonlinear too. For example, if the ith constraint is revised to $g_i(x) \leq e$, where e is a very small positive number, then the optimal value of $y(x)$ will increase approximately by the amount $e\lambda_i$ (except in cases where optimum x and λ are not continuous functions of e for infinitesimal values of e).

It should not be assumed that the Lagrangian multipliers technique will solve all problems of optimization of objective functions with constraints. The solution set may be neither a maximum nor minimum and the result must be subjected to sufficiency proofs or other tests. Also, size of the number of variables and constraints is a limit to the generalness of the solution.

12.3
GEOMETRIC PROGRAMMING

In engineering design problems such as

$$y = u_1 + u_2 + \cdots + u_n \tag{12-16}$$

where y = unit cost function,

$$u = \text{component costs}, \quad i = 1, \ldots, n$$

are frequent. The component costs are in turn expressed as a power function:

$$u_i = C_t t_1^{a_{i1}} t_2^{a_{i2}} \cdots t_m^{a_{im}} \tag{12-17}$$

where
C_t = positive constant
a_{ij} = arbitrary real exponents
$t_1, \ldots, t_m$ = design parameters presumed to be positive variables

Some design- and system-engineering problems can be represented in this fashion. For instance, linear programming problems have this particular polynomial form. Geometric programming, as an indirect solution method of nonlinear programming, concerns itself with problems of this nature.

The original work was motivated by the observation that the minimum value of a certain unconstrained nonlinear function could be found, almost by inspection, by exploiting the relationship that exists between the weighted arithmetic mean of a set of numbers and the corresponding geometric mean of the numbers (thus the origin of the name *geometric programming*). The generalization of the arithmetic-geometric mean inequality leads to

$$\sum_{i=1}^{N} C_i x_i \geq \prod_{i=1}^{N} x_i^{c_i}, \qquad C_i \leq 1 \text{ for all } i \tag{12-18}$$

where x_i is a set of any nonnegative numbers, and C_i is a set of nonnegative coefficients. Sometimes the maximum value of the geometric mean will be equal to the minimum value of the corresponding arithmetic mean. A distinguishing feature of geometric programming uses the geometric mean inequality law which in its simplest form is

$$\tfrac{1}{2}Z_1 + \tfrac{1}{2}Z_2 \geq Z_1^{1/2} Z_2^{1/2}$$

where Z_1 and Z_2 represent two positive numbers. To show this, note that

$$(Z_1 - Z_2)^2 \geq 0$$
$$Z_1^2 - 2Z_1 Z_2 + Z_2^2 \geq 0$$

which implies that after adding $4Z_1 Z_2$ to both sides

$$Z_1^2 + 2Z_1 Z_2 + Z_2^2 \geq 4Z_1 Z_2$$

A square root operation of the last inequality gives the original equation again. We have a strict equality if $Z_1 = Z_2$ and an inequality otherwise.

For the geometric programming approach [Equation (12-18)], the set of nonnegative coefficients, called λ_i, must satisfy a normality condition of

$$\sum_{i=1}^{N} \lambda_i = 1 \tag{12-19}$$

Define $\theta_i = \lambda_i x_i$, $i = 1, \ldots, N$, as a change of variable. Equation (12-18) becomes

$$\sum_{i=1}^{N} \theta_i \geq \prod_{i=1}^{N} \left(\frac{\theta_i}{\lambda_i} \right)^{\lambda_i} \tag{12-20}$$

where the left- and right-hand sides are called the primal and predual function.

A *posynomial* is defined as

$$g = \theta_1 + \theta_2 + \cdots + \theta_n \tag{12-21}$$

if

$$\theta_i = C_i \prod_{j=1}^{m} t_j^{a_{ij}}, \qquad C_i \geq 0 \text{ for all } i = 1, \ldots, N \qquad (12\text{-}22)$$

where the a_{ij}'s are real constants, m is the number of independent primal variables, and the t_j's at this point are the independent variables. If the primal function is a posynomial, the terms θ_i given by Equation (12-22) are substituted into the right-hand side of the geometric inequality (12-20) to obtain a predual function:

$$P(\bar{\lambda}, \bar{t}) = \prod_{i=1}^{N} \left(\frac{C_i}{\lambda_i}\right)^{\lambda_i} \prod_{j=1}^{m} t_j^{D_j} \qquad (12\text{-}23)$$

where the D_j's are linear combinations, or

$$D_j = \sum_{i=1}^{N} \lambda_i a_{ij}, \qquad j = 1, \ldots, m \qquad (12\text{-}24)$$

Importantly, the maximum value of the predual function is found by choosing the weights λ_i to have D_j equal zero; then the maximum value will be equal to the minimum value of the primal function. Thus, Equation (12-23) simplifies to a dual function

$$P(\bar{\lambda}) = \prod_{i=1}^{N} \left(\frac{C_i}{\lambda_i}\right)^{\lambda_i} \qquad (12\text{-}25)$$

With the values of λ_i determined, the dual function can be evaluated to give the minimum value of the primal function. With this minimum value, the optimum values of the independent variables can be found using

$$\theta_i = \lambda_i P(\bar{\lambda}) \qquad (12\text{-}26)$$

Consider again the standard tool life equation (8-5) on which the study of fundamental machining economics is based. This equation

$$y_0 = C_0 t_m + C_t \left(\frac{t_m}{T}\right) + C_0 t_c \left(\frac{t_m}{T}\right) + C_0 t_h$$

describes the average unit cost y of a single-point rough-turning operation as the sum of four component costs, namely the machining cost, the tool cost, the tool changing cost, and the handling cost. See Chapter 8 for notation. The machining time t_m is given as

$$t_m = \frac{L\pi D}{12Vf}$$

$$VT^n f^m = k$$

with the first constraint expressing length L, diameter D, velocity V, and feed f in consistent units, and the second as the modified Taylor tool life relationship between the variables with exponents n and m and constant k as empirical parameters.

Now the objective is to find the minimum cutting speed V. Rewriting by substituting in terms of V gives the single objective function

$$y_0 = C_{01} V^{-1} + C_{02} V^{-1+(1/n)} + C_0 t_h$$

where for convenience we have defined

$$C_{01} = C_0 \left(\frac{L\pi D}{12f}\right)$$

$$C_{02} = \frac{f^{m/n}}{k^{1/n}} [C_0 t_c + C_t] \frac{L\pi D}{12f}$$

Dropping, for the moment, the constant term $C_0 t_h$, the problem becomes

$$y_0 = C_{01} V^{-1} + C_{02} V^{-1+(1/n)}$$

This equation is not a polynomial (in the sense of an integral rational expression) since the exponents violate positive integers; however, the C_i must be positive and for this reason these functions are called posynomials (positive polynomials). The usual way for minimizing functions such as this involves the classic calculus of setting to zero the first partial derivative with respect to the independent variables. This produces nonlinear simultaneous equations. If these first derivatives vanish, we have satisfied the necessary condition for a minimum. Functions of higher-order partial derivatives are normally required to satisfy sufficient conditions. Frequently this causes serious difficulty. The procedure is demonstrated for y_0, which has two terms and one independent variable V. Therefore, using Equation (12-23), the predual equation is

$$P(\bar{\lambda}, \bar{V}) = \left(\frac{C_{01} V^{-1}}{\lambda_1}\right)^{\lambda_1} \left(\frac{C_{02} V^{1/n-1}}{\lambda_2}\right)^{\lambda_2}$$

and from Equations (12-19) and (12-24), respectively,

$$\lambda_1 + \lambda_2 = 1$$

$$-\lambda_1 + \lambda_2 \left(\frac{1}{n} - 1\right) = 0$$

from which it follows that

$$\lambda_1 = 1 - n$$

$$\lambda_2 = n$$

where n is the empirically determined exponent of the Taylor tool life equation. The optimum minimum from the dual is

$$y^* = \left(\frac{C_{01}}{1-n}\right)^{1-n} \left(\frac{C_{02}}{n}\right)^{n}$$

From this optimum y^* value, the corresponding optimum value V^* of the cutting speed is found by using Equation (12-26):

$$C_{01} V^{-1*} = (1-n) \left(\frac{C_{01}}{1-n}\right)^{1-n} \left(\frac{C_{02}}{n}\right)^{n}$$

$$V^* = \left(\frac{C_{01} n}{C_{02}(1-n)}\right)^{n}$$

Substituting back for C_{01} and C_{02} yields, in the original symbols,

$$V^* = \left\{\frac{C_0 n}{(f^{m/n}/k^{1/n})(C_0 t_c + C_t)(1-n)}\right\}^{n}$$

which is the operating design for the machining-economic problem and which corresponds to Equation (8-8). No calculus was used to obtain this result. A particular value can be determined by the appropriate substitution of empirical and cost data for equations.

Note that the above procedure applies to unconstrained posynomials inasmuch as substitution of constraints into the cost function results in a single function. Extensions given in the Selected References removed this posynomial restriction and permit the addition of constraints to give the customary appearance of the nonlinear mathematical program statement.

12.4
SUMMARY

Linear programming, the most-used optimization technique, was introduced. The computations that were shown were demonstrated for manual manipulation. Naturally, real problems are computer-solved, but easy problems must be viewed first. If the linearities are like the real thing, linear programming can be exploited.

The method of the Lagrangian undetermined multiplier has the advantage of being able to take the nonlinear classic optimization statement and bound the problem. Its usefulness is limited to small-scale problems and the insight that it gives to nonlinear programming.

Finally, an introduction to geometric programming was given. This method has the important estimating characteristics of being able to find the cost contribution of the terms of the objective function when $C(x)$ is expressed in a special structured way. Certain kinds of engineering design problems can be cost-optimized using this method.

SELECTED REFERENCES

A great variety of linear programming texts exists. A few are given here:

FABRYCKY, W. J., P. M. GHARE, and P. E. TORGERSEN: *Industrial Operations Research*, Prentice-Hall, Inc., Englewood Cliffs, N.J., 1972, pp. 441–500.

GARVIN, WALTER W.: *Introduction to Linear Programming*, McGraw-Hill Book Company, New York, 1960.

LLEWELLYN, ROBERT W.: *Linear Programming*, Holt, Rinehart and Winston, Inc., New York, 1964.

WAGNER, HARVEY M.: *Principles of Operations Research with Applications to Managerial Decisions*, Prentice-Hall, Inc., Englewood Cliffs, N.J., 1969, Chaps. 2–5, pp. 33–164.

A listing of references for the Lagrangian method includes

BAUMOL, WILLIAM J.: *Economic Theory and Operations Analysis*, Prentice-Hall, Inc., Englewood Cliffs, N.J., 1965.

GUE, R. L., and M. E. THOMAS: *Mathematical Methods in Operations Research*, The Macmillian Company, New York, 1968.

KREYSZIG, ERWIN: *Advanced Engineering Mathematics*, John Wiley & Sons, Inc., New York, 1963.

Benchmark references for geometric programming are found in

DUFFIN, R. J., E. L. PETERSON, and C. ZENER: *Geometric Programming*, John Wiley & Sons, Inc., New York, 1967.

WILDE, D. J., and C. S. BEIGHTLER: *Foundations of Optimization*, Prentice-Hall, Inc., Englewood Cliffs, N.J., 1967, Chap. IV, pp. 99–131.

ZENER, C.: "A Mathematical Aid in Optimizing Engineering Design," *Proceedings of the National Academy of Science*, 47 (1961), pp. 537–539.

QUESTIONS

1. What are the basic concepts employed in linear programming? If these are to be exploited, what kinds and quality of information are required?
2. What does nonnegativity imply in a realistic sense? Under what circumstances do you imagine that variables can assume negative values?
3. What kinds of solutions are found in solving sets of linear equations? If there are eight variables and four constraints, what are the number of possible corner solutions? Will these all be feasible?
4. Why are solutions with a minimizing objective function improper at the origin?
5. What is the connection between the linear programming and Lagrangian method of undetermined multiplier dual variables?

6. Will the Lagrangian method work with linear programs?

7. In a cost optimization of an engineering design, what are the functions required before geometric programming can be applied?

12-1. Solve the following linear programs by graphical methods.

(a) maximize $3x_1 + 5x_2$

subject to $2x_1 + 3x_2 \leq 18$

$2x_1 + x_2 \leq 12$

$3x_1 + 3x_2 \leq 24$

$x_{1,2} \geq 0$

(b) minimize $4x_1 + 6x_2$

subject to $2x_1 + x_2 \geq 10$

$2x_1 + 3x_2 \geq 18$

$x_1 + 3x_2 \geq 12$

$x_{1,2} \leq 0$

(c) maximize $x_1 + x_2$

subject to $x_1 \leq 6$

$x_2 \leq 8$

$x_1 + x_2 \leq 10$

$x_1, x_2 \geq 0$

(d) minimize $4x_1 + 6x_2$

subject to $2x_1 + x_2 \geq 8$

$x_1 + 4x_2 \geq 8$

$2x_1 + 3x_2 \geq 12$

$x_{1,2} \geq 0$

(e) maximize $37x_1 + 10x_2$

subject to $10x_1 + 5x_2 \leq 50$

$25x_1 + 9x_2 \leq 100$

$5x_1 + x_2 \leq 15$

$8x_1 + 20x_2 \leq 180$

$4x_1 \leq 10$

$x_{1,2} \geq 0$

12-2. A total profit function has been constructed as $\$37x_1 + \$10_2$ which is restricted by

$$10x_1 + 5x_2 \leq 50$$
$$4x_1 \leq 10$$
$$25x_1 + 9x_2 \leq 100$$
$$8x_1 + 20x_2 \leq 180$$
$$5x_1 + x_2 \leq 15$$

Set up a graphical solution.

12-3. Describe the feasibility space after you have plotted the constraints. All $x_{1,2} \geq 0$.

(a) $-x_1 + x_2 \leq 2$
$\quad\ x_1 - x_2 \leq 1$

(b) $-x_1 + x_2 \geq 2$
$\quad\ x_1 - x_2 \leq 1$

(c) $\quad\ x_1 + x_2 = 6$
$\quad\ x_1 \qquad \leq 4$
$\qquad\quad x_2 \leq 4$

12-4. What is the slope of the following objective functions?
(a) Isoprofit lines: $3x_1 + 5x_2$ and $x_1 + x_2$.
(b) Isocost lines: $4x_1 + 6x_2$ and $37x_1 - 10x_2$.

12-5. Convert the inequality constraint sets to an equality constraint set. Use the parts of Problem 12-1.

12-6. With the linear program

maximize $-3x_1 - 2x_2$

subject to $-x_1 + x_2 \leq 1$

$\qquad\qquad 6x_1 + 4x_2 \geq 24$

$\qquad\qquad x_1 \geq 0, \ x_2 \geq 3$

transform to

$$\sum_{j=1}^{n} c_j x_j$$

subject to

$$\sum_{j=1}^{n} a_{ij} x_j = b_i \qquad \text{for } i = 1, \ldots, m, \ b_i \geq 0 \text{ and } x_j \geq 0 \text{ for } j = 1, \ldots, n$$

12-7. Solve Problem 12-1(a) by the simplex procedure.

12-8. Solve Problem 12-1(c) by the simplex procedure.

12-9. Solve the primal given below by simplex practices.

maximize $5x_1 + 4x_2 + 3x_3$

subject to $x_1 + 0.5x_2 + 0.7x_3 \leq 20$

$\qquad\qquad 5x_1 + \qquad\qquad x_3 \leq 10$

$\qquad\qquad 0.2x_2 + 0.5x_3 \leq 14$

$\qquad\qquad\qquad x_{1,2,3} \geq 0$

12-10. (a) Solve the following primal:

maximize $3x_1 + 5x_2 - x_3$

subject to $2x_1 + 3x_2 + \frac{1}{2}x_3 \leq 18$

$\qquad\qquad 2x_1 + x_2 - 2x_3 \leq 12$

$\qquad\qquad 3x_1 + 3x_2 \qquad\quad \leq 24$

$\qquad\qquad\qquad x_{1,2,3} \geq 0$

(b) Solve the following primal:

maximize $x_1 + x_2 - x_3$

subject to $x_1 \qquad + x_3 \leq 6$

$\qquad\qquad x_2 - x_3 \leq 8$

$\qquad\qquad x_1 + x_2 - x_3 \leq 10$

$\qquad\qquad\qquad x_{1,2,3} \geq 0$

12-11. Using the Big M method, solve

(a) minimize $x_1 + 2x_2$

subject to $x_1 + x_2 \geq 4$

$x_1 + 3x_2 \geq 9$

$x_2 = 2$

$x_{1,2} \geq 0$

(b) minimize $-3x_1 - 2x_2$

subject to $-2x_1 - x_2 \geq -3$

$-x_1 - x_2 \geq -2$

$x_{1,2} \geq 0$

12-12. Solve Problem 12-1(b) by the penalty method.

12-13. Solve Problem 12-1(d) by the penalty method.

12-14. Solve Problem 12-1(e) by the penalty method.

12-15. Determine the initial iteration using the Big M method. Stop after you have identified the new basis for the second iteration.

minimize $x_1 - x_2$

subject to $x_1 - x_2 \leq 1$

$3x_1 + 4x_2 \geq 12$

$x_1 \geq 0$

$x_2 \geq 1$

12-16. Given the primal

maximize $7x_1 + 8x_2$

subject to $2x_1 + x_2 \leq 4$

$3x_1 + 5x_2 \leq 6$

$x_{1,2} \geq 0$

convert to the dual and solve.

12-17. Use the dual method to solve Problems 12-11(a) and 12-11(b).

12-18. In analyzing the costs of their new superbounce ball, the promoters were contemplating whether costs could be reduced using different combinations of raw materials. They decided that their ball should bounce to at least 75% of its original height. Raw material x was good for a 60% bounce, y for a 75% bounce, and z for a 90% bounce. Material x cost $0.50, y cost $0.85, and z cost $1 for like quantities. However, more than 50% of either x or y in the finished product would seriously reduce the bounce of the ball. What is the least expensive combination of raw materials?

12-19. Two different sizes of cans comprise the output of a can manufacturer. Due to a prolonged steel strike, the firm only has 10 million square feet of tin plate. Large cans use 1 square foot of tin, while smaller ones require only half as much. The profit on the large and small cans, respectively, is $0.005 and $0.003. During the month's interval until tin plate shipments are expected, the capacity for large cans is 8 million units and for small cans is 17.5 million units. Solve by graphical methods for a maximum profit.

12-20. Solve the following by using the Lagrangian method of undetermined multipliers. Show the optimal program and dual variable.
(a) $y = x_1x_2$ subject to $2x_1 + 2x_2 = c$, where c is any constant.
(b) Minimize $2x_1^2 + 3x_2^2$ subject to $x_1 + 3x_2 = 10$.
(c) Minimize $y = 3x_1^2 + 4x_2^2$ subject to $2x_1 - 3x_2 = 10$.
(d) Maximize $y = 2x_1x_2 + 3x_1x_2$

subject to $x_1^2 + x_2 = 3$

$x_1x_3 = 2$

(e) The cost-estimating relationship of an antenna used to receive signals from a telecommunications satellite is given by

$$C = \frac{20 \times 10^5}{D^{1/3}} \exp\left(\frac{D}{45}\right), \qquad 15 < D < 250$$

where D = diameter in feet and C = dollars. Find the minimum using differentiation. How does this differ from the Lagrangian method of undetermined multipliers?

12-21. Using the Lagrangian method of undetermined multipliers, find the firm's expansion model expressed in terms of its total expenditure (in dollars) on its input, given the production function

$$Q = 2 \log L + 4 \log K$$

and the input prices $P = 4$ and $P_c = 12$. The object is to maximize output Q subject to the expenditure constraint $4L + 12K = M$.

12-22. Find the optimal values for the function and variables using the geometric programming method:
   (a) Min $y = 2x + (1/x^2)$.
   (b) Min $y = c_1 x^{-1} + c_2 x^{\frac{1-n}{n}}$, $c_i > 0$, $0 < n < 1$.
   (c) Min $y = 4t_1 t_2^{-1} + 256 t_1^{-1} t_2^{-1} + 4t_2$.
   (d) Min $y = 40/x_1 x_2 x_3 + 40 x_2 x_3 + 20 x_1 x_3 + 10 x_1 x_2$.
   Solve these problems by differential calculus for a check.

12-23. A cost model for the optimal velocity of heat-transfer media for a heat exchanger with tubes is summarized by the following single variable model as

$$\frac{c}{q} = \frac{K_1}{v^{0.8}} + K_2 v^2$$

where c/q is cost per heat transfer area, $K_{1,2}$ are proportionality constants, and v is velocity in feet per second. Find the optimum velocity and c/q.

12-24. The horsepower requirement for a propeller-driven aircraft is given as

$$\text{hp} = \frac{\rho v^3 C_d S}{1100} + \frac{4\omega^2}{1100\pi e A R} \frac{1}{\rho v s}$$

but rewriting as $\text{hp} = Av^3 + (B/v)$, A and B are appropriate constants as defined between the two equations. Use normality and orthogonality concepts to determine optimum values. Typical values are $\rho = 0.002$, $C_d = 0.04$, $S = 300$, $\omega = 15,000$, and $eAR = 6$. Find the horsepower.

12-25. A five-sided box (open top) is to be built with a volume of 20 cubic feet. Material costs $10 per square foot, while joint welding is $8 per joint foot. What size box should be built? Describe the equations and solve by both the Lagrangian and geometric programming methods.

chapter 13

Many have been ruined by buying good pennyworths.

BENJAMIN FRANKLIN
Poor Richard's Almanac

MANAGEMENT OF COST ESTIMATING

Cost estimating is done for a number of purposes. One reason is to set a standard of performance, another is to provide a basis for comparing alternatives, but probably the most popular purpose is to determine a price either to pay for a certain benefit or to charge for a specific operation or product or project. In view of its importance, verification of cost estimating is mandatory, as profitable orders, contracts, and operations result from successful estimating. In this chapter, the word *management* implies analysis of successful and unsuccessful estimates, adjustment practices based on the estimate for bidding or pricing, and finding the amount of detail to use for an estimate. Normally, one would think of staffing a cost-estimating department, administration, training, and recruitment, as the material for a chapter on management. Indeed, cost control and accounting variances, critical path planning, and cost reduction tactics are also intentionally overlooked. As practices vary and many books on the general field of management and cost control are available that satisfy these general objectives, we choose to give our attention to the analysis of the estimate and its manipulation.

13.1 TRADE-OFF: INFORMATION VERSUS COST

This book has stressed that quality and quantity of information are crucial to estimating. If information is lacking, estimating methods accommodate to the precondition that data are limited. Even so, many hard-dollar decisions are made on this basis. As more information becomes available, estimating methods earn the appearance of *certified tools*, and faith and acceptance, justifiably so, improve. Decisions improve, errors are reduced, and a list of positive factors to encourage methods of detailed estimating can be written. If this is true, with what level of detail should the estimating department operate? Specifically, we do not answer this question. Broadly, the issues are made clearer by Figure 13.1.

The axes are not scaled in this figure other than to indicate direction of increasing or decreasing magnitude. Our experience indicates that more detail, rather than less, is warranted at the early decision point. This in no way indicates that computer programming is required or even essential. In the vast majority of early design decisions, the methods of estimating are adapted to the design, and the computer provides little service other than occasional arithmetic calculation. If the computer is used extensively in the estimating process, the estimating is likely to be routine and detailed.

Certainly, as detail increases, maintenance, application, and training cost increase. These costs of estimating may be significant. Sometimes only 1 out of 5 or 10 estimates or *bids* may be won. The lost bid costs are covered by the winning bid. Yet it takes exposure to win business, and if profits decline and business conditions become more competitive, the tendency is to increase the estimating activity. Estimating is not alone responsible for winning or losing bids, but as errors decline and as more detail becomes available, it is axiomatic that estimating is better able to reflect the strengths of an organization or expose its weaknesses.

13.2 ESTIMATORS ESTIMATE: MANAGEMENT CONTROLS

The goal of the estimating function is to produce estimates that are exact. While this is a commendable purpose, it is more realistic to say that the goal would have the estimate value fall within some acceptable range. This recognizes that

FIGURE 13.1

TRADE-OFF: INFORMATION VERSUS COST

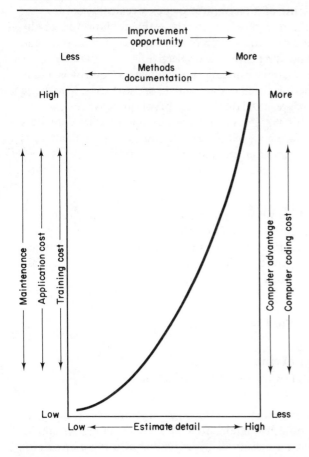

estimates are seldom true and advances the notion of a tolerance. What this tolerance should be is open to analysis and consideration. Factors such as cost of preparation, time available, impact on the organization, and data requirements bear on the selection of an estimate tolerance. Inasmuch as estimates are prior to the fulfillment of the design, a passage of time exists between the estimate and the historical determination of the actual value. In some situations, data may never be gathered to allow even a nominal comparison. At the other extreme an abundance may be on hand; yet due to accounting effects a tidy comparison may be impractical for the estimate. In practice a reconciliation between estimate and actual measure, while very desirable, is difficult to bring about.

There is yet another problem in assuring and controlling the value of the estimate. Estimators estimate, yet it is management that controls the elements of cost. Certainly estimators are a part of the general management team; however, the prerogatives of cost control, manning, budgets, and so forth are the responsibilities of nonestimators. One may argue that an estimator is responsible for forecasting the peccadilloes of management. Certainly estimators should try to be aware of these factors when costing designs; nonetheless the responsibility of cost control is either delegated or shared with others.

It may be argued that the estimate should reflect an upper limit of cost, and it need not have the objective of being an exact image of the final value. In these circumstances the estimate serves as a *target*, and accordingly management strives to bring the actual cost (less profit in the case of a product estimate) under the projected estimate. If in the case of a product estimate, the actual realization of the cost were always above the estimate, the firm could be in jeopardy. The philosophy that the estimate serves as an upper limit is appealing, although not always convincing. In the sense of a target, it provides a realistic goal for cost control. This may not be the best objective for the estimate, however. Faulty misinterpretation, biases, and distortion creep into estimates when they serve in this capacity. "Truth" in estimating is the best objective.

The dilemma that estimators estimate and managers control is not a sufficient reason to ignore performance of the estimate or estimators. Whether a direct or indirect comparison is resorted to, the principle of surveillance cannot be overlooked. Therefore, in this chapter we shall consider methods and practices necessary to measure the performance of the estimate and to determine a means to analyze the amount of information necessary for an estimate.

13.3
ANALYSIS OF SUCCESSFUL ESTIMATES[1]

Cost estimating is seldom done without an effort to check its effectiveness. A common method of verification is to find out what the actual costs and prices have been on past estimates and how these compare with the original estimates. That was done in the case displayed in Figure 13.2, which resulted from a study of 157 cost estimates for tools in a manufacturing plant.

An error is measured as

$$E = \left(\frac{c_e}{c_a} - 1 \right) \times 100 \tag{13-1}$$

where E = percentage error of estimate

c_e = estimate of cost, price, return or effectiveness

c_a = actual cost, price, return or effectiveness

Now if a firm determined this percentage error for a large number of estimates and plotted these findings on a histogram, we could, by assuming an infinite population of estimates, form a density distribution curve to these error estimates. In this actual study the deviations from true cost ranged from a low of 50% to a high of 450%, but the sum of the estimated amounts was less than 4% above the total actual cost. Obviously all the jobs that were estimated low did not yield as much return as was expected of them. The others brought in more than was expected because they actually cost less than what was estimated. A natural question is: "What can be learned from an analysis of this sort to help improve the cost-estimating practice?" Should some amount be added to each estimate to reduce the losses from the low ones? If so, how much? Obviously to do so would raise the high estimates and diminish their chances in competition. The low estimates cannot be identified beforehand or there would not be any low estimates. While analysis of this sort is useful, it does not give information about jobs that did not meet competition. The actual costs of unsuccessful bids were never determined. Study of estimating policies must consider all estimates, not just the ones that have produced orders.

In further discussion herein, the estimated cost c_e is understood to be the total price for a job. This includes the usual items of manufacturing cost, direct and indirect, overhead, contingencies, and an expected profit as depicted in

[1]Lawrence E. Doyle and Phillip F. Ostwald, "Closed Bidding Strategies to Improve the Performance of Cost Estimating," *SME Paper*, MM70–721, May 9, 1970.

Figure 9.1. The contingency factor allows for average unassignable extra costs that may occur. The total price is the firm price for the market, a quotation, or a bid. The actual cost c_a includes these same factors of cost except for contingencies because they are inherent in the actual performance for each job.

For simplicity the deviations in cost estimates will be expressed by the ratio c_e/c_a. This does not change the shape of Figure 13.2, to which a new scale

FIGURE 13.2

DISTRIBUTION OF ESTIMATES FOR VARIOUS
TOOLING JOBS

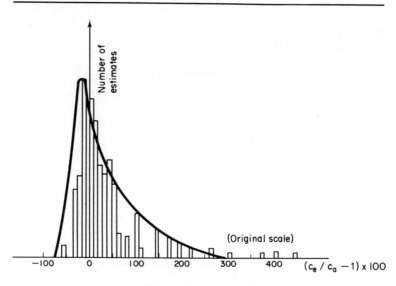

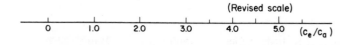

may be applied to show the deviations in terms of c_e/c_a. Obviously this ratio equals one where actual cost equals the estimated cost. This baseline is where the important break-even line originates. Those points to the left of the baseline on the horizontal axis are operationally undesirable because the firm will be operating at a loss. Depending on bidding policies and bidding objectives, the distribution of estimates is to the right of the break-even point. Long-range survival depends on that.

Now consider a hypothetical manufacturing plant and competitive situation in which estimates and bids are regularly made. For this situation we assume three distributions which describe our firm's bids and the competitors' bids. These three distributions, $f(v)$, $f(s)$, and $f(r)$, are shown in Figure 13.3.

The density curve $f(v)$ describes the distribution of all possible estimates competing for orders for our manufacturing plant. This density curve must withstand the test of competition if it is to generate successful bids. If it is a winning bid, it is, of course, lower than the lowest competing bid. Since the density curve $f(v)$ contains all bids made under the prevailing circumstances, it naturally contains those bids which are successful in generating winning orders. This leads to the density curve for the distribution of successful bids $f(s)$. Curve

$f(s)$ contains the distribution of bids that were successful on past jobs when plotted with respect to c_e/c_a. Its appearance would be similar to Figure 13.2. The firm must keep records and be aware of actual cost c_a for those won jobs. The distinction between $f(s)$ and $f(v)$ is that the actual cost c_a is unknown for the jobs that were unsuccessful in $f(v)$. Moreover, the actual cost of jobs won by a competitor remains unobtainable by our firm. Certainly, competitors do not like to release proprietary costs of this sort to each other. Nonetheless, there are ways of approximating competitor actual cost, but this discussion is deferred until later.

FIGURE 13.3

DISTRIBUTION OF SUCCESSFUL,
UNSUCCESSFUL, AND ANTICIPATED
INTENSITY OF ESTIMATES

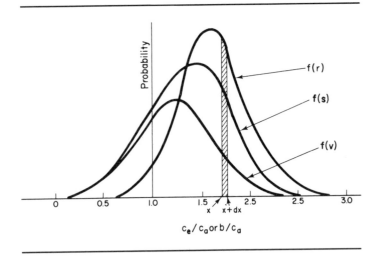

The density curve which describes the competitors' bidding practices is denoted by $f(r)$, where $r = b/c_a$. We let b stand for the lowest competitive bid and c_a stand for the actual cost for the same job in our plant. Curve $f(r)$ is the distribution of the intensity of competition because it shows our competitors' lowest bid compared to the actual cost in our plant.

Using these three curves, it is possible to determine the bid which will produce the winning bid for our firm. The problem, of course, arises in uncovering these curves. Now we describe a probabilistic model which will determine one curve by knowing two out of the three distributions. For the moment we assume that $f(v)$ and $f(r)$ exist. Later we shall discuss ways of evaluating them.

Now assume that our firm's proposed bid is X. The probability that our competitor's bid is in the region of X (i.e., region x to $x + dx$) is $f(r_x)\,dr$. The probability that our competitor's bid is below our bid of X and therefore successful is

$$P(X > x) = \int_{-\infty}^{x} f(r)\,dr \tag{13-2}$$

From probability then, the probability that our firm's bid is successful as compared to our competitor's bid is 1 minus the probability that our competitor's bid is successful. In other words,

$$P \text{ (our firm's bid successful)} = P \text{ (competitor's bid unsuccessful)}$$

$$= 1 - P \text{ (competitor's bid successful)}$$

$$= 1 - \int_{-\infty}^{x} f(r) \, dr$$

$$= \int_{x}^{\infty} f(r) \, dr \tag{13-3}$$

Now, the probability that our bid is in the region $x < X < x + dx$ is given by $P(x < X < x + dx) = f(v) \, dv$. Since the distribution $f(s)$ is the density function of our successful bids, the probability that our bid is successful is equal to $f(s) \, ds$ and equal to the probability that our bid is in the region of X times the probability our competitor's bid is not successful. Therefore, in the region of X,

$$f(s) \, ds = f(v) \, dv \int_{x}^{\infty} f(r) \, dr \tag{13-4}$$

Now, since $ds = dv$ for an incremental region,

$$f(s) = f(v) \int_{x}^{\infty} f(r) \, dr \tag{13-5}$$

Thus the density function $f(s)$ describing successful bids is given in terms of $f(v)$, the density function of all our firm's bids, and $f(r)$, the density function which describes the intensity of competition. The area under $f(s)$ represents the population of successful jobs, and proportionately the area under $f(v)$ is to the area under $f(s)$ as the total number of estimates made is to the number of successful estimates. By manipulating Equation (13-5), $f(v)$ can be determined if $f(s)$ and $f(r)$ are known. A count of the two classes of $f(s)$ and $f(r)$ will give an initial evaluation of $f(v)$ from $f(s)$.

The distribution $f(r)$ is difficult to evaluate directly. Values of b, low bids on past jobs, may be kept on record, particularly in an industry where the results of bidding are publicized. This is common for open competitive bidding for government contracts. When private companies are the customers, information about winning bids is often closed; however, only follow-up may be necessary to uncover this information. What is not obtainable generally is information on c_a, the actual cost of jobs won by competitors. However, what can be plotted from data generally available is the distribution b/c_e, the ratio of the winning bid to our cost estimate price. The ratio of the winning bid to estimated value is a random variable with a probability distribution independent of the amount involved. In other cases that may not be so, and a separate probability density distribution may have to be set up for different ranges of job values. Any data collected for an analysis will have to be tested for correlation between job size and the ratio b/c_e. A logarithmic-normal-type curve fits this distribution.

The distribution of $r = b/c_a$ may be estimated in several ways, depending on the information available. The ratios may be computed directly for jobs made in our plant. Ratios of b/c_e may be compiled for other jobs and for those job ratios of c_e/c_a generated by means of random numbers or Monte Carlo techniques. These two ratios are multiplied to obtain the ratio b/c_a for each job. Once $f(r)$ and $f(s)$ are estimated from past experience, the pattern of the total estimating practice, $f(v)$, may be constructed by computation or graphical manipulation of Equation (13-5). A final check of $f(v)$ is made to assure that it is in reasonable proportion in size to $f(s)$ as the total number of jobs studied is to the number of successful estimates.

Another approach to setting up the model to portray a given situation is to make an initial estimate of $f(v)$ on the basis of its included area in relation to

$f(s)$ and bias factors and random selection of values of c_a to be divided into c_e for the estimates not sold. This synthesized $f(v)$ may be combined with $f(s)$ in accordance with Equation (13-5) to yield a suitable $f(r)$.

After the model has been arranged to reflect reasonably well the performance of cost-estimating practice, it may be manipulated to indicate how the performance may be optimized.

13.4 OPTIMIZING COST-ESTIMATING PERFORMANCE

Cost-estimating performance can be changed in one or both of two ways. The entire distribution $f(v)$, and thus its mean, $\bar{v}$, can be moved to the right or left by multiplying some or all estimates by one or more factors. The spread of the distribution may be increased or decreased by exerting more or less effort in obtaining, refining, and testing basic information. These changes obviously alter the competitive situation and the net results achieved.

Figure 13.4 indicates in a general way some of the effects of variations in the mean and the spread of the total distribution of the ratio c_e/c_a for sets of cost estimates. These curves carry the same axis designations given previously. Table 13.1 lists relevant measurements from Figure 13.4. These include the areas under the curves $f(v)$ and $f(s)$ and the distances between means, $\bar{v}$ and $\bar{s}$. A yield q is defined as $f(s)/f(v)$ and is an indicator of the number of successful estimates obtained from the total number of estimates made in each case. An average margin $|\bar{s}|$ is defined as the distance between $\bar{s}$ and a baseline. With the baseline

FIGURE 13.4

HYPOTHETICAL COMBINATIONS OF COST AND BID RATIOS

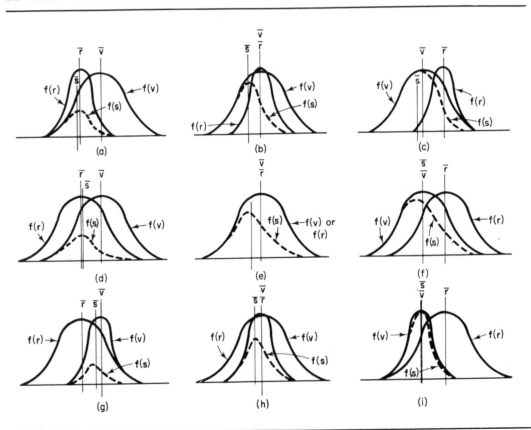

passing through the point $c_e/c_a = 1.0$, the distance from it to $\bar{s}$ is an indicator of the average excess of the sale price over actual cost. If we normalize $f(r)$ and require that the baseline lie a specified number of standard deviations from $\bar{r}$, we are able to determine the average margin. In Table 13.1 two conditions are considered. These stipulations require that c_e/c_a be -3σ and -1σ, respectively, below $\bar{r}$. These conditions imply that competition in the market place is related to actual costs because all distributions are on the same horizontal scale and 1.0 for c_e/c_a is also 1.0 for b/c_a. Total gain is the product of the margin (average excess realized over the base estimate) and the yield (proportional to the number of successful estimates).

In the example of Figure 13.4 and Table 13.1, the yield increases as the mean, $\bar{v}$, of the cost-estimating ratio, c_e/c_a, is decreased with respect to $\bar{r}$. The yield is also increased for all relative positions of $\bar{v}$ and $\bar{r}$ as the spread of $f(v)$ is reduced as compared to the spread of $f(r)$. The amount of gain depends largely on competitive conditions. If average low bids are close to actual costs, a low yield with an average high return on each job appears to be optimum.

TABLE 13.1

MEASUREMENTS FROM FIGURE 13.4

	Parts of Figure 13.4								
	(a)	(b)	(c)	(d)	(e)	(f)	(g)	(h)	(i)
Area $f(v)$	1.4	1.4	1.4	1.4	1.4	1.4	0.7	0.7	0.7
Area $f(s)$	0.28	0.61	1.02	0.34	0.68	1.02	0.21	0.39	0.60
Yield $= \dfrac{f(s)}{f(v)} = q$	0.20	0.44	0.73	0.24	0.49	0.73	0.30	0.56	0.86
$(\bar{v} - \bar{s})^a$	1.2	0.7	0.3	1.0	0.4	0	0.7	0.6	0
Average margin $= \|\bar{s}_1\|^b$	2.8	1.8	0.5	2.0	2.5	2.0	3.7	2.3	2.0
Total gain, $q\|\bar{s}_1\|$	0.6	0.8	0.4	0.5	1.2	1.5	1.1	1.3	1.7
Average margin $= \|\bar{s}_2\|^c$	0.8	-0.2	-1.5	0	0.5	0	1.7	0.3	0
Total gain, $q\|\bar{s}_2\|$	0.2	-0.1	-1.1	0	0.3	0	0.5	0.2	0

[a]In units of σ_v.
[b]For $\bar{r} - 1 = 3\sigma_r$.
[c]For $\bar{r} - 1 = \sigma_r$.

Changing the level of the average ratio c_e/c_a as discussed is equivalent to setting the bid or quotation price of jobs individually or as a whole to optimize profits. The ratio c_a/c_e can be used to maximize profits. Let x be the bid price for a job for which a cost of c_{e0} is estimated and for which the actual cost is as yet unknown and is designated by the variable c_a. The profit on the job will ultimately be $x - c_a = x - c_{e0}/v$. The factor v is the ratio c_e/c_a, as before. The probability that this ratio is between v and $v + dv$ is $f(v)\, dv$. Then, the expected profit for a bid of x is

$$E(x) = \int_0^\infty P(x)\left(x - \frac{c_{e0}}{v}\right) f(v)\, dv \qquad (13\text{-}6)$$

where $P(x)$ is the probability that x is the lowest bid.

Accordingly, it wins the contract. Since x and $P(x)$ are independent of v and

$$\int_0^\infty f(v)\, dv = 1$$

the expected profit equation becomes

$$E(x) = P(x)(x - c_0) \qquad (13\text{-}7)$$

where

$$c_0 = c_{e0} \int_0^\infty \frac{f(v)}{v} \, dv$$

$P(x)$ can be determined directly from a distribution of $u = b/c_e$ evaluated from past experience. Thus $P(x)$ is given by

$$P(x) = P\left(\frac{x}{c_{e0}} - \frac{b}{c_{e0}}\right) = \int_{x/c_{e0}}^\infty f(u) \, du \qquad (13\text{-}8)$$

$$E(x) = \int_{x/c_{e0}}^\infty (x - c_0) f(u) \, du \qquad (13\text{-}9)$$

Equations (13-7) and (13-8) are solved together for successive values of x. The curve for $E(x)$ versus x should generally have a form like that in Figure 13.5. The bid x_B that yields the maximum expected profit $E(x)$ is the optimum bid.

FIGURE 13.5

GRAPH FOR $E(x)$

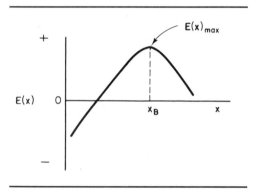

13.5
PRACTICAL DIFFICULTIES IN COMPARING ACTUAL TO ESTIMATED FACTS

How simple is it to obtain actual cost information to allow a *comparison to the estimate*? Success in finding actual values comparable to the estimate is mixed. A number of factors prevent a straightforward comparison. Often, accounting cost systems are developed in such a way that collecting and analyzing actual costs may not facilitate comparison. Actual cost systems allow little advantage because their information may be gathered too late. Actual costs that are gathered are used principally for income tax, ownership, and open disclosure laws. This information, despite its availability, may not meet the needs of estimating. Furthermore, it may have little resemblance to the way actual or *true costs* were incurred for the estimate. The depreciation account is an approximation, and will show the real decline in capital value consumption only after a sale or bankruptcy liquidation of the assets—hardly an action suitable for checking estimates. A comparison of actual costs to the estimate is a dilemma. In effect, the comparison remains an "estimate of actual costs following the design to the estimate of costs prior to the design." Explicit actual costs are difficult to obtain, but cost estimators need these facts to allow improvement and control. Usually, the first comparison is that of direct costs. Subsequent enlargement of data calls for ingenuity and careful analysis.

There is a cost associated with making a cost estimate and relying on the value of that estimate. Now we consider the problem of minimizing the total cost of making the estimate. Initially, several statements are made regarding the types of costs that are treated here. Statistical decision theory, as it pertains to our needs, is explained. The types of information and assumptions necessary for the application of statistical decision theory are briefly considered. Numerical methods are used to uncover a solution to several equations that are mentioned. Finally, an example is provided which uses arbitrary costs and other information. It is seen that emphasis is more on illustrating the optimization method and deriving certain specific relations of optimality than on recommending a general course of action.

Assume that the cost of estimating is the sum of a monotonic increasing function $C(M)$, which is the functional cost of making the estimate, and a monotonic decreasing function $C(E)$, which is the functional cost of making an error in the estimate. Both are dependent on the amount of detail of the estimate. This function is given by

$$C_T = C(M) + C(E) \tag{13-10}$$

where C_T = total cost of making estimate, dollars
$C(M)$ = functional cost of making estimate, dollars
$C(E)$ = functional cost of error in estimate, dollars

Figure 13.6 describes what we mean.

FIGURE 13.6

COST OF INCREASING DETAIL

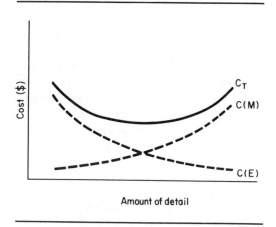

It should be added that the cost of estimating can be a significant factor. In some industries there is evidence to suggest that 90% of the estimates do not lead to contracts and sales. In these situations the cost of estimating must be applied to won jobs. For this reason, improvements in estimating are considered worthy problems, and economies that can be introduced into estimating and optimization of existing practices are met with interest.

[2]Jack C. Hanson and Phillip Ostwald, "Finding the Optimal Amount of Segmentation for Estimates," *Transactions of the AACE*, 1971, pp. 249–254.

While a variety of linear or nonlinear cost functions of n can be found for $C(M)$, we shall use a flat or fixed cost and a constant variable cost per unit, or

$$C(M) = C_f + C_v(n - 1)^a, \qquad n \geq 1 \qquad (13\text{-}11)$$

where C_f = fixed cost of estimating, dollars
$\qquad C_v$ = variable cost of estimating per element n
$\qquad n$ = number of elements or detail of estimate
$\qquad a$ = experience exponent

It is apparent that as more time and detail are put into the estimate the cost of making the estimate increases. It is appropriate here to enlarge on the nature of n. This variable could simply represent the number of segments or components of an estimate such as engineering, manufacturing, and marketing. In this case $n = 3$. The total value would be the sum of these three component estimates. In another case n could represent the number of hours spent in making the estimate. The measures of the amount of detail or segmentation can vary widely.

The fixed cost of an estimate includes space, utilities, training, documentation, and other expenses that are more or less fixed charges insofar as estimating is concerned. It would be necessary to arbitrarily classify the total cost-estimating department expenses into fixed and variable categories. When $n = 1$ the cost of making the estimate is merely equal to C_f. In many situations a preliminary estimate is all that is needed. Even when enlarged numbers of estimating elements are anticipated, a preliminary estimate is usually provided. We assume that a first estimate is always made. The variable cost C_v applies to second level or detailed estimates. As n increases, the variable cost increases. Fixed cost becomes unimportant in the optimization procedure as it is invariant to changes in n. Our question is to determine how much more information we should add before making an estimate decision.

13.6-2
Cost of Making a Wrong Decision

The second term of cost equation (13-10) is related to a potential error of an estimate. Whenever the estimate leads to the wrong decision there is a subsequent monetary loss. There are numerous ways to consider this loss. In one circumstance the cost of a wrong decision could result from an overbid in which we lost the bidding competition. In a second, we underbid, and although we won the job, our actual expenses deviated from forecast expenses. In the latter situation our profit may be less or it may not materialize at all. As has been implied, the accuracy of the estimate can be improved and can thus reduce the chance for making a wrong decision.

Before proceeding, let us define *value* and *cost* as now used. The *cost* of the estimate is the expense incurred by the makers of the estimate, while *value* is the monetary worth of the design (product, project, and so forth) that is being estimated. The value would include labor, material (direct and indirect), overhead, general and administrative expenses, and perhaps profit as fits the particular situation. This value we call V. The estimate value is V_e, while the actual value is V_a. The better the estimate, the closer V_e is to V_a.

The opportunity to make a decision implies the availability of at least two alternatives. For this dichotomy a criterion must be established to determine which of two policies will be implemented. In cost estimating the criterion is usually a break-even point of some type. This we define as the critical value, V_c. If we have made a decision and the actual value supports our decision, it is correct, and we have incurred no loss or cost of error. The cost of an incorrect decision would be the amount that could have been saved or earned had the correct decision been made. If by good fortune we know V_e to be V_a in advance and could compare it to the critical value, the correct policy could always be implemented. A happenstance like this is nonexistent in estimating. The only information known in advance is V_e. Thus the decision must be based on the estimate.

Since the estimate can be in error, we recognize two types of error which have corresponding and distinct costs. They are dependent on the critical and actual value and could be defined by different cost equations. For example, if the value of a project or product is greater than the critical value, the decision would be to follow a certain policy A; and consequently, if the value of the project is less than the critical value, the decision would be to follow alternative policy B. Table 13.2 describes what we have said.

TABLE 13.2

DECISION TABLE FOR GUIDING POLICY ACTIONS

		The actual value V_a implies	
		Policy A $V_a > V_c$ Buy decision	Policy B $V_a < V_c$ Make decision
The value of estimate V_e implies	Policy A $V_e > V_c$ Buy decision	Correct decision	Incorrect decision Type II error
	Policy B $V_e < V_c$ Make decision	Incorrect decision Type I error	Correct decision

In the case of a *make* or *buy* example, a type I error would be to make a product and have the actual value be greater than what it could have been purchased for. This would be when V_a is greater than the purchase price, V_c, but V_e was less than the purchase price. A type II error would be to buy a product when it actually would have been cheaper to make it. That is, V_a is less than the purchase price, V_c, when V_e is greater than the purchase price. A type I error occurs whenever V_e indicates that we make a product when in fact V_a contradicts our estimate and implies the alternative decision. A type II error occurs whenever V_e indicates that we buy the product when in fact V_a implies the make decision. In our specific example, policies A and B correspond to buy and make.

Now that both types of errors have been discussed, their corresponding costs may be written. As stated before, this is the cost that could have been saved had the correct decision been made. The cost of both a type I and a type II error would be the difference between what was actually paid for the item and what the actual cost would have been. For this make or buy hypothesis,

$$\text{cost of type I error} = V_a - V_c \quad \text{if } V_a \geq V_c, V_e \leq V_c \quad (13\text{-}12)$$
$$\text{cost of type II error} = V_c - V_a \quad \text{if } V_a \leq V_c, V_e \geq V_c \quad (13\text{-}13)$$

where $V_c =$ the point where our decision, based on value, is changed.

In actuality, Equations (13-12) and (13-13) can represent any type of a function, i.e., from a lost bid to an out of pocket loss. Other criteria and corresponding decisions leading to several types of errors are also possible.

Costs, as defined by Equations (13-12) and (13-13), are sometimes called conditional costs, as they occur under given conditions. A fundamental of statistical decision theory implies that the expected cost is equal to the product of the conditional cost and the probability that the condition will occur. For example, if the probability of cost were zero, the expected cost would also be zero. However, if there were a 50% chance of $1000 cost and a 50% chance of a $2000 gain, over the long run we would expect to gain an average of $500. In a one-time instance the loss or gain is either $1000 or $2000. Hence, the probability of the conditions implying a corresponding cost become important in a continuous estimating procedure.

It must be remembered that the value of the estimate, V_e, is an expected value, and this supposes that it is the mean of a probability distribution. That is, the value estimated has the highest probability of occurrence. Of course, a probability distribution has some measure of variance. This can most easily be seen in Figure 13.2.

The expected value of the estimate from a rough approximation is V_e, and the standard deviation can be determined from historical or confidence information, assuming that the current estimate has the same distribution as past estimates. The probability of the condition V_a occurring can be written as a function of the normal probability distribution equation, or

$$P(V_a) = \frac{1}{\sigma\sqrt{2\pi}} e^{-(1/2)[(V_a - V_e)/\sigma]^2} \tag{13-14}$$

To find the expected cost of a type I or type II error we take the sum of the products of the conditional loss times the probability of that loss occurring. The above equation has defined the probability as being a function of a random variable V_a. The costs of a type I and type II error are continuous over defined intervals, and the sum of the products of probability and conditional costs can be expressed by integrating over the appropriate ranges for each conditional cost. These expected costs would be

$$C_{\mathrm{I}} = \int_{V_c}^{\infty} C_2(V_a - V_c) \frac{1}{\sigma\sqrt{2\pi}} e^{-(1/2)[(V_a - V_e)/\sigma]^2} \, dV_a \tag{13-15}$$

$$C_{\mathrm{II}} = \int_{0}^{V_c} C_3(V_c - V_a) \frac{1}{\sigma\sqrt{2\pi}} e^{-(1/2)[(V_a - V_e)/\sigma]^2} \, dV_a \tag{13-16}$$

where C_{I} = expected cost of a type I error

C_{II} = expected cost of a type II error

V_c = point where decision based on value is changed

C_2, C_3 = slope of the cost functions

These curves are superimposed in Figure 13.7. The lower limit of integral (13-16) is zero instead of $-\infty$ to maintain the validity of the cost portion of the equation.

The total expected loss is not merely the sum of these two equations but is also dependent on the probability of the occurrence of the corresponding type I and type II errors. In our example of a make or buy decision, it can be seen that the probability of the occurrence of a type I error is the probability that the actual cost is greater than the critical value when the estimate is below that value. For a type II error the actual value is less than the critical value when the estimate is greater than that value. Since we are considering the normal distribu-

FIGURE 13.7

EXPECTED COST OF AN ERROR

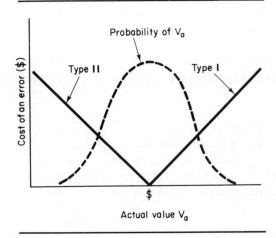

tion, these probabilities would be as follows:

$$P(\text{I}) = \int_{V_c}^{\infty} \frac{1}{\sigma\sqrt{2\pi}} e^{-(1/2)[(V_a - V_c)/\sigma]^2} \, dV_a \qquad (13\text{-}17)$$

$$P(\text{II}) = \int_{-\infty}^{V_c} \frac{1}{\sigma\sqrt{2\pi}} e^{-(1/2)[(V_a - V_c)/\sigma]^2} \, dV_a \qquad (13\text{-}18)$$

where $P(\text{I}) =$ probability of type I error $(V_a > V_c)$
$P(\text{II}) =$ probability of type II error $(V_a < V_c)$

The total expected cost is the sum of the expected cost of each type of error times the probability of the occurrence of that error and the cost of making the estimate, or

$$C_T = C(M) + C_{\text{I}}(P(\text{I})) + C_{\text{II}}(P(\text{II})) \qquad (13\text{-}19)$$

13.6-4
Finding the Standard Deviation

The variable in this expected cost equation is the standard deviation, and it can be manipulated by increasing or reducing the amount of detail. If a relationship between the standard deviation and n is known beforehand or can be estimated, the optimal amount of detail for an estimate can be determined.

If an estimate is divided into components, the total value of the estimate is the sum of the expected value of the components. The standard deviation, however, is more complex. The standard deviation for an estimate comprising two components is given by

$$\sigma_{1+2} = (\sigma_1^2 + \sigma_2^2 + 2r\sigma_1\sigma_2)^{1/2} \qquad (13\text{-}20)$$

where $r =$ correlation coefficient. Extensions of Equation (13-20) for n components are possible.

It is important to consider the nature of the standard deviation of the components and the correlation r. As n increases, we expect smaller and simpler elements. We presume they are easier to estimate than large and complex elements, and hence more precise. Continuing with this argument the smaller elements could be expected to have more than a proportionate drop in standard deviation. This means that if the value of the estimate were split into two equal components, the standard deviation for each component would be less than or equal to half the standard deviation or the total. If r were perfect, or equal to unity, the standard deviation of the components would be equal to half of the standard deviation of the total estimate. We could have

$$\sigma_1 + \sigma_2 \leq ((\tfrac{1}{2}\sigma)^2 + (\tfrac{1}{2}\sigma)^2 + 2(\tfrac{1}{2}\sigma)(\tfrac{1}{2}\sigma))^{1/2} \tag{13-21}$$

$$\sigma_{1+2} \leq (\tfrac{1}{4}\sigma^2 + \tfrac{1}{4}\sigma^2 + \tfrac{1}{2}\sigma^2)^{1/2} \tag{13-22}$$

$$\sigma_{1+2} \leq \sigma \tag{13-23}$$

This segmentation would have no benefit, as the standard deviation would not change. In reality r is not perfect and σ usually decreases more than proportionally, and the standard deviation reduces as a function of n. To determine which function should be used, we look to historical data for similar estimates and use them. In the absence of such data, the standard deviation of the function n may be approximated by

$$\sigma \approx \frac{K\sigma_f}{(n-1)^b} \tag{13-24}$$

where σ = standard deviation for estimate of n elements
    K = constant
    n = second level of detail, $n > 1$
    b = exponent
    σ_f = initial guess of standard deviation

By defining σ, the function C_T reduces to a single variable dependent on n. It then becomes the task to minimize this function with respect to n to determine the optimal amount of detail to minimize total costs.

13.6-5
Simplest Model for Optimum Detail

If V_e were to equal the critical value V_c, a special situation would be created. Under this condition integrals (13-17) and (13-18) reduce to simplified expressions. With a normal distribution the probability that the actual value is above the mean or below the mean is 50%, or

$$P(\text{I}) = \int_{V_c = V_e}^{\infty} \frac{1}{\sigma\sqrt{2\pi}} e^{-(1/2)[(V_a - V_e)/\sigma]^2} \, dV_a = \frac{1}{2} \tag{13-25}$$

and by symmetry,

$$P(\text{II}) = \frac{1}{2} \tag{13-26}$$

Integrals (13-15) and (13-16) simplify[3] when using (13-24) and $V_c = V_e$, or

$$
\begin{aligned}
C_\text{I} &= \int_{V_e = V_c}^{\infty} C_2(V_a - V_c)\frac{1}{\sigma\sqrt{2\pi}} e^{-(1/2)[(V_a - V_e)/\sigma]^2} \, dV_a \\
&= \frac{2}{5} C_2 \sigma \\
&= \frac{2}{5} C_2 K \frac{\sigma_f}{(n-1)^b}
\end{aligned} \tag{13-27}
$$

and similarly

$$C_\text{II} = \frac{2}{5} C_2 K \frac{\sigma_f}{(n-1)^b} \tag{13-28}$$

---

[3] Unit normal loss integral, Robert Schlaifer, *Introduction to Statistics for Business Decisions*, McGraw-Hill Book Company, New York, 1961, pp. 370–371.

where C_2 and $C_3 =$ linear slopes of conditional opportunity loss curves of Figure 13.7. These forms can be inserted into the general equation (13-19) to yield

$$C_T = C_f + C_v(n-1)^a + \tfrac{1}{2} \times \tfrac{2}{5}(C_2 + C_3)K\sigma_f(n-1)^b \qquad (13\text{-}29)$$

Differentiating Equation (13-29), we have

$$\frac{dC_T}{dn} = C_v a(n-1)^{a-1} - \frac{1b}{5}(C_2 + C_3)K\sigma_f(n-1)^{-b-1} \qquad (13\text{-}30)$$

Equating the derivative to zero will establish the equations for minimum cost:

$$n' = \left[\frac{b(C_2 + C_3)}{5aC_v}K\sigma_f\right]^{1/(a+b)} \qquad (13\text{-}31)$$

where $n = n' + 1$, which gives the optimal detail for this specific situation.

Now consider a shop which manufactures tools and cast products. A die for casting a part is necessary. The company can either make the die or buy it from a vendor. Obviously an estimate is required to determine which policy should be followed. The fixed cost C_f for a rough estimate is \$50. Each elemental extension for a detailed estimate is $C_v = $ \$5. Using Equation (13-11), we have $C(M) = 50 + 5(n-1)$, where we let $a = 1$.

The critical value is the purchase quotation of \$3000. If V_a is less than \$3000, the die should be made; conversely, if V_a exceeds the purchase price, the die should be purchased. A type I error is to buy the die when it is cheaper to make it; that is, V_a is less than \$3000 when the value of the estimate V_e is greater than \$3000. A type II error would be to make the die and have the actual value be greater than what it could have been purchased for. This would be when V_a is greater than \$3000 but the estimate V_e is less than \$3000.

Equation (13-19) can be written as

$$C_T = 50 + 5(n-1) + \int_{3000}^{\infty} (V_a - 3000)\frac{1}{\sigma\sqrt{2\pi}}e^{-(1/2)[(V_a - V_e)/\sigma]^2}\, dV_a$$

$$\times \int_{V_c}^{\infty} \frac{1}{\sigma\sqrt{2\pi}}e^{-(1/2)[(V_a - V_e)/\sigma]^2}\, dV_a$$

$$+ \int_0^{3000} (3000 - V_a)\frac{1}{\sigma\sqrt{2\pi}}e^{-(1/2)[(V_a - V_e)/\sigma]^2}\, dV_a$$

$$\times \int_0^{V_c} \frac{1}{\sigma\sqrt{2\pi}}e^{-(1/2)[(V_a - V_e)/\sigma]^2}\, dV_a$$

The preliminary estimate and the standard deviation must be provided, and they are estimated as $V_e = $ \$3000 and $\sigma_f = $ \$600. Defining Equation (13-24) as $\sigma = \sigma_f/n^{0.75}$, where $a = 1$, $b = 0.75$, and $K = 1$, computation can proceed. A computer analysis which briefly is a step-by-step approach provides the results shown in Table 13.3. Additionally, the simplest model (13-31) can be used for this same problem. As before we let the initial cost estimate be the same as the critical value, or $V_e = V_c$. For our example $V_e = $ \$3000, and the solution is

$$n' = \left[\frac{(0.75)(1+1)(600 \times 1)}{5 \times 1 \times 5}\right]^{1/(1+0.75)} = 8 \quad \text{and} \quad n = 9$$

Both the computer-iterative method and the simplest model yield the same optimal segmentation of detail. However, in using the computer-iterative method other variations are possible. For instance when $V_e = $ \$2750 and other data remain the same, we obtain Table 13.4. The number of segments are logically fewer and for this case $n = 8$. The stars in Tables 13.3 and 13.4 indicate optimum selection on the basis of total cost and net gain.

TABLE 13.3

RESULTS FOR V_e = \$3000 AND σ_f = \$600
WHERE OPTIMUM n = 9

Segments of Detail	Expected Loss	Estimating Cost	Total Cost	Reduction in Expected Loss	Increment Cost of Estimate	Net Gain
1	—	\$50	—	—	—	—
2	\$239.36	55	\$264.36	—	\$5	—
3	142.33	60	202.33	\$97.03	5	\$92.04
4	105.01	65	170.01	37.32	5	32.32
5	84.63	70	154.63	20.38	5	15.38
6	71.59	75	146.59	13.04	5	8.04
7	62.44	80	142.44	9.15	5	4.15
8	55.62	85	140.62	6.82	5	1.82
9	50.32	90	140.32*	5.30	5	0.30*
10	46.07	95	141.07	4.25	5	−0.75

TABLE 13.4

RESULTS OF V_e = \$2750 AND σ_f = \$600
WHERE OPTIMUM n = 8

Segments of Detail	Expected Loss	Estimating Cost	Total Cost	Reduction in Expected Loss	Increment Cost of Estimating	Net Gain
1	—	\$50	—	—	—	—
2	\$219.46	55	\$274.46	—	\$5	—
3	111.34	60	171.34	\$108.12	5	\$103.12
4	66.89	65	131.89	44.46	5	39.46
5	42.26	70	112.26	24.63	5	19.63
6	27.12	75	102.12	15.14	5	10.14
7	17.43	80	97.43	9.69	5	4.69
8	11.14	85	96.14*	6.29	5	1.29*
9	7.06	90	97.06	4.09	5	−0.91

The cost of making the estimate is an increasing function, while the cost of an error is a decreasing function. The minimum cost value for an n resides between 1 and a large value. The best method for finding optimum n depends on the nature of the equations involved. If the equations can be simplified, the method of setting the first derivative equal to zero and solving for n yields the optimal and minimum value. If the nature of the equations is more complex, it may be necessary to use a computer program and some method of iteration.

In a computer program the total cost equation (13-19) is evaluated for values of n beginning at $n = 2$ and increasing in steps of 1. The evaluation of the integrals may be done by using Simpson's rule or by other routines. The optimal value for n will be found when the net gain from additional estimate elements decreases to zero. A computer procedure can accommodate other functions to suit the situation. If the normal distribution were not a valid assumption, another distribution could be inserted. It is then necessary to relate the parameters of the distribution to n.

An assumption so far unstated requires that the estimating segments be of near equal value. If n is defined as a re-occurring breakdown, the application of the model should be used for each level. The value of the estimate and standard deviation will change and become more accurate. The selected number of estimating elements may also change. This is apparent if the different results of the previous examples are considered. If accuracy increases faster than was expected or if V_e moves further from the critical value, additional breakdowns may not be necessary. If the normal distribution is not a valid assumption, a proper distribution should be found, and this may entail difficulties in other applications.

13.7 SUMMARY

The estimated cost and true cost of a job are usually not the same. Their ratio is a variable that can be used to analyze and optimize the cost-estimating activity. The ultimate success of each cost estimate in generating an order or contract depends on its competitive position. Low estimates are likely to beat competitors bids but may lead to losses when the job is done. High estimates are profitable if sold but are more prone to be underbid by competitors. When the distribution of estimated costs is multiplied by the corresponding probabilities of competitive bids, the distribution of successful cost estimates ensues. These are the estimates that generate orders. The distribution of successful estimates and the bidding of competitors may be estimated from past experience. From these distributions a model is constructed of the cost-estimating practice. This model may be manipulated to show what changes in procedures may tend to optimize the practice.

The cost of making an estimate is an important feature in managing estimates. The optimal value of n determines where the benefits of further breakdown are exceeded by the additional cost of that breakdown. In the broadest sense this model can be used to determine the need for segmentation. This could be interpreted as determining when the estimated value is truly the best estimate of the actual value considering the ultimate use of the estimate. For a project or product having insignificant results only a few details may be warranted, but for a job which has important overtones a greater amount of segmentation is required so long as it remains in proportion to the cost of making the estimate.

SELECTED REFERENCES

Statistical decision theory can be studied in

JEDAMUS, PAUL, and ROBERT FRAME: *Business Decision Theory*, McGraw-Hill Book Company, New York, 1969.

SCHLAIFER, ROBERT: *Introduction to Statistics for Business Decisions*, McGraw-Hill Book Company, New York, 1961.

QUESTIONS

1. Discuss the trade-off question of quality and quantity of information versus cost. While a precise point cannot be identified in a general situation, what are some management guidelines to determine the amount of detail?
2. Should cost estimating be excused from accuracy in estimating because it does not cost-control the business?
3. Why is there so little performance checking of the cost estimate and the cost-estimating function? What can you suggest to deal with variations with estimators? Can you name statistical tests that deal with differences in estimators and or estimates?

4. The model for determining the amount of detail involves a statistical question where each item or detail is about the same size. What if they are not, as in the practical case?

5. How would you estimate the cost of estimating? What forms or procedures would you suggest?

6. The contradiction is sometimes stated that the preliminary estimate should have an abundance of data while detailed estimates need very little data, as early decisions are more critical. If early product cost estimates are the acute problem as opposed to cost-estimating on-going operations, products, and so forth, what new rules can you suggest for management of cost estimating?

APPENDIXES

APPENDIX I: VALUES OF THE STANDARD NORMAL DISTRIBUTION FUNCTION[a]

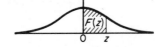

Areas Under the Normal Curve

$$F(z) = \int_0^z \frac{1}{\sqrt{2\pi}} e^{-z^2/2} \, dz$$

z	0.00	0.01	0.02	0.03	0.04	0.05	0.06	0.07	0.08	0.09
0.0	0.0000	0.0040	0.0080	0.0120	0.0159	0.0199	0.0239	0.0279	0.0319	0.0359
0.1	0.0398	0.0438	0.0478	0.0517	0.0557	0.0596	0.0636	0.0675	0.0714	0.0753
0.2	0.0793	0.0832	0.0871	0.0910	0.0948	0.0987	0.1026	0.1064	0.1103	0.1141
0.3	0.1179	0.1217	0.1255	0.1293	0.1331	0.1368	0.1406	0.1443	0.1480	0.1517
0.4	0.1554	0.1591	0.1628	0.1664	0.1700	0.1736	0.1772	0.1808	0.1844	0.1879
0.5	0.1915	0.1950	0.1985	0.2019	0.2054	0.2088	0.2123	0.2157	0.2190	0.2224
0.6	0.2257	0.2291	0.2324	0.2357	0.2389	0.2422	0.2454	0.2486	0.2518	0.2549
0.7	0.2580	0.2611	0.2642	0.2673	0.2704	0.2734	0.2764	0.2794	0.2823	0.2852
0.8	0.2881	0.2910	0.2939	0.2967	0.2995	0.3023	0.3051	0.3078	0.3106	0.3133
0.9	0.3159	0.3186	0.3212	0.3238	0.3264	0.3289	0.3315	0.3340	0.3365	0.3389
1.0	0.3413	0.3438	0.3461	0.3485	0.3508	0.3531	0.3554	0.3577	0.3599	0.3621
1.1	0.3643	0.3665	0.3686	0.3708	0.3729	0.3749	0.3770	0.3790	0.3810	0.3830
1.2	0.3849	0.3869	0.3888	0.3907	0.3925	0.3944	0.3962	0.3980	0.3997	0.4015
1.3	0.4032	0.4049	0.4066	0.4082	0.4099	0.4115	0.4131	0.4147	0.4162	0.4177
1.4	0.4192	0.4207	0.4222	0.4236	0.4251	0.4265	0.4279	0.4292	0.4306	0.4319
1.5	0.4332	0.4345	0.4357	0.4370	0.4382	0.4394	0.4406	0.4418	0.4430	0.4441
1.6	0.4452	0.4463	0.4474	0.4485	0.4495	0.4505	0.4515	0.4525	0.4535	0.4545
1.7	0.4554	0.4564	0.4573	0.4582	0.4591	0.4599	0.4608	0.4616	0.4625	0.4633
1.8	0.4641	0.4649	0.4656	0.4664	0.4671	0.4678	0.4686	0.4693	0.4699	0.4706
1.9	0.4713	0.4719	0.4726	0.4732	0.4738	0.4744	0.4750	0.4756	0.4762	0.4767
2.0	0.4772	0.4778	0.4783	0.4788	0.4793	0.4798	0.4803	0.4808	0.4812	0.4817
2.1	0.4821	0.4826	0.4830	0.4834	0.4838	0.4842	0.4846	0.4850	0.4854	0.4857
2.2	0.4861	0.4865	0.4868	0.4871	0.4875	0.4878	0.4881	0.4884	0.4887	0.4890
2.3	0.4893	0.4896	0.4898	0.4901	0.4904	0.4906	0.4909	0.4911	0.4913	0.4916
2.4	0.4918	0.4920	0.4922	0.4925	0.4727	0.4929	0.4931	0.4932	0.4934	0.4936
2.5	0.4938	0.4940	0.4941	0.4943	0.4945	0.4946	0.4948	0.4949	0.4951	0.4952
2.6	0.4953	0.4955	0.4956	0.4957	0.4959	0.4960	0.4961	0.4962	0.4963	0.4964
2.7	0.4965	0.4966	0.4967	0.4968	0.4969	0.4970	0.4971	0.4972	0.4973	0.4974
2.8	0.4974	0.4975	0.4976	0.4977	0.4977	0.4978	0.4979	0.4980	0.4980	0.4981
2.9	0.4981	0.4982	0.4983	0.4983	0.4984	0.4984	0.4985	0.4985	0.4986	0.4986
3.0	0.4987	0.4987	0.4987	0.4988	0.4988	0.4989	0.4989	0.4989	0.4990	0.4990
3.1	0.4990	0.4991	0.4991	0.4991	0.4992	0.4992	0.4992	0.4992	0.4993	0.4993

[a]This table gives the probability of a random value of a normal variate falling *in* the range $z = 0$ to $z = z$ (in the *shaded area in figure*). The probability of the same variate having a deviation greater than z is given by 0.5—probability from the table for the given z. The table refers to a single tail of the normal distribution; therefore the probability of a normal variate falling in the range $\pm z = 2 \times$ probability from the table for the given z. The probability of a variate falling outside the range $\pm z$ is $1 - 2 \times$ probability from the table for given z.

The values in this table were obtained by permission of authors and publishers from C. E. Weatherburn, *Mathematical Statistics*, Cambridge University Press, London, 1946.

APPENDIX II^a: VALUES OF t_α

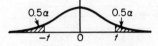

Degrees of Freedom ν	Probability α			
	0.10	0.05	0.01	0.001
1	6.314	12.706	63.657	636.619
2	2.920	4.303	9.925	31.598
3	2.353	3.182	5.841	12.941
4	2.132	2.776	4.604	8.610
5	2.015	2.571	4.032	6.859
6	1.943	2.447	3.707	5.959
7	1.895	2.365	3.499	5.405
8	1.860	2.306	3.355	5.041
9	1.833	2.262	3.250	4.781
10	1.812	2.228	3.169	4.587
11	1.796	2.201	3.106	4.437
12	1.782	2.179	3.055	4.318
13	1.771	2.160	3.012	4.221
14	1.761	2.145	2.977	4.140
15	1.753	2.131	2.947	4.073
16	1.746	2.120	2.921	4.015
17	1.740	2.110	2.898	3.965
18	1.734	2.101	2.878	3.922
19	1.729	2.093	2.861	3.883
20	1.725	2.086	2.845	3.850
21	1.721	2.080	2.831	3.819
22	1.717	2.074	2.819	3.792
23	1.714	2.069	2.807	3.767
24	1.711	2.064	2.797	3.745
25	1.708	2.060	2.787	3.725
26	1.706	2.056	2.779	3.707
27	1.703	2.052	2.771	3.690
28	1.701	2.048	2.763	3.674
29	1.699	2.045	2.756	3.659
30	1.697	2.042	2.750	3.646
40	1.684	2.021	2.704	3.551
60	1.671	2.000	2.660	3.460
120	1.658	1.980	2.617	3.373
∞	1.645	1.960	2.576	3.291

^{a}This table gives the values of t corresponding to various values of the probability α (level of significance) of a random variable falling inside the shaded areas in the figure, for a given number of degrees of freedom ν available for the estimation of error. For a one-sided test the confidence limits are obtained for $\alpha/2$. This table is taken from Table III of Fisher and Yates, *Statistical Tables for Biological, Agricultural, and Medical Research*, Oliver & Boyd Ltd., Edinburgh, 1963.

APPENDIX III: LEARNING CURVE FACTORS—
UNIT VALUES[a]

Unit Number	Slope Parameter, ϕ				
	0.70	0.75	0.80	0.85	0.90
1	1.0000	1.0000	1.0000	1.0000	1.0000
2	0.7000	0.7500	0.8000	0.8500	0.9000
3	0.5682	0.6338	0.7021	0.7729	0.8462
4	0.4900	0.5625	0.6400	0.7225	0.8100
5	0.4368	0.5127	0.5956	0.6857	0.7830
6	0.3977	0.4754	0.5617	0.6570	0.7616
7	0.3674	0.4459	0.5345	0.6337	0.7439
8	0.3430	0.4219	0.5120	0.6141	0.7290
9	0.3228	0.4017	0.4929	0.5974	0.7161
10	0.3058	0.3846	0.4765	0.5828	0.7047
11	0.2912	0.3696	0.4621	0.5699	0.6945
12	0.2784	0.3565	0.4493	0.5584	0.6854
13	0.2672	0.3449	0.4379	0.5481	0.6771
14	0.2572	0.3344	0.4276	0.5386	0.6695
15	0.2482	0.3250	0.4182	0.5300	0.6626
16	0.2401	0.3164	0.4096	0.5220	0.6561
17	0.2327	0.3085	0.4017	0.5146	0.6501
18	0.2260	0.3013	0.3944	0.5078	0.6444
19	0.2198	0.2946	0.3876	0.5014	0.6392
20	0.2141	0.2884	0.3812	0.4954	0.6342
21	0.2088	0.2826	0.3753	0.4898	0.6295
22	0.2038	0.2772	0.3697	0.4845	0.6251
23	0.1992	0.2722	0.3644	0.4794	0.6209
24	0.1949	0.2674	0.3595	0.4747	0.6169
25	0.1908	0.2629	0.3548	0.4702	0.6131
26	0.1870	0.2587	0.3503	0.4658	0.6094
27	0.1834	0.2546	0.3461	0.4617	0.6059
28	0.1800	0.2508	0.3421	0.4578	0.6026
29	0.1768	0.2472	0.3382	0.4541	0.5994
30	0.1738	0.2437	0.3346	0.4505	0.5963
31	0.1708	0.2405	0.3310	0.4470	0.5933
32	0.1681	0.2373	0.3277	0.4437	0.5905
33	0.1654	0.2343	0.3244	0.4405	0.5877
34	0.1629	0.2314	0.3213	0.4375	0.5851
35	0.1605	0.2286	0.3184	0.4345	0.5825
36	0.1582	0.2260	0.3155	0.4316	0.5800
37	0.1560	0.2234	0.3127	0.4289	0.5776
38	0.1539	0.2210	0.3100	0.4262	0.5753
39	0.1518	0.2186	0.3075	0.4236	0.5730
40	0.1498	0.2163	0.3050	0.4211	0.5708
41	0.1480	0.2141	0.3026	0.4187	0.5686
42	0.1461	0.2120	0.3002	0.4163	0.5666
43	0.1444	0.2099	0.2979	0.4140	0.5645
44	0.1427	0.2079	0.2958	0.4118	0.5626
45	0.1410	0.2060	0.2936	0.4096	0.5607

[a]Appendixes III and IV are from W. J. Fabrycky, P. M. Ghare, and P. E. Torgersen, *Industrial Operations Research*, Prentice-Hall, Inc., Englewood Cliffs, N.J., 1972.

APPENDIX III: LEARNING CURVE FACTORS—
UNIT VALUES (continued)

Unit Number	Slope Parameter, ϕ				
	0.70	0.75	0.80	0.85	0.90
46	0.1394	0.2041	0.2915	0.4075	0.5588
47	0.1379	0.2023	0.2895	0.4055	0.5570
48	0.1364	0.2005	0.2876	0.4035	0.5552
49	0.1350	0.1988	0.2857	0.4015	0.5534
50	0.1336	0.1972	0.2838	0.3996	0.5517
55	0.1272	0.1895	0.2753	0.3908	0.5438
60	0.1216	0.1828	0.2676	0.3829	0.5367
65	0.1167	0.1768	0.2608	0.3758	0.5302
70	0.1124	0.1715	0.2547	0.3693	0.5242
75	0.1084	0.1666	0.2491	0.3634	0.5188
80	0.1049	0.1622	0.2440	0.3579	0.5137
85	0.1017	0.1582	0.2393	0.3529	0.5090
90	0.0987	0.1545	0.2349	0.3482	0.5046
95	0.0960	0.1511	0.2308	0.3438	0.5005
100	0.0935	0.1479	0.2271	0.3397	0.4966
105	0.0912	0.1449	0.2235	0.3358	0.4929
110	0.0890	0.1421	0.2202	0.3322	0.4894
115	0.0870	0.1395	0.2171	0.3287	0.4861
120	0.0851	0.1371	0.2141	0.3255	0.4830
125	0.0834	0.1348	0.2113	0.3224	0.4800
130	0.0817	0.1326	0.2087	0.3194	0.4772
135	0.0801	0.1306	0.2062	0.3166	0.4744
140	0.0787	0.1286	0.2038	0.3139	0.4718
145	0.0772	0.1267	0.2015	0.3113	0.4693
150	0.0759	0.1250	0.1993	0.3089	0.4669
155	0.0746	0.1233	0.1972	0.3065	0.4646
160	0.0734	0.1217	0.1952	0.3042	0.4623
165	0.0723	0.1201	0.1933	0.3021	0.4602
170	0.0712	0.1184	0.1914	0.3000	0.4581
175	0.0701	0.1172	0.1896	0.2979	0.4561
180	0.0691	0.1159	0.1879	0.2960	0.4541
185	0.0681	0.1146	0.1863	0.2941	0.4522
190	0.0672	0.1133	0.1847	0.2922	0.4504
195	0.0663	0.1121	0.1831	0.2905	0.4486
200	0.0655	0.1109	0.1817	0.2887	0.4469
225	0.0616	0.1056	0.1749	0.2809	0.4390
250	0.0584	0.1011	0.1691	0.2740	0.4320
275	0.0556	0.0972	0.1640	0.2680	0.4258
300	0.0531	0.0937	0.1594	0.2626	0.4202
325	0.0510	0.0907	0.1554	0.2577	0.4151
350	0.0491	0.0879	0.1517	0.2532	0.4105
375	0.0474	0.0854	0.1484	0.2492	0.4062
400	0.0458	0.0832	0.1453	0.2454	0.4022
450	0.0431	0.0792	0.1399	0.2387	0.3951
500	0.0409	0.0758	0.1353	0.2329	0.3888

APPENDIX IV: LEARNING CURVE FACTORS—
CUMULATIVE VALUES

Unit Number	Slope Parameter, ϕ				
	0.70	0.75	0.80	0.85	0.90
1	1.0000	1.0000	1.0000	1.0000	1.0000
2	1.7000	1.7500	1.8000	1.8500	1.9000
3	2.2682	2.3838	2.5021	2.6229	2.7462
4	2.7582	2.9463	3.1421	3.3454	3.5562
5	3.1950	3.4590	3.7377	4.0311	4.3392
6	3.5927	3.9344	4.2994	4.6881	5.1008
7	3.9601	4.3803	4.8339	5.3218	5.8447
8	4.3031	4.8022	5.3459	5.9359	6.5737
9	4.6259	5.2039	5.8388	6.5333	7.2898
10	4.9317	5.5885	6.3153	7.1161	7.9945
11	5.2229	5.9581	6.7774	7.6860	8.6890
12	5.5013	6.3146	7.2267	8.2444	9.3744
13	5.7685	6.6595	7.6646	8.7925	10.0515
14	6.0257	6.9939	8.0922	9.3311	10.7210
15	6.2739	7.3189	8.5104	9.8611	11.3836
16	6.5140	7.6353	8.9200	10.3831	12.0397
17	6.7467	7.9438	9.3217	10.8977	12.6898
18	6.9727	8.2451	9.7161	11.4055	13.3342
19	7.1925	8.5397	10.1037	11.9069	13.9734
20	7.4066	8.8281	10.4849	12.4023	14.6076
21	7.6154	9.1107	10.8602	12.8921	15.2371
22	7.8192	9.3879	11.2299	13.3766	15.8622
23	8.0184	9.6601	11.5943	13.8560	16.4831
24	8.2133	9.9275	11.9538	14.3307	17.1000
25	8.4041	10.1904	12.3086	14.8009	17.7131
30	9.3051	11.4454	14.0199	17.0908	20.7267
35	10.1328	12.6175	15.6427	19.2940	23.6658
40	10.9025	13.7228	17.1934	21.4254	26.5425
45	11.6247	14.7727	18.6835	23.4958	29.3655
50	12.3070	15.7756	20.1216	25.5134	32.1416
55	12.9553	16.7382	21.5147	27.4847	34.8762
60	13.5743	17.6653	22.8678	29.4147	37.5735
65	14.1675	18.5611	24.1852	31.3077	40.2371
70	14.7378	19.4290	25.4708	33.1669	42.8699
75	15.2876	20.2717	26.7273	34.9955	45.4745
80	15.8191	21.0914	27.9573	36.7960	48.0530
85	16.3338	21.8904	29.1629	38.5705	50.6072
90	16.8333	22.6701	30.3460	40.3207	53.1388
95	17.3187	23.2811	31.5081	42.0484	55.1488
100	17.7912	24.1779	32.6509	43.7550	58.1399
150	21.9730	30.9338	43.2338	59.8901	82.1539
200	25.4833	36.8000	52.7203	74.7908	104.9614
300	31.3452	46.9418	69.6637	102.2341	148.1968
400	36.2640	55.7464	84.8495	127.5737	189.2588
500	40.5822	63.6741	98.8480	151.4560	228.7746

To Find	Given	Discrete Payments — Discrete Compounding	Discrete Payments — Continuous Compounding	Continuous Payments — Continuous Compounding
F	P	$F = P(1+i)^n = P(\ ^{F/P\,i,n})$	$F = Pe^{rn} = P(\ ^{F/P\,r,n})$	$F = Pe^{rn} = P(\ ^{F/P\,r,n})$
P	F	$P = F\dfrac{1}{(1+i)^n} = F(\ ^{P/F\,i,n})$	$P = F\dfrac{1}{e^{rn}} = F(\ ^{P/F\,r,n})$	$P = F\dfrac{1}{e^{rn}} = F(\ ^{P/F\,r,n})$
F	A	$F = A\left[\dfrac{(1+i)^n - 1}{i}\right] = A(\ ^{F/A\,i,n})$	$F = A\left[\dfrac{e^{rn} - 1}{e^r - 1}\right] = A(\ ^{F/A\,r,n})$	$F = \bar{A}\left[\dfrac{e^{rn} - 1}{r}\right] = \bar{A}(\ ^{F/\bar{A}\,r,n})$
A	F	$A = F\left[\dfrac{i}{(1+i)^n - 1}\right] = F(\ ^{A/F\,i,n})$	$A = F\left[\dfrac{e^r - 1}{e^{rn} - 1}\right] = F(\ ^{A/F\,r,n})$	$\bar{A} = F\left[\dfrac{r}{e^{rn} - 1}\right] = F(\ ^{\bar{A}/F\,r,n})$
P	A	$P = A\left[\dfrac{(1+i)^n - 1}{i(1+i)^n}\right] = A(\ ^{P/A\,i,n})$	$P = A\left[\dfrac{1 - e^{-rn}}{e^r - 1}\right] = A(\ ^{P/A\,r,n})$	$P = \bar{A}\left[\dfrac{e^{rn} - 1}{re^{rn}}\right] = \bar{A}(\ ^{P/\bar{A}\,r,n})$
A	P	$A \doteq P\left[\dfrac{i(1+i)^n}{(1+i)^n - 1}\right] = P(\ ^{A/P\,i,n})$	$A = P\left[\dfrac{e^r - 1}{1 - e^{-rn}}\right] = P(\ ^{A/P\,r,n})$	$\bar{A} = P\left[\dfrac{re^{rn}}{e^{rn} - 1}\right] = P(\ ^{\bar{A}/P\,r,n})$
A	g	$A = g\left[\dfrac{1}{i} - \dfrac{n}{(1+i)^n - 1}\right] = g(\ ^{A/G\,i,n})$	$A = g\left[\dfrac{1}{e^r - 1} - \dfrac{n}{e^{rn} - 1}\right] = g(\ ^{A/G\,r,n})$	

[a]Appendixes V—IX are from H. G. Thuesen, W. J. Fabrycky, and G. J. Thuesen, *Engineering Economy*, 4th ed.,
    Prentice-Hall, Inc., Englewood Cliffs, N.J., 1971.

[b]Notation: n = number of compounding periods
    i = effective interest rate per interest period
    r = nominal interest rate per year
    P = present sum of money
    F = future sum of money
    A = uniform series end-of-period cash flows
    $\bar{A}$ = amount of money flowing continuously and uniformly during period

	Single Payment		Equal Payment Series			
	Compound-amount factor	Present-worth factor	Compound-amount factor	Sinking-fund factor	Present-worth factor	Capital-recovery factor
n	To find F Given P F/P i, n	To find P Given F P/F i, n	To find F Given A F/A i, n	To find A Given F A/F i, n	To find P Given A P/A i, n	To find A Given P A/P i, n
1	1.050	0.9524	1.000	1.0000	0.9524	1.0500
2	1.103	0.9070	2.050	0.4878	1.8594	0.5378
3	1.158	0.8638	3.153	0.3172	2.7233	0.3672
4	1.216	0.8227	4.310	0.2320	3.5460	0.2820
5	1.276	0.7835	5.526	0.1810	4.3295	0.2310
6	1.340	0.7462	6.802	0.1470	5.0757	0.1970
7	1.407	0.7107	8.142	0.1228	5.7864	0.1728
8	1.477	0.6768	9.549	0.1047	6.4632	0.1547
9	1.551	0.6446	11.027	0.0907	7.1078	0.1407
10	1.629	0.6139	12.587	0.0795	7.7217	0.1295
11	1.710	0.5847	14.207	0.0704	8.3064	0.1204
12	1.796	0.5568	15.917	0.0628	8.8633	0.1128
13	1.886	0.5303	17.713	0.0565	9.3936	0.1065
14	1.980	0.5051	19.599	0.0510	9.8987	0.1010
15	2.079	0.4810	21.579	0.0464	10.3797	0.0964
16	2.183	0.4581	23.658	0.0423	10.8378	0.0923
17	2.292	0.4363	25.840	0.0387	11.2741	0.0887
18	2.407	0.4155	28.132	0.0356	11.6896	0.0856
19	2.527	0.3957	30.539	0.0328	12.0853	0.0828
20	2.653	0.3769	33.066	0.0303	12.4622	0.0803
21	2.786	0.3590	35.719	0.0280	12.8212	0.0780
22	2.925	0.3419	38.505	0.0260	13.1630	0.0760
23	3.072	0.3256	41.430	0.0241	13.4886	0.0741
24	3.225	0.3101	44.502	0.0225	13.7987	0.0725
25	3.386	0.2953	47.727	0.0210	14.0940	0.0710
26	3.556	0.2813	51.113	0.0196	14.3752	0.0696
27	3.733	0.2679	54.669	0.0183	14.6430	0.0683
28	3.920	0.2551	58.403	0.0171	14.8981	0.0671
29	4.116	0.2430	62.323	0.0161	15.1411	0.0661
30	4.322	0.2314	66.439	0.0151	15.3725	0.0651
31	4.538	0.2204	70.761	0.0141	15,5928	0.0641
32	4.765	0.2099	75.299	0.0133	15.8027	0.0633
33	5.003	0.1999	80.064	0.0125	16.0026	0.0625
34	5.253	0.1904	85.067	0.0118	16.1929	0.0618
35	5.516	0.1813	90.320	0.0111	16.3742	0.0611
40	7.040	0.1421	120.800	0.0083	17.1591	0.0583
45	8.985	0.1113	159.700	0.0063	17.7741	0.0563
50	11.467	0.0872	209.348	0.0048	18.2559	0.0548
55	14.636	0.0683	272.713	0.0037	18.6335	0.0537
60	18.679	0.0535	353.584	0.0028	18.9293	0.0528
65	23.840	0.0420	456.798	0.0022	19.1611	0.0522
70	30.426	0.0329	588.529	0.0017	19.3427	0.0517
75	38.833	0.0258	756.654	0.0013	19.4850	0.0513
80	49.561	0.0202	971.229	0.0010	19.5965	0.0510
85	63.254	0.0158	1245.087	0.0008	19.6838	0.0508
90	80.730	0.0124	1594.607	0.0006	19.7523	0.0506
95	103.035	0.0097	2040.694	0.0005	19.8059	0.0505
100	131.501	0.0076	2610.025	0.0004	19.8479	0.0504

	Single Payment		Equal Payment Series			
	Compound-amount factor	Present-worth factor	Compound-amount factor	Sinking-fund factor	Present-worth factor	Capital-recovery factor
n	To find F Given P $F/P \quad i,n$	To find P Given F $P/F \quad i,n$	To find F Given A $F/A \quad i,n$	To find A Given F $A/F \quad i,n$	To find P Given A $P/A \quad i,n$	To find A Given P $A/P \quad i,n$
1	1.100	0.9091	1.000	1.0000	0.9091	1.1000
2	1.210	0.8265	2.100	0.4762	1.7355	0.5762
3	1.331	0.7513	3.310	0.3021	2.4869	0.4021
4	1.464	0.6830	4.641	0.2155	3.1699	0.3155
5	1.611	0.6209	6.105	0.1638	3.7908	0.2638
6	1.772	0.5645	7.716	0.1296	4.3553	0.2296
7	1.949	0.5132	9.487	0.1054	4.8684	0.2054
8	2.144	0.4665	11.436	0.0875	5.3349	0.1875
9	2.358	0.4241	13.579	0.0737	5.7590	0.1737
10	2.594	0.3856	15.937	0.0628	6.1446	0.1628
11	2.853	0.3505	18.531	0.0540	6.4951	0.1540
12	3.138	0.3186	21.384	0.0468	6.8137	0.1468
13	3.452	0.2897	24.523	0.0408	7.1034	0.1408
14	3.798	0.2633	27.975	0.0358	7.3667	0.1358
15	4.177	0.2394	31.772	0.0315	7.6061	0.1315
16	4.595	0.2176	35.950	0.0278	7.8237	0.1278
17	5.054	0.1979	40.545	0.0247	8.0216	0.1247
18	5.560	0.1799	45.599	0.0219	8.2014	0.1219
19	6.116	0.1635	51.159	0.0196	8.3649	0.1196
20	6.728	0.1487	57.275	0.0175	8.5136	0.1175
21	7.400	0.1351	64.003	0.0156	8.6487	0.1156
22	8.140	0.1229	71.403	0.0140	8.7716	0.1140
23	8.954	0.1117	79.543	0.0126	8.8832	0.1126
24	9.850	0.1015	88.497	0.0113	8.9848	0.1113
25	10.835	0.0923	98.347	0.0102	9.0771	0.1102
26	11.918	0.0839	109.182	0.0092	9.1610	0.1092
27	13.110	0.0763	121.100	0.0083	9.2372	0.1083
28	14.421	0.0694	134.210	0.0075	9.3066	0.1075
29	15.863	0.0630	148.631	0.0067	9.3696	0.1067
30	17.449	0.0573	164.494	0.0061	9.4269	0.1061
31	19.194	0.0521	181.943	0.0055	9.4790	0.1055
32	21.114	0.0474	201.138	0.0050	9.5264	0.1050
33	23.225	0.0431	222.252	0.0045	9.5694	0.1045
34	25.548	0.0392	245.477	0.0041	9.6086	0.1041
35	28.102	0.0356	271.024	0.0037	9.6442	0.1037
40	45.259	0.0221	442.593	0.0023	9.7791	0.1023
45	72.890	0.0137	718.905	0.0014	9.8628	0.1014
50	117.391	0.0085	1163.909	0.0009	9.9148	0.1009
55	189.059	0.0053	1880.591	0.0005	9.9471	0.1005
60	304.482	0.0033	3034.816	0.0003	9.9672	0.1003
65	490.371	0.0020	4893.707	0.0002	9.9796	0.1002
70	789.747	0.0013	7887.470	0.0001	9.9873	0.1001
75	1271.895	0.0008	12708.954	0.0001	9.9921	0.1001
80	2048.400	0.0005	20474.002	0.0001	9.9951	0.1001
85	3298.969	0.0003	32979.690	0.0000	9.9970	0.1000
90	5313.023	0.0002	53120.226	0.0000	9.9981	0.1000
95	8556.676	0.0001	85556.760	0.0000	9.9988	0.1000
100	13780.612	0.0001	137796.123	0.0000	9.9993	0.1000

APPENDIX VI: 20% INTEREST FACTORS FOR ANNUAL COMPOUNDING INTEREST

	Single Payment		Equal Payment Series			
	Compound-amount factor	Present-worth factor	Compound-amount factor	Sinking-fund factor	Present-worth factor	Capital-recovery factor
n	To find F Given P F/P i,n	To find P Given F P/F i,n	To find F Given A F/A i,n	To find A Given F A/F i,n	To find P Given A P/A i,n	To find A Given P A/P i,n
1	1.200	0.8333	1.000	1.0000	0.8333	1.2000
2	1.440	0.6945	2.200	0.4546	1.5278	0.6546
3	1.728	0.5787	3.640	0.2747	2.1065	0.4747
4	2.074	0.4823	5.368	0.1863	2.5887	0.3863
5	2.488	0.4019	7.442	0.1344	2.9906	0.3344
6	2.986	0.3349	9.930	0.1007	3.3255	0.3007
7	3.583	0.2791	12.916	0.0774	3.6046	0.2774
8	4.300	0.2326	16.499	0.0606	3.8372	0.2606
9	5.160	0.1938	20.799	0.0481	4.0310	0.2481
10	6.192	0.1615	25.959	0.0385	4.1925	0.2385
11	7.430	0.1346	32.150	0.0311	4.3271	0.2311
12	8.916	0.1122	39.581	0.0253	4.4392	0.2253
13	10.699	0.0935	48.497	0.0206	4.5327	0.2206
14	12.839	0.0779	59.196	0.0169	4.6106	0.2169
15	15.407	0.0649	72.035	0.0139	4.6755	0.2139
16	18.488	0.0541	87.442	0.0114	4.7296	0.2114
17	22.186	0.0451	105.931	0.0095	4.7746	0.2095
18	26.623	0.0376	128.117	0.0078	4.8122	0.2078
19	31.948	0.0313	154.740	0.0065	4.8435	0.2065
20	38.338	0.0261	186.688	0.0054	4.8696	0.2054
21	46.005	0.0217	225.026	0.0045	4.8913	0.2045
22	55.206	0.0181	271.031	0.0037	4.9094	0.2037
23	66.247	0.0151	326.237	0.0031	4.9245	0.2031
24	79.497	0.0126	392.484	0.0026	4.9371	0.2026
25	95.396	0.0105	471.981	0.0021	4.9476	0.2021
26	114.475	0.0087	567.377	0.0018	4.9563	0.2018
27	137.371	0.0073	681.853	0.0015	4.9636	0.2015
28	164.845	0.0061	819.223	0.0012	4.9697	0.2012
29	197.814	0.0051	984.068	0.0010	4.9747	0.2010
30	237.376	0.0042	1181.882	0.0009	4.9789	0.2009
31	284.852	0.0035	1419.258	0.0007	4.9825	0.2007
32	341.822	0.0029	1704.109	0.0006	4.9854	0.2006
33	410.186	0.0024	2045.931	0.0005	4.9878	0.2005
34	492.224	0.0020	2456.118	0.0004	4.9899	0.2004
35	590.668	0.0017	2948.341	0.0003	4.9915	0.2003
40	1469.772	0.0007	7343.858	0.0002	4.9966	0.2001
45	3657.262	0.0003	18281.310	0.0001	4.9986	0.2001
50	9100.438	0.0001	45497.191	0.0000	4.9995	0.2000

APPENDIX VI: 30% INTEREST FACTORS FOR ANNUAL COMPOUNDING INTEREST

	Single Payment		Equal Payment Series			
n	Compound-amount factor	Present-worth factor	Compound-amount factor	Sinking-fund factor	Present-worth factor	Capital-recovery factor
	To find F Given P F/P i,n	To find P Given F P/F i,n	To find F Given A F/A i,n	To find A Given F A/F i,n	To find P Given A P/A i,n	To find A Given P A/P i,n
1	1.300	0.7692	1.000	1.0000	0.7692	1.3000
2	1.690	0.5917	2.300	0.4348	1.3610	0.7348
3	2.197	0.4552	3.990	0.2506	1.8161	0.5506
4	2.856	0.3501	6.187	0.1616	2.1663	0.4616
5	3.713	0.2693	9.043	0.1106	2.4356	0.4106
6	4.827	0.2072	12.756	0.0784	2.6428	0.3784
7	6.275	0.1594	17.583	0.0569	2.8021	0.3569
8	8.157	0.1226	23.858	0.0419	2.9247	0.3419
9	10.605	0.0943	32.015	0.0312	3.0190	0.3312
10	13.786	0.0725	42.620	0.0235	3.0915	0.3235
11	17.922	0.0558	56.405	0.0177	3.1473	0.3177
12	23.298	0.0429	74.327	0.0135	3.1903	0.3135
13	30.288	0.0330	97.625	0.0103	3.2233	0.3103
14	39.374	0.0254	127.913	0.0078	3.2487	0.3078
15	51.186	0.0195	167.286	0.0060	3.2682	0.3060
16	66.542	0.0150	218.472	0.0046	3.2832	0.3046
17	86.504	0.0116	285.014	0.0035	3.2948	0.3035
18	112.455	0.0089	371.518	0.0027	3.3037	0.3027
19	146.192	0.0069	483.973	0.0021	3.3105	0.3021
20	190.050	0.0053	630.165	0.0016	3.3158	0.3016
21	247.065	0.0041	820.215	0.0012	3.3199	0.3012
22	321.184	0.0031	1067.280	0.0009	3.3230	0.3009
23	417.539	0.0024	1388.464	0.0007	3.3254	0.3007
24	542.801	0.0019	1806.003	0.0006	3.3272	0.3006
25	705.641	0.0014	2348.803	0.0004	3.3286	0.3004
26	917.333	0.0011	3054.444	0.0003	3.3297	0.3003
27	1192.533	0.0008	3971.778	0.0003	3.3305	0.3003
28	1550.293	0.0007	5164.311	0.0002	3.3312	0.3002
29	2015.381	0.0005	6714.604	0.0002	3.3317	0.3002
30	2619.996	0.0004	8729.985	0.0001	3.3321	0.3001
31	3405.994	0.0003	11349.981	0.0001	3.3324	0.3001
32	4427.793	0.0002	14755.975	0.0001	3.3326	0.3001
33	5756.130	0.0002	19183.768	0.0001	3.3328	0.3001
34	7482.970	0.0001	24939.899	0.0001	3.3329	0.3001
35	9727.860	0.0001	32422.868	0.0000	3.3330	0.3000

APPENDIX VII: EFFECTIVE INTEREST RATES
CORRESPONDING TO NOMINAL RATE r

r	Compounding Frequency					
	Semi-annually $\left(1+\frac{r}{2}\right)^2 - 1$	Quarterly $\left(1+\frac{r}{4}\right)^4 - 1$	Monthly $\left(1+\frac{r}{12}\right)^{12} - 1$	Weekly $\left(1+\frac{r}{52}\right)^{52} - 1$	Daily $\left(1+\frac{r}{365}\right)^{365} - 1$	Continuously $\left(1+\frac{r}{\infty}\right)^{\infty} - 1$
1	1.0025	1.0038	1.0046	1.0049	1.0050	1.0050
2	2.0100	2.0151	2.0184	2.0197	2.0200	2.0201
3	3.0225	3.0339	3.0416	3.0444	3.0451	3.0455
4	4.0400	4.0604	4.0741	4.0793	4.0805	4.0811
5	5.0625	5.0945	5.1161	5.1244	5.1261	5.1271
6	6.0900	6.1364	6.1678	6.1797	6.1799	6.1837
7	7.1225	7.1859	7.2290	7.2455	7.2469	7.2508
8	8.1600	8.2432	8.2999	8.3217	8.3246	8.3287
9	9.2025	9.3083	9.3807	9.4085	9.4132	9.4174
10	10.2500	10.3813	10.4713	10.5060	10.5126	10.5171
11	11.3025	11.4621	11.5718	11.6144	11.6231	11.6278
12	12.3600	12.5509	12.6825	12.7336	12.7447	12.7497
13	13.4225	13.6476	13.8032	13.8644	13.8775	13.8828
14	14.4900	14.7523	14.9341	15.0057	15.0217	15.0274
15	15.5625	15.8650	16.0755	16.1582	16.1773	16.1834
16	16.6400	16.9859	17.2270	17.3221	17.3446	17.3511
17	17.7225	18.1148	18.3891	18.4974	18.5235	18.5305
18	18.8100	19.2517	19.5618	19.6843	19.7142	19.7217
19	19.9025	20.3971	20.7451	20.8828	20.9169	20.9250
20	21.0000	21.5506	21.9390	22.0931	22.1316	22.1403
21	22.1025	22.7124	23.1439	23.3153	23.3584	23.3678
22	23.2100	23.8825	24.3596	24.5494	24.5976	24.6077
23	24.3225	25.0609	25.5863	25.7957	25.8492	25.8600
24	25.4400	26.2477	26.8242	27.0542	27.1133	27.1249
25	26.5625	27.4429	28.0731	28.3250	28.3901	28.4025
26	27.6900	28.6466	29.3333	29.6090	29.6796	29.6930
27	28.8225	29.8588	30.6050	30.9049	30.9821	30.9964
28	29.9600	31.0796	31.8880	32.2135	32.2976	32.3130
29	31.1025	32.3089	33.1826	33.5350	33.6264	33.6428
30	32.2500	33.5469	34.4889	34.8693	34.9684	34.9859
31	33.4025	34.7936	35.8068	36.2168	36.3238	36.3425
32	34.5600	36.0489	37.1366	37.5775	37.6928	37.7128
33	35.7225	37.3130	38.4784	38.9515	39.0756	39.0968
34	36.8900	38.5859	39.8321	40.3389	40.4722	40.4948
35	38.0625	39.8676	41.1979	41.7399	41.8827	41.9068

	Single Payment		Equal Payment Series			
	Compound-amount factor	Present-worth factor	Compound-amount factor	Sinking-fund factor	Present-worth factor	Capital-recovery factor
n	To find F Given P F/P r,n	To find P Given F P/F r,n	To find F Given A F/A r,n	To find A Given F A/F r,n	To find P Given A P/A r,n	To find A Given P A/P r,n
1	1.051	0.9512	1.000	1.0000	0.9512	1.0513
2	1.105	0.9048	2.051	0.4875	1.8561	0.5388
3	1.162	0.8607	3.156	0.3168	2.7168	0.3681
4	1.221	0.8187	4.318	0.2316	3.5355	0.2829
5	1.284	0.7788	5.540	0.1805	4.3143	0.2318
6	1.350	0.7408	6.824	0.1466	5.0551	0.1978
7	1.419	0.7047	8.174	0.1224	5.7598	0.1736
8	1.492	0.6703	9.593	0.1043	6.4301	0.1555
9	1.568	0.6376	11.084	0.0902	7.0678	0.1415
10	1.649	0.6065	12.653	0.0790	7.6743	0.1303
11	1.733	0.5770	14.301	0.0699	8.2513	0.1212
12	1.822	0.5488	16.035	0.0624	8.8001	0.1136
13	1.916	0.5221	17.857	0.0560	9.3221	0.1073
14	2.014	0.4966	19.772	0.0506	9.8187	0.1019
15	2.117	0.4724	21.786	0.0459	10.2911	0.0972
16	2.226	0.4493	23.903	0.0418	10.7404	0.0931
17	2.340	0.4274	26.129	0.0383	11.1678	0.0896
18	2.460	0.4066	28.468	0.0351	11.5744	0.0864
19	2.586	0.3868	30.928	0.0323	11.9611	0.0836
20	2.718	0.3679	33.514	0.0298	12.3290	0.0811
21	2.858	0.3499	36.232	0.0276	12.6789	0.0789
22	3.004	0.3329	39.090	0.0256	13.0118	0.0769
23	3.158	0.3166	42.094	0.0238	13.3284	0.0750
24	3.320	0.3012	45.252	0.0221	13.6296	0.0734
25	3.490	0.2865	48.572	0.0206	13.9161	0.0719
26	3.669	0.2725	52.062	0.0192	14.1887	0.0705
27	3.857	0.2593	55.732	0.0180	14.4479	0.0692
28	4.055	0.2466	59.589	0.0168	14.6945	0.0681
29	4.263	0.2346	63.644	0.0157	14.9291	0.0670
30	4.482	0.2231	67.907	0.0147	15.1522	0.0660
31	4.711	0.2123	72.389	0.0138	15.3645	0.0651
32	4.953	0.2019	77.101	0.0130	15.5664	0.0643
33	5.207	0.1921	82.054	0.0122	15.7584	0.0635
34	5.474	0.1827	87.261	0.0115	15.9411	0.0627
35	5.755	0.1738	92.735	0.0108	16.1149	0.0621
40	7.389	0.1353	124.613	0.0080	16.8646	0.0593
45	9.488	0.1054	165.546	0.0061	17.4485	0.0573
50	12.183	0.0821	218.105	0.0046	17.9032	0.0559
55	15.643	0.0639	285.592	0.0035	18.2573	0.0548
60	20.086	0.0498	372.247	0.0027	18.5331	0.0540
65	25.790	0.0388	483.515	0.0021	18.7479	0.0533
70	33.115	0.0302	626.385	0.0016	18.9152	0.0529
75	42.521	0.0235	809.834	0.0012	19.0455	0.0525
80	54.598	0.0183	1045.387	0.0010	19.1469	0.0522
85	70.105	0.0143	1347.843	0.0008	19.2260	0.0520
90	90.017	0.0111	1736.205	0.0006	19.2875	0.0519
95	115.584	0.0087	2234.871	0.0005	19.3354	0.0517
100	148.413	0.0067	2875.171	0.0004	19.3728	0.0516

	Single Payment		Equal Payment Series			
	Compound-amount factor	Present-worth factor	Compound-amount factor	Sinking-fund factor	Present-worth factor	Capital-recovery factor
n	To find F Given P F/P r, n	To find P Given F P/F r, n	To find F Given A F/A r, n	To find A Given F A/F r, n	To find P Given A P/A r, n	To find A Given P A/P r, n
1	1.105	0.9048	1.000	1.0000	0.9048	1.1052
2	1.221	0.8187	2.105	0.4750	1.7236	0.5802
3	1.350	0.7408	3.327	0.3006	2.4644	0.4058
4	1.492	0.6703	4.676	0.2138	3.1347	0.3190
5	1.649	0.6065	6.168	0.1621	3.7412	0.2673
6	1.822	0.5488	7.817	0.1279	4.2901	0.2331
7	2.014	0.4966	9.639	0.1038	4.7866	0.2089
8	2.226	0.4493	11.653	0.0858	5.2360	0.1910
9	2.460	0.4066	13.878	0.0721	5.6425	0.1772
10	2.718	0.3679	16.338	0.0612	6.0104	0.1664
11	3.004	0.3329	19.056	0.0525	6.3433	0.1577
12	3.320	0.3012	22.060	0.0453	6.6445	0.1505
13	3.669	0.2725	25.381	,0394	6.9170	0.1446
14	4.055	0.2466	29.050	0.0344	7.1636	0.1396
15	4.482	0.2231	33.105	0.0302	7.3867	0.1354
16	4.953	0.2019	37.587	0.0266	7.5886	0.1318
17	5.474	0.1827	42.540	0.0235	7.7713	0.1287
18	6.050	0.1653	48.014	0.0208	7.9366	0.1260
19	6.686	0.1496	54.063	0.0185	8.0862	0.1237
20	7.389	0.1353	60.749	0.0165	8.2215	0.1216
21	8.166	0.1225	68.138	0.0147	8.3440	0.1199
22	9.025	0.1108	76.305	0.0131	8.4548	0.1183
23	9.974	0.1003	85.330	0.0117	8.5550	0.1169
24	11.023	0.0907	95.304	0.0105	8.6458	0.1157
25	12.183	0.0821	106.327	0.0094	8.7279	0.1146
26	13.464	0.0743	118.509	0.0084	8.8021	0.1136
27	14.880	0.0672	131.973	0.0076	8.8693	0.1128
28	16.445	0.0608	146.853	0.0068	8.9301	0.1120
29	18.174	0.0550	163.297	0.0061	8.9852	0.1113
30	20.086	0.0498	181.472	0.0055	9.0349	0.1107
31	22.198	0.0451	201.557	0.0050	9.0800	0.1101
32	24.533	0.0408	223.755	0.0045	9.1208	0.1097
33	27.113	0.0369	248.288	0.0040	9.1576	0.1092
34	29.964	0.0334	275.400	0.0036	9.1910	0.1088
35	33.115	0.0302	305.364	0.0033	9.2212	0.1085
40	54.598	0.0183	509.629	0.0020	9.3342	0.1071
45	90.017	.0111	846.404	0.0012	9.4027	0.1064
50	148.413	0.0067	1401.653	0.0007	9.4443	0.1059
55	244.692	0.0041	2317.104	0.0004	9.4695	0.1056
60	403.429	0.0025	3826.427	0.0003	9.4848	0.1054
65	665.142	0.0015	6314.879	0.0002	9.4940	0.1053
70	1096.633	0.0009	10417.644	0.0001	9.4997	0.1053
75	1808.042	0.0006	17181.959	0.0001	9.5031	0.1052
80	2980.958	0.0004	28334.430	0.0001	9.5052	0.1052
85	4914.769	0.0002	46721.745	0.0000	9.5064	0.1052
90	8103.084	0.0001	77037.303	0.0000	9.5072	0.1052
95	13359.727	0.0001	127019.209	0.0000	9.5076	0.1052
100	22026.466	0.0001	209425.440	0.0000	9.5079	0.1052

APPENDIX VIII: 20% INTEREST FACTORS FOR
CONTINUOUS COMPOUNDING INTEREST

	Single Payment		Equal Payment Series			
	Compound-amount factor	Present-worth factor	Compound-amount factor	Sinking-fund factor	Present-worth factor	Capital-recovery factor
n	To find F Given P F/P r,n	To find P Given F P/F r,n	To find F Given A F/A r,n	To find A Given F A/F r,n	To find P Given A P/A r,n	To find A Given P A/P r,n
1	1.221	0.8187	1.000	1.0000	0.8187	1.2214
2	1.492	0.6703	2.221	0.4502	1.4891	0.6716
3	1.822	0.5488	3.713	0.2693	2.0379	0.4907
4	2.226	0.4493	5.535	0.1807	2.4872	0.4021
5	2.718	0.3679	7.761	0.1289	2.8551	0.3503
6	3.320	0.3012	10.479	0.0954	3.1563	0.3168
7	4.055	0.2466	13.799	0.0725	3.4029	0.2939
8	4.953	0.2019	17.854	0.0560	3.6048	0.2774
9	6.050	0.1653	22.808	0.0439	3.7701	0.2653
10	7.389	0.1353	28.857	0.0347	3.9054	0.2561
11	9.025	0.1108	36.246	0.0276	4.0162	0.2490
12	11.023	0.0907	45.271	0.0221	4.1069	0.2435
13	13.464	0.0743	56.294	0.0178	4.1812	0.2392
14	16.445	0.0608	69.758	0.0143	4.2420	0.2357
15	20.086	0.0498	86.203	0.0116	4.2918	0.2330
16	24.533	0.0408	106.288	0.0094	4.3326	0.2308
17	29.964	0.0334	130.821	0.0077	4.3659	0.2291
18	36.598	0.0273	160.785	0.0062	4.3933	0.2276
19	44.701	0.0224	197.383	0.0051	4.4156	0.2265
20	54.598	0.0183	242.084	0.0041	4.4339	0.2255
21	66.686	0.0150	296.683	0.0034	4.4489	0.2248
22	81.451	0.0123	363.369	0.0028	4.4612	0.2242
23	99.484	0.0101	444.820	0.0023	4.4713	0.2237
24	121.510	0.0082	544.304	0.0018	4.4795	0.2232
25	148.413	0.0067	665.814	0.0015	4.4862	0.2229
26	181.272	0.0055	814.228	0.0012	4.4917	0.2226
27	221.406	0 0045	995.500	0.0010	4.4963	0.2224
28	270.426	0.0037	1216.906	0.0008	4.5000	0.2222
29	330.300	0.0030	1487.333	0.0007	4.5030	0.2221
30	403.429	0.0025	1817.632	0.0006	4.5055	0.2220
31	492.749	0.0020	2221.061	0.0005	4.5075	0.2219
32	601.845	0.0017	2713.810	0.0004	4.5092	0.2218
33	735.095	0.0014	3315.655	0.0003	4.5105	0.2217
34	897.847	0.0011	4050.750	0.0003	4.5116	0.2217
35	1096.633	0.0009	4948.598	0.0002	4.5125	0.2216
40	2980.958	0.0004	13459.444	0.0001	4.5152	0.2215
45	8103.084	0.0001	36594.322	0.0000	4.5161	0.2214
50	22026.466	0.0001	99481.443	0.0000	4.5165	0.2214

	Single Payment		Equal Payment Series			
	Compound-amount factor	Present-worth factor	Compound-amount factor	Sinking-fund factor	Present-worth factor	Capital-recovery factor
n	To find F Given P F/P r, n	To find P Given F P/F r, n	To find F Given A F/A r, n	To find A Given F A/F r, n	To find P Given A P/A r, n	To find A Given P A/P r, n
1	1.350	0.7408	1.000	1.0000	0.7408	1.3499
2	1.822	0.5488	2.350	0.4256	1.2896	0.7754
3	2.460	0.4066	4.172	0.2397	1.6962	0.5896
4	3.320	0.3012	6.632	0.1508	1.9974	0.5007
5	4.482	0.2231	9.952	0.1005	2.2205	0.4504
6	6.050	0.1653	14.433	0.0693	2.3858	0.4192
7	8.166	0.1225	20.483	0.0488	2.5083	0.3987
8	11.023	0.0907	28.649	0.0349	2.5990	0.3848
9	14.880	0.0672	39.672	0.0252	2.6662	0.3751
10	20.086	0.0498	54.552	0.0183	2.7160	0.3682
11	27.113	0.0369	74.638	0.0134	2.7529	0.3633
12	36.598	0.0273	101.750	0.0098	2.7802	0.3597
13	49.402	0.0203	138.349	0.0072	2.8004	0.3571
14	66.686	0.0150	187.751	0.0053	2.8154	0.3552
15	90.017	0.0111	254.437	0.0039	2.8266	0.3538
16	121.510	0.0082	344.454	0.0029	2.8348	0.3528
17	164.022	0.0061	465.965	0.0022	2.8409	0.3520
18	221.406	0.0045	629.987	0.0016	2.8454	0.3515
19	298.867	0.0034	851.393	0.0012	2.8487	0.3510
20	403.429	0.0025	1150.261	0.0009	2.8512	0.3507
21	544.572	0.0018	1553.689	0.0007	2.8531	0.3505
22	735.095	0.0014	2098.261	0.0005	2.8544	0.3503
23	992.275	0.0010	2833.356	0.0004	2.8554	0.3502
24	1339.431	0.0008	3825.631	0.0003	2.8562	0.3501
25	1808.042	0.0006	5165.062	0.0002	2.8567	0.3501
26	2440.602	0.0004	6973.104	0.0002	2.8571	0.3500
27	3294.468	0.0003	9413.706	0.0001	2.8574	0.3500
28	4447.067	0.0002	12708.174	0.0001	2.8577	0.3499
29	6002.912	0.0002	17155.241	0.0001	2.8578	0.3499
30	8103.084	0.0001	23158.153	0.0001	2.8580	0.3499
31	10938.019	0.0001	31261.237	0.0000	2.8580	0.3499
32	14764.782	0.0001	42199.257	0.0000	2.8581	0.3499
33	19930.370	0.0001	56964.038	0.0000	2.8582	0.3499
34	26903.186	0.0001	76894.409	0.0000	2.8582	0.3499
35	36315.503	0.0000	103797.595	0.0000	2.8582	0.3499

APPENDIX IX: FUNDS FLOW
CONVERSION FACTOR

r	$\dfrac{e^r - 1}{r}$ $(A/\bar{A}\ r)$
1	1.005020
2	1.010065
3	1.015150
4	1.020270
5	1.025422
6	1.030608
7	1.035831
8	1.041088
9	1.046381
10	1.051709
11	1.057073
12	1.062474
13	1.067910
14	1.073384
15	1.078894
16	1.084443
17	1.090028
18	1.095652
19	1.101313
20	1.107014
21	1.112752
22	1.118530
23	1.124347
24	1.130204
25	1.136101
26	1.142038
27	1.148016
28	1.154035
29	1.160094
30	1.166196
31	1.172339
32	1.178524
33	1.184751
34	1.191022
35	1.197335
36	1.203692
37	1.210093
38	1.216538
39	1.223027
40	1.229561

INDEX

Discounting:
  primary, 87
  system, 377-78
Distribution (*see also* Statistics):
  beta, 182
  cumulative probability, 180-81
  of estimates, 449
  Monte Carlo, 179-81
  normal, 182
  primary, 87
  secondary, 87
  skewed left, 182
  symmetric, 182
Distribution of costs, 86-91
Doubled quantities, 273-75
Dual, 431-34
  linear programming, 431
  variables, 436
Duty, 305
Dynamic curve (*see* Learning curve)
Dynamic programming, 394-96

E

Economics, laws, 4
Economic survival, governments, 3, 4
Effectiveness:
  Lagrange undetermined multipliers, 434
  system, 366-69
Elemental time, 71
Empirical distribution, 124 (*see also* Distribution)
Engineer, 6 (*see also* Estimator)
Engineering:
  changes, 30
  change orders, 280
  cost (definition), 52
  drawing, 30, 269
  materials, 96
Engineering change orders, 280
Engineering costs, 265, 268
  estimating, 267
  functional model, 269
Engineering economic method (*see* Return)
Engineering economy, 37
  formulas, 338
  tables, use of, 338
  models, 324-33
Engineering performance data (*see* Standard time data)

Errors, 42, 127, 174, 196
  of belief, 5
  blunders, 5
  conditional costs, 458-59
  of estimate, 448
  estimating performance, 41
  standard deviation, 459-60
  Type I or II, 457
Estimate:
  appropriate, 11
  error, 448 (*see also* Errors)
  operation, 8
  probability, 450-52
  product, 9
  project, 10
  system, 10
  successful, 448
Estimates, analysis of, 448
Estimating, 11
  assembly, 244-45
  cost of, 455-63
  cost of preparation, 5
  curves, 204
  engineering, 249
  everyday problem, 3
  errors, 5 (*see also* Errors)
  indirect labor, 245-48
  labor, 236-49
  optimal amount of detail, 455-63
  optimal performance, 452-63
  process, 19
  resources, 5
  sheet metal, 241-44
  truth, 448
Estimators:
  estimate, 446-48
  qualifications, 7
Exclusion chart, 171-73, 265
Expected cost, 182
Exponential smoothing, 142 (*see also* Smoothing)
Expected value method, 175-77
Expense budget, 92
Experience curve (*see* Learning curve)
Export duty, 305

F

Factor method, 196-201, 351
  time scale, 199
Failure, 3

Profit *(cont.)*

    marginal, 216
    maximum expected, 454
    maximum gross, 216
    maximum per unit, 216
Profitability, 264
Profit and loss statement, 60, 269-71
Profit engineering, 8
Profit volume analysis, 119, 291
Project, 10
    decision making, 336-37
    estimating examples, 349-53
    marginal analysis, 333-34
    return, 313-15
Project analysis, engineering economic,
    319-38
Project estimate, information, 316-19
Purchasing, assets, 346-48

R

Random variable, 181
Ratio method, 196
Raw material *(see* Material)
Recapitulation sheets, 42
Rectification, linear, 132
Regression:
    coefficients, 128
    computer, 139
    confidence limits, 129
    correlation, 135
    degrees of freedom, 130
    individual value limits, 129
    intercept, 131
    learning curve, 278
    least squares fit, exponential, 279
    linear rectification, 134
    multiple linear, 137-39
    multiple linear assumptions, 137
    normal equations, 129, 134
    polynomial, 133
    significance:
        intercept, 132
        slope, 131, 132
        standard time data, 203
Regression analysis, 126-39
Regression coefficients, calculation,
    linear model, 128
Relative frequency, 175
    curve, 125
Repairs, estimating, 299

Replacement, 343-46
Reserve for depreciation, 103
Return, 313-15
    average annual rate, 313-14
    average revenue, 211
    continuous interest model, 329
    engineering economic, 319-38
    maximum marginal, 216
    minimum attractive rate, 336-37
    minimum marginal, 216
    project criteria, 336-37
    purposes, 337-38
    rate method, 323-24
    sales revenue, 324-33
    salvage, 324-33
    total cost, 324-33
    yield, 337-38
Return, total, 216
Risk, 4, 175
Run time, 237

S

S curve, 153
Scientist, 6
Seasonal, 140
Seasonal factor, 148
    pattern, 147
Semi-fixed costs, 90
Sensitivity:
    factors, 436
    systems, 383-84
Setup, 237
Setup time standard, 281 *(see also* Time
    study)
Significant inventions, 152
Simplex method, 424-34 *(see also*
    Linear programming)
Simulation, 177-81
    deterministic vs. probabilistic, 177
Single point cutting tool model, 239
Sixth-tenth model, 201
Sizing and power law model, 338-43
Short term, 215
Slack variables, 422-23
Smoothing, 142 *(see also* Moving
    averages)
Spares, estimating, 299
Standard, 71, 205
Standard costs, 63
Standards, adjustment, 282